Lecture Notes on Mathematical Modelling in the Life Sciences

The rapid pace and development of new methods and techniques in mathematics and in biology and medicine creates a natural demand for up-to-date, readable, possibly short lecture notes covering the breadth and depth of mathematical modelling, mathematical analysis and numerical computations in the life sciences, at a high scientific level.

The volumes in this series are written in a style accessible to graduate students. Besides monographs, we envision the series to also provide an outlet for material less formally presented and more anticipatory of future needs due to novel and exciting biomedical applications and mathematical methodologies.

The topics in LMML range from the molecular level through the organismal to the population level, e.g. gene sequencing, protein dynamics, cell biology, developmental biology, genetic and neural networks, organogenesis, tissue mechanics, bioengineering and hemodynamics, infectious diseases, mathematical epidemiology and population dynamics.

Mathematical methods include dynamical systems, partial differential equations, optimal control, statistical mechanics and stochastics, numerical analysis, scientific computing and machine learning, combinatorics, algebra, topology and geometry, etc., which are indispensable for a deeper understanding of biological and medical problems.

Wherever feasible, numerical codes must be made accessible.

Arnaud Ducrot • Quentin Griette • Zhihua Liu
Pierre Magal

Differential Equations and Population Dynamics II

Advanced Approaches

Springer

Arnaud Ducrot
Laboratoire de Mathématiques
Appliquées du Havre
Université Le Havre Normandie
Le Havre, France

Quentin Griette
Laboratoire de Mathématiques
Appliquées du Havre
Université Le Havre Normandie
Le Havre, France

Zhihua Liu
School of Mathematical Sciences
Beijing Normal University
Beijing, China

Pierre Magal
Institut de Mathématiques de Bordeaux
University of Bordeaux
Bordeaux, France

ISSN 2193-4789 ISSN 2193-4797 (electronic)
Lecture Notes on Mathematical Modelling in the Life Sciences
ISBN 978-3-031-91411-9 ISBN 978-3-031-91409-6 (eBook)
https://doi.org/10.1007/978-3-031-91409-6

This Springer imprint is published by the registered company Springer Nature Switzerland AG
The registered company address is: Gewerbestrasse 11, 6330 Cham, Switzerland

Pierre Magal, who was the driving force behind this book and a mentor for the rest of us, sadly passed away before the publication. We hope that through this writing, some of his great passion for the science can be passed on to the future generations. Arnaud, Quentin and Zhihua.

Foreword

The second volume of the book devoted to Advanced Approaches to Differential Equations and Population Dynamics handles fundamental concepts such as notions of biological kinetics and attractor in population dynamics we will briefly present here in their historical context.

1 The notion of attractor

1.1 The general framework

The Ancients had already formalized, with the help of the magic arithmetic, two attractors that summarize the dynamics of a human life, that of homeostatic life with the maintenance of a living metabolism, and that of death where catabolism prevails. We presented in the Foreword to the first volume of this book the medieval manuscript entitled The Secret of Secrets, containing esoteric calculations on life and death attractors. The manuscript presented as a missive sent by Aristotle to Alexander was in fact coming from a tenth-century Arabic treatise, the Kitâb sirr al-'asrârL, one of the most widely read in the Middle Ages. In Figure 1 we see a table in which two parts are presented simultaneously to determine whether a person will live or die, based on the numerical value of his name and the frequency of his heart.

Fig. 1: Two charts from "The Secret of Secrets", the left (respectively right) corresponding to frailty and death (respectively healing and life). Central thumbnails, top: Arabic numerals, bottom: perfect Tutte's equilateral triangle dissection by equilateral sub-triangles.

The Secret of Secrets is in fact an astonishing mixture of simple rules from Tutte's triangle, ignored in Middle Ages, because it was discovered only in 2010 that it performed the smallest perfect dissection of an equilateral triangle by equilateral triangles W.T. Tutte had conjectured[1].

The numbers in the right table of Figure 1 are inside the external (11, 19, 20) and internal (2, 3, 7) white sub-triangles of Tutte's triangle, or are sums (modulo 30) of these numbers in the following series of steps considered as opportune moments ("χαιροσ") of human life (unlike the numbers in the left table considered as critical moments):

$$12 + 11 + 5, 12 + 11 + 3, 12 + 11, 11 + 11, 11 + 9, 11 + 8, 9 + 8, 9 + 7, 8 + 5, 8 + 3, 7, 3, 2, 1 \equiv 11 + 20$$

Unfortunately, not only was the justification for these abstruse calculations absent from the manuscript, but the notions of dynamics and stability were also absent and will be introduced in the following.

1.2 The definitions

The primitive notion of attractor was proposed by Buffon[2] under the name of "inner mold" ("moule intérieur"). Then, many other authors refined his primitive intuition: Birkhoff was the first to introduce the notion of center of attraction of a dynamical system[3]. He was followed by Smale[4], Bathia and Szegô[5], Thom[6] and Williams[7]. Then, the multiplication of numerical experiments on strange attractors has shown that the previous definitions were too restrictive. Using the notion of Bowen's shadow trajectory[8], Conley[9], Ruelle[10], Sinaï[11], Guckenheimer and Holmes[12], and Garrido and Simo[13] refined the first definitions. Then, Cosnard and

[1] Drápal, A. and Hämäläinen, C. An enumeration of equilateral triangle dissections. *Discr. Appl. Math.* **158**, 1479-1495 (2010).

[2] Leclerc de Buffon, G.L. *Histoire naturelle, générale et particulière, avec la description du Cabinet du Roy*. Imprimerie Royale, Paris, 1749.

[3] Birkhoff, G.D. *Dynamical Systems*. A.M.S. Colloq. Publ. **9**, A.M.S., New York, 1927.

[4] Smale, S. Differentiable dynamical Systems, *Bull. A.M.S. Soc.* **73**, 747-817 (1967).

[5] Bhatia, N.P. and Szegö, G.P. Dynamical Systems: stability theory and applications. *Lect. Notes in Maths* **35**, Springer, New York, 1967.

[6] Thom R. *Stabilité structurelle et morphogenèse*. InterEditions, Paris, 1972.

[7] Williams, R.F. Expanding attractors. *Publ. Math. I.H.E.S.* **43**, 1974, 169-203 (1974).

[8] Bowen R. (1978) ω-limit sets for Axiom A diffeomorphisms. *J. Diff. Equ.*, **18**, 333-339 (1978).

[9] Conley, C. Isolated invariant sets and the Morse index. *Reg. Conf. Series in Maths* **38**, A.M.S., Providence, 1978.

[10] Ruelle, D. Small random perturbations and the definition of attractors, *Comm. Math. Phys.* **82**, 137-151 (1981).

[11] Sinaï, Y.G. The stochasticity of dynamical Systems. *Sef. Math. Sov.* **1**, 100-119 (1981).

[12] Guckenheimer, G. and Holmes, P. *Nonlinear oscillations, dynamical systems, and bifurcations of vector fields*. Applied Math. Science **42**, Springer, New York, 1983.

[13] Garrido, L. and Simo, C. *Some ideas about strange attractors*. Lecture Notes in Physics **179**, Springer, New York, 1983.

Demongeot[14] proposed a more general approach to the concept of attractor, sought to broaden these all definitions by adding the following essential characteristics:

1) the basin of the attractor contains a strict neighborhood of it,
2) the attractor verifies a minimality condition,
3) it is generally assumed to be closed and invariant by the flow of the dynamics,

They can be translated mathematically into the following framework. Let consider a temporal set $T = \mathbb{N}$ or $\mathbb{R}^+$ and a state space $E \subset \mathbb{R}^n$ provided with a dynamic flow φ on $E \times T$ and a distance d: $\{\varphi(a,t)\}_{t\in T}$ denotes the trajectory starting at state a in E and being in state $\varphi(a,t)$ at time t. We will define two reciprocal operators L and B on the set of subsets of E. $L(a)$ denotes the limit set of the trajectory starting in a, made of the accumulation points of the trajectory $\{\varphi(a,t)\}_{t\in T}$, when t tends to infinity:

$$L(a) = \{y \in E\,;\, \forall \varepsilon > 0, \forall t \in T, \exists s(\varepsilon,t) \in T, \text{ such as } s(\varepsilon,t) > t \\ \text{and } d(\varphi(a, s(\varepsilon,t)), y) \leq \varepsilon\},$$

where $s(\varepsilon,t)$ is a time after t at which the trajectory starting in a is in a ball of radius ε centered in y. Considering a subset A of E, $L(A)$ is the union of all limit sets $L(a)$, for a belonging to A:

$$L(A) = \bigcup_{a\in A} L(a).$$

Conversely, $B(A)$ is the set of all initial conditions a outside A, whose limit set $L(a)$ is included in A (Figure 2). $B(A) = \{y \in E\backslash A; L(y) \subset A)\}$ is called the attraction basin of A. Then, A is an attractor if it verifies the three following conditions (cf. Figure 4):

i) $A = L(B(A))$.
ii) There is no set A' containing strictly A and whose any state is shadow-connected to A.
iii) There is no set A'' strictly contained in A verifying (i) and (ii).

The definition of the "shadow-connectivity" between subsets F_1 and F_2 of E, lies on the fact that there exists a "shadow-trajectory" between F_1 and F_2 (cf. Figure 4, where $F_1 = \{x\}$ and $F_2 = A$). The notion of "shadow-trajectory" between x and A has been defined first by R. Bowen: for any $\varepsilon > 0$, there is an integer $n(\varepsilon)$, times $t_{(1,\varepsilon)}, \ldots, t_{(n(\varepsilon),\varepsilon)}$ and states $y_{(1,\varepsilon)}, \ldots, y_{(n(\varepsilon),\varepsilon)} = y_\varepsilon$, such that all distances between successive states of the shadow-trajectory are less than ε:

$$d(\varphi(x, t(1,\varepsilon)), y(1,\varepsilon)) \leq \varepsilon, \ldots, \\ d(\varphi(y(k-1,\varepsilon), t(k,\varepsilon)), y(k,\varepsilon)) \leq \varepsilon, \ldots, \\ d(\varphi(y(n(\varepsilon)-1,\varepsilon), t(n(\varepsilon),\varepsilon)), y_\varepsilon) \leq \varepsilon), d(y_\varepsilon, A) \leq \varepsilon.$$

For all known dynamical systems, the definition of attractor by Cosnard and Demongeot remains available and is meaningful for any discrete or continuous time set T,

[14] Cosnard, M. and Demongeot, J. Attracteurs : Une approche déterministe. *Comptes Rendus Acad. Sci. Maths. Série I* **300**, 551-556 (1985).

when the dynamical system is autonomous (with respect to the time), contrary to the other previous definitions, which are all less general.

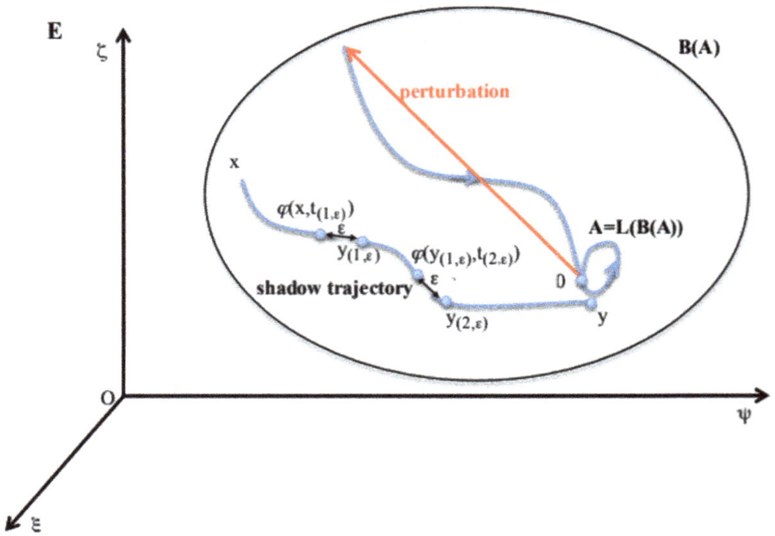

Fig. 2: Representation of the basin of attraction $B(A)$ of an attractor limit-cycle A, which contains exactly all the perturbation admissible, i.e., whose return trajectory goes back into A. An ε-shadow trajectory is represented inside $B(A)$ from the initial state x until the state y_ε such that $d(y_\varepsilon, A) \leq \varepsilon$.

1.3 Random and diffusion cases

In random case, the notion of confiner has been introduced to generalize that of attractor[15]. Eventually, Haraux has defined an attractor when the dynamics contains a diffusion[16] and in the present volume the authors have refined this definition for a PDE with a central manifold.

2 Practical use of the notion of attractor

In the applications, namely in epidemic diseases or in morphogenesis modelling, the notion of attractor is fundamental. René Thom called the growth of trees a "wave towards the light", maximizing the capture of solar energy through photosynthesis.

[15] Demongeot, J., Cosnard, M. and Jacob, C. Attractors and confiners: deterministic and stochastic approaches. In: Dynamical systems: a renewal of mechanism, eds S. Diner et al., World Sc. Pub., Singapour, 20-37 (1986).

[16] Haraux A. Attractors of asymptotically compact processes and applications to nonlinear partial differential equations. *Commun. Partial Differ. Equ.* **13**, 1383-1414 (1988).

A very simple modeling, valid for the growth of Araucaria can account for the main stages observed in this morphogenesis. The local apices, at the end of its branches, promote the synthesis in its meristems, of a plant hormone, the auxin, which stimulates local growth and promotes the synthesis of an inhibitor which diffuses, prohibiting the growth of secondary axillary buds. Any local apex therefore inhibits until the diffusion end the other buds on its branch. The modeling of this growth uses the variables X_0, cell mass of a branch apex, and X_1, mass of the primary axillary buds, which succeed it on the branch. The differential system linking X_0 and X_1 can be written as:

$$\begin{aligned} \frac{dX_0}{dt} &= s_0 X_0 + \frac{aX_0}{K_0 + X_0 + X_1} - d_0 X_0, \\ \frac{dX_1}{dt} &= s_1 X_1 + \frac{aX_1}{K_1 + X_0 + X_1} - d_1 X_1 \end{aligned} \tag{1}$$

The kinetics is said to be Michaelis-Menten with competitive inhibition, where the maximum speed a is proportional to the concentration of the auxin. The K_i's, constants modulating the speed at low mass X_i, the energy input si and the degradation rate d_i of X_i, define two steady states equations:

$$\begin{aligned} (d_0 - s_0)x_0 + (d_0 - s_0)x_1 + K_0(d_0 - s_0) - a = 0 \\ (d_1 - s_1)x_1 + (d_1 - s_1)x_0 + K_1(d_1 - s_1) - a = 0. \end{aligned} \tag{2}$$

Suppose that $K_0 = K_1 = K$, $s_0 = s_1 = s$ and $d_0 = d_1 = d$, then we obtain, by subtracting the second equation above from the first one: $(x_0+x_1)+K-\frac{a}{d-s} = 0$, hence the calculation of the two stable stationary states x^* and x^{**}, gives: $x_1^* = \frac{a}{d-s} - K$, $x_0^* = 0$ and $x_0^{**} = \frac{a}{d-s} - K$, $x_1^{**} = 0$.

We can show that, even if the auto-catalytic kinetics concerning the apex and its m axillary buds is strongly non-linear of Hill type: $\frac{aX_{in}}{K+\sum_{k=0}^{m} X_k^n}$, where n is the cooperativity parameter, the system is of gradient type, provided that we can replace the X_i's by their square root $Y_i = (X_i)^{1/2}$ and consider the potential P defined by:

$$P(y) = \sum_{k=1}^{m} (d_k - s_k)\frac{Y_k^2}{4} - \frac{a}{4n} \log\left(K + \sum_{k=0}^{m} Y_k^{2n}\right). \tag{3}$$

Then, the differential system governing the Y_k's can be written as:

$$\forall k = 1, \ldots, m, \ \frac{dY_k}{dt} = -\frac{\partial P}{\partial Y_k} \tag{4}$$

If the kinetics is more non-linear, e.g., allosteric[17], this result is still available by replacing the kinetic term by $a\frac{(1+X_i)^{n-1}+L_i c(1+cX_i)^{n-1}}{(1+X_i)^n+L_i(1+cX_i)^n}$, with the constant L_i equal to: $L_i = \underset{\prod_{k=1}^{n}}{k \neq i} \frac{(1+cX_k)^n}{(1+eX_k)^n}$, where $e < c < 1$. By changing the variables $Y_i = (X_i)^{1/2}$, the potential function P is given, if $e = c$, by the formula:

[17] Demongeot, J. and Laurent, M. Sigmoidicity in allosteric models. *Math. Biosciences* **67**, 1-17 (1983).

$$P(y) = \sum_{k=1}^{n}(d_k - s_k)\frac{Y_k^2}{4} - \frac{a}{4n}\log((1+Y_k^2)^n + c(1+cY_k^2)^n) \tag{5}$$

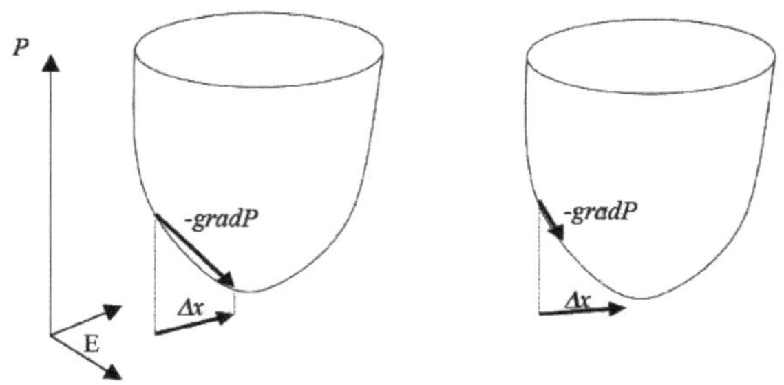

Potential automaton with Δx=-*gradP* (on the left) and automaton with a Lyapunov function decreasing on its trajectories (on the right).

Fig. 3: Illustration of the potential P gradient dynamic (left) and Liapunov functional P dynamic (right), where the energy P decreases on the trajectories, but without its gradient being equal to the velocity in the second case.

If the system is not potential (Figure 3), there is in general no algebraic solution to equations like (1), but a Liapunov functional can be used for ensuring the existence of a stable asymptotic solution as in the exemple of the predatory-prey system presented in chapter 6 of the present volume. If the system studied has a limit-cycle issued from a bifurcation, the Hodge decomposition[18] in a neighboorhood of this attractor can be used to give an approximate solution from a polynomial approximation of the potential $\mathbf{P}(x, y)$ and Hamiltonian $\mathbf{H}(x, y)$ parts of the flow:

$$\frac{dx}{dt} = -\frac{\partial \mathbf{P}(x,y)}{\partial x} + \frac{\partial \mathbf{H}(x,y)}{\partial y}, \quad \frac{dy}{dt} = -\frac{\partial \mathbf{P}(x,y)}{\partial y} - \frac{\partial \mathbf{H}(x,y)}{\partial x} \tag{6}$$

Here, the gradient of the Potential **P** represents the dissipative component of the velocity field normal to the limit-cycle, which vanishes on the attractor, and the derivate of the Hamiltonian **H** is the conservative component of the velocity tangential to the limit-cycle.

3 Conclusion

This second volume of the book devoted to Advanced Approaches to Differential Equations and Population Dynamics presents the same educational quality and same demonstrative force of the interest of differential methods in population dynamics modeling as the first volume. It will be very useful to master's students in mathematical biology, medical informatics and biomedical statistics, but also to any researcher,

[18] Demongeot, J and Minonzio, J.G. A signal processing tool adapted to the periodic biphasic phenomena: the Dynalet transform. *Mathematical Medicine & Biology* **42**, dqae025 (2025).

doctor, biologist, mathematician or computer scientist interested in the formalization of a rational and effective framework for solving biomedical modeling problems, in particular those relating to population dynamics, from the molecular level to the population level.

Final Tribute to Pierre Magal

Pierre Magal (1967-2024)

Pierre was a demanding teacher, for himself, for his students and for his friends, but everyone recognized a posteriori the validity of his ideas and his strength of conviction. His search for perfection was that of a transcendence that he sometimes touched upon, but which had to remain a perfectly pure, therefore inaccessible, goal. He was an upright, honest, creative man, who, like all innovative scientists, aroused either unconditional support or a jealous refusal to recognize his total mastery of the subject he had decided to investigate. This whole, sincere man was one of the best friends one could dream of and his friends could apply to him what his famous Bordeaux predecessor Michel de Montaigne said about his friend Etienne de La Boétie: "It's because it was him and because it was us."

Grenoble, France, February 24, 2025 *Jacques Demongeot (UGA, IUF)*

Foreword

"Mathematics is the key and door to the sciences."

Galileo Galilei

"Life preys upon life. This is biology's most fundamental fact."

Martin H. Fischer

In "Differential Equations and Population Dynamics II: Advanced Approaches", the authors continue their development of mathematical population theory in the context of ordinary differential equations. Their goal is to assemble the past and current fundamental results in this subject, and to provide their exposition to students and researchers. The presentation of their results gives great care to mathematical details and completeness. In addition, they provide extensive MATLAB computer codes for numerical simulations of applications of their results. Through this book, there are graphics illustrations of the mathematical ideas developed. The organization of this book is based on **Part I: Dynamical Systems in Population Dynamics**, and **Part II: Applications to Predator Prey Systems**.

Part I: Dynamical Systems in Population Dynamics.

- In **Semiflows, ω-limit Sets, α-limit Sets, Attraction, and Dissipation**, basic concepts of semi-flows, such as stability and instability of equilibria are explained. Examples are given for the logistic equation, periodic behavior equations, 2-dimensional Bernoulli-Verhulst equation, and Lorentz chaotic behavior. MATLAB computer codes are given for the graphic illustrations of these behaviors.
- In **Global Attractors and Uniform Persistence**, basic concepts, such as compact subsets in metric spaces, measures of non-compactness, spaces of continuous functions, and co-existence of steady states are developed. Examples of asymptotically smooth semiflows and κ-contracting mappings are given for illustration.
- In **Bifurcations**, bifurcating branches of equilibria and periodicity, are developed in finite and infinite dimensional cases. The Hopf Bifurcation Theorem is proven, and an application is given to pattern formation for a predator prey system.
- In **Center Manifold and Center Unstable Manifold Theory**, the smoothness and stability of center manifolds is developed. The results are applied to traveling waves.
- In **Normal Forms**, normal form theory is classified. The results are applied to Cauchy problems in Banach spaces for abstract non-densely defined operators, and bifurcations of age structured models.

Part II: Applications to Predator Prey Systems.

- In **A Holling's Predator Prey Model with Handling and Searching**, predation behavior is developed in systems of equations. The concepts of uniform persistence and extinction of the species are analyzed. Three-dimensional systems of K-competitive systems are described and applied to snowshoe hares and lynx systems. Illustrations are given with MATLAB codes and graphics.
- In **Hopf bifurcation for a Holling's predator-prey model with handling and searching predators**, a Holling's predator prey model is treated with linearization, Hopf bifurcation, and computation of the Taylor's expansion of the reduced system. Numerical simulations are given with MATLAB codes and graphics.
- In **Large Speed Traveling Waves for the Rosenzweig-MacArthur Model**, the analysis of traveling waves in the Rosenzweig-MacArthur model is developed with the center manifold theorem. Global attractors, interior attractors, and periodic orbits are analyzed. Numerical simulations are given with MATLAB codes and graphics.
- In **A Predator Prey Model with Infinitely Many Variants**, a generalized Cauchy problem for existence, uniqueness, persistence, and asymptotic behavior is developed. Examples are given to monovalent replacement dynamics and bivalent dynamics.

This book is a very valuable resource for important subjects in mathematical population dynamics. Its value will have lasting impact for investigators of the mathematical theory of the principal elements of this subject. This value is especially benefited by the contributions of the author Pierre Magal, who passed suddenly on February 20, 2024. Pierre was a leading researcher in mathematical biology, functional analysis, and differential equations. Pierre made many fundamental contributions to both theoretical and applied aspects of these subjects. Pierre was a great colleague, collaborator, and friend.

Nashville, Tennessee, USA, February 15, 2025 *Glenn Webb*

Preface

This second volume continues the exploration of dynamical systems applied to population dynamics. While the first volume,"Differential Equations and Population Dynamics I: Introductory Approaches", focused on mathematical modeling and laid the theoretical foundations of ordinary differential equations, as well as their applications to epidemiological models, this second volume introduces advanced tools and methods to analyze the dynamical behavior of ODE-based mathematical models, with a primary focus on predator-prey systems.

The book is divided into two parts. The first part, covering Chapters 1 to 5, is dedicated to dynamical systems in population dynamics.

In Chapter 1, we introduce fundamental concepts such as semiflows, ω-limit and α-limit sets, and dissipative semiflows. Chapter 2 focuses on the analysis of global attractors and uniform persistence. We present various results concerning metric spaces and their applications to population dynamics, highlighting compactness properties and ω-limit sets. These concepts are illustrated through both classical and modern models, providing a robust mathematical framework for studying the long-term dynamics of biological systems. Chapter 3 is dedicated to bifurcation analysis, while Chapter 4 focuses on the theory of center and unstable manifolds. We explore the Hopf bifurcation theorem. Additionally, Chapter 5 discusses normal forms, essential tools for the qualitative analysis of nonlinear differential systems, and specifically for studying the stability of bifurcated solutions.

The second part of this volume, covering Chapter 6 to 9, presents a series of applications, discussing recent developments in the application of ordinary differential equation models to predator-prey systems and ecological modeling. In Chapter 6, we present and analyze a handling-and-searching predator-prey system and study its connection to the usual Rosenzweig-MacArthur model when some parameters are small. Real-world applications are also provided. This system is further studied in Chapter 7, where Hopf bifurcation and normal form reduction are examined. Chapter 8 concerns the analysis of the Rosenzweig-MacArthur system with diffusion. After describing the global attractor of the ODE problem, we develop a suitable center manifold reduction for large-speed traveling wave solutions and prove the existence of large-speed traveling waves for the Rosenzweig-MacArthur problem with diffusion, connecting the predator-free equilibrium to a periodic wave train. Finally, Chapter 9 is devoted to the analysis of the long-term behavior of a predator-prey model with infinitely many variants. Here, we consider a discrete structure for the different variants that all feed and compete for a single resource.

Following the approach of the first volume, we have included concrete examples and mathematical illustrations drawn from population dynamics. Additionally, each chapter concludes with a section of remarks and notes, providing further reading suggestions and MATLAB codes for the numerical simulations presented. The book also offers online supplementary materials, including the MATLAB codes mentioned above, which are available for viewing and download on the Springer Nature GitHub repository : https://github.com/sn-code-inside/differential-equations-and-population-dynamics-II/

We hope that this second volume will provide readers with a deeper understanding of the complex dynamics governing biological and ecological systems.

Contents

Part I
Dynamical Systems in Population Dynamics

Chapter 1
Semiflows, ω-limit Sets, α-limit Sets, Attraction, and Dissipation

1.1 Introduction

Let (M, d) be a complete metric space. A typical example in population dynamics will be

$$M = \mathbb{R}_+^n = [0, \infty)^n.$$

A more general class of subset M are the intervals in $\mathbb{R}^n$, that is subsets of the following form

$$M = [c, \infty) = \{x \in \mathbb{R}^n : c_i \le x_i\},$$

$$M = [c, d] = \{x \in \mathbb{R}^n : c_i \le x_i \le d_i\},$$

or

$$M = (-\infty, d] = \{x \in \mathbb{R}^n : x_i \le d_i\},$$

where $c = (c_1, \cdots, c_n), d = (d_1, \cdots, d_n) \in \mathbb{R}^n$ and $c \le d$, that mens $c_i \le d_i$ for all components $i = 1, .., n$.

These subsets will be endowed with usual distance (induced by the norm $\|.\|$ of $\mathbb{R}^n$)

$$d(x, y) = \|x - y\|, \forall x, y \in M.$$

Then (M, d) is a complete metric space (since the subsets M are closed subsets of $\mathbb{R}^n$).

In order to consider both discrete and continuous time dynamical systems, the time will vary either in

$$I_+ = [0, +\infty) \text{ and } I = \mathbb{R} \text{ (\textbf{continuous time})}$$

or in

$$I_+ = \mathbb{N} \text{ and } I = \mathbb{Z} \text{ (\textbf{discrete time})}.$$

Below we shall use the notation I_+ to denote one of the above sets.

A. Ducrot et al., *Differential Equations and Population Dynamics II*,
Lecture Notes on Mathematical Modelling in the Life Sciences,
https://doi.org/10.1007/978-3-031-91409-6_1

Definition 1.1 Let $\{U(t)\}_{t\in I_+}$ be a family of continuous maps from the metric space M into itself parameterized by $t \in I_+$. We will say that U is a **continuous semiflow** on M if the following properties are satisfied

(i) $U(0)x = x, \forall x \in M$;
(ii) $U(t)U(s)x = U(t+s)x, \forall t, s \in I_+, \forall x \in M$;
(iii) The map $(t, x) \to U(t)x$ is continuous from $I_+ \times M$ into M.

Moreover we will say that U is a **continuous time semiflow** if $I_+ = [0, +\infty)$ and a **discrete time semiflow** if $I_+ = \mathbb{N}$.

Remark 1.2 Since we assumed that the map $x \to U(t)x$ continuous (for each $t \geq 0$), it follows that a discrete time semiflow is always continuous. That is to say that the property (iii) of Definition 1.1 is always satisfied for discrete time semiflow.

1.1.1 Discrete time semiflow and difference equation

Let $\{U(t)\}_{t\in\mathbb{N}}$ be a given discrete semiflow, that $I_+ = \mathbb{N}$. Define the map $T : M \to M$ by

$$T(x) = U(1)(x)$$

and the sequence $\{u_n\}_{n\in\mathbb{N}}$ as follows

$$u_n := U(n)(x), \forall n \geq 0.$$

Then by using the property (ii) in Definition 1.1, one has

$$u_n = U(n)(x) = U(1)(U(n-1)(x)) = T(u_{n-1}), \forall n \in \mathbb{N}, \text{ and } u_0 = x.$$

So we obtain the following difference equation

$$u_{n+1} = T(u_n), \forall n \in \mathbb{N}, \text{ with } u_0 = x \in M.$$

Let us also observe that

$$\begin{aligned} u_0 &= x \\ u_1 &= T(x) \\ u_2 &= T(T(x)) = T^2(x) \end{aligned}$$

so that we get

$$u_n = T^n(x), \forall n \in \mathbb{N},$$

where T^n is defined as the n−fold composition of T, namely

$$T^{n+1} = T \circ T^n, \forall n \in \mathbb{N} \text{ and } T^0 = I.$$

Conversely, assume that the map $T : M \to M$ is given. Then the semiflow $\{U(t)\}_{t\in\mathbb{N}}$ is defined by

$$U(n)(x) = T^n(x), \forall n \in \mathbb{N}, \forall x \in M.$$

1.1.2 Semiflow generated by the 1-dimensional logistic equation

Let us consider the family of maps $\{U(t)\}_{t \geq 0}$ defined on $M = \mathbb{R}_+$ by

$$U(t)x = \frac{e^{\lambda t} x}{1 + \kappa \int_0^t e^{\lambda \sigma} x d\sigma}, \forall t \geq 0, \forall x \geq 0. \tag{1.1}$$

Here $\lambda \in \mathbb{R}$ and $\kappa \geq 0$ are given parameters.

Lemma 1.3 *The family $\{U(t)\}_{t \geq 0}$ is a semiflow on the metric space $M = [0, +\infty)$.*

Remark 1.4 The semiflow $U(t)$ cannot be extended (backward in time) to a flow because the solution of the logistic equation $t \to U(t)x$ is blowing up for negative time (whenever $x > \lambda/\kappa$). But $U(t)$ restricted to $[0, \lambda/\kappa]$ defines a flow (whenever $\lambda > 0$).

Proof Let us verify that $\{U(t)\}_{t \geq 0}$ is a continuous semiflow on $\mathbb{R}_+$. Indeed, note that one has

$$U(0)x = x, \forall x \in \mathbb{R}_+.$$

Next let $t, s \geq 0$ be given. Then we have

$$\begin{aligned} U(t)U(s)x &= \frac{e^{\lambda t} \frac{e^{\lambda s} x}{1+\kappa \int_0^s e^{\lambda r} x dr}}{1 + \kappa \int_0^t e^{\lambda \sigma} \frac{e^{\lambda s} x}{1+\kappa \int_0^s e^{\lambda r} x dr} d\sigma} \\ &= \frac{e^{\lambda(t+s)} x}{1 + \kappa \int_0^s e^{\lambda r} x dr + \kappa \int_0^t e^{\lambda(\sigma+s)} x d\sigma} \end{aligned}$$

and by using a change of variable, we obtain that

$$\int_0^t e^{\lambda(\sigma+s)} x d\sigma = \int_s^{t+s} e^{\lambda r} x dr.$$

It follows that

$$U(t)U(s)x = \frac{e^{\lambda(t+s)} x}{1 + \kappa \int_0^{t+s} e^{\lambda r} x dr} = U(t+s)x,$$

so that U is a semiflow on $\mathbb{R}_+$. □

Remark 1.5 For given x, the map $t \to N(t) := U(t)x$ satisfies the logistic equation

$$N'(t) = \lambda N(t) - \kappa N^2(t), \quad N(0) = x. \tag{1.2}$$

We refer to Chapter 5 in Volume I for more results on such a problem.

1.1.3 Semiflow generated by a 2-dimensional Bernoulli-Verhulst equation

Let $\theta > 0$ be given. Consider the family of maps $V(t)$ for $t \geq 0$ on $\mathbb{R}^2$ defined by

$$V(t)X = \begin{cases} \left(\frac{U(t)\left(\|X\|_2^\theta\right)}{\|X\|_2^\theta}\right)^{\frac{1}{\theta}} e^{At}X, & \text{if } X = \begin{pmatrix} x \\ y \end{pmatrix} \neq 0, \\ 0, & \text{if } X = \begin{pmatrix} 0 \\ 0 \end{pmatrix}, \end{cases} \tag{1.3}$$

where $\{U(t)\}_{t\geq 0}$ denotes the semiflow defined by (1.1) while

$$\|X\|_2 = \left\| \begin{pmatrix} x \\ y \end{pmatrix} \right\|_2 = \sqrt{x^2 + y^2}$$

is the Euclidean norm. Note that we have (see Section 3.9.3 in Volume I)

$$e^{At}\begin{pmatrix} x \\ y \end{pmatrix} = \begin{pmatrix} \cos(\omega t)x - \sin(\omega t)y \\ \sin(\omega t)x + \cos(\omega t)y \end{pmatrix},$$

with

$$A = \begin{pmatrix} 0 & -\omega \\ \omega & 0 \end{pmatrix}.$$

Next we obtain the following result.

Lemma 1.6 *The family $\{V(t)\}_{t\leq 0}$ is a semiflow on $\mathbb{R}^2$.*

Proof Let us first observe that

$$\begin{aligned} \left\|e^{At}X\right\|_2^2 &= (\cos(\omega t)x)^2 - 2\cos(\omega t)x\sin(\omega t)y + (\sin(\omega t)y)^2 \\ &\quad + (\sin(\omega t)x)^2 + 2\sin(\omega t)x\cos(\omega t)y + (\cos(\omega t)y)^2 \\ &= x^2 + y^2 \\ &= \|X\|_2^2, \end{aligned}$$

and we deduce that e^{At} preserves the Euclidean norm of X

$$\left\|e^{At}X\right\|_2 = \|X\|_2, \forall t \in \mathbb{R}.$$

It follows that

$$V(t)\left(\mathbb{R}^2 \setminus \{0\}\right) \subset \mathbb{R}^2 \setminus \{0\}, \forall t \geq 0.$$

Let $X \neq 0$ be given. Next applying the Euclidean norm on both sides of (1.3), we deduce that

$$\|V(t)X\|_2^\theta = \frac{U(t)\left(\|X\|_2^\theta\right)}{\|X\|_2^\theta} \left\|e^{At}X\right\|_2^\theta = U(t)\left(\|X\|_2^\theta\right) \forall t \geq 0,$$

and

$$\begin{aligned}
V(t)V(s)X &= \left(\frac{U(t)\left(\|V(s)X\|_2^\theta\right)}{\|V(s)X\|_2^\theta}\right)^{\frac{1}{\theta}} e^{At} V(s)X \\
&= \left[U(t)\left(U(s)\left(\|X\|_2^\theta\right)\right)\right]^{\frac{1}{\theta}} e^{At} \frac{V(s)X}{\|V(s)X\|_2} \\
&= \left[U(t+s)\left(\|X\|_2^\theta\right)\right]^{\frac{1}{\theta}} e^{At} \frac{e^{As}X}{\|X\|_2} \\
&= V(t+s)X,
\end{aligned}$$

whenever $t \geq 0$ and $s \geq 0$. It follows that $\{V(t)\}_{t\geq 0}$ is a continuous semiflow. □

Remark 1.7 The map $t \to X(t) = V(t)X$ satisfies the 2-dimensional Bernoulli-Verhulst equation

$$X'(t) = \left[A + \frac{\lambda}{\theta} I\right] X(t) - \frac{\kappa}{\theta} \|X(t)\|_2^\theta X(t), \forall t \geq 0. \tag{1.4}$$

We refer to chapter 5 in volume I for more results.

1.1.4 Explicit formula for the semiflow of the Poincaré normal form

In the special case $\theta = 2$, we obtain

$$V(t)X = \begin{cases} \sqrt{U(t)\left(\|X\|_2^2\right)} \; e^{At}\left(\frac{X}{\|X\|}\right) & \text{if } X \neq 0, \\ 0 & \text{if } X = 0, \end{cases} \tag{1.5}$$

and from the above computation we deduce that $t \to V(t)X = \begin{pmatrix} x(t) \\ y(t) \end{pmatrix}$ satisfies the following system of ordinary differential equations

$$\begin{cases} x'(t) = \frac{\lambda}{2}x(t) - \omega y(t) - \frac{\kappa}{2}\left(x(t)^2 + y(t)^2\right)x(t), \\ y'(t) = \omega x(t) + \frac{\lambda}{2}y(t) - \frac{\kappa}{2}\left(x(t)^2 + y(t)^2\right)y(t). \end{cases} \tag{1.6}$$

System (1.6) is nothing but the so-called Poincaré normal form for Hopf bifurcation. We can therefore say that the Poincaré normal form is a special case of the 2-dimensional Bernoulli-Verhulst equation.

1.2 Stability of an Equilibrium

Let $\{U(t)\}_{t\in I_+}$ be a given continuous semiflow. Here recall that $I_+ = \mathbb{N}$ or $I_+ = [0, +\infty)$.

Definition 1.8 A point $\overline{x} \in M$ is an **equilibrium** for U if

$$U(t)\overline{x} = \overline{x}, \forall t \in I_+.$$

Set

$$B_M(x,\varepsilon) := \{y \in M : d(x,y) \leq \varepsilon\}.$$

Definition 1.9 We will say that an equilibrium $\overline{x} \in M$ is **stable** for U if for each $\varepsilon > 0$, there exists $\eta \in (0,\varepsilon]$ such that

$$U(t)B_M(\overline{x},\eta) \subset B_M(\overline{x},\varepsilon), \forall t \in I_+.$$

We will say that $\overline{x} \in M$ is **unstable** otherwise. That is to say that, there exist $\varepsilon > 0$, and sequences $(x_n)_{n\geq 0} \subset M$ and $(t_n)_{n\geq 0} \subset I_+$ with $x_n \to \overline{x}$ and $t_n \to +\infty$ such that

$$\|U(t_n)x_n - \overline{x}\| > \varepsilon.$$

We will say that $\overline{x} \in M$ is **asymptotically stable**, if $\overline{x}$ is stable for U and if there exists $\eta > 0$ such that for each $x \in B_M(\overline{x},\eta)$

$$\lim_{t\to+\infty} U(t)x = \overline{x}.$$

We will say that $\overline{x} \in M$ is **exponentially asymptotically stable**, if $\overline{x}$ is stable for U and if in addition we can find three constants $\eta > 0$, $\alpha > 0$, $M \geq 1$ such that

$$d(U(t)x,\overline{x}) \leq Me^{-\alpha t}d(x,\overline{x}), \forall t \in I_+, \forall x \in B_M(\overline{x},\eta).$$

Another equivalent definition of stability is the following one: the equilibrium $\overline{x} \in M$ is stable if for each neighborhood V of $\overline{x}$ in M (that is to say that V contains a ball $B_M(\overline{x},\varepsilon)$), we can find a neighborhood $W \subset V$ of $\overline{x}$ in M such that

$$U(t)W \subset V, \forall t \in I_+.$$

In this case, by considering the set

$$\widehat{W} := \bigcup_{t\in I_+} U(t)W,$$

we have $U(0)W = W$ (since $U(0) = I$), so that we get

$$W \subset \widehat{W} \subset V.$$

Hence $\widehat{W}$ is a neighborhood of $\overline{x}$ in M, (since W is a neighborhood of $\overline{x}$ in M), and we have

$$\begin{aligned} U(t)\widehat{W} &= U(t)\bigcup_{s\in I_+} U(s)W = \bigcup_{s\in I_+} U(t)U(s)W \\ &= \bigcup_{s\in I_+} U(s+t)W = \bigcup_{l\geq t} U(l)W. \end{aligned}$$

This yields

$$U(t)\widehat{W} \subset \widehat{W}, \forall t \in I_+,$$

and we obtain the following lemma.

Lemma 1.10 *The following properties are equivalent*

(i) *the point* $\overline{x} \in M$ *is a stable equilibrium for* U;
(ii) *For each neighborhood* V *of* $\overline{x}$ *in* M, *we can find neighborhood* $W \subset V$ *of* $\overline{x}$ *in* M, *such that*

$$U(t)W \subset V, \forall t \in I_+.$$

(iii) *For each neighborhood* V *of* $\overline{x}$ *in* M, *we can find a neighborhood* $W \subset V$ *of* $\overline{x}$ *in* M, *such that*

$$U(t)W \subset W, \forall t \in I_+.$$

We can observe that an equilibrium $\overline{x}$ is unstable for U if there exists $\varepsilon > 0$ such that for each $\eta > 0$ there exists a time $t = t(\eta) > 0$ in I_+ such that

$$U(t)B_M(\overline{x}, \eta) \not\subset B_M(\overline{x}, \varepsilon).$$

Therefore by considering a sequence $\eta_n = 1/(n+1) \to 0$ we obtain that for each integer $n \geq 0$ there exists a time $t_n \in I_+$ such that

$$U(t_n)B_M(\overline{x}, 1/(n+1)) \not\subset B_M(\overline{x}, \varepsilon).$$

Therefore we can find $x_n \in B_M(\overline{x}, 1/(n+1))$ with

$$d(U(t_n)x_n, \overline{x}) \geq \varepsilon.$$

Moreover, we must have $t_n \to +\infty$ since U is a continuous semiflow. Otherwise, we can find a sub-sequence $t_{n_p} \to \widehat{t}$ as $p \to +\infty$ for some $\widehat{t} \in I_+$, and by continuity of U, we deduce that

$$\lim_{p \to \infty} d(U(t_{n_p})x_{n_p}, \overline{x}) = d\left(U\left(\widehat{t}\right)\overline{x}, \overline{x}\right) = d(\overline{x}, \overline{x}) = 0,$$

which is impossible since $\varepsilon > 0$. Now due to the above analysis, we end-up with the following lemma.

Lemma 1.11 *An equilibrium point* $\overline{x} \in M$ *is unstable for* U *if and only if there exist* $\varepsilon > 0$ *and two sequences* $x_n \to \overline{x}$ *and* $t_n \to +\infty$ *such that*

$$d(U(t_n)x_n, \overline{x}) \geq \varepsilon, \forall n \geq 0.$$

Definition 1.12 Let A be a subset of M. We will say that A is **positively invariant** by U if

$$U(t)A \subset A, \forall t \in I_+.$$

The subset A is positively invariant if and only if for each $x \in A$

$$U(t)x \in A, \forall t \in I_+.$$

We will say that A is **negatively invariant** by U if

$$U(t)A \supset A, \forall t \in I_+.$$

The subset A is negatively invariant if and only if for each $y \in A$ and each $t \in I_+$, there exists $x \in A$ such that

$$U(t)x = y.$$

We will say that A is **invariant** by U if A is both positively and negatively invariant, that is

$$U(t)A = A, \forall tinI_+.$$

Example 1.13 Consider the map $T : [0, 1] \rightarrow [0, 1]$ defined by

$$T(x) = 4x(1 - x), \forall x \in [0, 1] .$$

Then T reaches its maximum on $[0, 1]$ at $x = 1/2$ and

$$T(1/2) = 1.$$

Therefore

$$T([0, 1]) = [0, 1] .$$

We deduce that $[0, 1]$ is invariant by the discrete time semiflow

$$U(n) = T^n, \forall n \geq 0.$$

But clearly the map T is not one to one, so the map $x \rightarrow U(n)x$ is not invertible.

Definition 1.14 We will say that $O_+ = \{u(t)\}_{t \in I_+} \subset M$ is a **positive orbit** of a semiflow U if

$$u(t) = U(t)u(0), \forall t \in I_+,$$

or, equivalently, if

$$u(t + s) = U(s)u(t), \forall t, s \in I_+.$$

We will say that $\{u(t)\}_{t \in I_+} \subset M$ is a **negative orbit** of U if

$$u(-t) = U(s)u(-t - s), \forall t, sinI_+.$$

Finally we will say that $O = \{u(t)\}_{t \in I} \subset M$ is a **complete orbit** of U if

$$u(t) = U(s)u(t - s), \forall s \geq 0, \forall t \in I.$$

We will say that an orbit (positive, negative or complete) **passes through** $x \in M$ **at time** $t = 0$ if $u(0) = x$.

Remark 1.15 Let $x \in M$ be given. Then there exists at most one positive orbit passing through x at time $t = 0$, which is

$$u(t) := U(t)x, \forall t \in I_+.$$

But in general there is no negative orbit passing through x at time 0. Since the map $U(t)$ is not always onto for $t > 0$. Moreover when there exists a negative orbit passing through x, the negative orbit is not necessarily unique since the map $U(t)$ is not always one to one for $t > 0$. As an example of non-unique negative orbit consider Example 1.13.

Remark 1.16 If $\{u(t)\}_{t\in I} \subset M$ is a complete orbit passing through x then the set

$$O := \bigcup_{t\in I} \{u(t)\}$$

satisfies

$$U(t)O = O, \forall t \in I_+.$$

This is an example of invariant set.

1.3 ω-Limit and α-Limit Sets

Definition 1.17 Let $x \in M$ be given. Let $\{u(t)\}_{t\in I_+} \subset M$ be a positive orbit of U passing through x at time 0. The ω**-limit set** of x is defined as

$$\omega(x) := \bigcap_{t\in I_+} \overline{\bigcup_{s\geq t} \{u(s)\}}.$$

Let $\{u(t)\}_{t\in -I_+}$ be a negative orbit of U passing through x at time 0. Then the α**-limit set** of x (with respect to this negative orbit) is

$$\alpha(x) := \bigcap_{t\in -I_+} \overline{\bigcup_{s\leq t} \{u(s)\}}.$$

The omega limit set satisfies

$$\begin{aligned}\omega(x) &= \left\{y \in M : \exists \{t_n\}_{n\in\mathbb{N}} \subset I_+ \to +\infty \text{ such that } \lim_{n\to+\infty} u(t_n) = y\right\}\\ &= \{y \in M : \forall t \in I_+, \forall \varepsilon > 0, \exists s \in I_+ \cap (t, +\infty) \text{ such that } d(u(s), y) \leq \varepsilon\}.\end{aligned}$$

Similarly the alpha limit set satisfies

$$\begin{aligned}\alpha(x) &= \left\{y \in M : \exists \{t_n\}_{n\in\mathbb{N}} \subset I_+ \to +\infty \text{ such that } \lim_{n\to+\infty} u(-t_n) = y\right\}\\ &= \{y \in M : \forall t \leq 0, \forall \varepsilon > 0, \exists s \in -I_+ \cap (-\infty, t) \text{ such that } d(u(s), y) \leq \varepsilon\}.\end{aligned}$$

Example 1.18 We have for example

$$\begin{aligned}u(t) = \cos(t) &\Rightarrow \bigcap_{t\geq 0} \overline{\bigcup_{s\geq t} \{u(s)\}} = [0, 1],\\ u(t) = t\cos(t) &\Rightarrow \bigcap_{t\geq 0} \overline{\bigcup_{s\geq t} \{u(s)\}} = \mathbb{R},\\ u(t) = t &\Rightarrow \bigcap_{t\geq 0} \overline{\bigcup_{s\geq t} \{u(s)\}} = \varnothing.\end{aligned}$$

So the omega limit sets can be compact, non compact or empty.

Definition 1.19 Let (M, d) be a metric space.

(i) A subset $C \subset M$ is **compact** if and only if any sequence in C has a sub-sequence which converges in C.
(ii) A subset $C \subset M$ is **relatively compact** if and only if $\overline{C}$ (the closure of C in (M, d)) is compact.

In the general case ω-limit set is non empty only whenever the positive orbit

$$O_+ = \{u(t)\}_{t \in I_+}$$

is relatively compact (i.e. its closure is compact).

Theorem 1.20 *Let $\{u(t)\}_{t \in I_+} \subset M$ be a positive orbit passing through $x \in M$ at time $t = 0$. Assume that the closure of this positive orbit*

$$\overline{\bigcup_{t \in I_+} \{u(t)\}}$$

is compact.

Then the ω-limit set satisfies the following properties:

(i) *$\omega(x)$ is a non empty compact subset of M;*
(ii) *$\omega(x)$ is invariant by U;*
(iii) $\lim_{t \to +\infty} d\left(u(t), \omega(x)\right) = 0$, *where*

$$d\left(x, B\right) := \inf_{y \in B} d(x, y).$$

Remark 1.21 If M is a closed subset of $\mathbb{R}^n$, and the metric d is induced by a norm on $\mathbb{R}^n$ (i.e. $d(x, y) = \|x - y\|$), then $\overline{\bigcup_{t \in I_+} \{u(t)\}}$ is compact if and only if the positive orbit $\bigcup_{t \in I_+} \{u(t)\}$ is a bounded set.

Before proving Theorem 1.20 we need the following lemma.

Lemma 1.22 *Let $B \subset M$ be given. Then the map $x \to d(x, B)$ is Lipschitz continuous. More precisely we have*

$$|d(x, B) - d(y, B)| \leq d\left(x, y\right), \forall x, y \in M.$$

Proof Let $x, y \in M$ and $z \in B$ be given. Then we have

$$d(x, B) \leq d(x, z) \leq d(x, y) + d(y, z).$$

Thus we get

$$d(x, B) \leq d(x, y) + d(y, B)$$

and the result follows. □

Proof (of Theorem 1.20) Define for $t \in I_+$

$$A_t := \overline{\bigcup_{s \geq t} \{u(s)\}}.$$

By assumption for each $t \in I_+$, the subset A_t is compact. Moreover the family $t \longrightarrow A_t$ is decreasing, that is to say that

$$t \geq s \Rightarrow A_t \subset A_s.$$

Proof of (i). Let $\{t_n\} \subset I_+ \longrightarrow +\infty$ be an increasing sequence and $x_n \in A_{t_n}, \forall n \geq 0$. Then since A_0 is compact, and the family $t \longrightarrow A_t$ is decreasing, we have

$$x_n \in A_0, \forall n \geq 0.$$

So, we can find a converging sub-sequence $\{x_n\}_{n \geq 0} \rightarrow z \in A_0$ (denoted for notational simplicity by the same index).

Moreover for each $t \in I_+$, we can find an integer $n_0 \geq 0$ such that $t_n \geq t, \forall n \geq n_0$, and since the family $t \longrightarrow A_t$ is decreasing

$$x_n \in A_t, \forall n \geq n_0.$$

But the subset A_t is closed by construction, therefore

$$z \in A_t, \forall t \in I_+.$$

hence $\omega(x)$ is non-empty (since $z \in \omega(x)$).

Next, consider a sequence in the ω-limit set

$$x_n \in \omega(x), \forall n \geq 0.$$

Then

$$x_n \in A_t, \forall n \geq 0, \forall t \in I_+.$$

Since A_0 is compact, we can find a converging sub-sequence, and since this sub-sequence belongs to each subset A_t (which is closed), we deduce that the limit of this converging sub-sequence belongs to $\omega(x)$. Therefore $\omega(x)$ is compact.

Proof of (ii). Observe that for each $t, s \in I_+$ one has

$$U(s)\left(\bigcup_{l \geq t} \{u(l)\}\right) = \bigcup_{l \geq t+s} \{u(l)\}. \tag{1.7}$$

This equality implies that

$$U(s)\left(\bigcup_{l \geq t} \{u(l)\}\right) \subset A_{t+s}$$

and since the map $x \longrightarrow U(s)x$ is continuous we obtain

$$U(s)\,(A_t) \subset A_{t+s}. \tag{1.8}$$

From (1.7), we also have

$$U(s)\,(A_t) \supset \bigcup_{l \geq t+s} \{u(l)\}$$

and since by assumption A_t is compact and $x \longrightarrow U(s)x$ is continuous, it follows that $U(s)\,(A_t)$ is compact and we obtain

$$U(s)\,(A_t) \supset A_{t+s}. \tag{1.9}$$

By combining the inclusions (1.8) and (1.9) we obtain

$$U(s)\,(A_t) = A_{t+s}, \forall t, s \in I_+.$$

The invariance of the ω-limit set follows from the following observation

$$U(s)\,(\omega(x)) = U(s)\left(\bigcap_{t \in I_+} A_t\right) = \bigcap_{t \in I_+} U(s)A_t = \bigcap_{t \in I_+} A_{t+s} = \omega(x).$$

Proof of (iii). Assume by contradiction that there exist $\varepsilon > 0$ and a sequence $(t_n)_{n \geq 0} \subset I_+$ with $t_n \longrightarrow +\infty$ such that

$$d\,(u(t_n), \omega(x)) \geq \varepsilon, \forall n \geq 0.$$

By compactness of A_0, we can find a converging sub-sequence (denoted with the same index) such that

$$\lim_{n \to +\infty} u(t_n) = z \in \omega(x).$$

By Lemma **1.22**, we deduce that

$$\varepsilon/2 \leq \lim_{n \to +\infty} d\,(u(t_n), \omega(x)) = d\,(z, \omega(x)) = 0,$$

which is a contradiction. □

The proof for alpha limit sets is similar to the proof of Theorem 1.20.

Theorem 1.23 *Let $\{u(t)\}_{t \in -I_+} \subset M$ be a negative orbit passing through $x \in M$ at time $t = 0$. Assume that $\overline{\bigcup_{t \in -I_+} \{u(t)\}}$ is compact. Then the α-limit set satisfies the following properties*

(i) *$\alpha(x)$ is a non empty compact subset of M;*
(ii) *$\alpha(x)$ is invariant by U.*
(iii) $\lim_{t \to -\infty} d\,(u(t), \alpha(x)) = 0.$

1.4 Heteroclinic and Homoclinic Orbits

In this section, as before we denote by $\{U(t)\}_{t\in I_+}$ a continuous semiflow on a metric space (M, d). Then we first have the following result.

Proposition 1.24 *Let $\{u(t)\}_{t\in I_+} \subset M$ be a positive orbit passing through $x \in M$ at time $t = 0$. Then we have*

$$\omega(x) = \{\overline{x}_+\} \Leftrightarrow \lim_{t\to+\infty} u(t) = \overline{x}_+.$$

Moreover in that case, $\overline{x}_+$ must be an equilibrium point of U.

Let $\{u(t)\}_{t\in -I_+} \subset M$ be a negative orbit passing through $x \in M$ at time $t = 0$. Then

$$\alpha(x) = \{\overline{x}_-\} \Leftrightarrow \lim_{t\to-\infty} u(t) = \overline{x}_-.$$

Moreover in that case, $\overline{x}_-$ must be an equilibrium of U.

Proof By using the definition of $\omega(x)$, we deduce that $\omega(x) = \{\overline{x}_+\}$ is equivalent to

$$\lim_{t\to+\infty} u(t) = \overline{x}_+.$$

Let $s \in I_+$ be given. Since $t \to u(t)$ is a positive orbit, we have

$$u(t+s) = U(s)u(t), \forall t \in I_+,$$

and by taking the limit when $t \to +\infty$ on both side, we obtain

$$\overline{x}_+ = U(s)\overline{x}_+.$$

By using the definition of $\alpha(x)$, we deduce that $\alpha(x) = \{\overline{x}_-\}$ is equivalent to

$$\lim_{t\to-\infty} u(t) = \overline{x}_-.$$

Let $s \in I_+$ be given. Since $t \to u(t)$ is a negative orbit, we have

$$u(t+s) = U(s)u(t), \forall t \in (-I_+) \cap (-\infty; -s],$$

and by taking the limit when $t \to -\infty$ on both side, we obtain

$$\overline{x}_- = U(s)\overline{x}_-.$$

This completes the proof of the result. □

Definition 1.25 Recall that $I = \mathbb{Z}$ if $I_+ = \mathbb{N}$ and $I = \mathbb{R}$ is $I_+ = [0, +\infty)$. A complete orbit $\{u(t)\}_{t\in I}$ is called a **heteroclinic orbit** if there exist $\overline{x}_{-\infty} \in M$ and $\overline{x}_{+\infty} \in M$ with $\overline{x}_{-\infty} \neq \overline{x}_{+\infty}$ satisfying

$$\lim_{t\to-\infty} u(t) = \overline{x}_{-\infty} \text{ and } \lim_{t\to+\infty} u(t) = \overline{x}_{+\infty}.$$

This is equivalent to say that both the omega limit set and alpha limit set are reduced to these single points, namely

$$\alpha(x) = \{\overline{x}_{-\infty}\} \text{ and } \omega(x) = \{\overline{x}_{+\infty}\}.$$

A complete orbit $\{u(t)\}_{t \in I}$ is called a **homoclinic orbit** if this orbit is not constant and there exists $\overline{x} \in M$ such that

$$\lim_{t \to +\infty} u(t) = \overline{x} \text{ and } \lim_{t \to +\infty} u(t) = \overline{x}.$$

This is equivalent to say that this orbit is not constant and the omega limit set and alpha limit set are reduced to the same single point, namely

$$\alpha(x) = \omega(x) = \{\overline{x}\}.$$

1.5 Attraction of Sets and Hausdorff Distance and Semi-Distance

Let (M, d) be a complete metric space. For any subsets A, $B \subset M$, we define Hausdorff's semi-distance of B to A as

$$\delta(B, A) := \sup_{x \in B} d(x, A),$$

where $d(x, A)$, the distance from x to A is given by

$$d(x, A) := \inf_{z \in A} d(x, z).$$

For each $\varepsilon > 0$, we define an **open ε-neighborhood of** A (see also Section 2.4 in Chapter 2 for more results) as

$$N(A, \varepsilon) := \{x \in M : \ d(x, A) < \varepsilon\},$$

and a **closed ε-neighborhood of** A as

$$\overline{N}(A, \varepsilon) := \{x \in M : \ d(x, A) \le \varepsilon\}.$$

Now since $x \longrightarrow d(x, A)$ is a continuous map, the set $N(A, \varepsilon)$ is an open neighborhood of A while $\overline{N}(A, \varepsilon)$ is a closed neighborhood of A. From these observations, it becomes clear that $\delta(B, A)$ is measuring the distance of B to A (and not the converse). Therefore δ is only a semi-distance (since $\delta(B, A) = 0$ does not imply $A = B$).

Remark 1.26 The open ball $B_M(x, \varepsilon)$ (respectively the closed ball $\overline{B}_M(x, \varepsilon)$) centered at x with radius $\varepsilon > 0$ satisfies

$$B_M(x, \varepsilon) = \{y \in M : \ d(y, x) < \varepsilon\} = N(\{x\}, \varepsilon),$$

and

$$\overline{B}_M(x,\varepsilon) = \{y \in M : d(y,x) \leq \varepsilon\} = \overline{N}(\{x\},\varepsilon).$$

Definition 1.27 The distance between two subsets $A, B \subset M$ is measured by using the so called **Hausdorff's distance** which is defined by

$$d_H(A,B) = \max(\delta(B,A), \delta(A,B)).$$

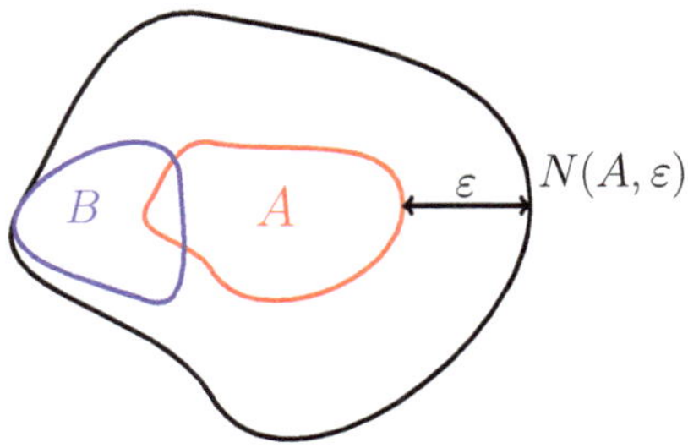

Fig. 1.1: *The figure illustrates the notion of Hausdorff's semi-distance of B to A. In the figure* $\varepsilon = \delta(B,A)$. *The black curve corresponds to the boundary of* $N(A,\varepsilon)$ *a* ε*-neighborhood of A.*

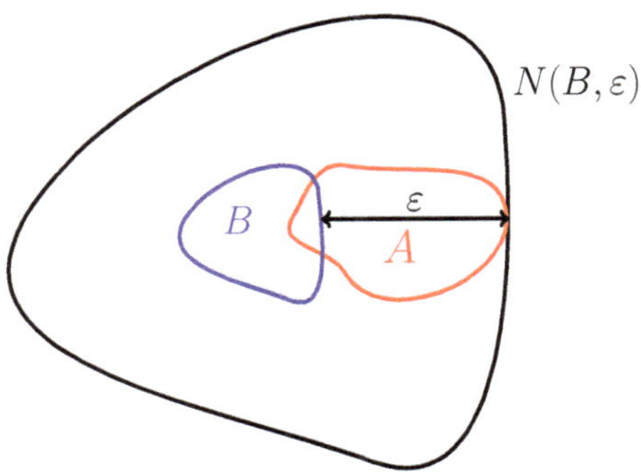

Fig. 1.2: *The figure illustrates the notion of Hausdorff's semi-distance of A to B. In the figure* $\varepsilon = \delta(A,B)$. *The black curve corresponds to the boundary of* $N(B,\varepsilon)$ *a* ε*-neighborhood of B.*

Definition 1.28 We say that a subset $A \subset M$ **attracts** a subset $B \subset M$ for a semiflow $\{U(t)\}_{t\in I_+}$ on M if

$$\lim_{t\to+\infty} \delta(U(t)B, A) = 0.$$

Remark 1.29 This means that for each $\varepsilon > 0$, we can find $t_0 = t_0(\varepsilon) > 0$ in I_+ (large enough) such that for each $t \geq t_0$, the subset $U(t)B = \{U(t)x : x \in B\}$ is included in $N(A,\varepsilon)$.

To illustrate the notion of attraction of sets, we first prove that positive orbit is attracted by the omega limit sets (whenever it is exists).

Theorem 1.30 *Let $\{u(t)\}_{t\in I_+} \subset M$ be a positive orbit passing through $x \in M$ at time $t = 0$. Assume that*

$$O_+(x) := \bigcup_{s\in I_+} \{u(s)\}$$

is relatively compact. Then $\omega(x)$ attracts $O_+(x)$ for U, that is,

$$\lim_{t\to+\infty} \delta\left(\bigcup_{s\geq t} \{u(s)\}, \omega(x)\right) = 0,$$

since

$$U(t)O_+(x) := \bigcup_{s\in I_+\cap[t,+\infty)} \{u(s)\}.$$

Proof We first observe that

$$U(t)O_+(x) = \bigcup_{s\in I_+\cap[t,+\infty)} \{u(s)\}, \forall t \in I_+.$$

Assume by contradiction that $\omega(x)$ does not attract $O_+(x)$ for U. Then we can find $\varepsilon > 0$ and a sequence $t_n \to +\infty$ such that

$$\delta(u(t_n), \omega(x)) \geq \varepsilon,$$

and we obtain a contradiction with Theorem **1.20**-(iii). □

Remark 1.31 If we consider the family of subsets $A_t := \bigcup_{s\geq t} \{u(s)\}$, then $\omega(x)$ attracts $O_+(x)$ for U if, and only if,

$$\lim_{t\to+\infty} \delta(A_t, \omega(x)) = 0.$$

Let $\{u(t)\}_{t\in -I_+} \subset M$ be a negative orbit passing through $x \in M$ at time $t = 0$. We can adapt this last notion of attractivity for the alpha limit sets. This can rewrites as follows: if the set

$$O_-(x) = \bigcup_{t\in -I_+} \{u(t)\}$$

is relatively compact, then

$$\lim_{t\to-\infty} \delta(B_t, \alpha(x)) = 0,$$

wherein

$$B_t = \bigcup_{s\in -I_+,\, s\leq t} \{u(s)\}.$$

The Hausdorff semi-distance measure the distance of B to A. Therefore if $B \subset A$ then $\delta(B, A) = 0$. Moreover, if $\varepsilon > 0$ then

$$\delta(B, A) = \varepsilon \Rightarrow B \subset \overline{N}(A, \varepsilon).$$

This means that

$$d(x, A) \leq \varepsilon, \quad \forall x \in B$$

and there exists a sequence $(x_n)_{n \geq 0} \subset B$ such that

$$\lim_{n \to \infty} d(x_n, A) = \varepsilon.$$

The Hausdorff distance can also be defined as follows

$$d_H(A, B) = \inf \{\varepsilon > 0 : A \subset N(B, \varepsilon) \text{ and } B \subset N(A, \varepsilon)\}.$$

Actually the Hausdorff distance is not a real distance, because we only have

$$d_H(A, B) = 0 \Leftrightarrow \overline{A} = \overline{B}.$$

But it becomes a true distance when restricted to the closed subsets.

Proposition 1.32 *Let (M, d) be a metric space. Then the set of closed subsets of M is a metric space endowed with the Hausdorff distance.*

Proof Due to the above analysis, it remains to prove the triangle inequality for the Hausdorff distance. Assume that

$$d_H(A, B) < \varepsilon \text{ and } d_H(B, C) < \varepsilon'.$$

From the proof of Lemma 1.22, we know that for each $x \in C$, $z \in B$,

$$d(x, A) \leq d(x, z) + d(z, A).$$

Let $x \in C$ be fixed. Since

$$d_H(B, C) < \varepsilon' \Rightarrow C \subset N(B, \varepsilon'),$$

we can choose $z \in B$ such that $d(z, x) \leq \varepsilon'$, and

$$d_H(A, B) < \varepsilon \Rightarrow B \subset N(A, \varepsilon).$$

We deduce that

$$d(x, A) \leq \varepsilon' + \varepsilon, \forall x \in C.$$

By taking the supremum with respect to x yields

$$\delta(C, A) \leq d_H(A, B) + d_H(B, C),$$

and by symmetry of the problem, we obtain

$$\delta(A, C) \leq d_H(A, B) + d_H(B, C).$$

We conclude that

$$d_H(C, A) \leq d_H(A, B) + d_H(B, C),$$

which completes the proof. □

1.6 Connectivity of ω-Limit Sets and α-Limit Sets

Internally chain transitive ω-limit sets and α-limit sets

The fundamental property of omega limit sets is the fact that each couple of points in a omega limit set can be almost connected by an orbit staying in a small neighborhood of the omega limit set. The following result is making this statement more precise.

Lemma 1.33 *Assume that the positive orbit $O_+(x)$ (respectively the negative orbit $O_-(x)$) is relatively compact. For each $a, b \in \omega(x)$ (respectively $a, b \in \alpha(x)$) and each $\varepsilon > 0$ we can find $a_\varepsilon, b_\varepsilon \in M$ and $t_\varepsilon > 0$ such that*

$$d(a, a_\varepsilon) \leq \varepsilon, \ d(b, b_\varepsilon) \leq \varepsilon,$$

$$U(t_\varepsilon)a_\varepsilon = b_\varepsilon,$$

and

$$U(t)a_\varepsilon \in N(\omega(x), \varepsilon), \forall t \in [0, t_\varepsilon],$$

which is equivalent to

$$d\left(U(t)a_\varepsilon, \omega\left(x\right)\right) \leq \varepsilon, \forall t \in [0, t_\varepsilon].$$

Proof We have $\delta(U(t)x, \omega(x)) \to 0$ as $t \to +\infty$. So we can find $t_0 > 0$ in I_+ such that

$$\delta(U(t)x, \omega(x)) \leq \varepsilon, \ \forall t \in I_+ \cap [t_0, +\infty).$$

Since a belongs to $\omega(x)$ we can find $t_1 > t_0$ such that $d(U(t_1)x, a) \leq \varepsilon$. Set $a_\varepsilon = U(t_1)x$. Then since b belongs to $\omega(x)$ we can find $t_2 > t_1 > t_0$ such that $d(U(t_2)x, b) \leq \varepsilon$. Set $b_\varepsilon = U(t_2)x$ and the result follows. □

Definition 1.34 Let A be a subset of M. We say that $a \in A$ **is chained to** $b \in A$ **in** A, if for each $t^\star > 0$, for each $\varepsilon > 0$, and each $\eta > 0$, there exist $\tau \in [t^\star, t^\star + \eta] \cap I_+$, and $x_1, x_2, ..., x_m \in A$ (with $m \geq 2$) such that

$$x_1 = a, \ x_m = b, \text{ and } d\left(U(\tau)x_i, x_{i+1}\right) \leq \varepsilon, \forall i = 1, ..., m-1.$$

We will say that A is **internally chain transitive**, if for each $a, b \in A$, a is chained to b in A.

Theorem 1.35 *Let $\{U(t)\}_{t \in I_+}$ be a continuous semiflow on (M, d). Then the omega limit set (respectively alpha limit set) of a relatively compact positive orbit (respectively negative orbit) is internally chain transitive.*

Before proving this theorem, let us state and prove the following lemma.

Lemma 1.36 *Let $\{U(t)\}_{t\in I_+}$ be a continuous semiflow on (M, d). Let C be a compact subset of M. Then for each $\varepsilon > 0$, and each $t^\star \in I_+$ (with $t^\star > 0$), there exists $\delta > 0$, such that $s \in [t^\star, t^\star + \delta] \cap I_+$, $u, v \in M$,*

$$d(u, C) \leq \delta, d(v, C) \leq \delta, \text{ and } d(u, v) \leq \delta \Rightarrow d(U(s)u, U(s)v) \leq \varepsilon.$$

Proof Assume by contradiction that there exists $\varepsilon > 0$, $t^\star \in I_+$ with $t^\star > 0$, and two sequences $u_n \in M$, and $v_n \in M$, and $s_n \in [t^\star, t^\star + 1/n] \cap I_+$, such that for each integer $n > 0$,

$$d(u_n, C) \leq 1/n, d(v_n, C) \leq 1/n, \text{ and } d(u_n, v_n) \leq 1/n,$$

and

$$d(U(s_n)u_n, U(s_n)v_n) \geq \varepsilon. \tag{1.10}$$

By definition of the distance $d(x, C)$, one can find two sequences $u_n^C \in C$ and $v_n^C \in C$ such that

$$d\left(u_n, u_n^C\right) \leq 2/n \text{ and } d\left(v_n, v_n^C\right) \leq 2/n.$$

By using the triangle inequality we obtain

$$d\left(u_n^C, v_n^C\right) \leq d\left(u_n^C, u_n\right) + d(u_n, v_n) + d\left(v_n, v_n^C\right) \leq 5/n.$$

Now, since C is compact, we can find some converging sub-sequences (still denoted with the same index), such that

$$u_n^C \to w, \text{ and } v_n^C \to w, \text{ as } n \to +\infty.$$

By using the continuity of the semiflow $(t, x) \to U(t)x$, we deduce from (1.10) that

$$0 = d\left(U(t^\star)w, U(t^\star)w\right) \geq \varepsilon > 0,$$

A contradiction which completes the proof of the lemma. □

Proof (of Theorem 1.35) Let us prove the result for omega limit set (the proof for alpha limit set is similar). Let $x \in M$ be given and assume that $\overline{\gamma^+(x)}$ is compact. Then $\omega(x)$ is nonempty, compact, invariant and

$$\lim_{t\to+\infty} d(U(t)x, \omega(x)) = 0.$$

Let $t^\star \in I_+$ (with $t^\star > 0$), $\varepsilon > 0$, and $\eta > 0$ be fixed. By continuity of U, and the compactness of $\omega(x)$, due to the above lemma we can find $\delta \in (0, \varepsilon/3) \cap (0, \eta)$, with the following property: If $s \in [t^\star, t^\star + \delta] \cap I_+$, $u, v \in M$,

$$d(u, \omega(x)) \leq \delta, d(v, \omega(x)) \leq \delta, \text{ and } d(u, v) \leq \delta \Rightarrow d(U(s)u, U(s)v) \leq \varepsilon/3.$$

Since

$$\lim_{t\to+\infty} d(U(t)x, \omega(x)) = 0,$$

we can find $t_1 \in I_+$ such that

$$d\left(U(t)x, \omega(x)\right) < \delta, \forall t \geq t_1.$$

Let $a, b \in \omega(x)$. Then we can find $t_a \geq t_1$ in I_+ such that

$$d\left(U(t_a)x, a\right) < \delta.$$

Let $k \in \mathbb{N}$, $k \geq 2$, such that $t^\star/k \leq \delta$. Then we can find $t_b \geq t_a + kt^\star$ in I_+ such that

$$d\left(U(t_b)x, b\right) < \delta.$$

However there exists $m \geq k+1$, such that $(m-1)\,t^\star \leq t_b - t_a < mt^\star$. We set $\tau = \dfrac{t_b - t_a}{m-1}$. Then by construction of $t^\star$, we have $\tau \in I_+$, and

$$\tau \in \left[t^\star, \left(1 + \frac{1}{m-1}\right)t^\star\right] \subset \left[t^\star, \left(1 + \frac{1}{k}\right)t^\star\right] \subset \left[t^\star, t^\star + \delta\right].$$

We set

$$y_1 = a, y_2 = U\left(\tau\right)U(t_a)x, \ldots, y_{m-1} = U\left((m-2)\tau\right)U(t_a)x, y_m = b.$$

Then we have

$$d\left(U\left(\tau\right)y_1, y_2\right) = d\left(U\left(\tau\right)a, U\left(\tau\right)U(t_a)x\right)$$

and since

$$a \in \omega(x), d\left(U(t_a)x, \omega(x)\right) < \delta, \text{ and } d\left(U(t_a)x, a\right) < \delta,$$

and $\tau \in [t, t+\delta]$, we deduce that

$$d\left(U\left(\tau\right)y_1, y_2\right) = d\left(U\left(\tau\right)a, U\left(\tau\right)U(t_a)x\right) \leq \varepsilon/3,$$

$$U\left(\tau\right)y_j = y_{j+1}, \forall j = 2, ..., m-2,$$

and

$$d\left(U\left(\tau\right)y_{m-1}, y_m\right) = d\left(U(t_b)x, b\right) \leq \delta \leq \varepsilon/3.$$

Now for each $j = 2, ..., m-1$, there exists $x_j \in \omega(x)$, such that $d\left(x_j, y_j\right) \leq \delta$. By setting $x_1 = a$ and $x_m = b$, we obtain for $j = 1, ..., m-1$,

$$\begin{aligned} d\left(U\left(\tau\right)x_j, x_{j+1}\right) &\leq d\left(U\left(\tau\right)x_j, U\left(\tau\right)y_j\right) + d\left(U\left(\tau\right)y_j, y_{j+1}\right) + d\left(y_{j+1}, x_{j+1}\right) \\ &\leq \varepsilon/3 + \varepsilon/3 + \delta \leq \varepsilon. \end{aligned}$$

The proof is completed. □

Invariantly connected ω-limit and α-limit Sets

Definition 1.37 A compact invariant set A is said to be **invariantly connected** if it is not the union of two nonempty disjoint compact invariant subsets. That is to say that if $A_1 \neq \emptyset$, and $A_2 \neq \emptyset$ are non empty and compact subsets satisfying

$$A = A_1 \cup A_2, \text{ with } A_1 \cap A_2 = \emptyset.$$

Then either A_1 or A_2 are not invariant by U. That is,

$$\text{either } U\left(t^\star\right) A_1 \neq A_1, \text{ or } U(t^\star)A_2 \neq A_2.$$

for some $t^\star > 0$.

Theorem 1.38 *Let $\{U(t)\}_{t\in I_+}$ be a continuous semiflow on a metric space (M, d). Then the omega limit set (respectively alpha limit set) of a relatively compact positive orbit (respectively negative orbit) is invariantly connected.*

Theorem 1.38 follows from Theorem 1.35 and the following lemma.

Lemma 1.39 *Let A be a compact subset of M which is invariant by U. If A is internally chain transitive then A is also invariantly connected.*

Proof We can prove the result by contradiction. Assume that A is the union of two disjoint closed invariant sets $A = A_1 \cup A_2$. We get a contradiction because the subsets A_1 and A_2 are invariant. So if we fix $\tau > 0$ then when $d(x, A_1) \leq \varepsilon$, then $d(U(\tau)x, A_2) > 2\varepsilon$ whenever $\varepsilon > 0$ is small enough. So a point of A_1 cannot be chained to a point of A_2. □

Connected ω-limit and α-limit sets

Definition 1.40 Let (M, d) be a metric space. Let $C \subset M$ be a subset of M. We will say that a pair of non empty subsets $A \subset C$ and $B \subset C$ is a **partition of** C if

$$A \cup B = C \text{ and } A \cap B = \emptyset.$$

Definition 1.41 Let (M, d) be a complete metric space. A subset C of M is said to be **connected** if there exists no partition of C in two subsets A and B which are both open subsets for the topology of (C, d). We will say that C is **disconnected** if C is not connected.

Remark 1.42 Recall that a subset is closed in (C, d) if its complementary set in C is open in (C, d). Therefore, in the above definition, it is equivalent to say that both subsets A and B are also closed in (C, d).

Assume that C is not connected. Then there exists a partition of C in two subsets A and B which are both open subsets for the topology of (C, d). Recall that a subset A is open in (C, d), if and only if for each $x \in A$, there exists $\varepsilon_x > 0$ such that

$$B_C(x, \varepsilon_x) = \{y \in C : d(x, y) \leq \varepsilon_x)\} \subset A,$$

where $B_C(x, \varepsilon_x)$ is the ball of center x and radius ε_x (in C).

Remark that

$$B_C(x, \varepsilon_x) = B_M(x, \varepsilon_x) \cap C,$$

where

$$B_M(x, \varepsilon_x) = \{y \in M : d(x, y) \leq \varepsilon_x\}.$$

So we must have

$$B_M(x, \varepsilon_x) \cap B = \emptyset.$$

Now we can define

$$U = \bigcup_{x \in A} B_M(x, \varepsilon_x),$$

and

$$V = \bigcup_{x \in B} B_M(x, \varepsilon_x),$$

where for each $x \in B$, ε_x is chosen small enough to guaranty

$$B_M(x, \varepsilon_x) \cap A = \emptyset.$$

We observe that U and V satisfy

$$U \cap C = A \text{ and } V \cap C = B$$

and

$$U \cap V = \emptyset.$$

Therefore we obtain the following proposition.

Proposition 1.43 *Let (M, d) be a complete metric space. A subset C is disconnected if there exist two open subsets $U \subset M$ and $V \subset M$, such that $C \subset U \cup V$,*

$$(C \cap U) \neq \emptyset \text{ and } (C \cap V) \neq \emptyset$$

and

$$U \cap V = \emptyset.$$

Definition 1.44 An **interval** in $\mathbb{R}$ is a subset I such that

$$a < c < b \text{ and } a, b \in I \Rightarrow c \in I.$$

Remark 1.45 This is definition can be extended to $\mathbb{R}^n$ endowed with some partial order $\leq$.

Theorem 1.46 *A connected set of real numbers is an interval.*

Remark 1.47 The converse is also true. An interval of real numbers is a connected set.

Proof Let C be a connected set in $\mathbb{R}$. Assume by contradiction that C is not an interval. Then we can find $a, b \in C$ and $c \notin C$ with $a < c < b$. The subsets $U = (-\infty, c)$ and $V = (c, \infty)$ are both open in $\mathbb{R}$, $C \subset U \cap V$ and

$$a \in C \cap U \neq \emptyset, \text{ and } b \in C \cap V \neq \emptyset.$$

Therefore $A = C \cap U$ and $B = C \cap V$ are both open in C. We deduce that C is disconnected. □

Theorem 1.48 *Let $T : M \to \widetilde{M}$ be a continuous map from a metric space (M, d) to a metric space $(\widetilde{M}, \widetilde{d})$. Then M is connected implies that $T(M)$ is connected.*

Proof Assume by contradiction that $T(M)$ is not connected. Then there exists $\widetilde{A}$ and $\widetilde{B}$ two open subset of $(T(M), \widetilde{d})$

$$\widetilde{A} \cup \widetilde{B} = T(M) \text{ and } \widetilde{A} \cap \widetilde{B} = \emptyset.$$

By continuity of T, we deduce that $A = T^{-1}\left(\widetilde{A}\right)$ and $B = T^{-1}\left(\widetilde{B}\right)$ are open in (M, d) and

$$A \cup B = M \text{ and } A \cap B = \emptyset.$$

This contradicts the fact that M is connected. The proof is completed. □

As a consequence of Proposition 1.43, Theorem 1.46, and Theorem 1.48, we obtain the following results that will be useful in the applications.

Theorem 1.49 *Let $M \subset X$ be a subset of Banach space $(X, \|.\|)$. Then we have the following properties:*

(i) *M is convex $\Rightarrow$ M is connected.*
(ii) *If M is connected and $x^* : X \to \mathbb{R}$ is a bounded linear map then*

$$I = \{x^*(x) : x \in M\} \subset \mathbb{R}$$

is an interval in $\mathbb{R}$.

As a consequence of the above theorem we have for example, a connected set in $\mathbb{R}^n$ becomes an interval when it is projected onto the axes.

Example of not connected ω-limit and α-limit sets: The ω-limit set (respectively the α-limit set) of a relatively compact positive orbit (respectively negative orbit) generated by a discrete time semiflow is not connected in general. Indeed, assume that $T : M \to M$ has a 2-periodic orbit

$$T(a) = b \text{ and } T(b) = a,$$

with

$$a \neq b.$$

If we define the complete orbit

$$u(n) = \begin{cases} a, & \text{if } n = 2k, \text{ for some integer } k \in \mathbb{Z}, \\ b, & \text{if } n = 2k + 1, \text{ for some integer } k \in \mathbb{Z}, \end{cases}$$

then the ω-limit set of the solution starting from a or b is

$$\omega(a) = \omega(b) = \{a, b\},$$

and

$$\alpha(a) = \alpha(b) = \{a, b\}.$$

This provides an example of disconnected ω-limit and α-limit sets.

The case of continuous time semiflow is different.

Theorem 1.50 *Let (M, d) be a complete metric space. The ω-limit set of a relatively compact orbit generated by a continuous time semiflow $\{U(t)\}_{t\in\mathbb{R}_+}$ is connected.*

Proof (of Theorem 1.50) Assume by contradiction that $\omega(x)$ is disconnected. Then there are disjoint open subsets U and V of M such that $U \cap \omega(x)$ and $V \cap \omega(x)$ are nonempty and $\omega(x) \subset U \cup V$. Let $a \in U \cap \omega(x)$ and $b \in V \cap \omega(x)$. Then we can have a sequence $t_1, t_2, \dots t_k \to \infty$ and a sequence $s_1, s_2, \dots s_k \to \infty$ (with $t_k \leq s_k$) such that $U(t_k)x \in U \to a$, and $U(s_k)x \in V \to b$. But $\{U(t)x : t \in [t_k, s_k]\}$ is a connected curve going from a point in U to a point in V. Therefore, there must be able to find $\tau_k \in (t_k, s_k)$ such that $U(\tau_k)x \in M \setminus (U \cup V)$. But the sequence $k \to U(\tau_k)x$ is relatively compact, so up to a sub-sequence (denoted with the same index) we can assume that

$$U(\tau_k)x \to c.$$

But by construction $M \setminus (U \cup V)$ is a closed subset, and it follows that

$$c \in M \setminus (U \cup V) \text{ and } c \in \omega(x).$$

We obtain a contradiction since $\omega(x) \subset (U \cup V)$, which completes the proof of the theorem. □

1.7 Dissipation and Absorbing Sets

Definition 1.51 A continuous semiflow $\{U(t)\}_{t\in I_+}$ on a metric space (M, d) is said to be **point dissipative (respectively compact dissipative, bounded dissipative)** if there exists a bounded set $B_0 \subset M$ attracting the points (respectively the compact subsets, the bounded subsets).

The notion of dissipative semiflow can be expressed by using one of the two following equivalent properties:

(i) There exists $B_0 \subset M$ a bounded subset such that

$$\lim_{t\to+\infty} \delta(U(t)B, B_0) = 0,$$

whenever B is a point (respectively a compact subset, a bounded subset).

(ii) For each $\varepsilon > 0$, and each subset $B \subset M$ that is a point (respectively a compact subset, a bounded subset), there exists $t_0 = t_0(\varepsilon, B) \in I_+$ with $t_0 > 0$ such that

$$U(t)B \subset \overline{N}(B_0, \varepsilon), \forall t \in I_+ \cap [t_0, +\infty),$$

where $\overline{N}(B_0, \varepsilon)$ is a closed ε-neighborhood of B_0 defined by

$$\overline{N}(B_0, \varepsilon) := \{x \in M : d(x, B_0) \leq \varepsilon\}.$$

Definition 1.52 A subset $B_0 \subset M$ is called **point absorbing, compact absorbing, bounded absorbing** if for each subset $B \subset M$ which is respectively a single point, a compact subset, a bounded subset, there exists $t_0 = t_0(B) \in I_+$ such that

$$U(t)B \subset B_0, \forall t \in I_+ \cap [t_0, +\infty).$$

A bounded absorbing subset is called **absorbing subset**.

1.8 Examples

1.8.1 Logistic equations: heteroclinic orbit

Consider the scalar logistic equation

$$N'(t) = N(t) - N(t)^2, \forall t \geq 0 \text{ and } N(0) = x. \tag{1.11}$$

The solution is explicitly given by the formula

$$N(t) = \frac{e^t x}{1 + \int_0^t e^l x dl}, \forall t \geq 0.$$

Define the maximal backward time of existence

$$\tau^-(x) = \inf\left\{t < 0 : 1 - \int_s^0 e^l x dl > 0, \forall s \in [t, 0]\right\}.$$

Then one has

$$\int_{-\infty}^0 e^l dl = 1,$$

therefore we get

$$\tau^-(x) = -\infty, \forall x \in [0, 1]$$

and the global solution is

$$N(t) = \frac{e^t x}{1 + \int_0^t e^\sigma x d\sigma}, \forall t \in \mathbb{R}. \tag{1.12}$$

It is clear that 0 and 1 are equilibrium solutions. Moreover for each $x \in (0, 1)$, the solution defined in (1.12) is a heteroclinic orbit and

$$\lim_{t \to -\infty} N(t) = 0 \text{ and } \lim_{t \to +\infty} N(t) = 1.$$

This is equivalent to say that

$$\alpha(x) = \{0\} \text{ and } \omega(x) = \{1\}.$$

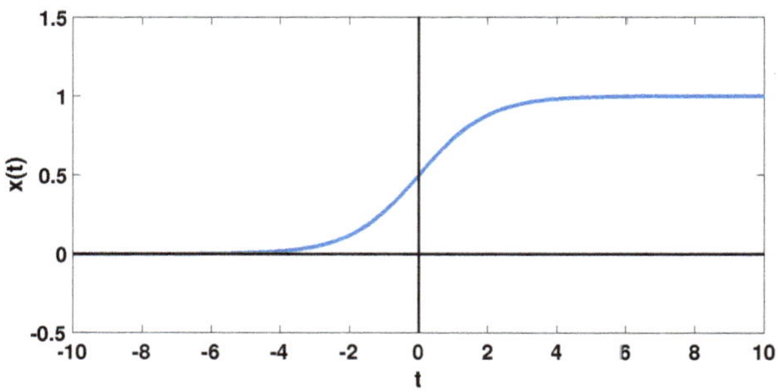

Fig. 1.3: *In this figure the blue curve represents the heteroclinic orbit* (1.6) *whenever* $x = 0.5$.

Dissipation: Consider the map

$$V(N) = (N - 1)^2.$$

We have

$$V'(N) = 2(N - 1)N' = 2(N - 1)N(1 - N) = -2N(1 - N)^2 \leq 0.$$

Theorem 1.53 *The semiflow*

$$U(t)x = \frac{e^t x}{1 + \int_0^t e^{\sigma} x d\sigma}, \forall t \in \mathbb{R}_+$$

of scalar logistic equation (1.11) *is bounded dissipative on* $\mathbb{R}_+$. *More precisely, each such that* $B_0 = [0, N_1]$ *(with* $N_1 > 2$*) is bounded absorbing set. That is to say that for each bounded set* $B \subset [0, +\infty)$, *there exists* $t_0 = t_0(B)$, *such that*

$$U(t)B \subset B_0, \forall t \geq t_0.$$

Proof We first observe that

$$\sup_{x \in [0,N_1]} V(x) = (N_1 - 1)^2, \forall N_1 > 2,$$

therefore

$$\left\{x \geq 0 : V(x) \leq (N_1 - 1)^2\right\} = [0, N_1], \forall N_1 > 2.$$

We choose $B_0 = \left\{x \geq 0 : V(x) \leq (N_1 - 1)^2\right\}$ for some $N_1 > 2$. Next, we observe that B_0 is positively invariant by U. Indeed, we have for each $x \in B_0$

$$V(U(t)x) \leq V(x) \leq \sup_{x \in B_0} V(x) = (N_1 - 1)^2.$$

Therefore

$$\sup_{x \in B_0} V(U(t)x) \leq (N_1 - 1)^2 \Rightarrow U(t)B_0 \subset B_0, \forall t \geq 0.$$

Assume by contradiction, that there exists a bounded set $B \subset [0, \infty)$, such that

$$\sup_{x \in B} V(U(t)x) \geq (N_1 - 1)^2, \forall t \geq 0.$$

Then we can construct a sequence $x_n \in B$ such that for each integer $n \geq 0$,

$$V(U(n)x_n) \geq (N_1 - 1)^2, \forall n \in \mathbb{N}.$$

Since the sequence x_n is bounded, we can find a sub-sequence (denoted with the same index) such that $x_n \to x_\infty$ and by continuity of U we deduce that

$$V(U(t)x_\infty) \geq (N_1 - 1)^2, \forall t \geq 0. \tag{1.13}$$

But since $N_1 \geq 2$, we must have

$$U(t)x_\infty > N_1, \forall t \geq 0. \tag{1.14}$$

Finally observe that

$$V'(U(t)x_\infty) = -U(t)x_\infty \times (1 - U(t)x_\infty)^2 = -U(t)x_\infty \times V\left(U(t)x_\infty\right),$$

so by integrating this formula and by using (1.14), we obtain

$$V\left(U(t)x_\infty\right) = e^{-\int_0^t U(\sigma)x_\infty d\sigma} V\left(U(t)x_\infty\right) \leq e^{-tN_1} V(x) \to 0, \text{ as } t \to +\infty.$$

We obtain a contradiction with (1.13). □

Theorem 1.54 *Consider the semiflow $U(t)$ of scalar logistic equation* (1.11) *restricted to* $(0, \infty)$. *Then the subset $B_0 = [N_0, N_1]$ (with $0 < N_0 < 1 < N_1$) is a compact absorbing set, but B_0 is not a bounded absorbing set. Moreover precisely*

(i) *The subset* $(0, 1]$ *attracts all the bounded subsets in* $(0, +\infty)$;
(ii) *The subset* $(0, 1]$ *is invariant by U. That is*

$$U(t)(0, 1] = (0, 1], \forall t \geq 0;$$

(iii) *The subset* $(0, 1]$ *is not compact in* $(0, +\infty)$*;*
(iv) *The subset* $(0, N_1]$ *is a bounded absorbing set in* $(0, +\infty)$.

1.8.2 Poincaré normal form: periodic orbit

Consider the Poincaré normal form

$$\begin{cases} x'(t) = \frac{\lambda}{2}x(t) - \omega y(t) - \frac{\kappa}{2}\left(x(t)^2 + y(t)^2\right)x(t) \\ y'(t) = \omega x(t) + \frac{\lambda}{2}y(t) - \frac{\kappa}{2}\left(x(t)^2 + y(t)^2\right)y(t) \end{cases} \tag{1.15}$$

From Section 1.1, we know that the semiflow generated by (1.15) is defined by

$$\begin{aligned} V(t)\begin{pmatrix} x \\ y \end{pmatrix} &= \sqrt{\frac{e^{\lambda t}\left(x^2 + y^2\right)}{1 + \kappa \int_0^t e^{\lambda s}\left(x^2 + y^2\right) ds}} \times \frac{1}{\sqrt{x^2 + y^2}} \\ &\quad \times \begin{pmatrix} \cos(\omega t)x - \sin(\omega t)y \\ \sin(\omega t)x + \cos(\omega t)y \end{pmatrix}, \end{aligned} \tag{1.16}$$

whenever $(x, y) \neq (0, 0)$ and

$$V(t)\begin{pmatrix} 0 \\ 0 \end{pmatrix} = \begin{pmatrix} 0 \\ 0 \end{pmatrix}.$$

Exercise 1.55 Derive the above formula by using the following changes of variables. Consider

$$r(t)^2 = x^2(t) + y(t)^2$$

and prove that

$$(r(t)^2)' = \lambda r(t)^2 - \kappa r(t)^4.$$

Consider

$$X(t) = \frac{x(t)}{\sqrt{x^2(t) + y^2(t)}}$$

and

$$Y(t) = \frac{y(t)}{\sqrt{x^2(t) + y^2(t)}}$$

whenever $(x, y) \neq (0, 0)$. Prove that

$$X' = -\omega Y, \text{ and } Y' = \omega X.$$

Figures 1.4, 1.5 and 1.6 illustrate the behavior of the solutions of the Poincaré normal form.

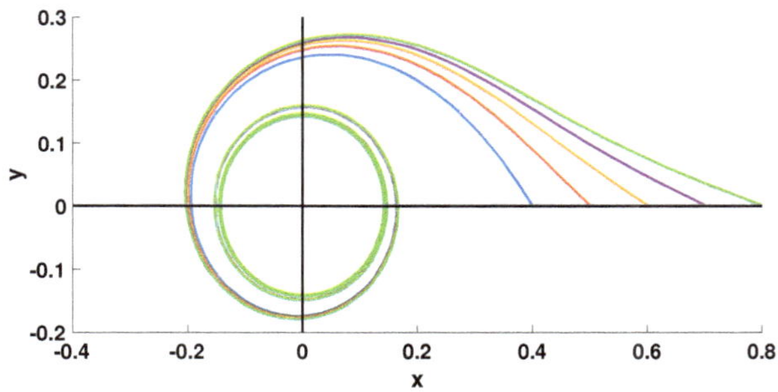

Fig. 1.4: *We plot some solutions of* (1.15) *in the phase plane* $(x(t), y(t))$ *whenever* $\lambda = 0.02$, $\omega = 0.1$ *and* $\kappa = 1$. *We choose several initial values where* $x = 0.2, 0.3, \ldots, 0.8$ *and* $y = 0$. *One may observe that the omega limit set of these solutions is the central circle, while the alpha limit set is empty since the norm of the solutions eventually blowup when the time goes to* $-\infty$.

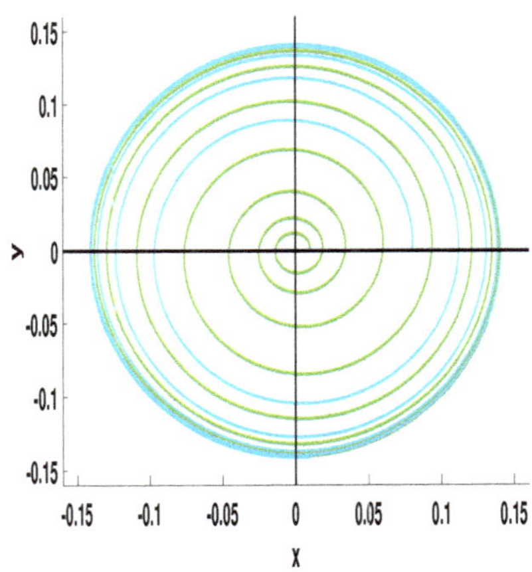

Fig. 1.5: *We plot some solutions of* (1.15) *in the phase plane* $(x(t), y(t))$ *whenever* $\lambda = 0.02$, $\omega = 0.1$ *and* $\kappa = 1$. *We choose several initial values, where* $x = 0.01, 0.08$ *and* $y = 0$. *The solutions are part of a complete orbits jointing the trivial equilibrium* 0 *to the circular periodic orbit.*

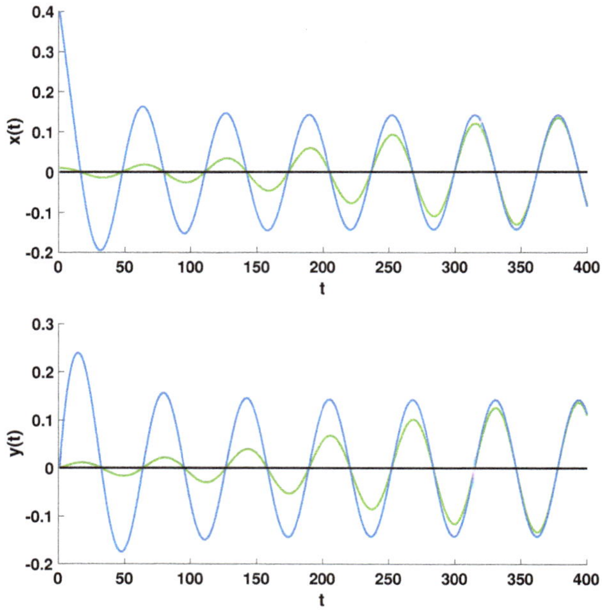

Fig. 1.6: *We plot* $(t, x(t))$ *(top) and* $(t, y(t))$ *(bottom) for two solutions of* (1.15) *whenever* $\lambda = 0.02$, $\omega = 0.1$ *and* $\kappa = 1$. *We choose several initial values where* $x = 0.01, 0.4$ *and* $y = 0$. *Both solutions coincide whenever they converge to the periodic orbit. That is because both solutions turn around* 0 *with the same rotation speed. This appears explicitly in the semiflow formula* (1.16) *(i.e. the rotation is guided by the linear term* $e^{At}X$*).*

1.8.3 Homoclinic orbit for a second order logistic equation

Consider the equation

$$x''(t) = -x(t)(1 - x(t)), \forall t \geq 0, \text{ with } x(0) = x_0 \text{ and } x'(0) = x'_0. \tag{1.17}$$

This equation can be regarded as the following system

$$\begin{cases} x'(t) = -y(t) \\ y'(t) = x(t)(1 - x(t)) \end{cases} \tag{1.18}$$

and the equilibria of (1.18) are

$$(x, y) = (0, 0) \text{ and } (x, y) = (1, 0).$$

System (1.18) is a Hamiltonian system which corresponds to a special case of the Bogdanov-Takens normal form (see [54] and [201]). The terminology Hamiltonian means that we have

$$x(1-x)x' = -yy'.$$

Therefore we obtain the following conservation property along the trajectories

$$\frac{x(t)^2}{2} - \frac{x(t)^3}{3} = -\frac{y(t)^2}{2} + h,$$

with a constant of integration $h \in \mathbb{R}$.

In order to study the trajectories passing through the points $y_0 = 0$ and x_0 between 0 and 1, we obtain the following condition for h

$$x_0^2\left(\frac{1}{2} - \frac{x_0}{3}\right) = h.$$

The function $x \to \frac{x^2}{2} - \frac{x^3}{3}$ is increasing between 0 and 1. Therefore h varies between 0 (for $x = 0$) and $\frac{1}{6}$ (for $x = 1$).

We deduce that the orbit passing through the point $(x, 0)$ (with $x \in [0, 1]$) satisfies

$$y = \pm\sqrt{2\left[h - \left(\frac{x^2}{2} - \frac{x^3}{3}\right)\right]},$$

where x takes some appropriate values in order for the quantity below the square root to be positive.

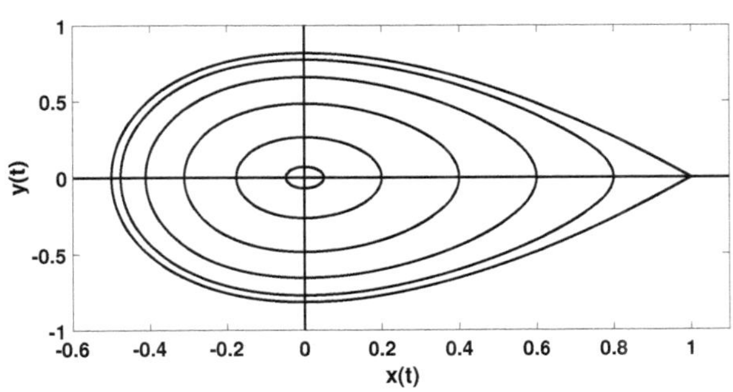

Fig. 1.7: *We plot the complete orbits* $(x(t), y(t))$ *passing respectively through* $x = 0.05, 0.2, 0.4, 0.6, 0.8$ *and* $y = 0$. *The last orbit is a homoclinic orbit passing through* $(x, y) = (-0.5, 0)$ *and the alpha limit set as well as the omega limit set is the equilibrium* $(x, y) = (1, 0)$.

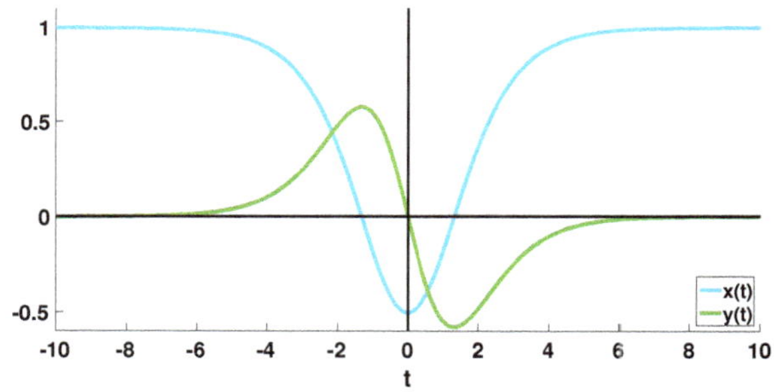

Fig. 1.8: *We plot the homoclinic orbits* $t \to (x(t), y(t))$ *which is the solution passing through* $(x, y) = (-0.5, 0)$. *We observe numerically the convergence to* $(1, 0)$ *when the time* t *goes to* $\pm\infty$.

1.8.4 Beverton and Holt discrete time model

The model of Beverton and Holt [29] was introduced in the context of fisheries in 1957. This model is the following

$$N(t+1) = \frac{\beta N(t)}{1 + \alpha N(t)}, \ \forall t \geq 0, \text{ with } N(0) = N_0 \geq 0, \tag{1.19}$$

where $N(t)$ denotes the number of individuals, $\beta > 0$ is the growth rate of the population and the term $1/(1 + \alpha N(t))$ (with $\alpha \geq 0$) describes the competition for food, cannibalism effect (indeed the adult fish eat the larvea when they reproduce) etc... The competition occurs between individuals of the same species. Therefore this effect is usually called *intra-specific competition*.

Semiflow: Recall that the semiflow $\{U(t)\}_{t \in \mathbb{N}}$ is defined by

$$U(t)(N_0) = T^t(N_0), \forall t \geq 0, \forall N_0 \geq 0,$$

where $T(N_0) = \frac{\beta N(t)}{1+\alpha N(t)}$ and T^t is defined by induction as follows

$$T^{t+1}(N_0) = T\left(T^t(N_0)\right) = T^t(T(N_0)), \forall t \in \mathbb{N},$$

and

$$T^0(N_0) = N_0.$$

Equilibria: It is readily checked that 0 is always an equilibrium of (1.19). The non-zero equilibrum satisfies

$$\frac{\beta}{1 + \alpha \overline{N}} = 1 \Leftrightarrow \overline{N} = \frac{\beta - 1}{\alpha}.$$

Dissipation property: We follow the construction of the Liapunov function by Fisher and Goh [112]. Define

$$V(N(t)) = |N(t) - \overline{N}|.$$

Then one has

$$\begin{aligned} V(N(t+1)) &= \left|N(t+1) - \overline{N}\right| = \left|\frac{\beta N(t)}{1+\alpha N(t)} - \frac{\beta - 1}{\alpha}\right| \\ &= \left|\frac{\alpha\beta N(t) - (\beta - 1)(1 + \alpha N(t))}{\alpha(1+\alpha N(t))}\right| \\ &= \left|\frac{\alpha N(t) - (\beta - 1)}{\alpha(1+\alpha N(t))}\right|, \end{aligned}$$

and we obtain

$$V(N(t+1)) = \frac{1}{(1+\alpha N(t))} V(N(t)). \tag{1.20}$$

We deduce that if we define $\{U(t)\}_{t\in\mathbb{N}}$ as the discrete time semiflow generated by (1.19) for each $t \in \mathbb{N}$, then we have

$$\begin{cases} V(U(t)N_0) < V(N_0), \text{ if } N_0 > 0, \text{ and } N_0 \neq \overline{N}, \\ V(U(t)N_0) = V(N_0), \text{ if } N_0 = 0, \text{ or } N_0 = \overline{N}. \end{cases} \tag{1.21}$$

Theorem 1.56 *The discrete time semiflow generated by* (1.19) *is bounded dissipative on* $[0, \infty)$. *More precisely, for each* $\gamma_0 > \overline{N}$ *the subset*

$$B_{\gamma_0} = \{x \geq 0 : V(x) \leq \gamma_0\}$$

is a bounded absorbing subset.

Proof We have for each $N_0 > 0$, with $V(N_0) \geq \overline{N}$,

$$V(U(t)N_0) < V(N_0), \forall t \geq 0,$$

and the result follows. □

Remark 1.57 One may observe that this model can be derived from the logistic equation (1.11) by using the following semi-implicit scheme

$$\frac{N(t+\Delta t) - N(t)}{\Delta t} = N(t)(\lambda - \chi N(t+\Delta t))$$

which is equivalent to

$$N(t+\Delta t) = \frac{(1+\Delta t\lambda)N(t)}{1+\Delta t\chi N(t)}.$$

The behavior of (1.19) is analogous to the behavior of the solution of the logistic equation (1.11) (see Figure 1.9).

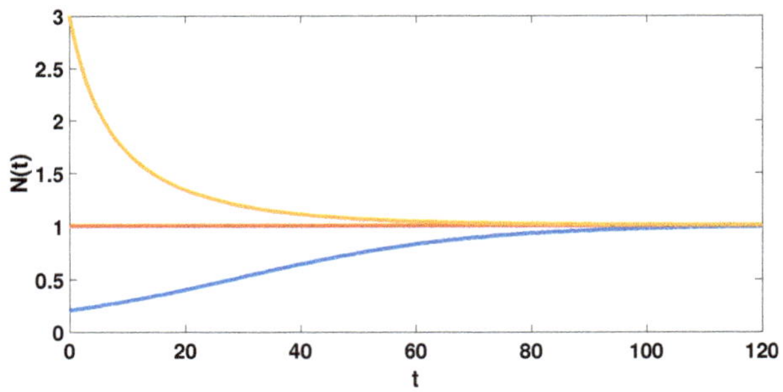

Fig. 1.9: *In this figure we plot three solutions of equation* (1.19) *with* $\alpha = 0.05$ *small and* $\lambda = 1 + \alpha$ *(similarly to the above approximation of the logistic equation). Then we choose three initial value* $N_0 = 0.2$ *(blue curve),* $N_0 = 1$ *(green curve) and* $N_0 = 3$ *(red curve). The green curve corresponds to the equilbrium* $N = 1$*. The second member of* (1.19) *being (concave) monotone increasing no other behavior can be observed whenever* $\lambda > 1$.

1.8.5 Ricker model: chaotic behavior

The first population dynamics model introduced in the context of fisheries to describe the reproduction of salmons in Canada was written by Ricker [280, 281] in 1954. The model reads as follows

$$N(t+1) = \beta N(t) \exp(-N(t)), \ \forall t \geq 0, \ \text{with } N(0) = N_0 \geq 0, \tag{1.22}$$

where $\beta > 0$ is the growth rate of the population and the term $\exp(-N(t))$ (with $\alpha \geq 0$) describes the intra-specific competition.

Semiflow: Recall that the semiflow $\{U(t)\}_{t \in \mathbb{N}}$ is defined by

$$U(t)(N_0) = T^t(N_0), \forall t \geq 0, \forall N_0 \geq 0,$$

where $T(N_0) = \beta N_0 \exp(-N_0)$ and T^t is defined by induction as follows

$$T^{t+1}(N_0) = T\left(T^t(N_0)\right) = T^t(T(N_0)), \forall t \in \mathbb{N},$$

and

$$T^0(N_0) = N_0.$$

Equilibria: It is readily checked that 0 is always an equilibrium of (1.22). The non-zero equilibrum satisfies

$$\beta \exp(-\overline{N}) = 1 \Leftrightarrow \overline{N} = \ln(\beta).$$

Dissipation property: We follow the construction of the Liapunov function by Fisher and Goh [112]. Define

$$V(N(t)) = \left(N(t) - \overline{N}\right)^2.$$

Then

$$\begin{aligned} V(N(t+1)) - V(N(t)) &= \left(N(t+1) - \overline{N}\right)^2 - \left(N(t) - \overline{N}\right)^2 \\ &= \left(\beta N(t)\exp(-N(t)) - \overline{N}\right)^2 - \left(N(t) - \overline{N}\right)^2 \\ &= (\beta N(t)\exp(-N(t)))^2 - N(t)^2 \\ &\quad -2\beta N(t)\exp(-N(t))\overline{N} + 2N(t)\overline{N} \\ &= [\beta\exp(-N(t)) - 1]\,[\beta\exp(-N(t)) + 1]\,N(t)^2 \\ &\quad -2N(t)\overline{N}\,[\beta\exp(-N(t)) - 1] \end{aligned}$$

and we obtain

$$V(N(t+1)) - V(N(t)) = h(N(t))\,[\beta\exp(-N(t)) - 1], \tag{1.23}$$

where

$$\begin{aligned} h(N(t)) &= \left\{[\beta\exp(-N(t)) + 1]\,N(t) - 2\overline{N}\right\}N(t) \\ &= \left\{N(t)\beta\exp(-N(t)) - \overline{N} + N(t) - \overline{N}\right\}N(t) \\ &= \left\{N(t)\exp(\overline{N} - N(t)) + \left[N(t) - 2\overline{N}\right]\right\}N(t) \end{aligned} \tag{1.24}$$

Proposition 1.58 *Assume that $\beta \in (1, e^2]$. We deduce that if we define $\{U(t)\}_{t\in\mathbb{N}}$ the discrete time semiflow generated by* (1.22) *for each $t \in \mathbb{N}$,*

$$V(U(t)N_0) \leq V(N_0), \forall t \in \mathbb{N}, \forall N_0 \geq 0. \tag{1.25}$$

Proof Let us prove that

$$h(N(t)) \leq 0, \forall N \leq \overline{N}$$

and

$$h(N(t)) \geq 0, \forall N \geq \overline{N}.$$

Indeed, we have

$$h(0) = 0, \text{ and } h(\overline{N}) = 0, \text{ and } h(N) > 0, \forall N \geq 2\overline{N}.$$

Consider $N \in (0, 2\overline{N})$ and $N \neq \overline{N}$. Then $h(N) = 0$ if

$$\begin{aligned} & N(t)\exp(\overline{N} - N(t)) + \left[N(t) - 2\overline{N}\right] = 0 \\ \Leftrightarrow\ & \exp(\overline{N} - N(t)) = 2\frac{\overline{N}}{N(t)} - 1 \\ \Leftrightarrow\ & \overline{N} - N(t) = \ln\left(2\frac{\overline{N}}{N(t)} - 1\right) \end{aligned}$$

which is equivalent to

$$\overline{N} = \frac{\ln\left(2\frac{\overline{N}}{N(t)} - 1\right)}{1 - \frac{N(t)}{\overline{N}}}. \tag{1.26}$$

First case: $N \in (0, \overline{N})$**:** Set $y = 1/(1 - N(t)/\overline{N}) > 1$. Then (1.26) gives

$$\begin{aligned}\overline{N} &= y\left\{\ln(1+1/y) - \ln(1-1/y)\right\} = y\left\{\sum_{n=1}^{\infty}(-1)^{n+1}\frac{y^{-n}}{n} + \sum_{n=1}^{\infty}\frac{y^{-n}}{n}\right\}\\ &= 2\sum_{n=0}^{\infty}\frac{y^{-2n}}{2n+1} > 2.\end{aligned}$$

Second case: $N \in (\overline{N}, 2\overline{N})$**:** Set $y = 1/(N(t)/\overline{N} - 1) > 1$. Then (1.26) gives

$$\overline{N} = y\left\{\ln(1+1/y) - \ln(1-1/y)\right\} > 2.$$

Finally $\beta \in (1, e^2]$ implies $\overline{N} \in (0, 2]$ we conclude that

$$h(N) \neq 0 \text{ and } N \in (0, 2\overline{N}) \Rightarrow N = \overline{N},$$

and the proof is completed. □

Theorem 1.59 *Assume that* $\beta \in (1, e^2]$. *The discrete time semiflow generated by* (1.22) *is bounded dissipative on* $[0, \infty)$. *More precisely, for each* $\gamma_0 > \overline{N}$ *the subset*

$$B_{\gamma_0} = \{x \geq 0 : V(x) \leq \gamma_0\}$$

is bounded absorbing subset.

Proof Let $U(t)$ be the discrete time semiflow generated by (1.22). We have

$$V(U(t)0) = V(0) = \overline{N},$$

and

$$V(U(t)\overline{N}) = V(\overline{N}) = 0,$$

and for each $N_0 \in (0, +\infty) \setminus \{\overline{N}\}$,

$$V(U(t)N_0) < V(N_0), \forall t \in \mathbb{N},$$

the result follows. □

Remark 1.60 An Hopf bifurcation for maps occurs at $\beta = e^2$ (i.e. the derivative of the map $x \to \beta x e^{-x}$) crosses -1 when β crosses e^{-2}). It means that periodic orbit with period 2 with appear. Then see Figure 1.10 and Figure 1.11 periodic orbits of period 4, 8, etc... appear. That is often called a period-doubling bifurcation. Chaos appears when β becomes even larger and gives rise to a periodic orbit of period 3.

Assume that $\beta > e^2$. We can use the fact that $x \to \beta x \exp(-x)$ is bounded to derive the dissipation property. Indeed, we have

$$(\beta x \exp(-x))' = \beta [1 - x] \exp(-x).$$

Therefore the map $x \to \beta x \exp(-x)$ reaches a maximum at $x = 1$, and this maximum is βe^{-1}. Define

$$W(N) = \max(N, \beta e^{-1}).$$

Let $U(t)$ be the discrete time semiflow generated by (1.22). Then we have

$$W(U(t)N_0) \leq W(N_0), \forall t \in \mathbb{N}, \forall N_0 \geq 0.$$

Theorem 1.61 *Assume that* $\beta \geq 0$. *The discrete time semiflow generated by* (1.22) *is bounded dissipative on* $[0, \infty)$.

Remark 1.62 Then by using a simple first order approximation $\exp(N(t)) = 1 + N(t) + h.o.t$, we can regard the model of Beverton and Holt as a first order approximation of the Ricker model. Nevertheless the behaviors of the two systems are very different.

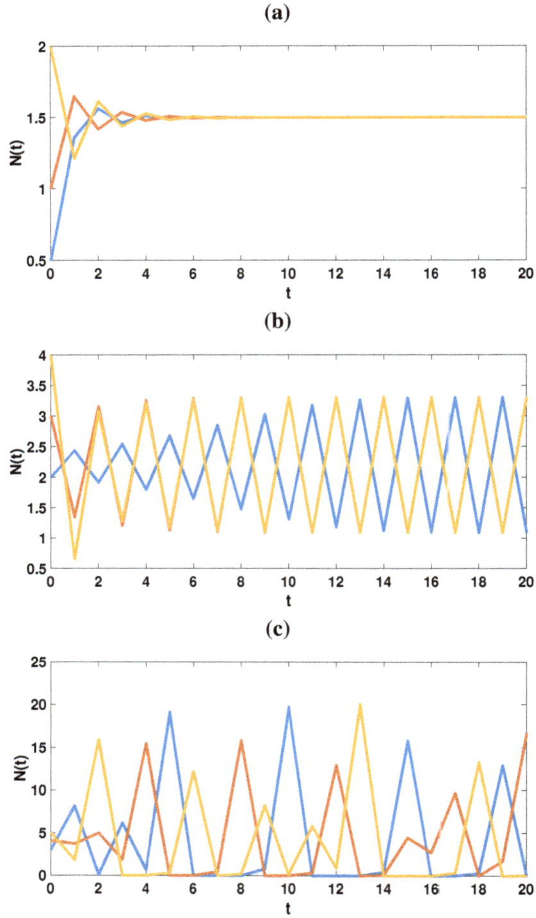

Fig. 1.10: *In Figures (a) (b) and (c) we plot three solutions of equation* (1.22). *In Figure (a), we fix* $\beta = e^{1.5}$, *we observe behavior that is similar to the behavior for the Beverton and Holt model (but oscillations around the positive equilibrium). In Figure (b) we fix* $\beta = e^{2.2}$, *the solutions converge to some periodic solutions. In Figure (c) we fix* $\beta = e^{4}$, *the behavior becomes chaotic (unpredictable).*

When the parameter β is large enough and a periodic orbit of period 3 appears, the system becomes chaotic (see Figures 1.10 and 1.11). In other words, period three implies chaos (Li and Yorke [215]).

This is the so called chaos of Sharkovskii [300, 301]. The result of Sharkovskii was rediscovered by Li and Yorke [215]. Both results of Sharkovskii [300, 301] and Li and Yorke show that when the system has a periodic solution of period 3 then periodic orbits of all periods exists. Therefore the system has infinitely many periodic orbits. This could serve as first step to characterize the chaos. But Li and

Yorke explored the dependency with respect to the initial condition in the following sense.

Theorem 1.63 (Li and Yorke [215]) *Let $T : I \to I$ be a continuous map on some interval $I \subset \mathbb{R}$. Assume that there exists $N_0 \geq 0$ such that*

$$T^3(N_0) \leq N_0 < T(N_0) < T^2(N_0)$$

or

$$T^3(N_0) \geq N_0 > T(N_0) > T^2(N_0).$$

Then

(i) *For each integer $k = 1, 2, \ldots$ there exists a periodic point in I having the period k.*

(ii) *There exists an uncountable set $S \subset I$ (containing no periodic points), which satisfies the following conditions:*

(a) *For every $p, q \in S$ with $p \neq q$,*

$$\limsup_{t \to \infty} |T^t(p) - T^t(q)| > 0,$$

and

$$\liminf_{t \to \infty} |T^t(p) - T^t(q)| = 0.$$

(b) *For every $p \in S$ and every periodic point $q \in I$,*

$$\limsup_{t \to \infty} |T^t(p) - T^t(q)| > 0.$$

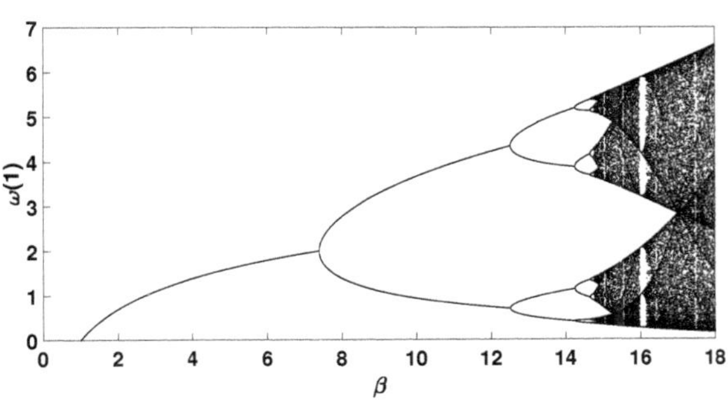

Fig. 1.11: *In this figure we plot for each β varying from* 0 *to* 18 *the omega limit set* $\omega(1)$ *of the solution of* (1.22) *starting from the initial value* $N_0 = 1$.

In Figure 1.11, we observe a cascade of bifurcations. The branch of single point corresponds to a positive equilibrium, which first bifurcate to a 2 periodic orbit. One may notice that the positive equilibrium exists for each β but becomes unstable after the bifurcation. Then we observe a period doubling bifurcation, we period orbits of period 2, 4, 8, etc. Periodic orbit of period 3 appears only when the parameter β becomes large enough, the chaos in the sense of Li and Yorke [215] appear.

1.8.6 Lorentz system: chaos and dissipation properties

The system of Lorenz [227] was introduced in his paper in 1963 as taken from Saltzman (1962, [292]) as a minimalist model of thermal convection in a box. This system is the following

$$\begin{cases} x' = \sigma(y-x), \\ y' = x(\rho - z) - y, \\ z' = xy - \beta z, \end{cases} \tag{1.27}$$

where $\sigma > 0, \rho > 0,$ and $\beta > 0$, and with initial value

$$x(0) = x_0, y(0) = y_0, \text{ and } z(0) = z_0.$$

In Figures 1.12-1.13, we plot two solutions of system (1.27) with the parameters values

$$\sigma = 10, \rho = 28, \text{ and } \beta = 8/3.$$

In Figures 1.12-1.13, we observe that starting from two very close initial values gives some very different trajectories. In Figure 1.13, the solution starting from the green dot (respectively from the black dot) end up at the yellow dot (respectively at the red dots) at time $t = 19$. So we can visualize the fact that changing a little bit the initial value may have a large impact on the trajectory.

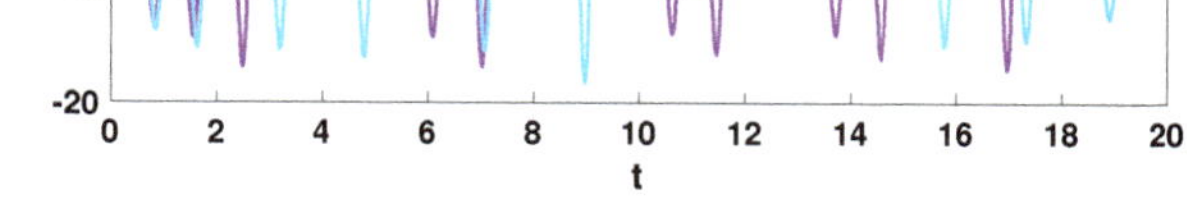

Fig. 1.12: *In this figure, we plot the first component* $t \to x(t)$ *of two solutions starting from the initial value from* $[x_0, y_0, z_0] = [10, 10, 10]$ *(light blue curve) and from* $[x_0, y_0, z_0] = [10, 10, 10.7]$ *(purple curve).*

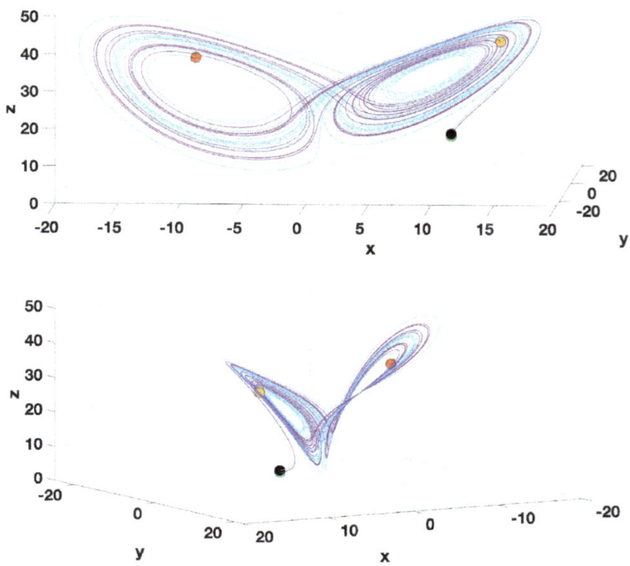

Fig. 1.13: *In this figure, the light blue curve (respectively the purple curve) corresponds to the solution* $t \longrightarrow (x(t), y(t), z(t))$ *starting from the initial value* $[x_0, y_0, z_0] = [10, 10, 10]$ *(green dot) (respectively from the initial value* $[x_0, y_0, z_0] = [10, 10, 10.7]$ *(black dot)). The two solutions are plotted on the time interval* $[0, 19]$.

Lorenz looks at his system as a simplified model for weather prediction. He is assuming that one loop corresponds to the calm weather and the other loop to tornados. The orbit generated by a slight disturbance of the initial distribution of the system looks very different. The shape of the omega limit set is very similar for each solution. The chaotic nature of this system is due to the "fractal" structure of the attractor in between the loops.

The frequency at which a solution passes from one loop to the next turns to be similar for each solution. Lorentz already mentioned this phenomenon during his presentation at the American Association for the Advancement of Science in 1972. During his presentation he says that:

Over the years minuscule disturbances neither increase nor decrease the frequency of occurrences of various weather events such as tornados; the most they may do is to modify the sequences in which they occur.

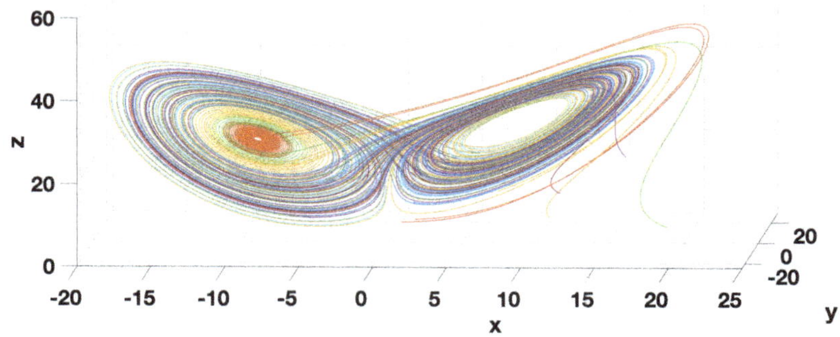

Fig. 1.14: *In this figure, we use 20 randomly chosen initial distributions.*

It turns out that the Lorenz attractor is an attractor in the sense developed in the Chapter 2. That is to say that, it is a compact invariant subset which attracts one of its neighborhood at each point of $\mathbb{R}^3$. The existence such a global attractor is a consequence of the dissipation of the system. So we now investigate the dissipation of the Lorentz system.

For the Lorenz system, the dissipativity can be studied by using the results of Leonov [209]. Let us consider

$$V(x, y, z) = \frac{1}{\sigma}x^2 + y^2 + z^2. \tag{1.28}$$

Then

$$\frac{dV(x(t), y(t), z(t))}{dt} = 2x(y-x) + 2yx(\rho - z) - 2y^2 + 2xyz - 2\beta z^2$$

and after simplification we obtain

$$\frac{dV(x(t), y(t), z(t))}{dt} = -2\left[x^2 + y^2 + \beta z^2\right] + 2(\rho+1)xy$$

but $2xy = x^2 + y^2 - (x-y)^2$ therefore

$$\frac{dV(x(t), y(t), z(t))}{dt} = -(1-\rho)x^2 - (1-\rho)y^2 - 2\beta z^2 - (\rho+1)(x-y)^2.$$

Lemma 1.64 *Assume that* $\rho \in (0, 1)$. *Then each solution of system* (1.27) *converges to* $0_{\mathbb{R}^3}$. *More precisely, we have the following estimation of the solution*

$$V(x(t), y(t), z(t)) \leq e^{-\delta t} V(x_0, y_0, z_0), \forall t \geq 0, \tag{1.29}$$

where

$$\delta := \min\{(1-\rho)\sigma, (1-\rho), 2\beta\}.$$

Proof By choosing the above constant $\delta > 0$, we have

$$\frac{dV(x(t), y(t), z(t))}{dt} \leq -\delta V(x(t), y(t), z(t))$$

and the result follows from the differential form of Gronwall's lemma (see Chapter 8 in Volume I). □

Theorem 1.65 *Assume that* $\rho \in (0, 1)$. *Then the semiflow* $U(t)$ *generated by the system* (1.27) *is bounded dissipative.*

Proof Let $\gamma_0 > 0$ be fixed. Let us define

$$B_{\gamma_0} = \left\{(x, y, z) \in \mathbb{R}^3 : V(x, y, z) \leq \gamma_0\right\}.$$

Then for each bounded subset $B \subset \mathbb{R}^3$ we have

$$\gamma := \sup_{(x,y,z)\in B} V(x, y, z) < \infty.$$

Let $t_0 = t_0(B) > 0$ such that

$$e^{-\delta t_0}\gamma \leq \gamma_0,$$

then by using the definition of B_{γ_0}, we deduce that

$$U(t)B \subset B_{\gamma_0}, \forall t \geq t_0,$$

and the proof is completed. □

Next we consider the functional

$$W(x, y, z) = y^2 + (z-\rho)^2. \tag{1.30}$$

Then

$$\begin{aligned}\frac{dW(x(t), y(t), z(t))}{dt} &= 2yx(\rho - z) - 2y^2 + 2(z-\rho)(xy - \beta z)\\ &= 2yx(\rho - z) - 2y^2 + 2xy(z-\rho) - 2\beta z(z-\rho)\end{aligned}$$

hence

$$\frac{dW(x(t), y(t), z(t))}{dt} = -2y^2 - 2\beta z(z-\rho).$$

Lemma 1.66 *Let* $\lambda \in (0, 2\min(1, \beta))$. *Then one has*

$$\frac{dW(x(t), y(t), z(t))}{dt} \leq -\lambda W(x(t), y(t), z(t)) + \chi, \tag{1.31}$$

where

$$\chi := [2\beta - \lambda]\left(\frac{\beta\rho}{\lambda - 2\beta}\right)^2.$$

Proof By using the fact that $\lambda < 2$ we deduce that

$$\begin{aligned}\frac{dW(x(t),y(t),z(t))}{dt} &+\lambda W(x(t),y(t),z(t)) \\ &= -2y^2 - 2\beta z(z-\rho) + \lambda y^2 + \lambda(z-\rho)^2 \\ &\leq [\lambda - 2\beta]\,(z-\rho)^2 - 2\beta\rho(z-\rho) \\ &= [\lambda - 2\beta]\left\{(z-\rho) - \frac{\beta\rho}{\lambda - 2\beta}\right\}^2 - [\lambda - 2\beta]\left(\frac{\beta\rho}{\lambda-2\beta}\right)^2.\end{aligned}$$

and since $\lambda < 2\beta$, inequality (1.31) follows. □

Theorem 1.67 *Let $\lambda \in (0, 2\min(1,\beta))$ be given. Then, for each $t \geq 0$, we have*

$$W(x(t),y(t),z(t)) \leq e^{-\lambda t} W(x_0,y_0,z_0) + \int_0^t e^{-\lambda(t-s)}\chi ds,$$

and

$$\limsup_{t\to+\infty} W(x(t),y(t),z(t)) \leq \frac{\chi}{\lambda}.$$

Moreover the semiflow $U(t)$ generated by the Lorenz system (1.27) *is bounded dissipative.*

Proof The proof is left as an exercise. (**Hint:** Start first with the two last components y and z use the same arguments than in the proof of Theorem 1.65. Then use the comparison principle for the x-equation to derive the dissipativity of the full system). □

1.9 Notes and Remarks

Existence and invariance and compactness of ω-limit sets

In this chapter, we presented some notions about continuous semiflow, omega and alpha limit sets and we refer to the books of LaSalle [207], Hale [135, 138]. As we will see in the next chapter the existence of ω-limit sets can be extended by replacing a point $x \in M$ by subset $B \subset M$ and replacing the orbit

$$t \to U(t)x.$$

by

$$t \to U(t)B.$$

we refer to Hale [138], Sell and You [298], Raugel [279] for more results.

Connectivity of ω-limit sets and chain recurrence

The notion of invariantly connected sets was introduced by LaSalle [207]. The notion of chain recurrence was originally introduced by Conley [63]. In the context of discrete dynamical systems generated by a continuous homeomorphism, Bowen [34] proved that the ω-limit sets are internally chain transitive when the orbit is precompact. Again for discrete dynamical systems, Robinson [282] proved the internal chain recurrence property for precompact orbits generated by continuous maps. In the context of asymptotically autonomous dynamical systems, Thieme [325, 327, 328] studied the long-time dynamics of solutions and Mischaikow, Smith and Thieme [258] discussed chain recurrence and Liapunov functions. Benaïm and Hirsch [25] introduced asymptotic pseudo-orbits in the context of continuous-time semiflows. Zhao [367, 368] discussed further properties of ω-limit sets for precompact orbits in the context of asymptotically autonomous dynamical systems. The notion of chain transitive sets was first introduced for discrete time semiflow by Hirsch, Smith, and Zhao [156], Zhao [372], and for continuous time semiflow by Magal [236]. We also refer to Raugel [279] for more results.

Dissipation in Beverton and Holt model and Ricker model

The Liapunov function for the Beverton and Holt model and Ricker model was introduced by Fisher and Goh [112]. We refer to their original articles for more results.

Chaos in one dimensional discrete time model

The understanding of one dimensional maps improved since Sharkovskii [300] and Li and Yorke [215]. We refer to [20] [100] [128] [216] [263] [353] for more results on the subject.

More about one dimensional discrete time model

The discrete time logistic equation was proposed by Verhulst [341] in 1838. That is the following difference equation

$$N(t+1) = \beta N(t)(1 - N(t)),, \forall t \geq 0, \text{ with } N(0) = N_0 \in [0,1],$$

where $\beta \in [0,4]$.

This equation also generates chaos in the sense of Li and Yorke [215] for $\beta = 4$.

Remark 1.68 The discrete logistic equation was introduced by Berkson [26] who invented the statistical logistic regression in 1944.

In connection with Ricker's model, the Beverton and Holt model is obtained using a first order (Talyor expansion) approximation (which is only valid whenever $\alpha N(t)$ is small enough)

$$\frac{\beta N(t)}{\exp(\alpha N(t))} \approx \frac{\beta N(t)}{1+\alpha N(t)}.$$

The significant difference between the Ricker's model and the Beverton and Holt model probably arise at the second order. It would be natural to consider the following extension of the above approximation

$$N(t+1) = \frac{\beta N(t)}{1+\alpha N(t) + (\alpha N(t))^2/2 + \ldots + (\alpha N(t))^m/m!},$$

as well as the higher order version of this model.

The discrete time logistic equation also follows from a first order approximation of the Ricker's model

$$\beta N(t) \exp(-\alpha N(t)) \approx \beta N(t)(1-\alpha N(t)).$$

An extended version of Beverton and Holt model is the cooperative model of Hill [154]

$$N(t+1) = \frac{\beta N(t)^n}{1+\alpha N(t)^n}.$$

Monod, Wyman and Changeux [262] also proposed an extended cooperative Hill's model which is called the Allosteric model

$$N(t+1) = \frac{\beta N(t)(1+N(t))^{n-1}}{1+\alpha(1+N(t))^n}.$$

Further application to embryology of such model can be found in the framework of gradient differential system in the paper of Demongeot, Glade and Forest [78] and Cinquin and Demongeot [62].

n-dimensional Ricker model

Ricker's model shows that salmons are first born into a river then swim to the sea. Female salmons spend several years in the ocean before returning to the river where they were born to reproduce. Therefore Ricker [280] considers an age-structured model of the following form

$$\begin{cases} N_1(t+1) = [\beta_1 N_1(t) + \ldots + \beta_m N_m(t)] \times \exp(-\alpha[\beta_1 N_1(t) + \ldots + \beta_m N_m(t)]) \\ N_2(t+1) = \pi_1 N_1(t) \\ \vdots \\ N_m(t+1) = \pi_{m-1} N_{m-1}(t), \end{cases}$$

where $N_1(t), \ldots, N_m(t)$ are the number of (female) individuals in each age classes going respectively from 1 to $m \geq 1$, β_i is the birth rate of the age class i, π_i is the probability to survive from age class i to the age class $i+1$. Ricker [280] was actually interested by the following special case

$$\beta_1 = \ldots = \beta_{m-1} = 0,$$

and

$$\beta_m > 0.$$

The age structured Ricker's model actually falls down into a more general class of age structured model so called density dependent model. Such a model was considered by Guckenheimer, Oster and Ipaktchi [129] and by Liu and Cohen [218] to quote a few.

Lorenz like attractor

An alternative to the Lorenz system is the model of Shimizu-Morioka [307]

$$\begin{cases} x' = y \\ y' = x(1-z) - \alpha y \\ z' = x^2 - \beta z \end{cases} \tag{1.32}$$

where $\beta > 0$, and $\alpha > 0$, and with initial value

$$x(0) = x_0, y(0) = y_0, \text{ and } z(0) = z_0.$$

Shilnikov [305] found some Lorenz like attractor for the parameters values

$$\alpha = 0.85, \text{ and } \beta = 0.5.$$

In Figures 1.15-1.16, we illustrate this result with a numerical simulation system (1.32) with the above parameters values.

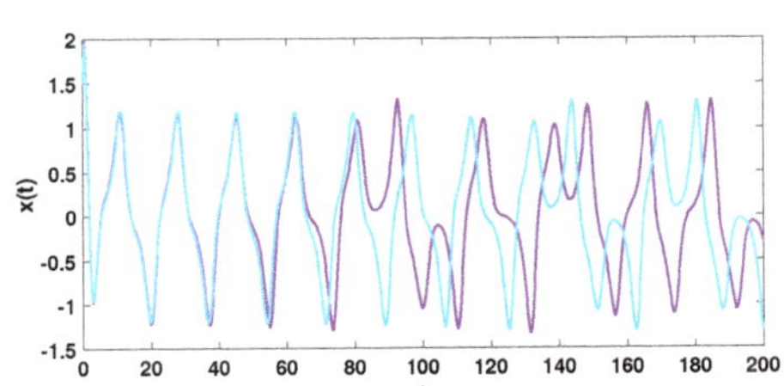

Fig. 1.15: *In this figure, we plot the first component $t \to x(t)$ of two solutions starting from the initial value from $[x_0, y_0, z_0] = [1.5, 1.5, 1.5]$ (light blue curve) and from $[x_0, y_0, z_0] = [1.5, 1.5, 1.605]$ (purple curve).*

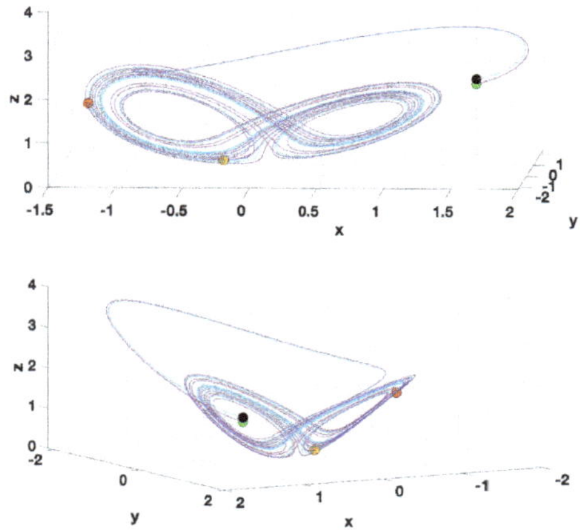

Fig. 1.16: *In this figure, the light blue curve (respectively the purple curve) corresponds to the solution* $t \to (x(t), y(t), z(t))$ *starting from the initial value* $[x_0, y_0, z_0] = [1.5, 1.5, 1.5]$ *(green dot) (respectively from the initial value* $[x_0, y_0, z_0] = [1.5, 1.5, 1.605]$ *(black dot)). The two solutions are plotted on the time interval* $[0, 200]$.

Remark 1.69 Shimizu-Morioka system can be regarded as the following second order ordinary differential equation

$$\begin{cases} x'' = x(1-z) - \alpha y \\ z' = x^2 - \beta z. \end{cases}$$

Lorenz like attractor in a multi-strain epidemic model

In this section, we present some multi-strain model simulations motivated by an application to the epidemiology of dengue fever. This chaotic numerical simulation was discovered by Aguiar, Kooi, and Stollenwerk [2]. We also refer to Aguiar et al. [1, 3] for more results going in that direction.

Consider an epidemic with two different strains, 1 and 2. Susceptibles to both strains (S) get infected with strain 1 (I_1 or I_{21}) or strain 2 (I_2 or I_{12}), with force of infection (β_1 and $\phi_1\beta_1$ respectively) and (β_2 and $\phi_2\beta_2$) respectively. They recover from infection with strain 1 (becoming R_1) or from strain 2 (becoming R_2), with recovery rate γ. In this recovered class, people have full and life-long immunity against the strain that they were exposed to and infected, and also a period of temporary cross-immunity against the other strain. After this, with rate α, they enter again in the susceptible classes (S_1 respectively S_2), where the index represents the first infection strain. Now, S_1 can be reinfected with strain 2 (becoming I_{12}), meeting I_2 with infection rate β_2 or meeting I_{12} with infection rate $\phi_2\beta_2$, secondary infected contributing differently to the force of infection than primary infected, and S_2 can be reinfected with strain 1 (becoming I_{21}) meeting I_1 or I_{21} with infections rates β_1 and $\phi_1\beta_1$ respectively.

The system introduced by Aguiar, Kooi, and Stollenwerk [2] is the following

$$\begin{cases} S' = -\beta_1 S (I_1 + \phi_1 I_{21}) - \beta_2 S (I_2 + \phi_2 I_{12}) + \mu (N - S), \\ I_1' = \beta_1 S (I_1 + \phi_1 I_{21}) - (\gamma + \mu) I_1, \\ I_2' = \beta_2 S (I_2 + \phi_2 I_{12}) - (\gamma + \mu) I_2, \\ R_1' = \gamma I_1 - (\alpha + \mu) R_1, \\ R_2' = \gamma I_2 - (\alpha + \mu) R_2, \\ S_1' = -\beta_2 S_1 (I_2 + \phi_2 I_{12}) + \alpha R_1 - \mu S_1, \\ S_2' = -\beta_1 S_2 (I_1 + \phi_1 I_{21}) + \alpha R_2 - \mu S_2, \\ I_{12}' = \beta_2 S_1 (I_2 + \phi_2 I_{12}) - (\gamma + \mu) I_{12}, \\ I_{21}' = \beta_1 S_2 (I_1 + \phi_1 I_{21}) - (\gamma + \mu) I_{21}, \\ R' = \gamma (I_{12} + I_{21}) - \mu R, \end{cases} \tag{1.33}$$

where $\beta_1 > 0$, $\beta_2 > 0$, and $\alpha > 0$, and with initial value

$$S(0) = S^0, I_1(0) = I_1^0, I_2(0) = I_2^0, R_1(0) = R_1^0, R_2(0) = R_2^0,$$
$$S_1(0) = S_1^0, S_2(0) = S_2^0, I_{12}(0) = I_{12}^0, I_{21}(0) = I_{21}^0, \text{ and } R(0) = R^0.$$

In all the figures, we use the following values of the parameters

$$N = 100, \mu = 1/65, \gamma = 52, \beta_1 = \beta_2 = 2\gamma,$$
$$\alpha = 2, \text{ and } \phi_1 = \phi_2 = \phi = 0.6.$$

In Figures 1.17-1.19 we plot two solutions of the system. The system seams to be very much initial condition dependent and the key to get some chaos is to use non symmetric initial percentage for each strain.

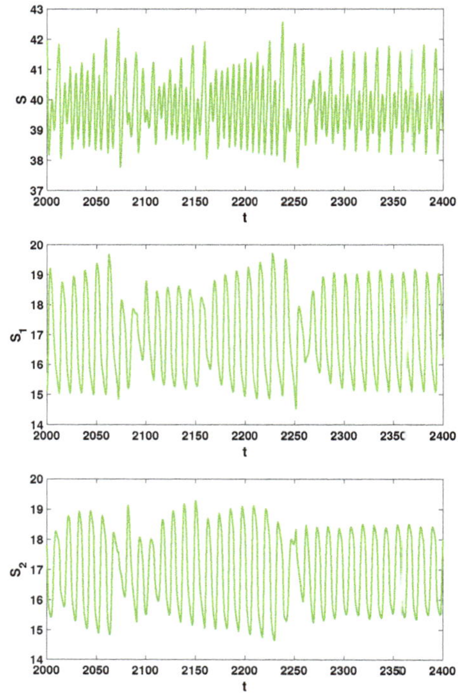

Fig. 1.17: *In this figure we use the initial value* $[S_0, I_1^0, I_2^0] = [70, 10, 20]$ *with the remaining components of the initial distribution equal to* 0. *In this figure we plot* $t \longrightarrow S(t)$, $t \longrightarrow S_1(t)$, *and* $t \longrightarrow S_2(t)$ *for* $t \in [2000, 2400]$.

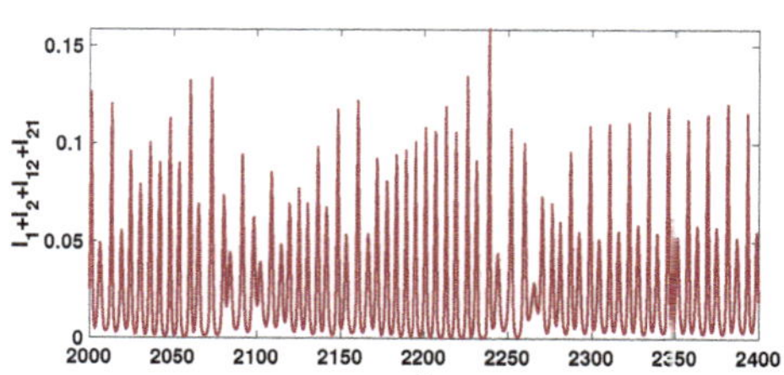

Fig. 1.18: *In this figure we use the initial value* $[S_0, I_1^0, I_2^0] = [70, 10, 20]$ *with the remaining components of the initial distribution equal to* 0. *In this figure we plot* $t \longrightarrow I_1(t) + I_2(t) + I_{12}(t) + I_{21}(t)$ *for* $t \in [2000, 2400]$.

In Figure 1.19, the global attractor is comparable to the Lorenz attractor because the solution jumps randomly from a region where the first strain is dominant to a region where the second strain is dominant back and forth. The changes from one dominant strain to the next are decided uncertainly along the axis $(S, 0, 0)$ where both strains are close to 0.

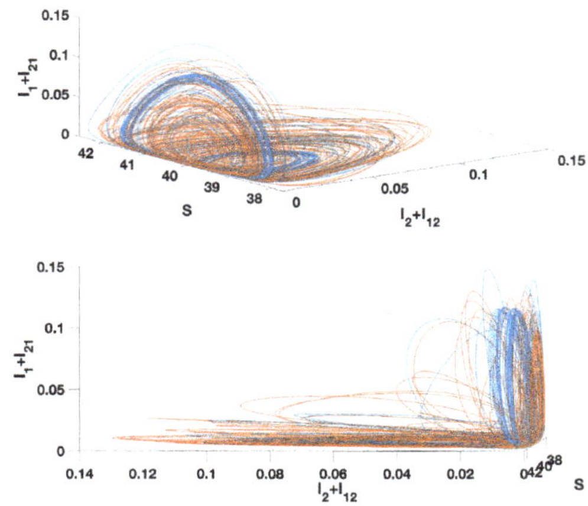

Fig. 1.19: *In this figure we use the initial value* $[S_0, I_1^0, I_2^0] = [70, 10, 20]$ *(blue curve)* $[S_0, I_1^0, I_2^0] = [70, 20, 10]$ *(orange curve) with the remaining components of the initial distribution equal to* 0*. The top and bottom figures corresponds to the same simulations but regarded from a different angle.*

Homoclinic orbits

Homoclinic trajectories are very important in the bifurcation theory and may induce chaos as in the system of Lorenz [128] [306] [350]. The existence of homoclinic orbits has been studied for Lorentz system by Leonov in [210], Leonov and Kuznetsov [211].

In the context of population dynamics, the existence of homoclinic orbits are usually induced by Bogdanov-Takens bifurcations (see Section 1.8.3 and Figure 1.8).

The existence homoclinic orbit have been obtained by Tang et al. [322] for an epidemic system of the form

$$\begin{cases} S' = b - \delta S - \kappa \dfrac{I^2 S}{1 + \alpha I^2} + \nu R \\ I' = \kappa \dfrac{I^2 S}{1 + \alpha I^2} - (\delta + \gamma) I \\ R' = \gamma I - \nu R. \end{cases}$$

A model without loss of immunity of the recovered (i.e. for $\nu = 0$ in the above model) was considered earlier by Xiao and Ruan [358].

Non-autonomous system

Here we briefly mention non-autonomous systems. An important example of such systems is given by the differential equation

$$x'(t) = F(\lambda(t), x(t))$$

where $t \longrightarrow \lambda(t)$ is a parameter which is time dependent, and which defines a continuous map from $\mathbb{R}$ to $\mathbb{R}^p$. Such a dynamical system can be studied by using the notion of skew product semiflow. We refer to Sell [297] and Sell and You [298] for a detailed presentation of skew product semi-flows.

Random dynamical systems

The theory presented here can be extended to random dynamical systems. We refer to Arnold et al [10] for a presentation of that topic.

Strange attractors

We refer to Lu, Wang, and Young [231] for more references and more results about strange attractors. In [231] they prove that a periodic perturbation of a system undergoing an Hopf bifurcation create sustained chaotic behavior. Specifically, strange attractors are shown to exist. The analysis is carried out for infinite dimensional systems.

Liapunov functions

Let $f : \mathbb{R}^n \longrightarrow \mathbb{R}^n$ be a continuously differentiable and Lipschitz continuous vector field and consider the equation

$$\begin{cases} u'(t) = f\big(u(t)\big), & t > 0, \\ u(0) = x \in \mathbb{R}^n. \end{cases} \tag{1.34}$$

We say that a continuously differentiable function $V : \mathbb{R}^n \longrightarrow \mathbb{R}$ is a **Liapunov function** associated with (1.34) if we have

$$\langle f(x), \nabla V(x)\rangle \leq 0 \text{ for all } x \in \mathbb{R}^n, \tag{1.35}$$

where $\langle \cdot, \cdot \rangle$ is the scalar product in $\mathbb{R}^n$ and

$$\nabla V(x) = \big(\partial_{x_1} V(x), \partial_{x_2} V(x), \ldots, \partial_{x_n} V(x)\big).$$

As a consequence of (1.35), the scalar function $t \mapsto V\big(x(t)\big)$ is nonincreasing along any orbit; indeed

$$\frac{\mathrm{d}}{\mathrm{d}t} V\big(u(t)\big) = \big\langle u'(t), \nabla V\big(u(t)\big)\big\rangle = \big\langle f\big(u(t)\big), \nabla V\big(u(t)\big)\big\rangle \leq 0 \text{ for all } t \geq 0. \tag{1.36}$$

Liapunov functions are generally hard to find in practice, but when they are known, they can be used to obtain informations about the global dynamics of the differential equations (1.34). A important particular case is when the strict inequality holds in (1.35) whenever $V(x) \neq 0$ (we say in this case that V is a **strict Liapunov function**) and moreover V has a unique global minimum; it is then elementary to show that any orbit converges to the global minimum of V.

In 1960, LaSalle [205, 206, 207] introduced a generalization of this principle to non-strict Liapunov function; this is often referred to as the LaSalle invariance principle.

Theorem 1.70 (LaSalle invariance principle) *Let $\Omega \subset \mathbb{R}^n$ be closed, bounded and positively invariant for* (1.34)*, and $V : \mathbb{R}^n \to \mathbb{R}$ a map satisfying* (1.35)*. Then the ω-limit set of any $x \in \Omega$ is included in the set of points $x \in \Omega$ where $\langle f(x), \nabla V(x)\rangle = 0$. That is*

$$\omega(x) \subset \big\{y \in \Omega : \big\langle f(y), \nabla V(y)\big\rangle = 0\big\} \qquad \textit{for all } x \in \Omega.$$

Moreover, the function $V(y)$ is constant for $y \in \omega(x)$.

Proof This proof is adapted from [206]. Let $x \in \Omega$ and $u(t)$ the solution of (1.34) starting from $u(0) = x$. We know that $t \mapsto V\big(u(t)\big)$ is nonincreasing, and since Ω is compact, V is bounded on Ω; therefore $V\big(u(t)\big)$ is bounded below and thus converges to a limit c. Thus $V \equiv c$ on $\omega(x)$ (thanks to the continuity of V), and since $\omega(x)$ is positively invariant, we have $\frac{\mathrm{d}}{\mathrm{d}t} V\big(v(t)\big) = \big\langle f\big(v(t)\big), \nabla V\big(v(t)\big)\big\rangle = 0$ for any orbit $\{v(t)\} \subset \omega(x)$. We conclude that $\big\langle f(y), \nabla V(y)\big\rangle = 0$ for any $y \in \omega(x)$, which finishes the proof of the Theorem. □

Chemostat models

A chemostat is an artificial device used to cultivate microorganisms in a tank containing a liquid medium to which nutrients are constantly added. A schematic representation of a chemostat is present in Figure 1.20 below. This device was first developed simultaneously by Monod [261] and Novick and Szilard [268]. Monod [261] also developed a mathematical model to describe the behavior of the cell population in his setting. Let us denote by $S(t)$ the concentration of nutrients in the tank at time $t > 0$ and $x(t)$ the concentration of cells. The model described by Monod can be written as follows:

$$\begin{cases} S' = D\left(S^{(0)} - S(t)\right) - \mu_0 \dfrac{S}{K_S + S}\dfrac{x}{R}, \\ x' = \left(\mu_0 \dfrac{S}{K_S + S} - D\right)x. \end{cases} \tag{1.37a}$$

with the initial data

$$S(0) = S_0 > 0, x(0) = x_0 > 0. \tag{1.37b}$$

Here $D > 0$ is the inflow and outflow rate (the volume of fluid going out of the tank by units of time), S_0 is the initial concentration of nutrients in the tank, $S^{(0)}$ is the concentration of nutrient of the fluid going in the tank, μ_0 is the maximal growth rate of the cells (in an environment saturated with nutrients), K_S is an affinity constant between the nutrient and the cells and R is the yield constant. The function defining the growth rate of the cells, $\mu(S) = \mu_0 \frac{S}{K_S+S}$, is often called *Monod function* as a tribute to the work of Monod [260]. An attempt to justify rigorously this model can be found in [226].

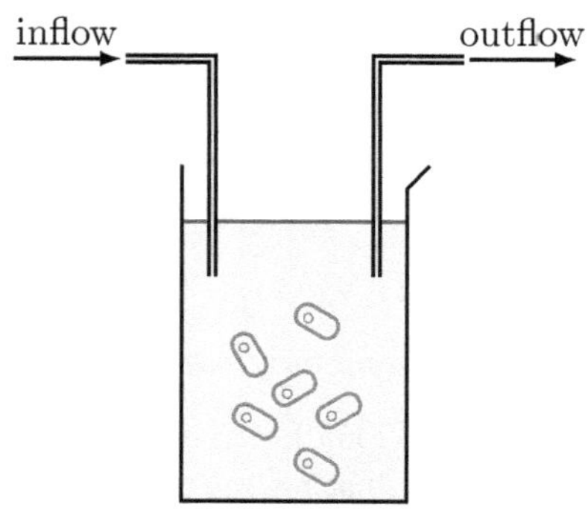

Fig. 1.20: *Schematic representation of a chemostat. The inflow of liquid contains nutrients, while the outflow takes out a mix of culture fluid, live and dead cells, and nutrients as well. Hence, there is a constant influx of nutrients, but at the same time, there is a permanent dilution of the cells due to the inflow and outflow of fluids.*

Taylor and Williams [323] proposed a generalization of the model (1.37) to account for multiple biological species competing for several resources in a chemostat. They argued that no more than n species can survive whenever n different resources are injected into the chemostat. This principle is called *competitive exclusion principle*; it is sometimes attributed to Gause [119]. We refer to the work of Hardin [149] for a review on the subject but note that this idea was present to some extent in the works of Lotka [230] and Volterra [344, 345]. Here, we present the model of Taylor and Williams [323], and the resolution method proposed by Hsu [167] with the help of a Liapunov function, simplifying the analysis of Hsu, Hubbell and Waltman [168].

The model with n species is the following.

$$\begin{cases} S'(t) = D\left(\Lambda - S(t)\right) - \sum_{i=1}^{n} \frac{\mu_i}{R_i} \frac{x_i(t)S(t)}{K_i + S(t)}, \\ x_1'(t) = \left(\mu_1 \frac{S(t)}{K_1 + S(t)} - D_1\right) x_1(t), \\ \vdots \\ x_n'(t) = \left(\mu_n \frac{S(t)}{K_n + S(t)} - D_n\right) x_n(t), \end{cases} \tag{1.38a}$$

with the initial data

$$S(0) = S_0 > 0, x_1(0) = x_1^0 \geq 0, \ldots, x_n(0) = x_n^0 \geq 0. \tag{1.38b}$$

We define

$$a_i := \frac{\mu_i - D_i}{K_i D_i} \tag{1.39}$$

the normalized growth rate of species i. Up to an index permutation, we assume that the sequence $(a_i)_{i=1,\ldots,n}$ is non-decreasing. That is,

$$a_1 \leq a_2 \leq \ldots \leq a_n. \tag{1.40}$$

The result of Hsu [167] is the following.

Theorem 1.71 *Suppose that* (1.40) *holds.*

1. *Suppose $a_i < \frac{1}{\Lambda}$ for some $i \in \{1, \ldots, n\}$, then*

$$x_i(t) \xrightarrow[t\to+\infty]{} 0.$$

2. *Suppose that $a_n \geq \frac{1}{\Lambda}$ and $a_n > a_{n-1}$, that is to say, a_n is the unique strict maximum of the sequence. Then*

$$S(t) \xrightarrow[t\to+\infty]{} S^* := \frac{1}{a_n}, \tag{1.41a}$$

$$x_n(t) \xrightarrow[t\to+\infty]{} x_n^* := \frac{R_n}{\mu_n S^*}\left(\Lambda - S^*\right)\left(K_n + S^*\right) \tag{1.41b}$$

$$x_i(t) \xrightarrow[t\to+\infty]{} 0, \qquad \forall i = 1, \ldots, n-1. \tag{1.41c}$$

3. *Suppose $a_n \geq \frac{1}{\Lambda}$ and we have $a_1 \leq a_2 \leq \ldots \leq a_{n-m} < a_{n-m+1} = \ldots = a_n$ for some $m \leq n$, that is to say, the m last equations have the same normalized growth rate which is strictly bigger than the others. Then, the ω-limit set of any initial data satisfying* (1.38b) *lies in the set*

$$S = S^* := \frac{1}{a_n} \tag{1.42}$$

and

$$(x_1, \ldots, x_n) = (0, \ldots, 0, x_{n-m+1}, \ldots, x_n), \tag{1.43}$$

with

$$(\Lambda - S^*)\, D = \sum_{j=n-m+1}^{n} \frac{\mu_j}{R_j} \frac{S^*}{K_j + S^*} x_j, \tag{1.44}$$

with $x_j \geq 0$, *and* $j = n - m + 1, \ldots, n$.

In this last case, the solutions are not necessarily converging.

In other words, the theorem states that the fate of n species competing for the same resource in a chemostat can be determined by looking at the normalized growth rate at equilibrium a_i defined in (1.39). If $a_i < \frac{1}{\Lambda}$ then the species cannot survive in any case (independently of the presence of another species). If at least a species survives, then the only species capable of maintaining their population are the ones that maximize the growth rate: any species that have a normalized growth rate below the maximal value will get extinct eventually.

Proof (Proof of Theorem 1.71) First we prove point 1. From the first line in (1.38a) we get $S'(t) \leq D\,(\Lambda - S)$ and therefore

$$S(t) \leq \Lambda + e^{-Dt}\,(S_0 - \Lambda)\,. \tag{1.45}$$

Next we rewrite the x_i equation in (1.38a) as

$$\frac{x_i'(t)}{x_i(t)} = \frac{\mu_i S(t)}{K_i + S(t)} - D_i,$$

then by integrating

$$\ln\left(x_i(t)\right) - \ln\left(x_i^0\right) = \int_0^t \frac{S(s)(\mu_i - D_i) - D_i K_i}{K_i + S(s)} \mathrm{d}s,$$

so finally

$$x_i(t) = x_i^0 \exp\left(\int_0^t \frac{S(s)(\mu_i - D_i) - D_i K_i}{K_i + S(s)} \mathrm{d}s\right). \tag{1.46}$$

If $\mu_i - D_i \leq 0$, then we have

$$0 \leq x_i(t) \leq x_i^0 \exp\left(\int_0^t \frac{-D_i K_i}{K_i + S(s)} \mathrm{d}s\right) \leq x_i^0 \exp\left(\frac{-D_i K_i}{K_i + \max\left(S_0, \Lambda\right)} t\right) \xrightarrow[t \to +\infty]{} 0, \tag{1.47}$$

so $x_i(t)$ converges to 0. If $\mu_i - D_i > 0$ but $a_n < \frac{1}{S_0}$, then rearranging the terms in (1.46) leads to

$$x_i(t) = x_i^0 \exp\left(\int_0^t (\mu_i - D_i) \frac{S(s) - \frac{D_i K_i}{\mu_i - D_i}}{K_i + S(s)} \mathrm{d}s\right) = x_i^0 \exp\left(\int_C^t \frac{\mu_i - D_i}{K_i + S(s)} \left(S(s) - \frac{1}{a_i}\right) \mathrm{d}s\right). \tag{1.48}$$

By assumption there exists $0 < \varepsilon \leq \frac{1}{a_i} - \Lambda$, and by (1.45) there is $t_0 > 0$ such that $S(s) \leq \Lambda + \frac{\varepsilon}{2}$ if $t \geq t_0$, therefore

$$
\begin{aligned}
x_i(t) &= x_i^0 \exp\left(\int_0^{t_0} \frac{\mu_i - D_i}{K_i + S(s)}\left(S(s) - \frac{1}{a_i}\right) \mathrm{d}s + \int_{t_0}^{t} \frac{\mu_i - D_i}{K_i + S(s)}\left(S(s) - \frac{1}{a_i}\right) \mathrm{d}s\right) \\
&\le x_i^0 \exp\left(\mathcal{I}_i + \int_{t_0}^{t} \frac{\mu_i - D_i}{K_i + \Lambda + \varepsilon/2}\left(-\frac{\varepsilon}{2}\right) \mathrm{d}s\right) \\
&= x_i^0 \exp\left(\mathcal{I}_i - \frac{\varepsilon}{2}\frac{\mu_i - D_i}{K_i + \Lambda + \varepsilon/2}(t - t_0)\right) \\
&\xrightarrow[t\to+\infty]{} 0,
\end{aligned}
$$

where $\mathcal{I}_i = \displaystyle\int_0^{t_0} \frac{\mu_i - D_i}{K_i + S(s)}\left(S(s) - \frac{1}{a_i}\right) \mathrm{d}s$ is a constant. Thus we have proved point 1.

Next we prove point 3. We define the following function, with S^* as in (1.41a) and x_n^* as in (1.41b).

$$
V(S, x_1, \ldots, x_n) = S - S^* - S^* \ln \frac{S}{S^*} + \sum_{j=1}^{n-1} c_j x_j + c_n \left[(x_n - x_n^*) - x_n^* \ln\left(\frac{x_n}{x_n^*}\right)\right]. \tag{1.49}
$$

We claim that V is a Liapunov function for (1.38) for suitable parameters c_i, $i = 1, \ldots, n$ that will be chosen later.

We have

$$
\begin{aligned}
\frac{\mathrm{d}}{\mathrm{d}t} V(S(t), x_1(t), \ldots, x_n(t)) &= S'(t)\left(1 - \frac{S^*}{S(t)}\right) + \sum_{j=1}^{n-1} c_j x_j'(t) + c_n x_n'(t)\left(1 - \frac{x_n^*}{x_n(t)}\right) \\
&= \left(D(\Lambda - S(t)) - \sum_{i=1}^{n} \frac{\mu_i}{R_i}\frac{x_i(t) S(t)}{K_i + S(t)}\right)\left(1 - \frac{S^*}{S(t)}\right) \\
&\quad + \sum_{j=1}^{n-1} c_j \left(\frac{\mu_j S(t)}{K_j + S(t)} - D_j\right) x_j(t) \\
&\quad + c_n \left(1 - \frac{x_n^*}{x_n(t)}\right) x_n(t) \left(\frac{\mu_n S(t)}{K_n + S(t)} - D_n\right) \\
&= \sum_{j=1}^{n-1} \left[c_j(\mu_j - D_j)\frac{S(t) - \frac{1}{a_i}}{K_j + S(t)} - \frac{\mu_j}{R_j}\frac{S(t) - S^*}{K_j + S(t)}\right] x_j(t) \\
&\quad + \left[D(\Lambda - S(t)) - \frac{\mu_n}{R_n}\frac{x_n(t) S(t)}{K_n + S(t)}\right] \frac{S(t) - S^*}{S(t)} \\
&\quad + c_n (x_n(t) - x_n^*)(\mu_n - D_n)\frac{S(t) - S^*}{K_n + S(t)},
\end{aligned}
$$

and by choosing $c_i = \frac{1}{R_i}\frac{\mu_i}{\mu_i - D_n}$ for $i = 1, \ldots, n$,

$$\frac{\mathrm{d}}{\mathrm{d}t}V\left(S(t),x_1(t),\ldots,x_n(t)\right)=\sum_{j=1}^{n-1}\left[\frac{\mu_j}{R_j}\frac{S^*-\frac{1}{a_i}}{K_j+S(t)}\right]x_j(t)$$
$$+\left[D\big(\Lambda-S(t)\big)-\frac{\mu_n}{R_n}\frac{x_n(t)S(t)}{K_n+S(t)}+c_n(\mu_n-D_n)\frac{(x_n(t)-x_n^*)S(t)}{K_n+S(t)}\right]\frac{S(t)-S^*}{S(t)}$$
$$=\sum_{j=1}^{n-1}\left[\frac{\mu_j}{R_j}\frac{S^*-\frac{1}{a_i}}{K_j+S(t)}\right]x_j(t)+\left[D\big(\Lambda-S(t)\big)-\frac{\mu_n}{R_n}\frac{x_n^*S(t)}{K_n+S(t)}\right]\frac{S(t)-S^*}{S(t)}$$
$$=\sum_{j=1}^{n-1}\left[\frac{\mu_j}{R_j}\frac{S^*-\frac{1}{a_i}}{K_j+S(t)}\right]x_j(t)$$
$$+\left[D\big(\Lambda-S(t)\big)-\frac{\mu_n}{R_n}\frac{x_n^*S(t)}{K_n+S(t)}-\left(D\big(\Lambda-S^*\big)-\frac{\mu_n}{R_n}\frac{x_n^*S^*}{K_n+S^*}\right)\right]\frac{S(t)-S^*}{S(t)}$$
$$=\sum_{j=1}^{n-1}\left[\frac{\mu_j}{R_j}\frac{S^*-\frac{1}{a_i}}{K_j+S(t)}\right]x_j(t)-D\frac{\big(S(t)-S^*\big)^2}{S(t)}+\frac{\mu_n}{R_n}x_n^*\frac{K_n\big(S^*-S(t)\big)}{\big(K_n+S(t)\big)\big(K_n+S^*\big)}\frac{S(t)-S^*}{S(t)},$$

so finally

$$\frac{\mathrm{d}}{\mathrm{d}t}V\left(S(t),x_1(t),\ldots,x_n(t)\right)$$
$$=\sum_{j=1}^{n-1}\left[\frac{\mu_j}{R_j}\frac{\frac{1}{a_n}-\frac{1}{a_i}}{K_j+S(t)}\right]x_j(t)-D\frac{\big(S(t)-S^*\big)^2}{S(t)}-\frac{\frac{\mu_n}{R_n}x_n^*K_n\big(S(t)-S^*\big)^2}{S(t)\big(K_n+S(t)\big)\big(K_n+S^*\big)}. \tag{1.50}$$

We see from (1.50) that $\frac{\mathrm{d}}{\mathrm{d}t}V\big(S(t),x_1(t),\ldots,x_n(t)\big)\le 0$ for all $t\ge 0$. Since the level sets of V in the positive quadrant $S>0,x_1>0,\ldots,x_n>0$ are bounded, it follows from the LaSalle invariance principle Theorem 1.70, that all orbits $(S(t),x_1(t),\ldots,x_n(t))$ included in the ω-limit set $\omega=\omega(S_0,x_1^0,\ldots,x_n^0)$ must satisfy $\frac{\mathrm{d}}{\mathrm{d}t}V\big(S(t),x_1(t),\ldots,x_n(t)\big)=0$. From (1.50), this means $S(t)\equiv S^*$ and $x_1(t)\equiv 0,\ldots,x_{n-m}(t)\equiv 0$. Then substituting $\frac{\mathrm{d}}{\mathrm{d}t}S(t)=0$ in (1.38a) we get

$$0=D\big(\Lambda-S^*\big)-\sum_{k=n-m+1}^{n}\frac{\mu_k}{R_k}\frac{S^*x_k(t)}{K_k+S^*},$$

which proves that ω is indeed included in M as in (1.42)-(1.44). Point 3 of Theorem 1.71 is proved. Now point 2 is proved by applying point 3 to the special case $m=1$ in which case M is reduced to a single point and Theorem 1.71 is proved. □

1.10 MATLAB Codes

1.10.1 Figure 1.3

```
L=10;
t=-L:0.1:L;
x0=0.5
x=exp(t)*x0./(1+(exp(t)-1)*x0);

aux=-L:0.1:L;
aux1=ones(size(aux));
aux2=-2:0.1:2;
hold on
plot(t,x,'Color','[    0    ,    0.4470    ,    0.7410]','LineWidth',5)
plot(aux,0*aux,'k',0*aux2,aux2,'k','LineWidth',3)
xlabel('t');
ylabel('N(t)');



xlabel('t')
ylabel('x(t)')

set(gca,'YLim',[-0.5  1.5])
set(gca,'XLim',[-L  L])

set(gca,'fontweight','bold','FontSize',30);
ax=gca;
hold off
```

1.10.2 Figure 1.4 and Figure 1.5 and Figure 1.6

```
close all
clear all;
figure(1)
dt=0.5
tspan = 1:dt:500;
hold on
y0 = [0.4;0];
[T,Y1] = ode45(@myfun,tspan,y0);

y0 = [0.5;0];
[T,Y2] = ode45(@myfun,tspan,y0);

```

```
y0 = [0.6;0];
[T,Y3] = ode45(@myfun,tspan ,y0);


y0 = [0.7;0];
[T,Y4] = ode45(@myfun,tspan ,y0);


y0 = [0.8;0];
[T,Y5] = ode45(@myfun,tspan ,y0);

plot(Y1(:,1),Y1(:,2),Y2(:,1),Y2(:,2),Y3(:,1),Y3(:,2)
    ,Y4(:,1),Y4(:,2),Y5(:,1),Y5(:,2),'linewidth',3);
sol = ode45(@myfun,tspan ,y0);

plot([-0.4,0.8],[0,0],'k','linewidth',3)

plot([0,0],[-0.2,0.3],'k','linewidth',3)

xlabel('x')
ylabel('y')

set(gca,'XLim',[-0.4,0.8])
set(gca,'YLim',[-0.2,0.3])
set(gca,'fontweight','bold','FontSize',30);
ax=gca;
hold off

figure(2)

dt=0.1
tspan = 1:dt:500;
y0 = [0.01;0];
[T,Y1] = ode45(@myfun,tspan ,y0);

tspan = 1:dt:300;
y0 = [0.08;0];
[T,Y2] = ode45(@myfun,tspan ,y0);
hold on

plot(Y1(:,1),Y1(:,2),'Color','[  0.4660   ,   0.6740
    ,  0.1880]','linewidth',3);

plot(Y2(:,1),Y2(:,2),'Color','[    0.3010  ,   0.7450
    ,   0.9330]','linewidth',3);
```

```
plot([-0.2,0.2],[0,0],'k','linewidth',3)

plot([0,0],[-0.2,0.2],'k','linewidth',3)
xlabel('x')
ylabel('y')

set(gca,'XLim',[-0.16,0.16])
set(gca,'YLim',[-0.16,0.16])
set(gca,'fontweight','bold','FontSize',30);
ax=gca;
hold off


figure(3)

dt=0.1
tspan = 1:dt:400;
y0 = [0.01;0];
[T,Y1] = ode45(@myfun,tspan,y0);

tspan = 1:dt:400;
y0 = [0.4;0];

[T,Y2] = ode45(@myfun,tspan,y0);
hold on

plot(T,Y1(:,1),'Color','[ 0.4660 , 0.6740 ,
    0.1880]','linewidth',3);

plot(T,Y2(:,1),'Color','[ 0 , 0.4470 ,
    0.7410]','linewidth',3);
plot(T,0*T,'k','linewidth',3)

xlabel('t')
ylabel('x(t)')

set(gca,'fontweight','bold','FontSize',30);
ax=gca;
hold off

figure(4)

dt=0.1
tspan = 1:dt:400;
y0 = [0.01;0];
[T,Y1] = ode45(@myfun,tspan,y0);
```

```
tspan = 1:dt:400;
y0 = [0.4;0];

[T,Y2] = ode45(@myfun,tspan,y0);
hold on

plot(T,Y1(:,2),'Color','[ 0.4660 , 0.6740 , 0.1880]','linewidth',3);

plot(T,Y2(:,2),'Color','[ 0 , 0.4470 , 0.7410]','linewidth',3);
plot(T,0*T,'k','linewidth',3)


xlabel('t')
ylabel('y(t)')


set(gca,'fontweight','bold','FontSize',30);
ax=gca;
hold off

function dy = myfun(t,y)
lambda = 0.02;
omega = 0.1;
kappa = 1;
dy = zeros(2,1);
dy(1) =(lambda/2)*y(1)-omega*y(2)-(kappa/2)*(y(1)*y(1)+y(2)*y(2))*y(1);
dy(2) = omega*y(1)+(lambda/2)*y(2)-(kappa/2)*(y(1)*y(1)+y(2)*y(2))*y(2);
dy=dy(:);
end
```

1.10.3 Figure 1.7

In this program we use the matlab solver *roots(p)* which compute the roots a polynomial $p(x) = a_n x^n + \ldots + a_0$ which is defined in Matlab as $[a_n\, a_{n-1}\, \ldots\, a_0]$.

```
clf;
aux2=0.05
h=(aux2^2)/2-(aux2^3)/3;
p = [(1/3) -(1/2) 0 h]
r = roots(p)
aux1=r(3)
x1=aux1:0.0001:aux2;
```

```
y1=2*(h- x1.^2/2+x1.^3/3).^0.5;
z1=-y1;

plot(x1,y1,'k',x1,-y1,'k','LineWidth',3)
hold on;

aux2=0.2;
h=(aux2^2)/2-(aux2^3)/3;
p = [(1/3) -(1/2) 0 h];
r = roots(p);
aux1=r(3);
x1=aux1:0.0001:aux2;
y1=2*(h- x1.^2/2+x1.^3/3).^0.5;
z1=-y1;

plot(x1,y1,'k',x1,-y1,'k','LineWidth',3)
hold on;


aux2=0.4;
h=(aux2^2)/2-(aux2^3)/3;
p = [(1/3) -(1/2) 0 h];
r = roots(p);
aux1=r(3);
x1=aux1:0.0001:aux2;
y1=2*(h- x1.^2/2+x1.^3/3).^0.5;
z1=-y1;

plot(x1,y1,'k',x1,-y1,'k','LineWidth',3);

aux2=0.6;
h=(aux2^2)/2-(aux2^3)/3;
p = [(1/3) -(1/2) 0 h];
r = roots(p);
aux1=r(3);
x1=aux1:0.0001:aux2;
y1=2*(h- x1.^2/2+x1.^3/3).^0.5;
z1=-y1;


plot(x1,y1,'k',x1,-y1,'k','LineWidth',3)

aux2=0.8;
h=(aux2^2)/2-(aux2^3)/3;
p = [(1/3) -(1/2) 0 h];
r = roots(p);
```

```
aux1=r(3);
x1=aux1:0.0001:aux2;
y1=2*(h- x1.^2/2+x1.^3/3).^0.5;
z1=-y1;


plot(x1,y1,'k',x1,-y1,'k','LineWidth',3)


aux2=1;
h=(aux2^2)/2-(aux2^3)/3;
p = [(1/3) -(1/2) 0 h];
r = roots(p);
aux1=r(3);
x1=aux1:0.0001:aux2;
y1=2*(h- x1.^2/2+x1.^3/3).^0.5;
z1=-y1;
plot(x1,y1,'k',x1,-y1,'k','LineWidth',3)


aux= -2:0.1:2;
plot(aux,0*aux,'k',0*aux,-aux,'k','LineWidth',3)
xlabel('x(t)');
ylabel('y(t)');
set(gca,'YLim',[-1 1])
set(gca,'XLim',[-0.6 1.1])
set(gca,'fontweight','bold','FontSize',30);
ax=gca;
```

1.10.4 Figure 1.8

In this program we need to increase the precision of the MATLAB ODE solver in order to get the convergence of the solution when t goes to $\pm\infty$.

```
clf;
X0=[-0.5 0];
t0=0:0.002:10;

options=odeset('RelTol',10^(-10));
[t1,x1]=ode45(@Model1,t0,X0,options);
hold on;
h=plot(t1,x1(:,1),'Color','[ 0.3010 , 0.7450 , 0.9330]','LineWidth',5);
h=plot(t1,x1(:,2),'Color','[ 0.4660 , 0.6740 , 0.1880]','LineWidth',5);
```

```
X0=[-0.5  0];
t0 =0:0.002:10;

options=odeset('RelTol',10^(-10));
[t2 ,x2]=ode45(@Model2,t0 ,X0,options);


h=plot(-t2 ,x2(:,1),'Color','[    0.3010 ,    0.7450    ,
    0.9330]','LineWidth',5);
h=plot(-t2 ,x2(:,2), 'Color','[ 0.4660   ,    0.6740     ,
    0.1880]','LineWidth',5);


aux=-10:0.1:10;
plot(aux,0*aux,'k',0*aux,aux,'k','LineWidth',3)
h=legend('x(t)','y(t)','Location','southeast');
xlabel('t');
set(gca,'YLim',[-0.6  1.1])
set(gca,'XLim',[-10  10])
set(gca,'fontweight','bold','FontSize',30);
ax=gca;


function dy=Model1(t,y)
dy=zeros(2,1);
dy(1)=-y(2);
dy(2)=y(1).*(1-y(1));
end

function dy=Model2(t,y)
dy=zeros(2,1);
dy(1)=y(2);
dy(2)=-y(1).*(1-y(1));
end
```

1.10.5 Figure 1.9

```
alpha=0.05;
lambda=1+alpha;
n=120; % number of time steps
x1=zeros(n+1,1);
x2=zeros(n+1,1);
x3=zeros(n+1,1);
```

```
t=zeros(n+1,1);
x1(1)=0.2;
x2(1)=1;
x3(1)=3;
t(1)=0;
for i=1:n
     x1(i+1)=lambda*x1(i)/(1+alpha*x1(i));
     x2(i+1)=lambda*x2(i)/(1+alpha*x2(i));
     x3(i+1)=lambda*x3(i)/(1+alpha*x3(i));
     t(i+1)=i;
end


clf;
plot(t,x1,t,x2,t,x3,'LineWidth',5)
xlabel('t');
ylabel('N(t)');
set(gca,'XLim',[0 n])
set(gca,'fontweight','bold','FontSize',30);
ax=gca;
```

1.10.6 Figure 1.10

```
close all;
clear all;
figure(1)
beta=exp(1.5);
n=20; % number of time steps
x1=zeros(n+1,1);
x2=zeros(n+1,1);
x3=zeros(n+1,1);
t=zeros(n+1,1);
x1(1)=0.5;
x2(1)=1;
x3(1)=2;
t(1)=0;
for i=1:n
     x1(i+1)=beta*x1(i)*exp(-x1(i));
     x2(i+1)=beta*x2(i)*exp(-x2(i));
     x3(i+1)=beta*x3(i)*exp(-x3(i));
     t(i+1)=i;
end
plot(t,x1,t,x2,t,x3,'LineWidth',5)
xlabel('t');
ylabel('N(t)');
set(gca,'XLim',[0 n])
```

```
set(gca,'fontweight','bold','FontSize',30);
ax=gca;


figure(2)
beta=exp(2.2);
n=20; % number of time steps
x1=zeros(n+1,1);
x2=zeros(n+1,1);
x3=zeros(n+1,1);
t=zeros(n+1,1);
x1(1)=2;
x2(1)=3;
x3(1)=4;
t(1)=0;
for i=1:n
    x1(i+1)=beta*x1(i)*exp(-x1(i));
    x2(i+1)=beta*x2(i)*exp(-x2(i));
    x3(i+1)=beta*x3(i)*exp(-x3(i));
    t(i+1)=i;
end


clf;
plot(t,x1,t,x2,t,x3,'LineWidth',5)
xlabel('t');
ylabel('N(t)');
set(gca,'XLim',[0 n])
set(gca,'fontweight','bold','FontSize',30);
ax=gca;

figure(3)
beta=exp(4);
n=20; % number of time steps
x1=zeros(n+1,1);
x2=zeros(n+1,1);
x3=zeros(n+1,1);
t=zeros(n+1,1);
x1(1)=3;
x2(1)=4.1;
x3(1)=5;
t(1)=0;
for i=1:n
    x1(i+1)=beta*x1(i)*exp(-x1(i));
    x2(i+1)=beta*x2(i)*exp(-x2(i));
    x3(i+1)=beta*x3(i)*exp(-x3(i));
```

```
        t(i+1)=i;
end


clf;
plot(t,x1,t,x2,t,x3,'LineWidth',5)
xlabel('t');
ylabel('N(t)');
set(gca,'XLim',[0 n])
set(gca,'fontweight','bold','FontSize',30);
ax=gca;
```

1.10.7 Figure 1.11

```
clear all;
close all;
n=10000;
p=50;
j=1;
for beta=1:0.005:exp(log(18))
x=1;


        for i=1:n
            x=beta*x*exp(-x);
            if (i >=n-p)
            aux(1,j)=beta;
            aux(2,j)=x;
            j=j+1;
            end
        end
end

clf;
plot(aux(1,:),aux(2,:),'k.','markersize', 2)
set(gca, 'xlim', [0 18]);
xlabel('\beta');
ylabel('\omega(1)');

set(gca,'fontweight','bold','FontSize',30);
ax=gca;
```

1.10.8 Figure 1.12 and Figure 1.13

```
close all;
clear all;
```

```
sigma=10;
rho=28;
beta=8/3

eps = 1e-15;


Tmax=19;


figure(1)

T = [0 Tmax];


initV = [10 10 10];


options = odeset('RelTol',eps,'AbsTol',[eps eps eps
    /10]);
[T,X] = ode45(@(T,X) F(T, X, sigma, rho, beta), T,
    initV, options);



plot3(X(:,1),X(:,2),X(:,3),'Color','[ 0.4940 ,
    0.1840 , 0.5560]','LineWidth', 1);

view
    ([19.1287420824236,-356.3357721553517,80.00898161095051])

hold on;

plot3(X(1,1),X(1,2),X(1,3),'ok', 'MarkerSize', 20, '
    MarkerFaceColor', 'g');

plot3(X(end,1),X(end,2),X(end,3),'ok', 'MarkerSize',
     20, 'MarkerFaceColor', '[ 0.9290 , 0.6940 ,
    0.1250]');

T = [0 Tmax];

initV = [10 10 10.7];
```

```
[T,X] = ode45(@(T,X) F(T, X, sigma, rho, beta), T, initV, options);


plot3(X(:,1),X(:,2),X(:,3),'Color','[ 0.3010 , 0.7450 , 0.9330 ]','LineWidth', 1);


plot3(X(1,1),X(1,2),X(1,3),'ok', 'MarkerSize', 20, 'MarkerFaceColor', 'k');

plot3(X(end,1),X(end,2),X(end,3),'ok', 'MarkerSize', 20, 'MarkerFaceColor', '[ 0.8500 , 0.3250 , 0.0980]');


grid;
xlabel('x'); ylabel('y'); zlabel('z');
set(gca, 'FontSize', 30);
set(gca, 'FontWeight', 'bold');

figure(2)

T = [0 Tmax];

initV = [10 10 10];

[T,X] = ode45(@(T,X) F(T, X, sigma, rho, beta), T, initV, options);

plot3(X(:,1),X(:,2),X(:,3),'Color','[ 0.4940 , 0.1840 , 0.5560]','LineWidth', 1);

hold on;

plot3(X(1,1),X(1,2),X(1,3),'ok', 'MarkerSize', 20, 'MarkerFaceColor', 'g');

plot3(X(end,1),X(end,2),X(end,3),'ok', 'MarkerSize', 20, 'MarkerFaceColor', '[ 0.9290 , 0.6940 , 0.1250]');

T = [0 Tmax];

initV = [10 10 10.7];

```

```
[T,X] = ode45(@(T,X) F(T, X, sigma, rho, beta), T, initV, options);

plot3(X(:,1),X(:,2),X(:,3),'Color','[ 0.3010 , 0.7450 , 0.9330 ]','LineWidth', 1);

view([183.3188733682035,331.5379873898024,112.19634248895])

plot3(X(1,1),X(1,2),X(1,3),'ok', 'MarkerSize', 20, 'MarkerFaceColor', 'k');

plot3(X(end,1),X(end,2),X(end,3),'ok', 'MarkerSize', 20, 'MarkerFaceColor', '[ 0.8500 , 0.3250 , 0.0980]');

grid;
xlabel('x'); ylabel('y'); zlabel('z');
set(gca, 'FontSize', 30);
set(gca, 'FontWeight', 'bold');


figure(3)
T = [0 Tmax];

initV = [10 10 10];


options = odeset('RelTol',eps,'AbsTol',[eps eps eps/10]);
[T,X] = ode45(@(T,X) F(T, X, sigma, rho, beta), T, initV, options);

plot(T,X(:,1),'Color','[ 0.4940 , 0.1840 , 0.5560]','LineWidth', 3);

hold on;
T = [0 Tmax];

initV = [10 10 10.7];


options = odeset('RelTol',eps,'AbsTol',[eps eps eps/10]);
```

```
[T,X] = ode45(@(T,X) F(T, X, sigma, rho, beta), T,
    initV, options);

plot(T,X(:,1),'Color','[ 0.3010 , 0.7450 ,
    0.9330 ]','LineWidth', 3);

xlabel('t'); ylabel('x(t)');
set(gca, 'FontSize', 30);
set(gca, 'FontWeight', 'bold');



function dx = F(T, X, sigma, rho, beta)
    dx = zeros(3,1);

    dx(1) = sigma*(X(2) - X(1));
    dx(2) = rho*X(1) - X(1)*X(3)- X(2);
    dx(3) = X(1)*X(2) - beta*X(3);

end
```

1.10.9 Figures 1.17-1.19

```
close all;
clear all;




Tmax=2800;


T =[0 Tmax];
initV=zeros(1,10);

initV(1,1) =70; % S at t=0
initV(1,2) =10; % I_1 at t=0
initV(1,3) =20; % I_2 at t=0


eps = 1e-40;
options = odeset('RelTol',eps,'AbsTol',[eps eps eps
    eps eps eps eps eps eps eps]);

[T,X] = ode45(@(T,X) F(T, X), T, initV, options);

```

```
figure(1)
    plot(T,X(:,1),'Color','[    0.4660    ,    0.6740    ,
        0.1880]','LineWidth', 3);
    xlabel('t'); ylabel('S');
    set(gca, 'FontSize', 30);
    set(gca, 'FontWeight', 'bold');
    xlim([2000,2400])

figure(2)
    plot(T,X(:,6),'Color','[    0.4660    ,    0.6740    ,
        0.1880]','LineWidth', 3);
    xlabel('t'); ylabel('S_1');
    set(gca, 'FontSize', 30);
    set(gca, 'FontWeight', 'bold');
    xlim([2000,2400])

figure(3)
    plot(T,X(:,7),'Color','[    0.4660    ,    0.6740    ,
        0.1880]','LineWidth', 3);
    xlabel('t'); ylabel('S_2');
    set(gca, 'FontSize', 30);
    set(gca, 'FontWeight', 'bold');
    xlim([2000,2400])
figure(4)
    plot(T,X(:,2)+X(:,3)+X(:,8)+X(:,9),'Color','
        [0.6350    ,    0.0780    ,  0.1840]','LineWidth',
        3);
    xlabel('t'); ylabel('I_1+I_2+I_{12}+I_{21}');
    set(gca, 'FontSize', 30);
    set(gca, 'FontWeight', 'bold');
    xlim([2000,2400])

figure(5)
    n=find(T>=2000);
    plot3(X(n(1):end,3)+X(n(1):end,8), X(n(1):end,1)
        ,X(n(1):end,2)+X(n(1):end,9),'Color','[  0    ,
         0.4470    ,    0.7410]','LineWidth', 1);


    hold on;

    initV=zeros(1,10);
    initV(1,1) =70; % S at t=0
    initV(1,2) =20; % I_1 at t=0
    initV(1,3) =10; % I_2 at t=0

```

```
    [T1,X1] = ode45(@(T,X) F(T, X), T, initV,
        options);
    n=find(T1>=2000);
    plot3(X1(n(1):end,3)+X1(n(1):end,8), X1(n(1):end
        ,1),X1(n(1):end,2)+X1(n(1):end,9),'Color','[
         0.8500 ,     0.3250  , 0.0980]','LineWidth',
        1);
    grid;
    ylabel('S'); xlabel('I_2+I_{12}'); zlabel('I_1+
        I_{21}');
    set(gca, 'FontSize', 30);
    set(gca, 'FontWeight', 'bold');


figure(6)

    plot3(X(n(1):end,3)+X(n(1):end,8), X(n(1):end,1)
        +X(n(1):end,6)+X(n(1):end,7),X(n(1):end,2)+X(
        n(1):end,9),'Color','[ 0   ,   0.4470   ,
        0.7410]','LineWidth', 1);

    hold on;


    plot3(X1(n(1):end,3)+X1(n(1):end,8), X1(n(1):end
        ,1)+X1(n(1):end,6)+X1(n(1):end,7),X1(n(1):end
        ,2)+X1(n(1):end,9),'Color','[   0.8500 ,
        0.3250   , 0.0980]','LineWidth', 1);
    grid;
    ylabel('S+S_1+S_2'); xlabel('I_2+I_{12}');
        zlabel('I_1+I_{21}');
    set(gca, 'FontSize', 30);
    set(gca, 'FontWeight', 'bold');
function dx = F(T, X)

N=100;
mu=1/65;
gamma= 52  ;
beta1= 2*gamma;
beta2= 2*gamma;
alpha= 2  ;
phi=0.6;
phi1=phi;
phi2=phi;

    dx = zeros(10,1);
```

```
dx(1) =- beta1/N*X(1)*(X(2)+phi1*X(9)) - beta2/N
    *X(1)*(X(3)+phi2*X(8))+mu*(N-X(1)); % S-
    equation

dx(2) = beta1/N*X(1)*(X(2)+phi1*X(9))- (gamma+mu
    )*X(2);                              % I_1
    equation
dx(3) = beta2/N*X(1)*(X(3)+phi2*X(8)) - (gamma+
    mu)*X(3);                            % I_2
    equation

dx(4)=gamma*X(2)-(alpha+mu)*X(4);

    % R_1 equation
dx(5)=gamma*X(3)-(alpha+mu)*X(5);

    % R_2 equation

dx(6)= - beta2/N*X(6)*(X(3)+phi2*X(8))+alpha*X
    (4)-mu*X(6);                         % S_1
    equation
dx(7)= - beta1/N*X(7)*(X(2)+phi1*X(9))+alpha*X
    (5)-mu*X(7);                         % S_2
    equation

dx(8)= beta2/N*X(6)*(X(3)+phi2*X(8))- (gamma+mu
    )*X(8);                              % I_12
    equation
dx(9)= beta1/N*X(7)*(X(2)+phi1*X(9))- (gamma+mu
    )*X(9);                              % I_21
    equation

dx(10)=gamma*(X(8)+X(9))-mu*X(10);

    % R equation
end
```

Chapter 2
Global Attractors and Uniform Persistence

2.1 Interior Global Attractor for an Elementary Example

To illustrate some ideas developed in this chapter, let us consider the difference equation

$$U(t+1)x = \frac{\lambda U(t)x}{1+U(t)x}, \forall t \in \mathbb{N}, \text{ and } U(0)x = x \in [0,\infty),$$

where $\lambda > 1$ is a given parameter.

We define the set $M = \mathbb{R}_+$, which is endowed with the distance induced by the absolute value in $\mathbb{R}$, namely

$$d(u,v) = |u-v|, \quad \forall (u,v) \in M.$$

Then the set $A = [0, \lambda - 1]$ (where $\overline{U} = \lambda - 1$ is the positive equilibrium of the above equation) is compact and

$$\lim_{t\to\infty} \delta(U(t)B, A) = 0,$$

whenever $B \subset \mathbb{R}_+$. Here recall that $\delta(\cdot,\cdot)$ denotes the Hausdorff semi-distance.

The subset A corresponds to the global attractor for the semiflow $U = \{U(t)\}_{t\in\mathbb{N}}$ in $\mathbb{R}_+$.

Next, we observe that the set

$$M_0 = (0, +\infty)$$

is positively invariant by U. But the existence of a global attractor for U restricted on the set M_0 is more delicate because the metric space (M_0, d) is not complete.

If we consider the positive interior equilibrium

$$\overline{U} = \lambda - 1 > 0,$$

we would like to define

A. Ducrot et al., *Differential Equations and Population Dynamics II*,
Lecture Notes on Mathematical Modelling in the Life Sciences,
https://doi.org/10.1007/978-3-031-91409-6_2

$$A_0 = \{\overline{U}\}$$

as the global attractor for U restricted to M_0. This is delicate, because the interior attractor A_0 will attract the compact subsets of M_0 (in general), and possibly not the bounded subsets of M_0 (in general).

2.2 Positive Orbit for a Set

Let (M, d) be a complete metric space. Let $\{U(t)\}_{t\geq 0}$ be a semiflow on this metric space (M, d). In Chapter 1, we described the ω-limit set for a point $x \in M$. Recall that this ω-limit set is defined as

$$\omega(x) = \bigcap_{t\geq 0} \overline{\bigcup_{s\geq t} U(s)x}.$$

The idea of global attractor is to understand the asymptotic properties of $t \rightarrow U(t)B$ when t goes to infinity and B is a suitable subset of M. To that aim we need to extend the notion ω-limit for the trajectory of a point x to the trajectory of a subset B of M. So for a subset $B \subset M$, we will consider $\omega(B)$ the set defined by

$$\omega(B) = \bigcap_{t\geq 0} \overline{\bigcup_{s\geq t} U(s)B}.$$

2.3 Examples of Metric Spaces

In practice the notion of metric space is introduced in order to keep a notion of distance between points for subsets of Banach spaces that are not necessarily vector spaces. Very often, a metric space will be a subset M of a Banach space $(X, \|.\|)$ endowed with the distance induced by the norm of X, that is

$$d(x, y) = \|x - y\|, \quad \forall x, y \in M.$$

Definition 2.1 A metric space (M, d) is **complete**, if any Cauchy sequence $\{u_n\}_{n\geq 0} \subset M$ converges in M. Recall that $\{u_n\}_{n\geq 0}$ in M is a Cauchy sequence if for each $\varepsilon > 0$ there exists an integer $r \in \mathbb{N}$, such that

$$d(u_m, u_p) \leq \varepsilon, \forall m, p \geq r.$$

In other words, the metric space (M, d) is complete if for each Cauchy sequence $\{u_n\}_{n\geq 0}$ in M, there exists $u \in M$ such that

$$d(u_m, u) \rightarrow 0, \text{ as } m \rightarrow \infty.$$

A metric space (M, d) is **not complete** otherwise. That is, if there exists a Cauchy sequence in (M, d) that is not convergent in M.

Example 2.2 (of metric space in $\mathbb{R}$) Consider the metric space $M_0 = (0, 1] \subset \mathbb{R}$ endowed with the metric $d_0(x, y) = |x - y|$. Then (M_0, d_0) is closed (because the empty set $\emptyset$ is open). If we consider the sequence

$$x_n = \frac{1}{2^n} \text{ for } n \geq 0,$$

then for any $n \in \mathbb{N}$ and $q \in \mathbb{N}$ one has

$$|x_n - x_{n+q}| = x_n - x_{n+q} \leq x_n = \frac{1}{2^n}.$$

Therefore $\{x_n\}_{n \in \mathbb{N}}$ is a Cauchy sequence. But this sequence converges to 0, which does not belong to M_0. Hence (M_0, d_0) is not a complete metric space.

Example 2.3 (of metric space in $\mathbb{R}^2$) Consider

$$M_1 = [0, 1] \times [0, 1], \text{ with } d_1(x, y) = ||x - y||_{\mathbb{R}^2}$$

which is a closed subset. Therefore, (M_1, d_1) will give a complete metric (i.e. any Cauchy sequence in (M_1, d_1) converges to an element of M_1).

An example of non complete metric space is given by

$$M_2 = M_1 \setminus \{(0, 0)\} = \{y \in M_1 : y \neq 0\},$$

which is not closed in $\mathbb{R}^2$. Then (M_2, d_1) is not a complete metric space. Indeed, let us consider a sequence

$$(x_n, y_n) \in (0, 1] \times (0, 1] \rightarrow (0, 0) \notin M_2, \text{ as } n \rightarrow \infty.$$

Then $\{(x_n, y_n)\}_{n \geq 0}$ is a Cauchy sequence in (M_2, d_1), but $\{(x_n, y_n)\}_{n \geq 0}$ does not converge in M_2.

Exercise 2.4 Prove the following properties.

(i) The map

$$d_2(x, y) = \left| \frac{1}{||x||_{\mathbb{R}^2}} - \frac{1}{||y||_{\mathbb{R}^2}} \right| + ||x - y||_{\mathbb{R}^2}, \forall x, y \in M_2,$$

is a distance on M_2.

(ii) M_2 endowed with the distance d_2 is a complete metric space.
(**Hint:** The proof should become clear at the end of this chapter.)

Example 2.5 (of metric space in $L^1(0, 1)$) We can also consider some infinite dimensional example such as

$$M_3 = \left\{ u \in L^1(0,1) : 0 \leq u \leq 1 \right\},$$

endowed with

$$d_3(u,v) = \|u - v\|_{L^1} = \int_0^1 |u(\sigma) - v(\sigma)| d\sigma, \ \forall (u,v) \in M_3.$$

Then since M_3 is closed in $L^1(0,1)$, the metric space (M_3, d_3) is complete. Next, if we consider

$$M_4 = M_3 \setminus \{0_{L^1}\},$$

then (M_4, d_3) is not a complete metric space. Indeed, let us consider a sequence

$$u_n \in L^1(0,1),$$

such that

$$\|u_n - 0_{L^1(0,1)}\|_{L^1} = \int_0^1 |u_n(\sigma)| d\sigma \to 0, \text{ as } n \to \infty.$$

Then one has

$$u_n \to 0_{L^1(0,1)} \notin M_4, \text{ as } n \to \infty,$$

where the limit is understood in $L^1(0,1)$. As a consequence $\{u_n\}_{n \geq 0}$ is a Cauchy sequence in (M_4, d_3), but $\{u_n\}_{n \geq 0}$ is not convergent in M_4.

Exercise 2.6 Prove the following properties.

(i) The map

$$d_4(u,v) = \left| \frac{1}{\|u\|_{L^1}} - \frac{1}{\|v\|_{L^1}} \right| + \|u - v\|_{L^1}, \forall u, v \in M_4,$$

is a distance on M_4.

(ii) M_4 endowed with the distance d_4 is a complete metric space.

(**Hint:** The proof should become clear at the end of this chapter.)

2.4 Neighborhood and ε-Neighborhood of a Subset

Throughout this part, (M, d) denotes a metric space.

Definition 2.7 Let $A \subset M$ be a subset. Then a subset $N \subset M$ (which is not necessarily open or closed in M) is called a **neighborhood of** A, if and only if for each $x \in A$, there exists $\varepsilon = \varepsilon(x) > 0$ (depends on x in general) such that

$$B_M(x, \varepsilon) := \{y \in M : d(x,y) < \varepsilon\} \subset N.$$

We will say that N is an **open neighborhood of** A whenever N is open, and we will say that N is an **closed neighborhood of** A whenever N is closed.

Recall that the distance between a point x and a subset $A \subset M$ is defined by

$$d(x, A) := \inf_{y \in A} d(x, y).$$

The quantity $d(x, A)$ is measuring the distance of x to A. Therefore if $x \in A$ then $d(x, A) = 0$. Moreover, if $d(x, A) = \varepsilon \geq 0$, this means that we can find a sequence $y_n \in A$ such that

$$\lim_{n \to \infty} d(x, y_n) = \varepsilon.$$

In particular $d(x, A) = 0$ does not imply that $x \in A$. But $d(x, A) = 0$ implies that $x \in \overline{A}$.

By Lemma 1.22, we know that the map $x \to d(x, A)$ is Lipschitz continuous. It follows that for each $\varepsilon > 0$, the subset

$$N(A, \varepsilon) := \{x \in M : d(x, A) < \varepsilon\},$$

is an open neighborhood of A (since it is the inverse image of $(-\infty, \varepsilon)$ by a continuous map), and the subset

$$\overline{N}(A, \varepsilon) := \{x \in M : d(x, A) \leq \varepsilon\},$$

is a closed neighborhood of A (since it is the inverse image of $(-\infty, \varepsilon]$ by a continuous map).

Definition 2.8 The subsets $N(A, \varepsilon)$ and $\overline{N}(A, \varepsilon)$ are called **ε-neighborhood of** A. The subset $N(A, \varepsilon)$ is called an **open ε-neighborhood of** A, and the subset $\overline{N}(A, \varepsilon)$ is called a **closed ε-neighborhood of** A.

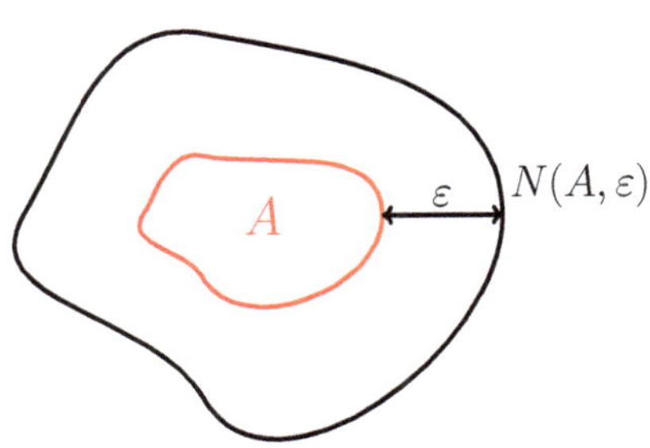

Fig. 2.1: *This figure illustrates the notion of ε-neighborhood of a subset A.*

The neighborhood $N(A, \varepsilon)$ is very specific. Indeed, for each $x \in A$, we have

$$B_M(x, \varepsilon) \subset N(A, \varepsilon),$$

where $\varepsilon > 0$ becomes independent of x. One can compare with Definition 2.7 where the value of ε may depend on x.

2.5 Compact Subsets in (Non Complete) Metric Spaces: Definitions and Main Results

Let (M, d) be a metric space.

Definition 2.9 A subset of O of M is **open**, if for each $x \in O$, there exists $\varepsilon > 0$ (depending on x) such that

$$B_M(x, \varepsilon) = \{y \in M : d(x, y) < \varepsilon\} \subset O.$$

A subset $\mathcal{F}$ is **closed** if its complementary in M is open. That is

$$M \setminus \mathcal{F} = \{y \in M : y \notin \mathcal{F}\}$$

is an open subset.

Definition 2.10 Let C be a subset of M. We will say that

(i) A family of subsets $\{\mathcal{F}_\alpha\}_{\alpha \in E}$ is a **covering** of C if

$$C \subset \bigcup_{\alpha \in E} \mathcal{F}_\alpha.$$

(ii) A family of subsets $\{\mathcal{F}_\alpha\}_{\alpha \in E}$ is an **open covering** of C, if $\{\mathcal{F}_\alpha\}_{\alpha \in E}$ is a covering of C, and for each $\alpha \in E$, the subset $\mathcal{F}_\alpha$ is an open subset of (M, d).

(iii) A family of subsets $\{\mathcal{F}_\alpha\}_{\alpha \in E}$ is a **finite covering** of C, if $\{\mathcal{F}_\alpha\}_{\alpha \in E}$ is a covering of C, and E is finite.

(iv) Suppose $\{\mathcal{F}_\alpha\}_{\alpha \in E}$ is a covering of C, and $E' \subset E$. Then $\{\mathcal{F}_\alpha\}_{\alpha \in E'}$ is a **sub-covering** of C (extracted from $\{\mathcal{F}_\alpha\}_{\alpha \in E}$), if

$$C \subset \bigcup_{\alpha \in E'} \mathcal{F}_\alpha.$$

Definition 2.11 Let C be a subset of M. We will say that

(i) The subset C is **compact** if any open covering of C admits a finite sub-covering of C. That is for each family $\{\mathcal{F}_\alpha\}_{\alpha \in E}$ such that

$$C \subset \bigcup_{\alpha \in E} \mathcal{F}_\alpha,$$

there exists a finite subset $E' \subset E$ such that

$$C \subset \bigcup_{\alpha \in E'} \mathcal{F}_\alpha.$$

(ii) The subset C is **sequentially compact** if each sequence $\{u_n\}_{n\in\mathbb{N}}$ admits a convergent sub-sequence $\{u_{n_p}\}_{p\in\mathbb{N}}$ converging to an element $x \in C$.
(iii) The subset C is **relatively compact**, if its closure $\overline{C}$ is compact.
(iv) The subset C is **totally bounded (or precompact)**, if for each $r > 0$, C has a finite covering with balls of radius r.

Theorem 2.12 (Corollary 3.8 p.34 in [203]) *Assume that (M, d) is a metric space (complete or not). Then we have the following*

(i) *A subset C is compact if and only if it is sequentially compact.*
(ii) *A subset C is compact if and only if (C, d) is a complete metric space and is totally bounded.*

Example 2.13 (of non compact subset which is totally bounded and closed) Consider the metric space $M = (0, 1] \subset \mathbb{R}$ endowed with the metric $d(x, y) = |x-y|$. Consider the set sequence

$$C := \bigcup_{p\in\mathbb{N}} \{x_p\},$$

where we have

$$x_n = \frac{1}{2^n}, \forall n \in \mathbb{N}.$$

(i) **The subset C is not compact.**
Indeed, we have for any $n \geq 1$

$$x_{n+1} = \frac{x_n}{2}, \text{ and } x_{n-1} = 2x_n.$$

Define

$$r_n = \frac{x_n - x_{n+1}}{2} = \frac{1}{2^{n+2}},$$

so that we have

$$B_M(x_n, r_n) \cap C = \{x_n\}.$$

It follows that $\mathcal{F} = \{\mathcal{F}_n\}_{n\geq 0}$ the family defined by

$$\mathcal{F}_n = B_M(x_n, r_n)$$

is an open covering for the set

$$C = \bigcup_{p\in\mathbb{N}} \{x_p\}.$$

But we can find no finite sub-covering of C extracted from $\mathcal{F}$ (since each element of $\mathcal{F}$ covers exactly one element of C). Therefore C is not compact.

(ii) **C is totally bounded.**
Indeed, if we consider any covering $\mathcal{G} = \{\mathcal{G}_n\}_{n\geq 0}$ the family defined by

$$\mathcal{G}_n = B(x_n, r), \forall n \in \mathbb{N},$$

for some $r > 0$.

Then setting $n_0 = n_0(r) > 0$ such that $x_{n_0} < r$, we have

$$C \subset \bigcup_{n=1,\dots,n_0} \{\mathcal{G}_n\},$$

and C is totally bounded.

(iii) **C is a closed subset in (M, d).**
Indeed, the complementary subset $M \setminus C$ is open (M, d), because we can find an open ball around each point of $M \setminus C$ included in $M \setminus C$.

Therefore C is closed, totally bounded, and C is not compact in (M, d).

The above example together with the following theorem explain why the notion of totally bounded sets is used only in complete metric spaces. This will be crucial (later in this chapter) when we will consider the measures of non compactness.

Theorem 2.14 (Corollary 3.9 p.35 in [203]) *Assume that (M, d) is a complete metric space. If C is totally bounded, then the subset $\overline{C}$ is compact.*

2.6 Cantor's Diagonal Process

In order to explain the diagonal process, we consider an example of sequence with double indexation

$$(n, m) \in \mathbb{N} \times \mathbb{N} \to u_n^m \in [0, 1].$$

This double indexation can be regarded as a sequence $m \to u^m$, where each element u^m is itself a sequence $n \to u_n^m$ with values in $[0, 1]$.

Claim 2.15 *There exists an increasing sequence $p \to m_p \in \mathbb{N}$, and there exists a sequence $n \to \overline{u}_n \in [0, 1]$, such that*

$$\lim_{p\to\infty} u_n^{m_p} = \overline{u}_n, \ \forall n \in \mathbb{N}.$$

Proof (of the claim) The principle of this proof is schematically represented in Figure 2.2.

Step $n = 0$: Let us first consider the sequence $m \to u_0^m \in [0, 1]$. Since $[0, 1]$ is compact in $\mathbb{R}$, we can find a convergent sub-sequence. That is, we can find a strictly increasing sequence of integer $p \to m_p^0 \in \mathbb{N}$ (with $m_p^0 \to \infty$), and $\overline{u}_0 \in [0, 1]$, such that

$$\lim_{p\to\infty} u_0^{m_p^0} = \overline{u}_0.$$

Step $n = 1$: We consider the sub-sequence $p \to u_1^{m_p^0} \in [0, 1]$. We can extract a sub-sequence $p \to m_p^1 \in \mathbb{N}$ (with $m_p^1 \to \infty$), and $\overline{u}_1 \in [0, 1]$, such that

$$\lim_{p\to\infty} u_1^{m_p^1} = \overline{u}_1.$$

Here to extract a sub-sequence from $p \longrightarrow m_p^0 \in \mathbb{N}$ means that, for each $p \in \mathbb{N}$,

$$m_p^1 \in \bigcup_{\widehat{p}\in\mathbb{N}} \{m_{\widehat{p}}^0\},$$

and

$$\lim_{p\to\infty} m_p^1 = \infty.$$

This process is represented by the diagonal red arrows in Figure 2.2.

Due to this construction, we keep the convergence from the previous sub-sequence. That is

$$\lim_{p\to\infty} u_0^{m_p^1} = \overline{u}_0.$$

We can keep the diagonal elements from the previous column without changing the convergence properties,

$$m_0^1 = m_0^0.$$

This process is represented by the horizontal red arrows in Figure 2.2.

Step $n = k + 1$**:** We consider the sub-sequence $p \longrightarrow u_{k+1}^{m_p^k} \in [0, 1]$. We can extract a sub-sequence $p \longrightarrow m_p^{k+1} \in \mathbb{N}$ (i.e. $m_p^{k+1} \longrightarrow \infty$ as $p \longrightarrow \infty$) and $\overline{u}_{k+1} \in [0, 1]$, such that

$$\lim_{p\to\infty} u_{k+1}^{m_p^{k+1}} = \overline{u}_{k+1}.$$

Here extracting a sub-sequence from $p \longrightarrow m_p^{k+1} \in \mathbb{N}$ means that, for each $p \in \mathbb{N}$,

$$m_p^{k+1} \in \bigcup_{\widehat{p}\in\mathbb{N}} \{m_{\widehat{p}}^k\},$$

and

$$\lim_{p\to\infty} m_p^{k+1} = \infty.$$

This process is represented by the diagonal red arrows in Figure 2.2.

Due to this construction, we keep the convergence from the previous sub-sequences, that is

$$\lim_{p\to\infty} u_0^{m_p^{k+1}} = \overline{u}_0, \lim_{p\to\infty} u_1^{m_p^{k+1}} = \overline{u}_1, \ldots, \lim_{p\to\infty} u_k^{m_p^{k+1}} = \overline{u}_k.$$

We can keep the above diagonal elements from the previous column without changing the convergence properties,

$$m_p^{k+1} = m_p^k, \forall p = 0, 1, \ldots, p - 1.$$

This process is represented by the horizontal red arrows in Figure 2.2.

Final step: By induction arguments we can construct such a sequence for each integer $k \in \mathbb{N}$, and we obtain

$$\lim_{p \to \infty} u_k^{m_p^k} = \overline{u}_k, \forall k \in \mathbb{N}.$$

Now we choose the diagonal element in Figure 2.2 (in green). That is we consider the sequence

$$\widehat{m}_p = m_p^p, \forall p \in \mathbb{N}.$$

Then by construction the sequence $p \to \widehat{m}_p$ is a sub-sequence of any sequence $p \to m_p^k$ for $k \in \mathbb{N}$. That is for each $p \in \mathbb{N}$, and $k \in \mathbb{N}$,

$$\widehat{m}_p \in \bigcup_{\widehat{p} \in \mathbb{N}} \{m_{\widehat{p}}^k\},$$

and

$$\lim_{p \to \infty} \widehat{m}_p = \infty.$$

Therefore, we obtain

$$\lim_{p \to \infty} u_k^{\widehat{m}_p} = \overline{u}_k, \forall k \in \mathbb{N},$$

and the claim is proved. □

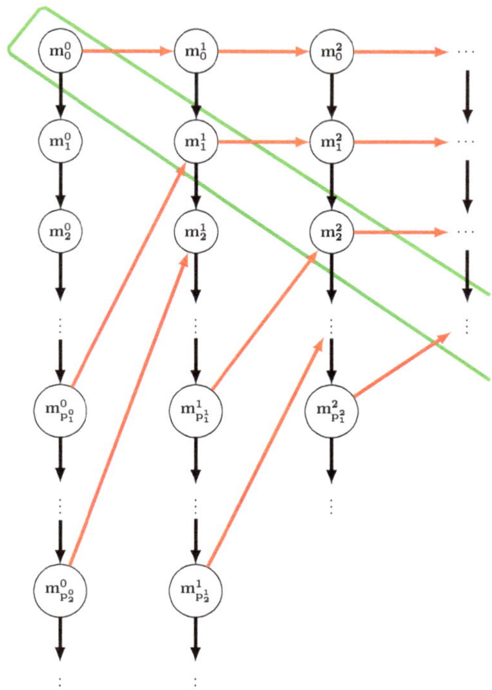

Fig. 2.2: *The figure gives a schematic representation of the diagonal process. The columns represent the index of each sub-sequence. The red arrows indicate where the values are taken in the previous column (i.e. on the left hand side). One must realize that all the indexes of a given column appear in the previous one (i.e., on the left-hand side). That is, each column is extracted from the previous one. The coefficients above the diagonal are unchanged. The diagonal process consists in choosing the diagonal elements.*

Example 2.16 (of sequence that is not converging uniformly with respect to $n \in \mathbb{N}$**)** The convergence in the Claim 2.15 is called **local convergence** in $n \in \mathbb{N}$. This means that the convergence holds simultaneously only for a finite number of $n \in \mathbb{N}$.

To convince ourselves that this convergence can not be uniform, let us consider the following example

$$u_n^p = \frac{n+1}{n+1+p+1}, \quad (n,p) \in \mathbb{N}^2.$$

Then one has

$$\lim_{p\to\infty} u_n^p = 0, \forall n \geq 0,$$

while we have

$$\|u^p\|_\infty = \sup_{n\in\mathbb{N}} u_n^p = \lim_{n\to\infty} \frac{n+1}{n+1+p+1} = 1, \forall p \in \mathbb{N}.$$

Therefore we have

$$\lim_{p\to\infty} \|u^p\|_\infty = 1 \neq 0.$$

This example shows that the convergence in the Claim 2.15 is not (in general) uniform with respect to $n \in \mathbb{N}$.

As a corollary of the previous diagonal extraction process, we obtain the following corollary.

Corollary 2.17 *The space of sequences*

$$M = \{u = \{u_n\}_{n\in\mathbb{N}} : u_n \in [0,1], \forall n \in \mathbb{N}\},$$

endowed with the distance

$$d(u,v) = \sup_{n\in\mathbb{N}} \frac{|u_n - v_n|}{2^n}, \ \forall (u,v) \in M,$$

is sequentially compact.

Proof Let $m \to u^m$ be a sequence of elements of M. That is for each $m \in \mathbb{N}$, u^m is a sequence of the form

$$n \to u_n^m \in [0,1].$$

By using the Claim 2.15, there exists a increasing sequence $p \to m_p \in \mathbb{N}$, and there exists a sequence $n \to \overline{u}_n \in [0,1]$, such that

$$\lim_{p\to\infty} u_n^{m_p} = \overline{u}_n, \forall n \in \mathbb{N}.$$

Let us prove that

$$\lim_{p\to\infty} \sup_{n\in\mathbb{N}} \frac{|u_n^{m_p} - \overline{u}_n|}{2^n} = 0.$$

Let $\varepsilon > 0$ be fixed. Since $u_n^m \in [0,1]$, and $\overline{u}_n \in [0,1]$, we deduce that

$$\frac{|u_n^{m_p} - \overline{u}_n|}{2^n} \leq \frac{2}{2^n}, \quad \forall n \geq 0.$$

Therefore, for $n_0 \in \mathbb{N}$ large enough, we have

$$\frac{|u_n^{m_p} - \overline{u}_n|}{2^n} \leq \frac{2}{2^n} \leq \varepsilon, \forall n \geq n_0.$$

It follows that

$$\sup_{n\in\mathbb{N}} \frac{|u_n^{m_p} - \overline{u}_n|}{2^n} \leq \max\left(\max_{n=0,\ldots,n_0} \frac{|u_n^{m_p} - \overline{u}_n|}{2^n}, \varepsilon \right).$$

Since

$$\lim_{p\to\infty} |u_n^{m_p} - \overline{u}_n| = 0, \forall n = 0, \dots, n_0,$$

one deduces that there exists an integer $p_0 \in \mathbb{N}$ such that

$$\sup_{n\in\mathbb{N}} \frac{|u_n^{m_p} - \overline{u}_n|}{2^n} \leq \varepsilon, \forall p \geq p_0.$$

The proof of the corollary is completed. □

2.7 Spaces of Continuous Functions: Compactness Properties and Diagonal Process

We introduce the Arzelà–Ascoli theorem, which is proved in [203, Theorem 3.1 in Chapter III].

Theorem 2.18 (Arzelà–Ascoli theorem) *Let X be a compact subset of a metric space (M, d). Let $(F, \|.\|_F)$ be a Banach space. Let Φ be a subset of $C(X, F)$, the space of continuous maps from X into F. Then Φ is relatively compact in $C(X, F)$ (endowed with the distance $d(u, v) = \sup_{x\in X} \|u(x) - v(x)\|_F$) if and only if*

(i) *Φ is equi-continuous, that is, for each $\varepsilon > 0$, there exists $\eta > 0$, such that*

$$d(x, y) \leq \eta \Rightarrow \|u(x) - u(y)\|_F \leq \varepsilon, \forall u \in \Phi.$$

(ii) *For each $x \in X$, the set $\{u(x) : u \in \Phi\}$ is relatively compact in F.*

To illustrates the above result, we will prove the following typical consequence of Arzelà–Ascoli theorem.

Lemma 2.19 *Let $I \subset \mathbb{R}$, $J_0 \subset \mathbb{R}$, and $J_1 \subset \mathbb{R}$ be three closed and bounded intervals. Then the set*

$$C := \{u \in C^1(I, J_0) \ : \ u(x) \in J_0, \text{ and } u'(x) \in J_1, \text{ for all } x \in I\}$$

is relatively compact in $C(I, J_0)$ endowed with the **supremum topology** *(i.e., $d(u, v) = \sup_{x\in I} |u(x) - v(x)|$). That is, for each sequence $\{u^m\}_{m\in\mathbb{N}} \subset C$, there exists a sub-sequence $p \to m_p \in \mathbb{N}$ (i.e., strictly increasing sequence of integers) and a continuous function $\bar{u} \in C(I, J_0)$, such that*

$$\lim_{p\to\infty} \sup_{x\in I} |u^{m_p}(x) - \bar{u}(x)| = 0.$$

Proof (of Lemma 2.19) To prove Lemma 2.19 we show that the Arzelà-Ascoli Theorem 2.18 can be applied to the set C. First we remark that for all $x \in I$ one has $\{u(x) \ : \ u \in C\} \subset J_0$ and J_0 is compact in $\mathbb{R}$, therefore condition (ii) in Theorem 2.18 is satisfied.

Next we show that the condition (i) holds. We let

$$D := \sup\{|y| \ : \ y \in J_1\},$$

so that

$$|u'(x)| \le D \text{ for all } u \in C \text{ and } x \in I.$$

Then by the mean value theorem, we have

$$|u(x) - u(y)| \le |x - y| \sup_{z \in I} |u'(z)| \le D|x - y| \text{ for all } u \in C \text{ and } x, y \in I.$$

In particular for each $\varepsilon > 0$ we can find

$$\eta := \frac{\varepsilon}{D},$$

such that

$$|x - y| \le \eta \Rightarrow |u(x) - u(y)| \le D|x - y| \le D\frac{\varepsilon}{D} = \varepsilon \text{ for all } u \in C.$$

Thus, by the Arzelà-Ascoli Theorem 2.18, C is relatively compact in $C(I, J_0)$. In other words for each sequence $\{u^m\}_{m\in\mathbb{N}} \subset C$, there exists a function $\bar{u} \in C(I, \mathbb{R})$ and a sub-sequence $p \to m_p \in \mathbb{N}$ such that

$$\lim_{p\to+\infty} d(u^{m_p}, \bar{u}) = \lim_{p\to+\infty} \sup_{x\in I} |u^{m_p}(x) - \bar{u}(x)| = 0.$$

Lemma 2.19 is proved. □

Theorem 2.20 *Let $J_0 \subset \mathbb{R}$, and $J_1 \subset \mathbb{R}$ be two closed and bounded intervals. Consider the subset of $C(\mathbb{R}, \mathbb{R})$*

$$M := \{u \in C(\mathbb{R}, \mathbb{R}) \, : \, u(x) \in J_0, \text{ for all } x \in \mathbb{R}\},$$

endowed with the distance associated to the ***local uniform topology****, that is the distance defined by*

$$d(u, v) = \sum_{n=1}^{+\infty} \frac{1}{2^n} \sup_{x\in[-n,n]} |u(x) - v(x)|, \quad \forall (u, v) \in M.$$

Then (M, d) is a complete metric space. Moreover, the set

$$C := \{u \in C^1(\mathbb{R}, \mathbb{R}) \, : \, u(x) \in J_0 \text{ and } u'(x) \in J_1 \text{ for all } x \in \mathbb{R}\},$$

is relatively compact in (M, d). That is, for each sequence $\{u^m\}_{m\in\mathbb{N}} \subset C$, there exists a sub-sequence $p \to m_p \in \mathbb{N}$ (i.e., strictly increasing sequence of integers) and a continuous function $\bar{u} \in C(\mathbb{R}, J_0)$, such that

$$\lim_{p\to+\infty} \sum_{n=1}^{+\infty} \frac{1}{2^n} \sup_{x\in[-n,n]} |u^{m_p}(x) - \bar{u}(x)| = 0.$$

Proof Let $\{u^m\}_{m\ge0}$ be a sequence in C. We use a diagonal process.

Step 1: Compactness on $[-1, 1]$.

Consider the restriction of $\{u^m\}_{m\in\mathbb{N}}$ to the interval $[-1,1]$. By Lemma 2.19, there exists a sub-sequence $\{u^{m^1_p}\}_{p\in\mathbb{N}}$ and a continuous function $\bar{u}_1 : [-1,1] \to \mathbb{R}$ such that

$$\lim_{p\to+\infty} \sup_{x\in[-1,1]} |u^{m^1_p}(x) - \bar{u}_1(x)| = 0.$$

Step 2: Compactness in $[-2,2]$.
By applying Lemma 2.19 to the sequence $\{u^{m^1_p}\}_{p\in\mathbb{N}}$ restricted to the interval $[-2,2]$, we get that there exists a sub-sequence $p \to m^2_p$ extracted from $p \to m^1_p$ and a continuous function $\bar{u}_2 \in C([-2,2],\mathbb{R})$ such that

$$\lim_{p\to+\infty} \sup_{x\in[-2,2]} |u^{m^2_p}(x) - \bar{u}_2(x)| = 0.$$

Since m^2_p is extracted from m^1_p, the restriction of $u^{m^2_p}$ to $[-1,1]$ converges to $\bar{u}_1$. Therefore

$$\bar{u}_2(x) = \bar{u}_1(x) \text{ for all } x \in [-1,1].$$

Step 3: Construction of $\{u^{m^n_p}\}_{p\in\mathbb{N}}$.
By induction, assume that we have constructed sequences $\{m^k_p\}_{p\in\mathbb{N}}$ for $k = 1,\dots,n$, such that each $\{m^k_p\}_{p\in\mathbb{N}}$ is extracted from $\{m^{k-1}_p\}_{p\in\mathbb{N}}$, and there exists $\bar{u}_k \in C([-k,k],\mathbb{R})$ such that

$$\sup_{x\in[-k,k]} |u^{m^k_p}(x) - \bar{u}_k(x)| \xrightarrow[p\to+\infty]{} 0 \text{ for all } k = 1,\dots,n.$$

By applying the Arzelà-Ascoli theorem to the sequence $\{u^{m^n_p}\}_{p\in\mathbb{N}}$ restricted to $[-(n+1),(n+1)]$, we find a sub-sequence $\{m^{n+1}_p\}_{p\in\mathbb{N}}$ extracted from $\{m^n_p\}_{p\in\mathbb{N}}$ and a function $\bar{u}_{n+1} \in C([-(n+1),n+1],\mathbb{R})$ such that

$$\lim_{p\to+\infty} \sup_{x\in[-(n+1),n+1]} |u^{m^{n+1}_p}(x) - \bar{u}_{n+1}(x)| = 0.$$

Since m^{n+1}_p is extracted from m^n_p, the restriction of $u^{m^n_p}$ to $[-n,n]$ converges to $\bar{u}_n$. Therefore

$$\bar{u}_{n+1}(x) = \bar{u}_n(x) \text{ for all } x \in [-n,n].$$

The sequence $\{u^{m^n_p}\}_{p\in\mathbb{N}}$ is thus constructed for all $n \in \mathbb{N}$.

Step 4: Diagonal process.
Now we choose $\widehat{m}_n := m^n_n$ for all $n \in \mathbb{N}$ and $\bar{u}(x) := \bar{u}_n(x)$ for all $x \in [-n,n]$ and $n \geq 1$. Observe that, by construction, the sequence $\{\widehat{m}_p\}_{p\geq n}$ is extracted from $\{m^n_p\}_{p\in\mathbb{N}}$ for all $n \in \mathbb{N}$. Therefore

$$\lim_{p\to+\infty} \sup_{x\in[-n,n]} |u^{\widehat{m}_p}(x) - \bar{u}(x)| = 0, \text{ for all } n \in \mathbb{N}.$$

Define $D := \sup_{x\in J_0} |x|$. Then, since $u^{\widehat{m}_p} \in C(\mathbb{R}, J_0)$ and $\bar{u} \in C(\mathbb{R}, J_0)$, we have

$$\sup_{x\in[-n,n]} |u^{\widehat{m}_p}(x) - \bar{u}(x)| \leq 2D.$$

Thus we have

$$\begin{aligned}
d(u^{\widehat{m}_p}, \bar{u}) &= \sum_{n=1}^{+\infty} \frac{1}{2^n} \sup_{x\in[-n,n]} |u^{\widehat{m}_p}(x) - \bar{u}(x)| \\
&\leq \sum_{n=1}^{n_0} \frac{1}{2^n} \sup_{x\in[-n,n]} |u^{\widehat{m}_p}(x) - \bar{u}(x)| + \sum_{n=n_0+1}^{+\infty} \frac{1}{2^n} \times 2D \\
&= \sum_{n=1}^{n_0} \frac{1}{2^n} \sup_{x\in[-n,n]} |u^{\widehat{m}_p}(x) - \bar{u}(x)| + \frac{2D}{2^{n_0+1}} \frac{1}{1-\frac{1}{2}}.
\end{aligned}$$

Let $\varepsilon > 0$ be given. Then for $n_0 \geq 0$ sufficiently large we have

$$\frac{2D}{2^{n_0+1}} \frac{1}{1-\frac{1}{2}} = \frac{2D}{2^{n_0}} \leq \frac{\varepsilon}{2},$$

and we obtain for all $n \geq n_0$,

$$d(u^{\widehat{m}_p}, \bar{u}) \leq \sum_{n=1}^{n_0} \frac{1}{2^n} \sup_{x\in[-n,n]} |u^{\widehat{m}_p}(x) - \bar{u}(x)| + \frac{\varepsilon}{2}.$$

Now we can find $n_1 \geq n_0$ such that $p \geq n_1$ with n_1 sufficiently large we have

$$\sup_{x\in[-n_0,n_0]} |u^{\widehat{m}_p}(x) - \bar{u}(x)| \leq \frac{\varepsilon}{2n_0}.$$

Therefore for all $p \geq n_1$ we have

$$d(u^{\widehat{m}_p}, \bar{u}) \leq n_0 \frac{\varepsilon}{2n_0} + \frac{\varepsilon}{2} = \varepsilon,$$

and since $\varepsilon > 0$ is arbitrary, we have proved that $u^{\widehat{m}_p}$ converges to $\bar{u}$ as $p \to +\infty$ for the topology induced by d. Theorem 2.20 is proved. □

Remark 2.21 Instead of the distance d, we could have used any other distance of the form

$$d^\star(u, v) = \sup_{x\in\mathbb{R}} \rho(x)|u(x) - v(x)|,$$

where $\rho : \mathbb{R} \to (0, \infty)$ is any continuous function, satisfying

$$\lim_{|x|\to\infty} \rho(x) = 0.$$

In that case, the conclusion of Theorem 2.20 still holds, that is, for each sequence $\{u^m\}_{m\in\mathbb{N}} \subset C$, there exists a sub-sequence $p \to m_p \in \mathbb{N}$ (i.e., strictly increasing sequence of integers) and a continuous function $\overline{u} \in C(\mathbb{R}, J_0)$, such that

$$\lim_{p\to\infty} d^\star(u^{m_p}, \overline{u}) = 0.$$

2.8 Compactness Properties for Families of Subsets

In this section, we prepare some key ingredients for the global attractor theory. We first recall that, in any metric space (M, d) (complete or not), a subset C of M is compact if and only if any sequence in C has a converging sub-sequence in C, that means that for each sequence $\{x_n\}_{n\geq 0} \subset C$, we can find a sub-sequence $\left\{x_{n_p}\right\}_{p\geq 0}$ (that is we can find a strictly increasing sequence of integer $p \to n_p \in \mathbb{N}$) and there exists $x \in C$, such that

$$\lim_{p\to\infty} x_{n_p} = x.$$

Definition 2.22 Let $I \subset [0, +\infty)$ be an unbounded set. Let $\{A_t\}_{t\in I}$ be a family of non-empty subsets of M. We say that $\{A_t\}_{t\in I}$ is **point-wise sequentially compact** if and only if for each sequence $\{t_n\}_{n\geq 0} \subset I \to +\infty$, and each sequence $\{x_n\}_{n\geq 0}$ satisfying

$$x_n \in A_{t_n}, \forall n \geq 0,$$

the sequence $\{x_n\}_{n\in\mathbb{N}}$ has a convergent sub-sequence.

Remark 2.23 Every family $\{A_t\}_{t\in I} \subset \mathbb{R}^n$ such that $A_t \subset B_{\mathbb{R}^n}(0, r), \forall t \in I$ (for some $r > 0$) is point-wise sequentially compact. This example provides an example of such a class of point-wise sequentially compact family of sets. The same property holds true in locally compact metric spaces.

Definition 2.24 Let $I \subset [0, +\infty)$ be an unbounded set, let $\{A_t\}_{t\in I}$ be a family of non-empty subsets of M. We will say that $\{A_t\}_{t\in I}$ is **decreasing** if and only if

$$t \geq s \Rightarrow A_t \subset A_s.$$

Recall that in any metric space (M, d) (complete or not) a subset C of M is compact if and only if every sequence has a convergent sub-sequence whose limit is in C.

Theorem 2.25 *Let $I \subset [0, +\infty)$ be unbounded, let $\{A_t\}_{t\in I}$ be a family of non-empty subsets of M. Assume that $\{A_t\}_{t\in I}$ is point-wise sequentially compact. Then the set*

$$A_\infty := \bigcap_{t\geq 0} \overline{\bigcup_{s\geq t} A_s}$$

is compact and non-empty and one has

$$\delta(A_t, A_\infty) \to 0, \; as \; t \to +\infty,$$

where $\delta(\cdot, \cdot)$ denotes the Hausdorff semi-distance. Moreover, if we assume in addition that A_t is a decreasing family of closed of subsets of M, then we have

$$A_\infty = \bigcap_{t \geq 0} A_t.$$

Proof Set $B_t = \overline{\bigcup_{s \geq t} A_s}, \forall t \geq 0$. Let $\{t_n\}_{n \geq 0} \subset I$ with $t_n \to \infty$, and let $\{x_n\}_{n \geq 0} \subset M$ be a sequence such that

$$x_n \in B_{t_n}, \forall n \geq 0.$$

We claim that $\{x_n\}_{n \geq 0}$ has a convergent sub-sequence. Indeed, there exists for each $n \geq 0$, $y_n \in \bigcup_{s \geq t_n} A_s$, such that

$$d(y_n, x_n) \leq \frac{1}{n+1}.$$

For each $n \geq 0$, there exists $s_n \geq t_n$ such that $y_n \in A_{s_n}$, and by assumption $\{y_n\}_{n \geq 0}$ has a convergent sub-sequence. Consider a sub-sequence $\{y_{n_p}\}_{p \geq 0}$ and $x \in M$ with $y_{n_p} \to x$ as $p \to \infty$. Then we have

$$d(x, x_{n_p}) \leq d(x, y_{n_p}) + d(y_{n_p}, x_{n_p}) \to 0, \text{ as } p \to +\infty.$$

Moreover, since $A_t \subset B_t$, we deduce that

$$\delta(A_t, A_\infty) \leq \delta(B_t, A_\infty), \forall t \geq 0.$$

Hence we can replace A_t by B_t and we can always assume that A_t is a non-increasing family of closed and non-empty subsets of M. We set $A_\infty = \bigcap_{t \geq 0} A_t$. Then it is clear that A_∞ is closed since it is an intersection of closed subsets.

Let us prove that A_∞ is not empty. To do so, by taking $t_n \to \infty$ and $x_n \in A_{t_n}$, and by choosing a sub-sequence we can assume that $x_n \to x$. Let $t \in I$ be given. Then for all $n \in \mathbb{N}$ with $t_n \geq t$, we have

$$x_n \in A_{t_n} \subset A_t$$

and since A_t is closed, we deduce that

$$x_n \to x \in A_t, \forall t \geq 0.$$

Hence we get

$$x \in A_\infty.$$

Moreover, if we take $\{y_n\}_{n \geq 0} \subset A_\infty$, then by induction we can find a sequence $\{t_n\}_{n \geq 0} \subset I$, and a sequence $\{x_n\}_{n \geq 0} \subset M$, such that $x_n \in A_{t_n}$ for all $n \geq 0$, $t_n \to +\infty$, and $d(y_n, x_n) \to 0$, as $n \to +\infty$. By assumption we can find a sub-sequence $\{x_{n_p}\}_{p \geq 0}$ such that $x_{n_p} \to y$ as $p \to +\infty$. Now we also has $y_{n_p} \to y$ as $p \to +\infty$. Since A_∞ is closed we end-up with $y \in A_\infty$ and it follows that A_∞ is compact.

Assume by contradiction that there exist $\varepsilon > 0$, and a sequence $\{t_n\}_{n \geq 0}$ such that $t_n \to \infty$ and $\delta(A_{t_n}, A_\infty) \geq \varepsilon, \forall n \geq 0$. Then for each $n \geq 0$, we can find $x_n \in A_{t_n}$, such that $d(x_n, A_\infty) \geq \varepsilon/2$. By assumption we can find a sub-sequence $\{x_{n_p}\}_{p \geq 0}$

with $x_{n_p} \to x$. But by construction we have $x \in A_\infty$ and

$$0 = d(x, A_\infty) \geq \varepsilon/2,$$

and we have obtained a contradiction. □

Definition 2.26 We say that a subset $C \subset M$ **attracts** a family $\{A_t\}_{t\in I}$, if

$$\lim_{t\to\infty} \delta(A_t, C) = 0.$$

That is equivalent to say that for each $\varepsilon > 0$, we can find $t_0 = t_0(\varepsilon) \geq 0$, such that

$$A_t \subset N(C, \varepsilon), \forall t \geq t_0.$$

We deduce the following result from Theorem 2.25.

Theorem 2.27 *The two following assertions are equivalent:*

(i) *The family $\{A_t\}_{t\in I}$ is point-wise sequentially compact.*
(ii) *The family $\{A_t\}_{t\in I}$ is attracted by some compact subset $C \subset M$.*

Moreover, if (i) *or* (ii) *is satisfied, then the set*

$$A_\infty = \bigcap_{t\geq 0} \overline{\bigcup_{s\geq t} A_s}$$

is the smallest compact subset $C \subset M$ attracting the family $\{A_t\}_{t\in I}$.

Definition 2.28 Let $I \subset [0, +\infty)$ be an unbounded set and let $\{A_t\}_{t\in I}$ be a family of non-empty subsets of M. We say that $\{A_t\}_{t\in I}$ is **Hausdorff sequentially compact** if and only if for each sequence $\{t_n\}_{n\geq 0} \subset I \to +\infty$, there exist a sub-sequence $\{t_{n_p}\}_{p\geq 0}$ and a subset $\widehat{A}_\infty \subset M$ satisfying

$$d_H\left(A_{t_{n_p}}, \widehat{A}_\infty\right) \to 0, \text{ as } p \to +\infty.$$

Recall that $d_H(\cdot, \cdot)$ denotes the Hausdorff distance as defined in Chapter 1. The convergence of the family A_t to $\widehat{A}_\infty$ for the Hausdorff distance means that for each $\varepsilon > 0$, we have both

$$A_t \subset N\left(\widehat{A}_\infty, \varepsilon\right) \tag{2.1}$$

and

$$\widehat{A}_\infty \subset N\left(A_t, \varepsilon\right) \tag{2.2}$$

whenever $t > 0$ is large enough.

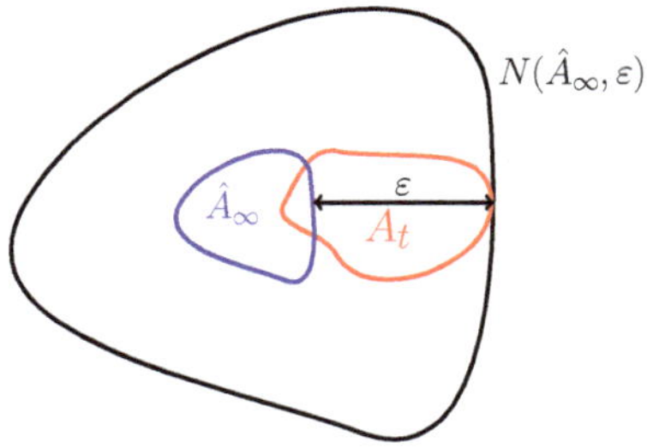

Fig. 2.3: *This figure illustrates the inclusion* (2.1). *When* ε *goes to* 0, *the set* A_t *approaches* $\widehat{A}_\infty$.

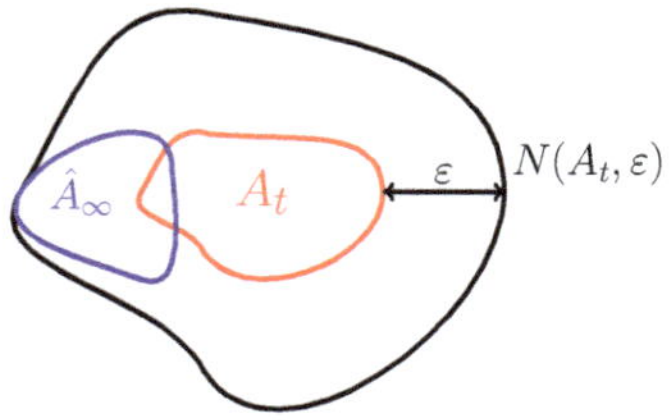

Fig. 2.4: *This figure illustrates the inclusion* (2.2). *When* ε *goes to* 0, *the set* $\widehat{A}_\infty$ *approaches* A_t.

In our previous convergence results (see Theorem 2.25), the property (2.2) was missing. This property can be interpreted by saying that every point of the limit set A_∞ is approached by some point of A_t. The following result ensures that the convergence is also true in the sense of the Hausdorff distance, but only for a subsequence A_{t_n} and a smaller subset $\widehat{A}_\infty \subset A_\infty$.

Theorem 2.29 *Let* $I \subset [0, +\infty)$ *be an unbounded set and let* $\{A_t\}_{t\in I}$ *be a family of non-empty closed subsets of* M. *If the family* $\{A_t\}_{t\in \bar{I}}$ *is point-wise sequentially compact, then* $\{A_t\}_{t\in I}$ *is Hausdorff sequentially compact. More precisely, for each* $\{t_n\}_{n\geq 0} \subset I \to +\infty$, *there exists a sub-sequence* $\left\{t_{n_p}\right\}_{p\geq 0}$ *such that*

$$d_H\left(A_{t_{n_p}}, \widehat{A}_\infty\right) \to 0, \text{ as } p \to +\infty,$$

where the set $\widehat{A}_\infty \subset A_\infty$ *is defined by*

$$\widehat{A}_\infty = \bigcap_{p\geq 0}\overline{\bigcup_{q\geq p} A_{t_{n_q}}}.$$

Proof Set

$$A_\infty = \bigcap_{t\geq 0}\overline{\bigcup_{s\geq t} A_s}.$$

Then by Theorem 2.25 we know that A_∞ is compact, and

$$\delta(A_t, A_\infty) \to 0, \text{ as } t \to +\infty.$$

Since A_∞ is compact, for each integer $m \geq 0$, there exists $\left\{x_1^m, ..., x_{k_m}^m\right\} \subset A_\infty$, such that

$$A_\infty \subset \bigcup_{i=1}^{k_m} B\left(x_i^m, \frac{1}{2^{m+1}}\right).$$

Let $\{t_n\}_{n\geq 0} \subset I$ be a sequence with $t_n \to +\infty$. By using a diagonal process, we can find a sequence of integers $\{l_m\}_{m\geq 0}$, such that $\left\{y_1^m, ..., y_{l_m}^m\right\} \subset \left\{x_1^m, ..., x_{k_m}^m\right\}$ and for any integer p and m with $p \geq m$

$$A_{t_{n_p}} \subset \bigcup_{i=1}^{l_m} B\left(y_i^m, \frac{1}{2^m}\right), \tag{2.3}$$

and

$$A_{t_{n_p}} \cap B\left(y_i^m, \frac{1}{2^m}\right) \neq \emptyset, \forall i = 1, ..., l_m. \tag{2.4}$$

Set

$$\widehat{A}_\infty = \bigcap_{p\geq 0}\overline{\bigcup_{q\geq p} A_{t_{n_q}}}.$$

Due to Theorem 2.25, we know that $\widehat{A}_\infty$ is a non-empty compact set, and

$$\delta(A_{t_{n_p}}, \widehat{A}_\infty) \to 0, \text{ as } p \to +\infty.$$

Moreover, by using (2.3) and (2.4), we deduce that for each integer $m \geq 0$,

$$\widehat{A}_\infty \subset \bigcup_{i=1}^{l_m} \overline{B}\left(y_i^m, \frac{1}{2^m}\right),$$

and

$$\widehat{A}_\infty \cap \overline{B}\left(y_i^m, \frac{1}{2^m}\right) \neq \emptyset, \forall i = 1, ..., l_m.$$

Let $x \in \widehat{A}_\infty$ be given. Then for each $m \geq 0$, there exists $i_0 \in \{1, ..., l_m\}$, such that

$$x \in \overline{B}\left(y_{i_0}^m, \frac{1}{2^m}\right) \text{ and } A_{t_{n_p}} \cap B\left(y_{i_0}^m, \frac{1}{2^m}\right) \neq \emptyset, \forall p \geq m.$$

Hence we get

$$d\left(x, A_{t_{n_p}}\right) \leq \frac{1}{2^{m-1}}, \forall p \geq m.$$

We conclude that

$$\delta\left(\widehat{A}_\infty, A_{t_{n_p}}\right) \leq \frac{1}{2^{m-1}}, \forall p \geq m,$$

and the proof is completed. □

Example 2.30 (Discrete time example) Consider the following example $\{A_n\}_{n\in\mathbb{N}}$ with

$$A_{2n} = \{0\}, \forall n \in \mathbb{N},$$

and

$$A_{2n+1} = \{1\}, \forall n \in \mathbb{N}.$$

Then

$$A_\infty = \{0, 1\},$$

and

$$\delta(A_n, A_\infty) = 0, \forall n \geq 0,$$

implies

$$\lim_{n\to\infty} \delta(A_n, A_\infty) = 0.$$

Moreover we have

$$\delta(A_\infty, A_n) = \sup_{x\in\{0,1\}} d(x, A_n) = 1, \forall n \in \mathbb{N}.$$

Therefore, the Hausdorff distance satisfies

$$d_H(A_n, A_\infty) = 1, \forall n \geq 0.$$

Moreover if we consider constant sub-sequence $n \to A_{2n}$ then

$$\widehat{A}_\infty = \{0\},$$

and if we consider constant sub-sequence $n \to A_{2n+1}$ then

$$\widehat{A}_\infty = \{1\}.$$

Example 2.31 (Continuous time example) Consider the following example $\{A_t\}_{t\in\mathbb{R}_+}$ with

$$A_t = \{\cos(t)\}, \forall t \in \mathbb{R}_+.$$

Then we have

$$A_\infty = [-1, 1],$$

and

$$\delta(A_\infty, A_t) = \sup_{x\in[-1,1]} d(x, A_t) = \max\left(|1 - \cos(t)|, |1 + \cos(t)|\right) \geq 1, \forall t \in \mathbb{R}_+.$$

Therefore, the Hausdorff distance satisfies

$$d_H(A_t, A_\infty) \geq 1, \forall t \geq 0.$$

Moreover if we consider constant sub-sequence $n \longrightarrow A_{t+2n\pi}$ then we get

$$A_{t+2n\pi} = \{\cos(t)\}, \forall n \geq 0,$$

and we obtain for this sequence $\widehat{A}_\infty = \{\cos(t)\}$.

2.9 Measure of Non-Compactness

Definition 2.32 The **Kuratowski measure of non-compactness**, κ, is defined by

$$\kappa(B) = \inf\{r > 0 : B \text{ has a finite cover with ball of radius } r\},$$

for any bounded set B of M. We set $\kappa(B) = +\infty$, whenever B is unbounded.

Remark 2.33 From Example 2.13, we can see that the measure of non-compactness is not measuring the compactness in non complete metric spaces.

From this definition, we deduce that

$$\kappa(B) = \kappa(\overline{B}),$$

for any subset B of M.

The following lemma can be reformulated by saying that (M, d) is a complete metric space if and only if each totally bounded subset of M is relatively compact.

Theorem 2.34 *The following properties are equivalent*

(i) *(M, d) is complete metric space.*
(ii) *For each subset C of M, $\kappa(C) = 0$ implies that $\overline{C}$ is compact.*

Proof **Proof of (ii)⇒(i).** Let $\{x_n\}_{n\geq 0}$ be a Cauchy sequence. Then for each $\varepsilon > 0$, there exists $n_0 \geq 0$, such that

$$d(x_{n_0}, x_{n_0+l}) \leq \varepsilon, \forall l \geq 0.$$

Hence we obtain

$$\kappa(\bigcup_{n\geq 0}\{x_n\}) = \kappa(\bigcup_{n\geq n_0}\{x_n\}) \leq \varepsilon.$$

Hence

$$\kappa(\bigcup_{n\geq 0}\{x_n\}) = 0,$$

and by using (ii), we deduce that

$$\overline{\bigcup_{n\geq 0}\{x_n\}}$$

is compact. By compactness, we deduce that $\{x_n\}_{n\geq 0}$ has a sub-sequence converging in M. But, by using again the fact that $\{x_n\}_{n\geq 0}$ is a Cauchy sequence, we conclude that the sequence $\{x_n\}_{n\geq 0}$ converges in M.

Proof of (i)⇒(ii). Let C be closed subset of M, and assume that $\kappa(C) = 0$. Let $\{x_n\}_{n\geq 0}$ be a sequence in C. For each integer $m \geq 0$, let $\left\{z_1^m, ..., z_{k_m}^m\right\} \subset M$ be such that

$$C \subset \bigcup_{i=1}^{k_m} B\left(z_i^m, \frac{1}{2^m}\right),$$

and by using a diagonal process, we can find a sub-sequence $\left\{x_{n_m}\right\}_{m\geq 0}$ and a sequence of integer $\{i_m\}_{m\geq 0}$ such that

$$x_{n_p} \in B\left(z_{i_m}^m, \frac{1}{2^m}\right), \forall p \geq m.$$

So

$$d\left(z_{i_m}^m, z_{i_{m+1}}^{m+1}\right) \leq d\left(z_{i_m}^m, x_{n_{m+1}}\right) + d\left(x_{n_{m+1}}, z_{i_{m+1}}^{m+1}\right) \leq \frac{1}{2^m} + \frac{1}{2^{m+1}} \leq \frac{1}{2^{m-1}}.$$

We deduce that for all $m \geq 0$, and $l \geq 0$,

$$\begin{aligned} d\left(z_{i_m}^m, z_{i_{m+l}}^{m+l}\right) &\leq d\left(z_{i_m}^m, z_{i_{m+1}}^{m+1}\right) + \ldots + d\left(z_{i_{m+l-1}}^{m+l-1}, z_{i_{m+l}}^{m+l}\right) \\ &\leq \frac{1}{2^{m-1}} + \frac{1}{2^m} + \ldots. + \frac{1}{2^{m+l-1}} \\ &\leq \frac{1}{2^{m-1}}\left[1 + \frac{1}{2} + \ldots. + \frac{1}{2^l}\right] \leq \frac{1}{2^{m-2}}. \end{aligned}$$

This implies that $\left\{z_{i_m}^m\right\}_{m\geq 0}$ is a Cauchy sequence. So by using (i), one deduce that $z_{i_m}^m \to x \in C$, as $m \to +\infty$, and since

$$d\left(x_{n_p}, z_{i_m}^m\right) \leq \frac{1}{2^m}, \forall p \geq m,$$

the result follows. □

Proposition 2.35 *The following statements are valid*

(i) *Let (M, d) be complete metric space. Let $I \subset [0, +\infty)$ be an unbounded set and let $\{A_t\}_{t\in I}$ be a decreasing family of non-empty closed subsets of M. Assume that*

$$\kappa(A_t) \to 0, \text{ as } t \to +\infty.$$

Then the set

$$A_\infty = \bigcap_{t \geq 0} A_t$$

is non-empty and compact, and one has

$$\delta(A_t, A_\infty) \to 0, \ as\ t \to +\infty.$$

(ii) *For each $A \subset M$, and $B \subset M$, we have*

$$\kappa(B) \leq \kappa(A) + \delta(B, A).$$

Proof **Proof of (i)**. Assume first that $\kappa(A_t) \to 0$ as $t \to +\infty$. Let $\{t_n\}_{n\geq 0} \subset I$ and $\{x_n\}_{n\geq 0} \subset M$ be two sequences such that $t_n \to \infty$ and $x_n \in A_{t_n}, \forall n \geq 0$. Then one has

$$\kappa\left(\bigcup_{n\geq 0} \{x_n\}\right) = \kappa\left(\bigcup_{n\geq p} \{x_n\}\right) \leq \kappa(A_{t_p}), \forall p \geq 0,$$

so that

$$\kappa\left(\bigcup_{n\geq 0} \{x_n\}\right) = 0.$$

By Theorem 2.34 we conclude that $\overline{\bigcup_{n\geq 0} \{x_n\}}$ is compact. We deduce that the family $\{A_t\}_{t\in I}$ is point-wise sequentially compact and the result follows from Theorem 2.25.

Proof of (ii). Let $\varepsilon > 0$ be given. Then there exists $\{x_1, ..., x_N\} \subset M$, such that

$$A \subset \bigcup_{n=1,\ldots,N} B\left(x_n, \kappa(A) + \varepsilon/2\right) \tag{2.5}$$

Let $x \in B$ be given. Then we have $d(x, A) \leq \delta(B, A)$ so that there exists $y \in A$, such that

$$d(x, y) \leq \delta(B, A) + \varepsilon/2.$$

But by (2.5), there exists $n \in \{1, ..., N\}$ such that

$$d(x_n, y) \leq \kappa(A) + \varepsilon/2.$$

This yields

$$d(x, x_n) \leq d(x, y) + d(y, x_n) \leq \kappa(A) + \delta(B, A) + \varepsilon.$$

Hence we get

$$B \subset \bigcup_{n=1,\ldots,N} B\left(x_n, \kappa(A) + \delta(B, A) + \varepsilon\right)$$

and

$$\kappa(B) \leq \kappa(A) + \delta(B, A) + \varepsilon, \forall \varepsilon > 0.$$

The result follows by letting $\varepsilon \to 0$. □

2.10 Omega-Limit Sets of a Subset

Let (M, d) be a given metric space and $U : I \times M \to M$ be a semiflow on (M, d). Here the set I is $[0, \infty)$ or $\omega\mathbb{N}$, for some $\omega > 0$.

Definition 2.36 For each subset $B \subset M$, we denote by

$$\gamma^+(B) = \bigcup_{t \geq 0} U(t)(B)$$

the **positive orbit of** B **for** U, and by

$$\omega(B) = \bigcap_{t \geq 0} \overline{\bigcup_{s \geq t} U(s)(B)}$$

the ω**-limit set of** B for U.

Definition 2.37 We say that a subset $A \subset M$ **attracts** a subset $B \subset M$ for U, if

$$\lim_{t \to \infty} \delta(U(t)B, A) = 0.$$

Lemma 2.38 *Assume that U is a continuous semiflow. If $B \subset M$ is positively invariant for U so is $\overline{B}$.*

Proof Assume that $B \subset M$ is positively invariant for U, that is to say that

$$U(t)B \subset B, \forall t \geq 0.$$

Let $x_n \in B \to x \in \overline{B}$ be a convergent sequence. Then since U is a continuous semiflow

$$U(t)x_n \in B \to U(t)x \in \overline{B},$$

and we deduce that $\overline{B}$ is positively invariant for U. □

The theory of attractors is based on the following fundamental result, which is related to Hale [138, Lemmas 2.1.1 and 2.1.2].

Proposition 2.39 (Omega-limit sets) *Assume that the semiflow $\{U(t)\}_{t \in I}$ is continuous on (M, d). Let B be a subset of M, which is attracted by a compact subset $C \subset M$ for U. Then one has*

(i) *$\omega(B)$ is a non-empty and compact set which attracts B for U, that is*

$$\lim_{t \to +\infty} \delta(U(t)B, \omega(B)) = 0.$$

(ii) *$\omega(B)$ invariant for U, namely*

$$U(t)\omega(B) = \omega(B), \forall t \geq 0.$$

Proof **Proof of (i)**. Set

$$A_t = \overline{\bigcup_{s\geq t} U(s)B}, \forall t \geq 0.$$

It is clear that the family $\{A_t\}_{t\in I}$ is decreasing and is attracted by C. By Theorem 2.27, we deduce that $\omega(B) = \bigcap_{t\geq 0} A_t$ is non-empty, compact, and attracts $\{A_t\}_{t\in I}$. Therefore $\omega(B)$ attracts B for U.

Proof of (ii). Let $\tau \geq 0$. We have

$$U(\tau)\left(\bigcup_{s\geq t} U(s)B\right) = \bigcup_{s\geq t+\tau} U(s)B, \forall t \geq 0. \tag{2.6}$$

Therefore, we obtain

$$\bigcup_{s\geq t+\tau} U(s)B \subset \overline{U(\tau)A_t}, \forall t \geq 0,$$

hence

$$A_{t+\tau} \subset \overline{U(\tau)A_t}, \forall t \geq 0. \tag{2.7}$$

By using again (2.6), we have

$$U(\tau)\left(\bigcup_{s\geq t} U(s)B\right) \subset A_{t+\tau}, \forall t \geq 0,$$

and since $U(\tau)$ is continuous,

$$U(\tau)A_t \subset A_{t+\tau}, \forall t \geq 0. \tag{2.8}$$

We deduce from (2.8) that

$$U(\tau)\omega(B) \subset U(\tau)A_t \subset A_{t+\tau}, \forall t \geq 0,$$

hence

$$U(\tau)\omega(B) \subset \omega(B), \forall t \geq 0. \tag{2.9}$$

To prove the converse inclusion, let $y \in \omega(B)$. Since the family A_t is decreasing, we can find a sequence $t_n \to \infty$ and $y_n \in A_{t_n+\tau}$ such that $y_n \to y$ as $n \to \infty$. By using (2.7) we deduce that we can find a sequence $x_n \in A_{t_n}$ such that

$$d(y_n, U(\tau)x_n) \leq \frac{1}{2^n}.$$

But we can also find a convergent sub-sequence

$$x_{n_p} \to x \in \omega(B).$$

Now by using the continuity of $U(\tau)$, it follows that $U(\tau)x = y$. It follows that

$$\omega(B) \subset U(\tau)\omega(B), \tag{2.10}$$

and the results follows from (2.9) and (2.10).

2.11 Global Attractors

Let (M, d) be a complete metric space. Let $U : I \times M \to M$ be a continuous semiflow on M with $I = \omega\mathbb{N}$ or $\mathbb{R}$.

Definition 2.40 The semiflow U is said to be **point (compact, bounded) dissipative** if there is a bounded set B_0 in M such that B_0 attracts each point (compact set, bounded set) in M;

Definition 2.41 The semiflow U is said to be **asymptotically smooth** if for any nonempty closed bounded set $B \subset M$ which is positively invariant for U (i.e. $U(t)B \subset B, \forall t \geq 0$), there is a compact set $C \subset M$ such that C attracts B for U (i.e. $\delta(U(t)B, C) \to 0$, as $t \to \infty$).

Lemma 2.42 *The semiflow U is asymptotically smooth if and only if*

$$\lim_{t \to \infty} \kappa\left(U(t)B\right) = 0,$$

for any nonempty closed bounded subset $B \subset M$ which is positively invariant for U.

Proof ($\Leftarrow$): Let $B \subset M$ be a nonempty bounded subset which is positively invariant for U. Assume that

$$\lim_{t \to \infty} \kappa\left(U(t)B\right) = 0.$$

Then we have

$$\lim_{t \to \infty} \kappa\left(\overline{U(t)B}\right) = 0.$$

By using Proposition 2.35-(i) (applied to the family $A_t = \overline{U(t)B}$), we deduce that $A_\infty = \bigcap_{t \geq 0} \overline{U(t)B}$ is non-empty and compact, and satisfies

$$\delta(U(t)B, A_\infty) \leq \delta(A_t, A_\infty) \to 0, \text{ as } t \to +\infty.$$

($\Rightarrow$): Conversely assume that there exists a compact subset $C \subset M$ such that

$$\delta(U(t)B, C) \to 0, \text{ as } t \to \infty.$$

Then by Proposition 2.35-(ii), we have

$$\kappa(U(t)B) \leq \kappa(C) + \delta(U(t)B, C) = \delta(U(t)B, C) \to 0, \text{ as } t \to \infty,$$

and the proof is completed. □

Definition 2.43 A positively invariant subset $A \subset M$ for U is said to be **stable** if for any neighborhood V of A, there exists a neighborhood $W \subset V$ of A such that

$$U(t)W \subset V, \forall t \geq 0.$$

We say that A is **globally asymptotically stable** for U if, in addition, A attracts points of M for U, that is

$$\lim_{t\to\infty} d\left(U(t)x, A\right) = 0, \forall x \in M.$$

Lemma 2.44 *A subset A is stable for U if and only if for each neighborhood V of A, there exists $W \subset V$ a neighborhood of A satisfying*

$$U(t)W \subset W, \forall t \geq 0.$$

Proof ($\Rightarrow$): Assume that $A \subset M$ is stable for U. Let V be neighborhood A. Then there exists a neighborhood $W \subset V$ of A such that

$$U(t)W \subset V, \forall t \geq 0.$$

Set

$$\widehat{W} = \bigcup_{s\geq 0} U(s)W.$$

Then $\widehat{W}$ is a neighborhood of A (since it contains W) and

$$U(t)\widehat{W} = \bigcup_{s\geq 0} U(s+t)W \subset \bigcup_{s\geq 0} U(s)W = \widehat{W}.$$

($\Leftarrow$): This implication is straightforward. □

Lemma 2.45 *Assume that A is a compact subset of M and V is a neighborhood of A. Then there exists $\varepsilon > 0$ such that*

$$N(A, \varepsilon) \subset V.$$

Proof Assume by contradiction that

$$N\left(A, \frac{1}{n+1}\right) \not\subset V, \forall n \in \mathbb{N}.$$

This implies that there exists $x_n \in A$ such that

$$B_M\left(x_n, \frac{1}{n+1}\right) \not\subset V, \forall n \in \mathbb{N}.$$

But since A is compact, we can find a converging sub-sequence $x_{n_p} \to \overline{x} \in A$. Moreover for each $p \in \mathbb{N}$ we can find $y_{n_p} \in B_M\left(x_{n_p}, \dfrac{1}{n_p+1}\right)$, and

$$y_{n_p} \notin V.$$

Hence we have for all $p \in \mathbb{N}$

$$d\left(\overline{x}, y_{n_p}\right) \leq d\left(\overline{x}, x_{n_p}\right) + d\left(x_{n_p}, y_{n_p}\right),$$

and letting $p \to \infty$ we obtain

$$\lim_{p\to\infty} d\left(\overline{x}, y_{n_p}\right) = 0.$$

Since $\overline{x} \in A$ and $y_{n_p} \notin V$, we obtain a contradiction with the fact that V is neighborhood of A. □

The following result can be found in the book of Hale [138, Theorem 2.2.5].

Proposition 2.46 (Stability) *Let $A \subset M$ be a compact subset which is positively invariant for U. If A attracts the compact subsets of one of its neighborhoods, then A is stable.*

Proof Let W be a neighborhood of A and assume that A attracts every compact subset of W for U. Assume by contradiction that A is not stable. To that aim assume that there exists a neighborhood V of A, such that for each neighborhood V' of A, with $V' \subset V$, there exists $t = t(V') > 0$, such that

$$U(t)V' \not\subset V.$$

Since A is compact and $V \cap W$ is a neighborhood of A, we can find $\varepsilon > 0$, such that

$$N(A, \varepsilon) \subset V \cap W.$$

Moreover by construction for each integer $m \geq 0$, we can find $t_m \geq 0$, such that

$$U(t_m)\left(N\left(A, \frac{\varepsilon}{m+1}\right)\right) \not\subset V.$$

So for each $m \geq 0$, we can find $x_m \in N\left(A, \frac{\varepsilon}{m+1}\right)$, such that

$$U(t_m)x_m \notin V.$$

Since A is compact, possibly along a sub-sequence, we can always assume that

$$x_m \to x \in A, \text{ as } m \to +\infty.$$

Moreover since

$$C = \{x_m : m \geq 0\} \cup \{x\},$$

is compact and is a subset of W, there exists $t_0 > 0$, such that

$$U(t)C \subset N(A, \varepsilon) \subset V, \forall t \geq t_0.$$

Hence

$$t_m \in I \cap [0, t_0], \forall m \geq 0,$$

and we have by construction

$$d(U(t_m)x_m, A) \geq \varepsilon, \forall m \geq 0. \tag{2.11}$$

Since A is positively invariant for the semiflow U, we deduce that for all $m \geq 0$

$$d\left(U(t_m)x_m, A\right) \leq d\left(U(t_m)x_m, U(t_m)x\right) \leq \sup_{t \in I \cap [0,t_0]} d\left(U(t)x_m, U(t)x\right).$$

Since U is continuous, U becomes uniformly continuous on the compact subset $(I \cap [0, t_0]) \times C$. Therefore we get

$$U(t)x_m \xrightarrow[m \to +\infty]{} U(t)x \in A, \text{ uniformly in } t \in I \cap [0, t_0],$$

and for all $m \geq 0$ large enough

$$d\left(U(t_m)x_m, A\right) \leq \varepsilon/2,$$

a contradiction with (2.11). □

As a direct consequence of Proposition 2.46, we deduce the following result.

Proposition 2.47 *Let $A \subset M$ be a compact subset which is positively invariant for U. The following properties are equivalent*

(i) *A attracts the compact subsets of M;*
(ii) *A is stable and attracts the points of M;*

Definition 2.48 A nonempty, compact and invariant subset $A \subset M$ is said to be

(i) an **attractor for** U if A attracts one of its neighborhoods;
(ii) a **global attractors for** U if A is an attractor that attracts every point in M;
(iii) a **strong global attractor for** U if A attracts every bounded subset of M.

We remark that the notion of attractor and global attractor was used in [156, 244, 298, 372]. The strong global attractor was defined as global attractor in [138, 324]. As we will see in the following the notion of strong global attractor is not applicable (in general) in the context of the uniform persistence.

A discrete time version of the following theorem was proved by Magal and Zhao [244]. The proof of the following Theorem is inspired by [138, Theorem 2.4.2, Lemmas 2.4.4 and 2.4.5]. This result is also similar to the result about the existence of global attractors in [298, Theorem 23.12] for continuous time semiflows.

Theorem 2.49 (Global attractor in M**)** *Let U be a continuous semiflow on a complete metric space (M, d). Assume that*

(i) *U is point dissipative;*
(ii) *U is asymptotically smooth;*
(iii) *The positive orbits for U of compact subsets of M are bounded.*

Then U has a global attractor $A \subset M$. Moreover, A attracts any subset $B \subset M$ with an eventually bounded positive orbit, that is to say that A attract any subset $B \subset M$ such that

$$\gamma^+(U(t_0)(B)) = \bigcup_{t \geq t_0} U(t)B,$$

is bounded for some $t_0 \geq 0$ (large enough).

Remark 2.50 As a direct consequence of Proposition 2.46, the subset A is a stable, compact and invariant for U in (M, d).

Proof Assume that (i) is satisfied. Since U is point dissipative, we can find a closed and bounded subset B_0 in M such that for each $x \in M$, there exists $t_0 = t_0(x) \in I$ such that

$$U(t)x \in B_0, \forall t \geq t_0.$$

Define

$$J(B_0) = \{y \in B_0 : U(t)y \in B_0, \forall t \geq 0\}.$$

Then,

$$U(t)J(B_0) \subset J(B_0), \forall t \geq 0,$$

and for every $x \in M$, there exists $t_0 = t_0(x) \geq 0$,

$$U(t_0)x \in J(B_0).$$

Since $J(B_0)$ is non-empty, closed and bounded, and U is asymptotically smooth, Proposition 2.39 implies that $\omega(J(B_0))$ is compact, invariant and attracts the points of M.

Assume, in addition, that (ii) is satisfied. We claim that there exists an $\varepsilon > 0$ such that

$$\gamma^+(N(\omega(J(B_0)), \varepsilon))$$

is bounded. Assume, by contradiction, that

$$\gamma^+\left(N\left(\omega(J(B_0)), \frac{1}{n+1}\right)\right)$$

is unbounded for each $n > 0$.

Let $z \in M$ be fixed. Then we can find a sequence $x_n \in N\left(\omega(J(B_0)), \frac{1}{n+1}\right)$ and a sequence $t_n \geq 0$ such that

$$d(z, U(t_n)x_n) \geq n, \ \forall n \geq 0.$$

Since $\omega(J(B_0))$ is compact, we can find a convergent sub-sequence $x_{n_p} \to x \in \omega(J(B))$ as $p \to +\infty$. Since the set

$$H := \{x_{n_p} : p \geq 0\} \cup \{x\}$$

is compact, Assumption (ii) implies that $\gamma^+(H)$ is bounded, and we obtain a contradiction.

Therefore the orbit $D = \overline{\gamma^+(N(\omega(J(B_0)), \varepsilon))}$ is bounded for $\varepsilon > 0$ small enough. Then D is closed, bounded and positively invariant for U. Since $\omega(J(B_0))$ attracts points of M for U, and

$$\omega(J(B_0)) \subset N(\omega(J(B_0)), \varepsilon) \subset \mathring{D},$$

we deduce that $N(\omega(J(B_0)), \varepsilon)$ is an absorbing set of M. It follows that for each $x \in M$, there exists $t_0 = t_0(x) \geq 0$ such that

$$U(t)x \in N(\omega(J(B_0)), \varepsilon) \subset \mathring{D}, \forall t \geq t_0.$$

Claim 2.51 *Since U is a continuous semiflow, it follows that for each compact subset C of M, there exists an integer $t_0 \geq 0$ such that*

$$U(t_0)C \subset D.$$

Proof (Proof of the claim:) Assume by contradiction that

$$U(t)C \not\subset D, \forall t \geq 0.$$

Then there exists a sequence $\{t_n\}_{n \geq 0}$ and a sequence $\{x_n\}_{n \geq 0} \in C$ such that $t_n \to \infty$ and

$$U(t_n)x_n \notin D, \forall n \geq 0.$$

Moreover, the subset D is positively invariant by U, so we deduce that

$$U(t)x_n \notin D, \forall t \in [0, t_n], \forall n \in \mathbb{N}. \tag{2.12}$$

But since C is compact, we can assume that $x_n \to x \in C$. By construction there exists $t_0 > 0$ such that

$$U(t)x \in \mathring{D}, \forall t \geq t_0.$$

But the inverse image $U(t_0)^{-1}\mathring{D}$ is open (since U is continuous) and contains x. Now since $x_n \to x$, we deduce that there exists an integer $n_0 \geq 0$ such that

$$x_n \in U(t_0)^{-1}\mathring{D}, \forall n \geq n_0.$$

Therefore

$$U(t_0)x_n \in \mathring{D}, \forall n \geq n_0,$$

and we obtain a contradiction with (2.12). The proof of the claim is completed. □

By using Claim 2.51, we deduce that

$$A = \omega(D)$$

attracts every compact subset of M.

Fix a bounded neighborhood V of A. Since U is continuous, we can apply Proposition 2.46, it follows that A is stable, and hence, there is a neighborhood W of A such that $U(t)(W) \subset V$, $\forall t \geq 0$. Clearly, the set

$$W' := \overline{\cup_{t \geq 0} U(t)W}$$

is a bounded neighborhood of A, and

$$U(t)W' \subset W', \forall t \geq 0.$$

Since U is asymptotically smooth, there is a compact set $J \subset W'$ such that J attracts W'. By Proposition 2.39, $\omega(W')$ is non-empty, compact, invariant for U, and attracts W'. Since A attracts $\omega(W')$, we deduce that

$$\omega(W') \subset A.$$

We conclude that A is a global attractor for U.

To prove the last part of the theorem, without loss of generality we assume that B is a bounded subset of M and $\gamma^+(B)$ is bounded. We set

$$K = \overline{\gamma^+(B)}.$$

Then we have

$$U(t)K \subset K, \forall t \geq 0.$$

Since K is closed and bounded and U is asymptotically smooth, we deduce that there exists a compact C which attracts K for U. Note that

$$U(t)B \subset U(t)\gamma^+(B) \subset U(t)K, \ \forall t \geq 0.$$

Thus, C attracts B for U. By Proposition 2.39, we deduce that $\omega(B)$ is non-empty, compact, invariant for U and attracts B. Since A is attracts the compact subset of M for U, we end-up with

$$\omega(B) \subset A,$$

and A attracts B for U. □

2.12 Uniform Persistence and Global Attractors

Let (M, d) be a complete metric space, and $\rho : M \to [0, +\infty)$ a continuous function. We decompose M into the open subset

$$M_0 := \{x \in M : \rho(x) > 0\},$$

and the closed subset

$$\partial M_0 := \{x \in M : \rho(x) = 0\}.$$

Then by using the definition of M_0 and ∂M_0, we deduce that

$$M = M_0 \cup \partial M_0,$$

and

$$M_0 \cap \partial M_0 = \emptyset.$$

Throughout this section, we assume that $\{U(t)\}_{t \in I}$ is a continuous semiflow, and

$$U(t)M_0 \subset M_0, \forall t \geq 0.$$

Definition 2.52 The region M_0 is called the **interior region**, and ∂M_0 is called the **boundary region**.

Remark 2.53 The boundary region ∂M_0 does not correspond (in general) to the topological notion of boundary of M_0 in the sense of the metric space (M, d) (i.e. we may have $\partial M_0 \not\subset \overline{M}_0$ in general).

Remark 2.54 In most of the examples, the boundary region ∂M_0 is also positively invariant by U. But this assumption is not needed to prove the existence of a global attractors in M_0.

Example 2.55 (of function $\rho(x)$) The function $\rho(x)$ measure the distance of x to the boundary region ∂M_0. A typical example of function ρ is the following

$$\rho(x) = d(x, \partial M_0) = \inf_{y \in \partial M_0} d(x, y).$$

The idea in the following definition is to consider ∂M_0 as part infinite.

Definition 2.56 A subset $B \subset M_0$ is said to be **ρ-strongly bounded** if B is bounded in (M, d), and

$$\inf_{x \in B} \rho(x) > 0.$$

Example 2.57 (of function $\rho(x)$) If

$$M = L^1_+(0, 1),$$

and

$$\partial M_0 = \{0_{L^1}\}, \text{ and } M_0 = L^1_+(0, 1) \setminus \{0_{L^1}\}.$$

There are many ways to measure the distance of a non-negative function u to the null function 0_{L^1}.

We can take

$$\rho(u) = \int_0^1 \chi(\sigma)u(\sigma)d\sigma$$

where $\chi \in C_+([0, 1], \mathbb{R})$ and

$$\chi(x) > 0, \forall x \in (0, 1).$$

So in particular we allow $\chi(0)$ and $\chi(1)$ to be $0_{\mathbb{R}}$.

Then the distance

$$d(u, \{0_{L^1}\}) = \|u\|_{L^1} = \int_0^1 u(\sigma)d\sigma,$$

which corresponds only to the case $\chi \equiv 1$. We will discuss this more in the section devoted to the examples at the end of this chapter.

Remark 2.58 If we change the function ρ, we will change the ρ-bounded sets.

Definition 2.59 (i) The semiflow $\{U(t)\}_{t\in I}$ is said to be **ρ-uniformly persistent** if there exists $\varepsilon > 0$ such that

$$\liminf_{t\to+\infty} \rho\left(U(t)x\right) \geq \varepsilon, \forall x \in M_0.$$

(ii) The semiflow $\{U(t)\}_{t\in I}$ is said to be **weakly ρ-uniformly persistent** if there exists $\varepsilon > 0$ such that

$$\limsup_{t\to+\infty} \rho\left(U(t)x\right) \geq \varepsilon, \forall x \in M_0.$$

(iii) The set ∂M_0 is said to be **ρ-ejective for** U if there exists $\varepsilon > 0$ such that for every $x \in M$ with $0 < \rho(x) < \varepsilon$, there is $t_0 = t_0(x) \geq 0$ such that

$$\rho\left(U(t_0)x\right) \geq \varepsilon.$$

For a given open subset $M_0 \subset M$, let $\partial M_0 := M \setminus M_0$. Then if $\partial M_0 \neq \emptyset$, we can use the continuous function $\rho : M \to [0, \infty)$ defined by $\rho(x) = d\left(x, \partial M_0\right), \forall x \in M$, to obtain the usual definition of persistence.

Proposition 2.60 *Assume that there is a compact subset C of M which attracts every point in M for U. Then the following statements are equivalent:*

(i) *U is weakly ρ-uniformly persistent;*
(ii) *U is ρ-uniformly persistent;*
(iii) *∂M_0 is ρ-ejective for U.*

Proof The observations (i)⇔(iii) and (ii)⇒(i) are obvious. Let us prove that (i)⇒(ii). Let $\varepsilon > 0$ be fixed such that

$$\limsup_{t\to+\infty} \rho\left(U(t)x\right) \geq \varepsilon, \forall x \in M_0. \tag{2.13}$$

Then for each $x \in M_0$, and each $t \geq 0$, there exists $r \geq 0$ such that $\rho\left(U(t+r)x\right) \geq \varepsilon/2$. Assume that U is not ρ-uniformly persistent. Then we can find a sequence $\{x_m\}_{m\geq 0} \subset M_0$ such that

$$\liminf_{t\to+\infty} \rho\left(U(t)x_m\right) \leq \frac{1}{m+1}, \forall m \geq 0. \tag{2.14}$$

By using the fact that C attracts every point of M and (2.13), we deduce that there exists $t_m^\star \geq 0$ such that

$$d(U(t)x_m, C) \leq \frac{1}{m+1}, \forall t \geq t_m^\star, \text{ and } \rho\left(U(t_m^\star)x_m\right) \geq \varepsilon/2.$$

By using (2.14) we deduce that there exists $l_m > 0$ such that

$$\rho\left(U(t_m^\star + l_m)x_m\right) \leq \frac{2}{m+1}.$$

Assuming that $\varepsilon/2 > \dfrac{2}{m+1}$ (which is true for all m large enough), we deduce that there exists $t_m \geq t_m^\star$ such that

$$\rho\left(U(t_m)x_m\right) \geq \varepsilon/2, \text{ and } \rho\left(U(t_m+l)x_m\right) \leq \varepsilon/2, \ \forall l \in (0, l_m].$$

We conclude that there exist $l_m > 0$ and $t_m \geq 0$ such that

$$\begin{aligned} &d(U(t_m)x_m, C) \leq \frac{1}{m+1}, \\ &\rho\left(U(t_m)x_m\right) \geq \varepsilon/2, \\ &\rho\left(U(t_m+l)x_m\right) \leq \varepsilon/2, \ \forall l \in (0, l_m], \end{aligned}$$

and

$$\rho\left(U(t_m+l_m)x_m\right) \leq \frac{2}{m+1}. \tag{2.15}$$

Since C is compact, by taking a sub-sequence that we denote with the same index, we can always assume that $y_m = U(t_m)x_m \to y \in C$. Since ρ is continuous and U is a continuous semiflow, we deduce that

$$\rho\left(y\right) \geq \varepsilon/2, \text{ and } \rho\left(U(l)y\right) \leq \varepsilon/2, \ \forall l \in \left[0, l^\star\right),$$

where $l^\star = \liminf_{m\to+\infty} l_m$. Since $H = \{y_m : m \geq 0\} \cup \{y\}$ is compact, and $(t,x) \to U(t)x$ is a continuous map, we deduce that for each $\tau > 0$, $\bigcup_{t\in[0,\tau]\cap I} U(t)H$ is a compact subset which is included in M_0. Hence for each $\tau > 0$, $\bigcup_{t\in[0,\tau]\cap I} U(t)H$ is ρ-bounded. By using (2.15), we deduce that

$$l^\star = +\infty.$$

Otherwise, assume by contradiction that we can find a sub-sequence $l_{m_p} \to \widehat{l}$ as $p \to \infty$. By continuity of U, and by (2.15), we deduce that

$$\rho(U(l_{m_p})y_{m_p}) \to \rho(U(\widehat{l})y) = 0,$$

and since $\rho\left(y\right) \geq \varepsilon/2$, we obtain a contradiction with the fact that M_0 is positively invariant by U.

We conclude that

$$\limsup_{t\to+\infty} \rho\left(U(t)y\right) \leq \varepsilon/2 < \varepsilon,$$

which contradicts (2.13). □

We note that the concept of general ρ-persistence was used in [317, 330, 372]. It was also shown in [330] that the ρ-uniform persistence implies the weak ρ-uniform persistence for non-autonomous semiflows under appropriate conditions. The following result shows that the notion of ρ-uniform persistence is independent of the choice of continuous function ρ.

Proposition 2.61 *Let $\xi : M \to [0, +\infty)$ be a continuous function such that $\partial M_0 = \{x \in M : \xi(x) = 0\}$. Assume that there is a compact subset C of M which attracts every point in M for U. Then U is ρ-uniformly persistent if and only if U is ξ-uniformly persistent.*

Proof It suffices to prove that ρ-uniform persistence implies ξ-uniform persistence since the problem is symmetric. Let us first remark that U is ρ-uniformly persistent if and only if there exists $\varepsilon > 0$ such that

$$\inf_{x \in M_0} \inf_{y \in \omega(x)} \rho(y) \geq \varepsilon,$$

where $\omega(x)$ is the ω-limit set of the positive orbit of x. Define

$$A_\omega = \bigcup_{x \in M_0} \omega(x) \text{ and } V = \{y \in M : \rho(y) \geq \varepsilon\}.$$

Then we have

$$\inf_{x \in M_0} \inf_{y \in \omega(x)} \rho(y) = \inf_{x \in A_\omega} \rho(x) \geq \varepsilon.$$

Clearly, $A_\omega \subset C$, so $\overline{A_\omega}$ is compact. Since $A_\omega \subset V \subset M_0$ and V is closed, we deduce that $\overline{A_\omega} \subset V \cap C \subset M_0$. As a consequence $\overline{A_\omega} \subset M_0$ is compact, and hence, there exists $\eta > 0$ such that $\inf_{x \in \overline{A_\omega}} \xi(x) \geq \eta$, which implies that U is ξ-uniformly persistent. □

Definition 2.62 Let A be a nonempty subset of M. A is said to be **ejective** for U if there exists a neighborhood V of A such that for every $x \in V \setminus A$, there is $t_0 = t_0(x) \geq 0$ such that

$$U(t_0)x \in M \setminus V.$$

Proposition 2.63 *Assume that $\partial M_0 \neq \emptyset$ and that there is a compact subset C of M which attracts every point in M for U. Then the following statements are equivalent*

(i) *U is ρ-uniformly persistent;*
(ii) *∂M_0 is ejective for U.*

Proof **Proof of** (i) $\Rightarrow$ (ii). Assume that (i) is true. Let $\varepsilon > 0$ be fixed such that

$$\limsup_{t \to +\infty} \rho\left(U(t)x\right) \geq \varepsilon, \forall x \in M_C.$$

Then it is clear that ∂M_0 is ejective for U, with $V = \{x \in M : \rho\left(x\right) \leq \varepsilon/2\}$.

Proof of (ii) $\Rightarrow$ (i). Conversely, assume that ∂M_0 is ejective for U. Let V be a neighborhood of ∂M_0 such that for every $x \in M_0 \cap V$, there is $n_0 = n_0(x) \geq 0$ such that $U(t_0)x \in M \setminus V$. By Proposition 2.61, it is sufficient to prove that U is ρ-uniformly persistent when $\rho\left(x\right) = d(x, \partial M_0)$. Assume, by contradiction, that U is not ρ-uniformly persistent. Then for each $n \geq 1$, there exists $x_n \in M_0$, such that

$$\limsup_{t \to +\infty} \rho\left(U(t)x_n\right) \leq \frac{1}{n}.$$

By the attractivity of C, it follows that for each $n \geq 1$, there exists $t_n \geq 0$ such that each $y_n := U(t_n)x_n \in M_0$ satisfies

$$d\left(U(s)y_n, C\right) \leq \frac{2}{n}, \text{ and } d\left(U(s)y_n, \partial M_0\right) \leq \frac{2}{n}, \forall s \geq 0.$$

Since C is compact and V is a neighborhood of ∂M_0, there exists $\delta > 0$ such that

$$\{x \in M : d\left(x, C\right) \leq \delta, \text{ and } d\left(x, \partial M_0\right) \leq \delta\} \subset V.$$

Let $n_0 \geq 2/\delta$ be fixed. Then we have $y_{n_0} \in M_0$, and

$$d\left(U(s)y_{n_0}, C\right) \leq \delta, \text{ and } d\left(U(s)y_n, \partial M_0\right) \leq \delta, \forall s \geq 0.$$

Thus, we obtain

$$y_{n_0} \in M_0 \cap V, \text{ and } U(s)y_{n_0} \in V, \forall s \geq 0,$$

a contradiction with the property (ii). □

Observe that M_0 is an open subset in (M, d). In order to make M_0 become a complete metric space, we define a new metric function d_0 on M_0 by

$$d_0(x, y) = \left|\frac{1}{\rho\left(x\right)} - \frac{1}{\rho\left(y\right)}\right| + d(x, y), \quad \forall x, y \in M_0. \tag{2.16}$$

Lemma 2.64 *(M_0, d_0) is a complete metric space.*

Proof It is easy to see that d_0 is a distance. Let $\{x_n\}_{n\geq 0}$ be a Cauchy sequence in (M_0, d_0). Since $d(x, y) \leq d_0(x, y)$, $\forall x, y \in M_0$, we deduce that $\{x_n\}_{n\geq 0}$ is a Cauchy sequence in (M, d), and there exists $x \in M$, such that $d(x_n, x) \to 0$ as $n \to +\infty$. To prove that $d_0(x_n, x) \to 0$ as $n \to +\infty$, it is sufficient to show that $x \in M_0$. Given $\varepsilon > 0$, since $\{x_n\}_{n\geq 0}$ is a Cauchy sequence in (M_0, d_0), there exists $n_0 \geq 0$ such that $d_0(x_n, x_p) \leq \varepsilon$, $\forall n, p \geq n_0$. In particular, we have $d_0(x_n, x_{n_0}) \leq \varepsilon, \forall n \geq n_0$. Then we get

$$\left|\frac{1}{\rho\left(x_n\right)} - \frac{1}{\rho\left(x_{n_0}\right)}\right| \leq \varepsilon, \forall n \geq n_0.$$

Hence there exists $r > 0$ such that $\inf_{n\geq 0} \rho\left(x_n\right) \geq r$. Since ρ is continuous and $d(x_n, x) \to 0$ as $n \to +\infty$, we deduce that $\rho\left(x\right) \geq r$, and hence $x \in M_0$. Thus the metric space (M_0, d_0) is complete. □

We denote for each couple of subsets $A, B \subset M$,

$$\delta\left(B, A\right) = \sup_{x\in B} \inf_{y\in A} d(x, y),$$

and if $A, B \subset M_0$, we denote

$$\delta_0\left(B, A\right) = \sup_{x\in B} \inf_{y\in A} d_0(x, y).$$

Here δ corresponds to the Hausdorff semi-distance in (M, d) while δ_0 is the Hausdorff semi-distance in (M_0, d_0).

Lemma 2.65 *The following two statements are valid:*

(i) *Let* $\{B_t\}_{t\in I}$ *be a family of subsets of* M_0, *where* I *is a unbounded subset of* $[0,+\infty)$. *If* $A \subset M_0$ *is compact in* (M,d) *and* $\lim_{t\to\infty}\delta(B_t,A) = 0$, *then* $\lim_{t\to\infty}\delta_0(B_t,A) = 0$.
(ii) *If* U *is asymptotically smooth in* (M,d), *then* U *is asymptotically smooth in* (M_0,d_0).

Proof **Proof of (i).** Let

$$\kappa := \frac{1}{2}\inf_{x\in A}\rho(x) > 0.$$

Assume by contradiction that

$$\limsup_{t\to+\infty}\delta_0(B_t,A) > \varepsilon > 0.$$

Then we can find a sequence $\{t_p\}_{p\geq 0} \subset I$ such that $t_p \to +\infty$ as $p \to +\infty$ and a sequence $\{x_{t_p}\}_{p\geq 0} \in B_{t_p} \subset M_0$ such that

$$d_0\left(x_{t_p},A\right) \geq \varepsilon/2, \forall p \geq 0.$$

Since A attracts B_t in (M,d), we deduce that

$$d\left(x_{t_p},A\right) \to 0 \text{ as } p \to +\infty.$$

So without loss of generality, we can assume that there exists $x \in A$ such that $d(x_{t_p},x) \to 0$ as $p \to +\infty$. Since ρ is continuous and $\rho(x) > \kappa$, there exists $p_0 \geq 0$ such that $\rho(x_{t_p}) \geq \kappa, \forall p \geq p_0$. Thus, we have

$$0 < \varepsilon/2 \leq d_0\left(x_{t_p},A\right) \leq d_0\left(x_{t_p},x\right) \leq \kappa^{-2}\left|\rho(x_{t_p})-\rho(x)\right| + d(x_{t_p},x) \to 0 \text{ as } p \to +\infty,$$

a contradiction.

Proof of (ii). Let B be a bounded subset in (M_0,d_0) such that

$$U(t)B \subset B, \forall t \geq 0.$$

Since U is asymptotically smooth in (M,d), there exists a compact subset $C \subset M$ which attracts B for U in (M,d). That is

$$\lim_{t\to\infty}\delta\left(U(t)B,C\right) = 0.$$

The subset $C_0 = C \cap \overline{B} \subset M_0$ is compact in (M,d), and attracts B for U in (M,d). It easily follows that C_0 is also compact in (M_0,d_0). Since C_0 attracts B for U in (M,d), the property (i) implies that C_0 attracts B for U in (M_0,d_0). □

Before going to the main result of this section the following lemma provides sufficient conditions for the positive orbit of a compact set to be bounded.

Lemma 2.66 *Assume that U is continuous and point dissipative. Let C be a compact subset of M. Assume that for every bounded sequence $\{x_n\}_{n\geq 0}$ in $\gamma^+(C)$, there exists $t_0 \geq 0$, and there exists a sub-sequence $\left\{x_{n_p}\right\}_{p\geq 0}$, such that the two following properties are satisfied:*

(i) $\left\{U(t_0)x_{n_p}\right\}_{p\geq 0}$ *converges;*

(ii) *The subset* $\bigcup_{p\geq 0}\bigcup_{0\leq t\leq t_0}\left\{U(t)x_{n_p}\right\}$ *is bounded.*

Then $\gamma^+(C)$ is bounded in M.

Proof Since U is point dissipative, we can choose a bounded and open subset V of M such that for each $x \in M$ there exists $\tau(x) \geq 0$ such that

$$U(t)x \in V, \ \forall t \geq \tau(x).$$

For each $x \in C$ we can find

$$\widehat{\tau}(x) = \inf\{t > 0 : U(t)x \in V\}.$$

We claim that

$$\sup_{x\in C} \widehat{\tau}(x) < \infty. \tag{2.17}$$

Assume by contradiction that, there exists a sequence $x_n \in C$ such that

$$\widehat{\tau}(x_n) \to \infty.$$

Since C is compact we can assume (by taking a sub-sequence) that $x_n \to x_\infty \in C$, and since

$$\widehat{\tau}(x_\infty) < \infty,$$

and $x \to U(\tau(x_\infty))x$ is continuous and V is open, we deduce that for all n large enough

$$U(\tau(x_\infty))x_n \in V,$$

and we obtain a contradiction with $\widehat{\tau}(x_n) \to \infty$. So we deduce (2.17) holds.

Let $r > \sup_{x\in C} \widehat{\tau}(x)$. Let $x \in C$, by using the definition of $\widehat{\tau}(x)$, we deduce that there exists $s = s(x) \in I \cap [0, r]$, such that

$$U(s)x \in V.$$

Let $z \in M$ be fixed. Assume, by contradiction, that $\gamma^+(C)$ is unbounded. Then there exists a sequence $\left\{x_p\right\}$ in $\gamma^+(C)$ such that

$$x_p = U\left(\tau_p\right) z_p, \text{ with } z_p \in C, \text{ and } \lim_{p\to\infty} d(z, x_p) = \infty.$$

Since U is continuous and V is bounded, we can assume that

$$\lim_{p\to\infty} \tau_p = \infty, \text{ and } \tau_p > r, \, x_p \notin \overline{V}, \ \forall p \geq 1.$$

For each $z_p \in C$, there exists $s_p \leq r$ such that

$$U(s_p)z_p \in V.$$

Since $x_p = U(\tau_p)(z_p) \notin V$, there exists $\delta_p \in [s_p, \tau_p)$ (that is the last time before τ_p such that the orbit starting from x_p belongs to $\overline{V}$) such that

$$y_p = U(\delta_p)z_p \in \overline{V}, \quad \text{and } U(l)y_p \notin V, \ \forall l \in I \cap \left(0, l_p\right], \text{ with } l_p = \tau_p - \delta_p.$$

Clearly,

$$x_p = U(l_p)y_p, \ \forall p \geq 1,$$

and $\{y_p\}_{p\in\mathbb{N}}$ is a bounded sequence in $\gamma^+(C)$.

By using assumptions (i) and (ii), we deduce that there exists $t_0 \in I$, and taking a sub-sequence, we have

$$\left\{U\left(t_0\right) y_p\right\}_{p\geq 0} \rightarrow \widehat{x} \in M.$$

and

$$\bigcup_{p\geq 0, t\in[0,t_0]\cap I} \left\{U(t)y_p\right\}$$

is bounded.

Since $\lim_{p\to\infty} d(z, x_p) = \infty$, we deduce that for all $p \geq 0$ large enough

$$l_p > t_0.$$

By construction the subset

$$H = \left\{U\left(t_0\right) y_p : p \geq 0\right\} \cup \{\widehat{x}\}$$

is compact, and since U is a continuous semiflow, the subset

$$\bigcup_{s\in I\cap[0,t]} U(s)H$$

is bounded for each $t \in I$, and we deduce that $l_p \rightarrow +\infty$, as $p \rightarrow +\infty$.

Now since U is continuous, and V is open, we deduce that

$$U(t)\widehat{x} \notin V, \forall t > 0,$$

which contradicts the definition of V (the fact that any orbit starting from a single point x ultimately belongs to V). □

The main result of this section is the following theorem.

Theorem 2.67 (Global attractor in M_0) *Assume that U is asymptotically smooth, ρ-uniformly persistent, and U has a global attractor A in M. Then U has a global attractor A_0 in (M_0, d). Moreover, for each subset B of M_0, if there exists $t \geq 0$ such that $\gamma^+\left(U(t)\left(B\right)\right)$ is ρ-strongly bounded, then A_0 attracts B for U.*

Proof Since U is point dissipative and ρ -uniformly persistent, U is point dissipative in (M_0, d_0). Moreover, Lemma 2.65 implies that U is asymptotically smooth in (M_0, d_0). It is clear that U is continuous in (M_0, d_0). Let C be a compact subset in (M_0, d_0), and $\{x_n\}_{n\geq 0}$ a bounded sequence in $\gamma^+(C)$ in (M_0, d_0). Then $x_n = U(t_n)z_n$, $z_n \in C$, $\forall n \geq 1$, and the sequence $\{x_n\}_{n\geq 0}$ is ρ-strongly bounded in (M, d). If $\{t_n\}_{n\geq 0}$ is bounded by $T > 0$, since $\overline{\bigcup_{t\in[0,T]\cap I} U(t)C}$ is a compact subset of M_0, we deduce that $\{x_p\}$ has a convergent sub-sequence in (M_0, d_0). If $\{t_n\}_{n\geq 0}$ is unbounded, by taking a sub-sequence, we can assume that $t_n \to +\infty$, as $n \to +\infty$. Since C is also compact in (M, d), we have $\lim_{t\to\infty} \delta(U(t)(C), A) = 0$. Thus, $\{x_n\}_{n\geq 0}$ has a convergent sub-sequence $x_{n_k} \to x$ in (M, d) as $k \to \infty$. By the continuity of ρ and the ρ-strong boundedness of $\{x_n\}_{n\geq 0}$, it follows that $\rho(x) > 0$, i.e., $x \in M_0$, and hence, $x_{p_k} \to x$ in (M_0, d_0) as $k \to \infty$. Thus, Lemma 2.66 (with $t_0 = 0$) implies that positive orbits of compact sets are bounded for U in (M_0, d_0). Then the conclusion for U in (M_0, d) follows from Theorem 2.49, as applied to U in (M_0, d_0). □

As a consequence of Proposition 2.46, we deduce that A_0 is a stable set for U with respect to the original distance d.

Corollary 2.68 (Stability of the global attractor in M_0**)** *Assume that A_0 is a global attractor for U in (M_0, d). Then A_0 is a stable invariant subset for U in (M, d).*

Proof Since A_0 is compact, we must have

$$\eta := \inf_{x\in A_0} \rho(x) > 0.$$

The result follows from the fact that A_0 attracts the compact subsets of

$$\{x \in M : \rho(x) \geq \eta/2\}$$

which is a neighborhood of A_0. □

2.13 Coexistence Steady States

In this section, we establish the existence of coexistence steady state (i.e., the fixed point in M_0) for uniformly persistent dynamical systems.

Definition 2.69 A map $T : M \to M$ is said to be **κ-condensing** if T takes bounded sets to bounded sets and

$$\kappa(T(B)) < \kappa(B).$$

for any nonempty bounded set $B \subset M$ with $0 < \kappa(B) < +\infty$.

We assume that M is a closed and convex subset of a Banach space $(X, \|\cdot\|)$, that $\rho : M \to [0, +\infty)$ is a continuous function such that $M_0 = \{x \in M : \rho(x) > 0\}$ is nonempty and convex, and that $T : M \to M$ is a continuous map with $T(M_0) \subset M_0$. We set $\partial M_0 := M \setminus M_0$.

Assume that $T : M_0 \to M_0$ has a global attractor A_0. It follows that for every compact set $K \subset M_0$, there exists an open neighborhood of K which is attracted by A_0. This property of A_0 is enough for the arguments in the proof of [366, Theorem

2.3] (see also [372, Theorem 1.3.6]) instead of the property that A_0 attracts ρ-strongly bounded sets in M_0. Thus, the proof of [366, Theorem 2.3] actually implies the following fixed point theorem.

Theorem 2.70 *Assume that T is κ-condensing. If $T : M_0 \to M_0$ has a global attractor A_0. Then T has a fixed point $x_0 \in A_0$.*

We also refer to Nussbaum [269] and to Hale and Lopes [142] for more results.

2.14 Two Examples

In this section, we provide two examples of discrete and continuous-time semiflows which admit global attractors, but no strong global attractors in the complete metric spaces (M_0, d_0) introduced in Section 2.11. These examples show that our notion of global attractors is needed.

2.14.1 Asymptotically smooth semiflows on (M_0, d_0)

The example of this subsection is taken from Magal and Zhao [244]. Let $C([0,1],\mathbb{R})$ be endowed with the usual norm $\|\varphi\|_\infty = \sup_{a\in[0,1]} |\varphi(a)|$. Let $M := C_+([0,1],\mathbb{R})$ be the set of the nonegative and continuous functions endowed with the metric $d(\varphi,\psi) = \|\varphi - \psi\|_\infty$, and $T : M \to M$ be the map defined by

$$T(\varphi) = \delta \frac{\mathcal{F}_\beta(\varphi)}{1 + \mathcal{F}_\beta(\varphi)} 1_{[0,1]},$$

where $1_{[0,1]}(a) = 1, \forall a \in [0,1]$, and $\mathcal{F}_\beta(\varphi) = \int_0^1 \beta(a)\varphi(a)da, \forall \varphi \in M$. We assume that

(A1) $\delta > 1, \beta \in C([0,1],\mathbb{R}), \int_0^1 \beta(a)da = 1, \beta(a) > 0, \forall a \in [0,1)$, *and* $\beta(1) = 0$.

Consider the following discrete time system on M:

$$u_{n+1} = T(u_n), \forall n \geq 0, \text{ and } u_0 \in M.$$

It is easy to see that the map T is continuous, and maps bounded sets into compact sets of M. Note that $T(M) \subset [0,\delta]\, 1_{[0,1]} = \{\alpha 1_{[0,1]} : \alpha \in [0,\delta]\}$ is bounded. So T is compact and point dissipative, and has a strong global attractor in M. Set

$$\partial M_0 = \{0\}, \text{ and } M_0 = M \setminus \{0\}, \text{ and } \rho(\varphi) = \|\varphi\|_\infty, \ \forall \varphi \in M.$$

Clearly, $T(M_0) \subset M_0$, $T(\partial M_0) \subset \partial M_0$, and the fixed points of T are 0 and $\overline{u} = (\delta - 1)\, 1_{[0,1]}$. Then it is easy to see that for each $\varphi \in M_0$, $T^m(\varphi) \to \overline{u}$, as $m \to +\infty$. So T is ρ-uniformly persistent. Let $\overline{\alpha} = (\delta - 1)$ and $B := \{\varphi \in M : \|\varphi\|_\infty = \overline{\alpha}\}$. Since $\beta(1) = 0$, we have $\mathcal{F}_\beta(B) = (0, \overline{\alpha}]$. Moreover, $T(B) = \{\alpha 1_{[0,1]} : \alpha \in (0, \overline{\alpha}]\}$, and $T^n(B) = T(B)$, $\forall n \geq 1$. Thus, there exists no compact subset in M_0 that attracts B for T. In particular, there is no strong global attractor for $T : (M_0, d_0) \to (M_0, d_0)$, where d_0 is defined as in (2.16).

2.14.2 κ-contracting maps on (M_0, d_0)

In this subsection, we construct κ-contracting maps on (M_0, d_0) such that they admit a global attractor, but no strong global attractor.

We set

$$X = L^1((0, +\infty), \mathbb{R}) \times \mathbb{R}, \ X_+ = L^1_+((0, +\infty), \mathbb{R}) \times \mathbb{R}_+,$$

and endow X with the product norm $\|(\varphi, y)\| = \|\varphi\|_{L^1} + |y|$.

Define $1_{[0,1]} \in L^1((0, +\infty), \mathbb{R})$ by

$$1_{[0,1]}(l) = \begin{cases} 1 \text{ if } l \in (0, 1), \\ 0 \text{ if } l \in [1, \infty). \end{cases}$$

Let a, b and c be three real numbers. Define $T : X_+ \to X_+$ by $T(\varphi, y) = (T_1(\varphi, y), T_2(\varphi, y))$ with

$$\begin{aligned} T_1(\varphi, y) &= a\varphi(\cdot + 1) + \left[a \int_0^1 \varphi(l)dl + c\frac{\int_0^1 \varphi(l)dl}{1+\|(\varphi, y)\|} \right] 1_{[0,1]}, \\ T_2(\varphi, y) &= ay + b\frac{\|(\varphi, y)\|}{1+\|(\varphi, y)\|}. \end{aligned}$$

We assume that

(A3) $a \in (0, 1)$, $b > 0$, $c > 0$, $\sqrt{a} < a + b < 1$, and $a + c > 1$.

Consider the discrete time system

$$x_{n+1} = T(x_n), \forall n \geq 0, \text{ and } x_0 \in X_+.$$

It is easy to see that $T^n(0, y) \to 0$, as $n \to +\infty$. Clearly, T is not uniformly persistent for $X_+ \setminus \{0\}$. We will find a closed subset M of X_+ such that it contains 0 and is positively invariant for T, and show that T is uniformly persistent for $M \setminus \{0\}$.

Lemma 2.71 *There exists a non-decreasing and right-continuous function* $f : \mathbb{R}_+ \to \mathbb{R}_+$ *such that* $f(0) = 0, f(x) > 0, \forall x > 0$, $\lim_{x \to 0} f(x) = 0$, *and the set*

$$M := \left\{ (\varphi, y) \in X_+ : y \leq f(\|\varphi\|_{L^1}) \right\}$$

is positively invariant for T.

Proof Along this proof we make use of the notation $x = (x_1, x_2)$ to denote the components of a point $x \in \mathbb{R}^2$. We define $F : \mathbb{R}^2_+ \to \mathbb{R}^2_+$ by

$$F(x_1, x_2) = \left(ax_1, ax_2 + b\frac{x_1 + x_2}{1 + x_1 + x_2}\right), \quad \forall x = (x_1, x_2) \in \mathbb{R}_+^2.$$

Then F is non-decreasing on $\mathbb{R}_+^2$. Set

$$\chi(t) = (ta + (1 - t), 1), \ \forall t \in [0, 1].$$

By induction, we define $\chi : \mathbb{R}_+ \to \mathbb{R}_+^2$ by

$$\chi(t) = F(\chi(t - 1)), \ \forall t \in (n, n + 1], \ \forall n \geq 1.$$

Note that $\chi(1)_1 = F(\chi(0))_1$ and $a < 1$. Then the function $t \to \chi(t)_1$ is strictly decreasing and continuous. Since $F(1, 1) \leq (a, 1)$, the function $t \to \chi(t)_2$ is non-increasing and left continuous. Moreover, since $a+b < 1$, we have $\lim_{t\to+\infty} \chi(t) = 0$. We further set

$$\chi(t) = (1 - t, 1), \ \forall t \in (-\infty, 0].$$

Since $\chi(t)_1$ is strictly decreasing in $t \in \mathbb{R}$, we can define

$$f(x) = \begin{cases} \chi(\chi(x)_1^{-1})_2, & \text{if } x > 0, \\ 0, & \text{if } x = 0. \end{cases}$$

It is easy to see that f has the desired properties.

Let $D := \{(x_1, x_2) \in \mathbb{R}_+^2 : x_2 \leq f(x_1)\}$. Since f is non-decreasing and right-continuous, it easily follows that D is closed. Now we show that $F(D) \subset D$. Let $x = (x_1, x_2) \in D$, then $x_2 \leq f(x_1)$. If $x_1 = 0$, there is nothing to prove because $F(0) = 0$. Assume that $x_1 > 0$, then there exists $t \in \mathbb{R}$ such that $\chi(t)_1 = x_1$, and hence, $x_2 \leq f(x_1) = \chi(t)_2$. Clearly, $x = (x_1, x_2) \leq \chi(t)$, and $F(x) \leq F(\chi(t))$. In the case where $t \geq 0$, we have

$$\chi(t + 1)_1 = F(\chi(t))_1 = F(x)_1,$$

and hence,

$$f(F(x)_1) = \chi(t + 1)_2 = F(\chi(t))_2 \geq F(x)_2,$$

which implies that $F(x) \in D$. In the case where $t \leq 0$, we have

$$x_1 \geq F(x)_1 = F(\chi(t))_1 = a\chi(t)_1 = a(1 - t) \geq a = \chi(1)_1,$$

and hence, there exists $s \in [t, 1]$ such that $\chi(s)_1 = F(x)_1$. It then follows that

$$f(F(x)_1) = \chi(s)_2 = 1 \geq F(\chi(t))_2 \geq F(x)_2,$$

which implies that $F(x) \in D$. This proves that $F(D) \subset D$.

Finally, we prove that $T(M) \subset M$. For any $(\varphi, y) \in M$, we have $(\|\varphi\|_{L^1}, y) \in D$, and hence, the positive invariance of D for F implies that $F(\|\varphi\|_{L^1}, y)_2 \leq f(F(\|\varphi\|_{L^1}, y)_1)$. Note that $\|T_1(\varphi, y)\| \geq a\|\varphi\|_{L^1} = F(\|\varphi\|_{L^1}, y)_1$ and $T_2(\varphi, y) = F(\|\varphi\|_{L^1}, y)_2$. By the monotonicity of f, it then follows that

$$T_2(\varphi, y) = F(\|\varphi\|_{L^1}, y)_2 \leq f(F(\|\varphi\|_{L^1}, y)_1) \leq f(\|T_1(\varphi, y)\|),$$

which implies that $T(\varphi, y) \in M$. Thus, M is positively invariant for T. $\square$

Now we consider $T : M \to M$, where M is endowed with the usual distance $d(x, \widehat{x}) = \|x - \widehat{x}\|$. We set

$$\partial M_0 = \{0\}, \ M_0 = M \setminus \{0\}, \text{ and } \rho(x) = \|x\|.$$

Since T is the sum of a compact operator and a linear operator with norm being a, we have $\kappa(T(B)) \leq a\kappa(B)$ for any bounded set $B \subset M$. Thus, T is κ-contraction. Moreover, for each $x \in M$, we have $\|T(x)\| \leq a\|x\| + b + c$, and hence

$$\|T^n(x)\| \leq a^n \|x\| + \left(\sum_{i=0}^{n-1} a^i\right)(b + c), \ \forall n \geq 1.$$

It then follows that $B = \left\{x \in M : \|x\| \leq \frac{b+c}{1-a}\right\}$ is positively invariant for T, and attracts every bounded subset of M for T. So $T : (M, d) \to (M, d)$ has a strong global attractor.

Let $\varepsilon > 0$ be fixed such that $a + \frac{c}{1+\varepsilon} > 1$. We claim that

$$\limsup_{n\to\infty} \|T^n x\| \geq \varepsilon, \ \forall x = (\varphi, y) \in M_0.$$

Assume, by contradiction, that $\limsup_{n\to\infty} \|T^n x\| < \varepsilon$ for some $x = (\varphi, y) \in M_0$. We set $(\varphi_n, y_n) = T^n x$, $\forall n \geq 0$. By the definition of M, we have $\varphi \in L^1_+((0, +\infty), \mathbb{R}) \setminus \{0\}$. It then follows that there exists $n_0 \geq 0$ such that $\int_0^1 \varphi_{n_0}(l)dl > 0$ and

$$\int_0^1 \varphi_{n+1}(l)dl \geq \left(a + \frac{c}{1+\varepsilon}\right) \int_0^1 \varphi_n(l)dl, \forall n \geq n_0.$$

Thus, we obtain

$$\int_0^1 \varphi_n(l)dl \to +\infty, \text{ as } n \to +\infty,$$

a contradiction. By Proposition 2.60, we conclude that T is ρ-uniform persistence. Since $T : (M, d) \to (M, d)$ has a global attractors, it follows from Theorem 2.67 that $T : (M_0, d_0) \to (M_0, d_0)$ has a global attractor.

To avoid possible confusion, we denote by κ_0 the Kuratowski measure of non-compactness on the complete metric (M_0, d_0). We now consider $T : (M_0, d_0) \to (M_0, d_0)$. Let $\varepsilon > 0$ be fixed such that

$$\sqrt{a} < d := a + \frac{b}{1+\varepsilon} < 1.$$

Then for each $x \in M$, we have

$$\|T(x)\| \geq a\|x\| + b\frac{\|x\|}{1+\|x\|} \geq d \min(\varepsilon, \|x\|).$$

Let $B \subset M_0$ be a ρ-bounded set. We set $\rho_0 = \inf_{x\in B} \rho(x)$. Then for each $x \in B$, we obtain

$$\|T(x)\| \geq a\|x\| + b\frac{\|x\|}{1+\|x\|} \geq d\min(\varepsilon, \|x\|).$$

By induction, it follows that

$$\rho\left(T^n(x)\right) \geq d^n \min\left(\varepsilon, \rho_0\right), \quad \forall n \geq 1, \ \forall x \in B.$$

Thus, for each $x, y \in B$, we have

$$\begin{aligned} d_0\left(T^n(x), T^n(y)\right) &= \left|\frac{1}{\rho\left(T^n(x)\right)} - \frac{1}{\rho\left(T^n(y)\right)}\right| + \left\|T^n(x) - T^n(y)\right\| \\ &\leq \left[\frac{1}{\rho\left(T^n(x)\right)\rho\left(T^n(y)\right)} + 1\right]\left\|T^n(x) - T^n(y)\right\| \\ &\leq \left[\frac{1}{d^{2n}\min\left(\varepsilon, \rho_0\right)^2} + 1\right] d(T^n(x), T^n(y)), \end{aligned}$$

and hence,

$$\begin{aligned} \kappa_0\left(T^n(B)\right) &\leq \left[\frac{1}{d^{2n}\min\left(\varepsilon, \rho_0\right)^2} + 1\right]\kappa\left(T^n(B)\right) \\ &\leq a^n\left[\frac{1}{d^{2n}\min\left(\varepsilon, \rho_0\right)^2} + 1\right]\kappa(B). \end{aligned}$$

Since $d > \sqrt{a}$, we obtain $\kappa_0\left(T^n(B)\right) \to 0$ as $n \to +\infty$. So $T : (M_0, d_0) \to (M_0, d_0)$ is κ_0-contracting.

It remains to show that $T : (M_0, d_0) \to (M_0, d_0)$ has no strong global attractor. Let $\delta > 0$ be fixed, and consider the ρ-strongly bounded set

$$B_\delta = \{x \in M : \rho(x) = \delta\}.$$

For each $m \geq 0$, we set $x^m := (\varphi^m, 0)$ with $\varphi^m = \delta 1_{[m,m+1]}$, and

$$x_n^m := \left(\varphi_n^m, y_n^m\right) = T^n\left(x^m\right), \quad \forall n \geq 0.$$

Then for each $m \geq 1$ and each $n \in \{0, ..., m-1\}$, we have $\int_0^1 \varphi_n^m(l)dl = 0$, and hence,

$$\begin{cases} \varphi_{n+1}^m(\cdot) = a\varphi_n^m(\cdot + 1) + a\int_0^1 \varphi_n^m(l)dl 1_{[0,1]}(\cdot) \\ y_{n+1}^m = a y_n^m + b\frac{\|x_n^m\|}{1+\|x_n^m\|}. \end{cases}$$

Thus, for each $m \geq 1$ and each $n \in \{0, ..., m-1\}$, we obtain

$$\left\|x_{n+1}^m\right\| \leq (a+b)\left\|x_n^m\right\| \leq (a+b)^n \delta.$$

It follows that $\inf_{x \in B_\delta} \rho\left(T^n(x)\right) \to 0$, as $n \to +\infty$. So the κ_0-contracting map $T : (M_0, d_0) \to (M_0, d_0)$ has a global attractor, but no strong global attractor.

2.15 Notes and Remarks

This chapter is concerned with global attractors. For more reading on this wide topic, we refer the reader to the books of Temam [324], Hale [138], Sell and You [298], Raugel [279], Chueshov [61], Robinson [283], Cholewa, Dlotko, and Chafee [52], Carvalho Langa and Robinson [45], Zhao [372], Cheban [46], Qin and Su [276] and the references cited therein.

So far, the only book similar to this chapter is the book of Zhao [372], where only the discrete-time case is considered. The novelty in this chapter is that we provide a global attractors theory in the context of uniform persistence for both discrete and continuous time dynamical systems.

Chains and uniform persistence

Uniform persistence (permanence) has been extensively studied for both continuous- and discrete-time dynamical systems. We refer the reader to Waltman [346], Hutson and Schmitt [181], and to Hofbauer and Sigmund [158] for surveys and reviews on the topic. We also refer to Magal and Zhao [244], Thieme [330], Zhao [369, 370], Smith and Thieme [314], Smith and Zhao [317] and references cited therein for further results and extensions.

Here we recall some more general results taken from Hale and Waltman [145], and Hofbauer and So [159].

Let (M, d) be a complete metric space, and $\rho : M \to [0, +\infty)$ be a continuous function. To decompose the set M we define the open subset

$$M_0 := \{x \in M : \rho(x) > 0\},$$

and the closed subset

$$\partial M_0 := \{x \in M : \rho(x) = 0\},$$

so that $M = M_0 \cup \partial M_0$. To state our results, we assume the following:

Assumption 2.72 Let $\{U(t)\}_{t \in I}$ be a continuous semiflow where the set of index $I = \mathbb{R}$ or $\mathbb{Z}$. We assume that it satisfies the following property

(i) $U(t)M_0 \subset M_0$, and $U(t)\partial M_0 \subset \partial M_0, \forall t \geq 0$.
(ii) U has a global attractor, denoted by A, in M.

Lemma 2.73 *Let Assumption 2.72 be satisfied. Then the set A_∂ given by*

$$A_\partial = A \cap \partial M_0$$

is a global attractor for the semiflow U_∂, the semiflow U restricted at ∂M_0.

Definition 2.74 Let $C \subset M$ be a compact invariant set. We recall that the *stable set (or attracting set)* of C is denoted by $W^s(C)$ and is defined by

$$W^s(C) = \{x \in M : \omega(x) \neq \emptyset \text{ and } \omega(x) \subset C\},$$

or equivalently

$$W^s(C) = \left\{x \in M : \lim_{t \to +\infty} d(U(t)x, C) = 0\right\}.$$

The *unstable set (or repelling set)* of C is defined by

$$W^u(C) = \{x \in M : \text{ there exists a backward orbit } \gamma^-(x) \text{ such that } \alpha_\gamma(x) \neq \emptyset \text{ and } \alpha_\gamma(x) \subset C\},$$

where $\alpha_\gamma(x)$ is the alpha limit set corresponding to the specific negative orbit $\gamma^-(x)$. That is also equivalent to say that

$$W^u(C) = \{x \in M : \text{ there exists a complete orbit } t \to u(t) \text{ passing through } x \text{ such that } \lim_{t \to -\infty} d(u(t), C) = 0\}.$$

Using the above definitions, we have the following representation for the global attractor A.

Theorem 2.75 *Let Assumption 2.72 be satisfied. Assume in addition that U has a global attractor A_0 in M_0. Then*

$$A = A_0 \cup W^u(A_\partial)$$

Proof See Theorem 3.2 p. 391 in [145] □

Definition 2.76 A nonempty invariant subset C of M is called an *isolated invariant set* if it is the maximal invariant set of a neighborhood of itself. The neighborhood is called an *isolating neighborhood.*

Definition 2.77 Let C_1 and C_2 be two isolated invariant sets (not necessarily distinct). The subset C_1 is said to be *chained* to C_2, and we write $C_1 \to C_2$, if there exists an element $x \in M$ such that

$$x \notin C_1 \cup C_2 \text{ and } x \in W^u(C_1) \cap W^s(C_2).$$

If we assume in addition that C_1 and C_2 are two compact subsets, $C_1 \to C_2$ is also equivalent to say that there exists a complete orbit $t \to u(t)$ such that

$$u(0) = x \notin C_1 \cup C_2,$$

with

$$\lim_{t \to -\infty} d(u(t), C_1) = 0, \text{ and } \lim_{t \to +\infty} d(u(t), C_2) = 0.$$

A finite sequence $C_1, C_2, \ldots, C_k$ of isolated invariant sets will be called a *chain* if $C_1 \to C_2 \to \ldots \to C_k$ (with $C_1 \to C_1$, if $k = 1$). The chain will be called a *cycle* if $C_k = C_1$.

A particular invariant set of interest is given by

$$\widehat{A}_\partial = \bigcup_{x \in A_\partial} \omega(x) \subset \partial M_0.$$

Definition 2.78 $\widehat{A}_\partial$ is *isolated* if there exists a covering $C = \{C_i\}_{i=1,\ldots,n}$ of $\widehat{A}_\partial$ by pairwise disjoint, compact, isolated invariant sets $C_1, \ldots, C_n$ for U_∂ (i.e. U restricted to ∂M_0) such that each C_i is also an isolated invariant set for U (i.e. U on M). The collection C is called an *isolated covering*. The set $\widehat{A}_\partial$ will be called *acyclic* if there exists some isolated covering $C = \{C_1, \cdots, C_n\}$ of $\widehat{A}_\partial$ such that no subset of the C_i forms a cycle. An isolated covering C satisfying this condition will be called *Morse decomposition* or *acyclic*.

Using the above definition, the main result recalled in this section reads as follows.

Theorem 2.79 *Let Assumption 2.72 be satisfied. Assume in addition that* $\widehat{A}_\partial$ *is isolated and has an acyclic covering* $C = \{C_1, \cdots, C_k\}$. *Then the semiflow* U *is* ρ*-uniformly persistent if and only if*

$$W^s(C_i) \cap M_0 = \emptyset, \forall i = 1, \ldots, k.$$

We refer to Theorem 4.2 p. 393 in [145] for the proof of the above theorem. This theorem generalizes earlier result on strong repeller, that roughly speaking corresponds to the above result with $\rho(x) = d(x, \partial M_0)$, where d denotes the distance function. We refer to Hirsch, Smith and Zhao [156] and Smith and Zhao [317] for more results and details. Theorem 2.79 has been extended to non-autonomous semiflows by Thieme in [329, 330]. The notion of generalized distance function ρ has been considered by Thieme in [330], where the concept of uniform ρ-persistence was introduced for nonautonomous semiflows. General results about uniform persistence have been obtained earlier by Hale and Waltman [145], Thieme [326] for autonomous semiflows, and Freedman and So [115], see also Hofbauer and So [159] for continuous functions.

Various concepts of persistence were used in the literature. We refer the reader to Hutson and Schmitt [181], Cantrell, Cosner and Hutson [43], Hutson and Zhao [373], Cosner [65], Cantrell and Cosner [41], Hutson and Mischaikow [179], Smith and Zhao [311, 316], and Ruan and Zhao [291]. We also refer to Thieme [330], Smith and Zhao [317] and Zhao [371] for further applications with different choice of $\rho-$functions.

Section 2.12 mostly follows Magal and Zhao [244], while Section 2.13 is adapted from Zhao [366] and Magal and Zhao [244]. Note that a class of continuous Kolmogorov like maps on the $m-$dimensional positive cone $\mathbb{R}_+^m$ has been studied by Hutson and Moran in [180]. The authors proved that the existence of a compact attracting set in the interior $int(\mathbb{R}_+^m)$ ensures the existence of a fixed point in $int(\mathbb{R}_+^m)$. Next applying Theorem 2.70 to the Poincaré map associated with a periodic semiflow, may be useful to derive the existence of a periodic orbit, and as a special case that of periodic coexistence solutions for differential equations with periodic coefficients. We refer the reader to Freedman and Yang [359, Theorem 4.11] who proved the existence of interior periodic solutions for periodic, dissipative, and uniformly persistent systems of ODEs. We also refer to Zanolin [362, Lemma 1] for the existence of interior periodic orbits of periodic and uniformly persistent ODE of Kolmogorov type. See also Hutson [178] for the analysis of autonomous Kolmogorov systems of ODEs and autonomous differential equations with time delay. As far as positive equilibrium are concerned, Hofbauer [157] proved an index theorem for dissipative ODE system, that ensures the existence of a positive equilibrium (see also [158]). Cantrell, Cosner and Hutson [43, Theorem 6.2] also prove the existence of coexistence steady state for an autonomous Kolmogorov reaction-diffusion problem with two species. See also Ducrot and Madec [93] for a spatially heterogeneous resource-consumer reaction-diffusion system.

There are also various results on uniform persistence with parameters and robustness. For results in these directions, one may refer to Hirsch, Smith and Zhao [156] and Smith and Zhao [316, 317]. We also refer to Hutson [177] and Schreiber [296] who studied the robustness of permanence for perturbed ordinary differential systems.

More properties on the measure of non-compactness

In this part, we collect more properties of the measure of non-compactness. Let $(X, \|.\|)$ be a Banach space. Recall that for each bounded set $B \subset X$, the quantity

$$\kappa(B) = \inf\{\varepsilon > 0 : B \text{ can be covered by a finite number of balls of radius } \leq \varepsilon\}$$

is the Kuratowski measure of non-compactness.

We now list various properties of the Kuratowski measure of non-compactness in a Banach space X. We refer the reader to Deimling [75], Martin [248], and Sell and You [298, Lemma 22.2].

Theorem 2.80 *Let $(X, \|.\|)$ be a Banach space and $\kappa(.)$ the measure of non-compactness defined as above. Then for any bounded subsets B and $\widehat{B}$ of X, we have the following properties:*

(i) $\kappa(B) = 0$ *if and only if* $\overline{B}$ *is compact;*

(ii) $\kappa(B) = \kappa\left(\overline{B}\right)$*;*

(iii) $\kappa(\overline{\text{co}}(B)) = \kappa(B)$*;*

(iv) *If* $B \subset \widehat{B}$ *then* $\kappa(B) \leq \kappa\left(\widehat{B}\right)$*;*

(v)

$$\kappa(\lambda B) \leq |\lambda| \kappa(B),$$

where

$$\lambda B = \{\lambda x : x \in B\}$$

and

$$\kappa\left(B + \widehat{B}\right) \leq \kappa(B) + \kappa\left(\widehat{B}\right),$$

where

$$B + \widehat{B} = \left\{x + y : x \in B, y \in \widehat{B}\right\}.$$

Let $L \in \mathcal{L}(X)$ be a given linear bounded operator on X. The *essential semi-norm* $\|L\|_{\text{ess}}$ of L is defined by

$$\|L\|_{\text{ess}} = \kappa\left(L\left(B_X(0,1)\right)\right),$$

where $B_X(0,1) = \{x \in X : \|x\|_X \leq 1\}$. Then the essential semi-norm satisfies the following properties.

Proposition 2.81 *For each pair of bounded linear operators $L, \widehat{L} \in \mathcal{L}(X)$, we have the following properties:*

(i) $\|L\|_{\text{ess}} = 0$ *if and only if L is compact (i.e. L maps bounded subsets into relatively compact subsets);*
(ii) $\|\lambda L\|_{\text{ess}} \leq |\lambda| \, \|L\|_{\text{ess}}, \forall \lambda \in \mathbb{C}$;
(iii) $\left\|L + \widehat{L}\right\|_{\text{ess}} \leq \|L\|_{\text{ess}} + \left\|\widehat{L}\right\|_{\text{ess}}$;
(iv) $\left\|L\widehat{L}\right\|_{\text{ess}} \leq \|L\|_{\text{ess}} \left\|\widehat{L}\right\|_{\text{ess}}$;
(v) $\|L\|_{\text{ess}} \leq \|L\|_{\mathcal{L}(X)}$;

The above proposition ensures that $\|\cdot\|_{\text{ess}}$ is a semi-norm on $\mathcal{L}(X)$, that becomes a norm on the quotient space $\mathcal{L}(X)/\mathcal{K}(X)$ where $\mathcal{K}(X)$ denotes the bilateral ideal of the compact operators on X.

Asymptotic smoothness

In view of the properties of the measure of non-compactness recalled above, as an application, we derive sufficient conditions for the asymptotic smoothness of a semiflow generated by a semilinear abstract Cauchy problem on a Banach space $(X, \|\cdot\|)$ of the form

$$\frac{du(t)}{dt} = Au(t) + F(u(t)), \; t > 0, \; u(0) = x \in X. \tag{2.18}$$

Here we assume that the following properties hold.

Assumption 2.82 The map $F : X \to X$ is Lipschitz continuous on bounded sets. We also assume F is completely continuous, that is to say that the closure $\overline{F(B)}$ is compact for each bounded subset B of X. In other words,

$$\kappa(F(B)) = 0 \text{ for any bounded set } B \subset X.$$

We assume that the linear operator $A : D(A) \subset X \to X$ is the infinitesimal generator of a strongly continuous semigroup on X, denoted by $\{T_A(t)\}_{t \geq 0} \subset \mathcal{L}(X)$.

The maximal semiflow generated by (2.18) is denoted by U. If it is globally defined as the fixed point for $x \in X$ of

$$U(t)x = T_A(t)x + \int_0^t T_A(t-s)F(U(s)x)ds, \; t \geq 0.$$

Now according to Definition 2.41 to investigate the asymptotic smoothness of this semiflow, let $B \subset X$ be a bounded set such that

$$U(t)B \subset B, \forall t \geq 0.$$

According to the above assumption, F maps bounded sets into relatively compact subsets of X. Then by using Mazur's theorem we have for all $t \geq 0$ and $x \in X$:

$$\int_0^t T_A(t-s)F(U(s)x)ds \in t \times \overline{\text{co}}(\{T_A(t-s)F(U(s)x) : s \in [0,t]\}).$$

This implies that for any $t \geq 0$

$$\kappa\left(\int_0^t T_A(t-s)F(U(s)B)ds\right) \leq t \times \kappa\left(\overline{\text{co}}\left(\{T_A(t-s)F(x) : x \in B, s \in [0,t]\}\right)\right) = 0,$$

Since the measure of compactness is additive, namely $\kappa(B_1 + B_2) \leq \kappa(B_1) + \kappa(B_2)$), we end-up with

$$\kappa(U(t)B) = \kappa(T_A(t)B), \ \forall t \geq 0.$$

Now since $\kappa(L(B)) \leq \|L\|_{\mathcal{L}(X)}\kappa(B)$ as soon as $L \in \mathcal{L}(X)$, we deduce the following proposition, which extends some ideas originally proved by Webb in [348].

Proposition 2.83 *Assume that $F : X \to X$ is a map on a Banach space that is Lipschitz continuous on the bounded sets of X and F is completely continuous, that is to say that the closure $\overline{F(B)}$ is compact for each bounded subset $B \subset X$.*

Then the semiflow $t \to U(t)$ is asymptotically smooth in X if we furthermore impose that

$$\|T_A(t)\|_{\mathcal{L}(X)} \to 0 \text{ as } t \to \infty.$$

We refer to Magal and Thieme [242] for more results going in this direction.

Existence of strong global attractors

This subsection is devoted to strong global attractors as defined in Definition 2.48. To that aim recall the following definitions.

Definition 2.84

(i) A map $T : M \to M$ is **compact** if for any bounded set $B \subset M$, $\overline{T(B)}$ is compact.

(ii) A semiflow U is said to be **eventually compact** there exists $t_0 > 0$, such that $U(t_0)$ is compact.

We complete this section with a variant of [138, Theorems 2.4.6 and 2.4.7] on the existence of strong global attractors. We also refer to Billotti and LaSalle [31] for other results.

Theorem 2.85 (Strong global attractor) *Let U be a continuous semiflow on a metric space (M, d). Assume that U is point dissipative on M, and assume that one of the two following conditions holds:*

(i) *U is eventually compact;*
(ii) *U is asymptotically smooth, and for each bounded set $B \subset M$, there exists $t_0 = t_0(B) \geq 0$ such that $\gamma^+(U(t_0)(B))$ is bounded.*

Then U has a strong global attractor $A \subset M$.

Proof The conclusion in case (ii) is an immediate consequence of Theorem 2.49. In the case of (i), since $U(t_0)$ is compact for some $t_0 \in I \setminus \{0\}$, it is sufficient to show that for each compact subset $C \subset M$,

$$\bigcup_{t \geq 0} U(t)(C)$$

is bounded. By applying Lemma 2.66 to the discrete time semiflow $\{V(t)\}_{t \in \mathbb{N}}$, defined by

$$V(n) = U(t_0)^n, \forall n \geq 0,$$

we deduce that for each compact subset $C \subset M$,

$$\bigcup_{n \geq 0} V(n)(C)$$

is bounded. So Theorem 2.49 implies that V has a global attractor $\widetilde{A} \subset M$. We set

$$\widetilde{B} = \bigcup_{0 \leq t \leq t_0} U(t)\left(\widetilde{A}\right).$$

Since U is a continuous, the subset $\widetilde{B}$ is compact, and $\widetilde{B}$ attracts every compact subset of M for U, and hence, the result follows from Theorem 2.49. □

Connected global attractors

We are interested in the connectedness of the global attractor. The goal is to give a generalized version of the connectness results in Gobbino and Sardella [122].

Definition 2.86 A subset C of M is said to be **connected** if there exists no partition of C in two subsets A and B that are both open and closed. That is to say that if for each $A, B \subset C$, such that

$$A \neq \emptyset, \ B \neq \emptyset, \ C = A \cup B, \text{ and } A \cap B = \emptyset,$$

we have

$$\left(\overline{A} \cap B\right) \cup \left(A \cap \overline{B}\right) \neq \emptyset.$$

If C is closed and not connected, there exist A and B two non empty subset of C, such that

$$C = A \cup B,\ A \cap B = \emptyset, \text{ and } \left(\overline{A} \cap B\right) \cup \left(A \cap \overline{B}\right) = \emptyset.$$

Assume for example that $A \neq \overline{A}$. Then $\overline{A} \subset C$, and $\left(\overline{A} \cap B\right) = \emptyset$. So $C = A \cup B \cup \left(\overline{A} \setminus A\right)$, and $C \neq A \cup B$. So $A = \overline{A}$. We conclude that if C is closed and not connected, there exist A and B two nonempty closed subset of C, such that

$$C = A \cup B, \text{ and } A \cap B = \emptyset.$$

Lemma 2.87 *Let $\{U(t)\}_{t\in I}$ be a continuous semiflow, and let A be a nonempty compact subset of M invariant for U. If there exists a connected subset C of M, such that $A \subset C$, and A attracts C for U, then A is connected.*

Proof Assume that A is not connected then we can find A_1 and A_2 two disjoint nonempty compact subsets of A such that $A = A_1 \cup A_2$. Since A_1 and A_2 are compact, we can find $\varepsilon > 0$, such that $\overline{N}(A_1, \varepsilon) \cap \overline{N}(A_2, \varepsilon) = \emptyset$. Since A attracts C for U there exists $t > 0$, such that $U(t)C \subset \overline{N}(A_1, \varepsilon) \cup \overline{N}(A_2, \varepsilon)$, and since $A \subset C$, and A is invariant for U, we deduce that $U(t)C \cap \overline{N}(A_i, \varepsilon) \neq \emptyset, \forall i = 1, 2$. But since $x \to U(t)x$ is continuous, we deduce that $\widehat{C} = U(t)C$ is connected. But $\widehat{C} = C_1 \cup C_2$ with $C_i = U(t)C \cap \overline{N}(A_i, \varepsilon), \forall i = 1, 2$. By construction we have

$$\left(C_1 \cap \overline{C_2}\right) \cup \left(\overline{C_1} \cap C_2\right) \subset \overline{N}(A_1, \varepsilon) \cap \overline{N}(A_2, \varepsilon) = \emptyset,$$

we obtain a contradiction with the fact that $\widehat{C}$ is connected. □

Theorem 2.88 *Let $\{U(t)\}_{t\in I}$ be a continuous semiflow on M a closed and convex subset of a Banach space $(X, \|\cdot\|)$. Assume that U has a global attractor $A \subset M$. Then A is connected.*

Proof Since A is compact, $\overline{\text{co}}(A)$ (the closed convex hull of A) is also compact. Moreover $A \subset \overline{\text{co}}(A)$, and since A attracts the compact subset of M by A the result follows from Lemma 2.87. □

The following result has been proved first by Zhao [366, Lemma 2.1].

Lemma 2.89 (Zhao) *Let M be a closed subset of a Banach space $(X, \|\cdot\|)$. Assume that $\rho : M \to [0, +\infty)$ is a continuous map, and assume that*

$$M_0 = \{x \in M : \rho(x) > 0\}$$

is nonempty and convex. Let $A \subset M_0$ be a compact subset, then

$$\overline{\text{co}}(A) \subset X_0.$$

Theorem 2.90 *Let $\{U(t)\}_{t\in I}$ be a continuous semiflow on M a closed subset of a Banach space $(X, \|\cdot\|)$. Assume that $\rho : M \to [0, +\infty)$ is a continuous map, and assume that*

$$M_0 = \{x \in M : \rho(x) > 0\}$$

is nonempty and convex. Assume that U has a global attractor $A \subset M_0$. Then A is connected.

Upper semi-continuity of global attractors

We now investigate the dependence to a parameter of the global attractors. The results presented in this subsection are mostly taken from Magal [236]. We also refer to [279]. Let (Λ, d_Λ) be a metric space, and let $\lambda_0 \in \Lambda$ be fixed. Let $\{U_\lambda(t)\}_{t\in I}$ (with $I = \mathbb{N}$ or $I = [0, \infty)$) be a family of semiflow on M parametrized by $\lambda \in \Lambda$. Let $\{A_\lambda\}_{\lambda\in\Lambda}$ be a family of compact subset of M. We will say that $\{A_\lambda\}_{\lambda\in\Lambda}$ is **upper semi-continuous at** $\lambda_0 \in \Lambda$, if and only if

$$\delta\left(A_\lambda, A_{\lambda_0}\right) \to 0 \text{ as } \lambda \to \lambda_0.$$

For each $\lambda \in \Lambda$, and each $\eta \geq 0$, we set

$$\overline{B}_\Lambda(\lambda, \eta) = \{\alpha \in \Lambda : d_\Lambda(\lambda, \alpha) \leq \eta\}.$$

We first have the following result.

Proposition 2.91 *We assume that for each $(t, x_0) \in I \times M$, the map $(\lambda, x) \to U_\lambda(t)x$ is continuous at (λ_0, x_0), and for each $\lambda \in \Lambda$, A_λ is invariant for U_λ, and A_{λ_0} is a global attractors for U_{λ_0}. Then the following statement are equivalents:*

(i) *$\{A_\lambda\}_{\lambda\in\Lambda}$ is upper semi-continuous at $\lambda_0 \in \Lambda$.*
(ii) *There exists a compact subset $C \subset M$, such that $\delta(A_\lambda, C) \to 0$ as $\lambda \to \lambda_0$.*

Proof (i)⇒(ii) is obvious with $C = A_{\lambda_0}$.

(ii)⇒(i). Conversely, assume that there exists a compact subset C of M such that $\delta(A_\lambda, C) \to 0$ as $\lambda \to \lambda_0$. We argue by contradiction that there exist $\varepsilon > 0$, and a sequence $\lambda_n \to \lambda_0$, as $n \to +\infty$, such that

$$\delta\left(A_{\lambda_n}, A_{\lambda_0}\right) \geq \varepsilon, \forall n \geq 0.$$

Then by using Theorem 2.29, we find a sub-sequence $\lambda_{n_p} \to \lambda_0$, and a compact subset $\widehat{A}_\infty$ such that $d_H(A_{\lambda_{n_p}}, \widehat{A}_\infty) \to 0$, as $p \to +\infty$. By construction we have $\delta\left(\widehat{A}_\infty, A_{\lambda_0}\right) \geq \varepsilon$. Since for each $p \geq 0$, $A_{\lambda_{n_p}}$ is invariant for $U_{\lambda_{n_p}}$, and since for each $t \in I$, and each $x_0 \in M$, the map $(\lambda, x) \to U_\lambda(t)x$ is continuous at (λ_0, x_0), we deduce that $\widehat{A}_\infty$ is invariant for U_{λ_0}. Finally since A_{λ_0} is a global attractor for U_{λ_0}, we deduce that $\widehat{A}_\infty \subset A_{\lambda_0}$, so $\delta\left(\widehat{A}_\infty, A_{\lambda_0}\right) = 0$, a contradiction. □

Proposition 2.92 *We assume that for each $\lambda \in \Lambda$, A_λ is invariant for U_λ, and there exists a subset B of M such that:*

a) A_{λ_0} *attracts* B *for* U_{λ_0}.
b) $A_\lambda \subset B, \forall \lambda \in \Lambda$.
c) For each $t \geq 0$, $U_\lambda(t)x \to U_{\lambda_0}(t)x$ *as* $\lambda \to \lambda_0$ *uniformly in* $x \in B$.
Then $\{A_\lambda\}_{\lambda\in\Lambda}$ *is upper semi-continuous at* $\lambda_0 \in \Lambda$.

Proof We have for each $t \geq 0$, and each $\lambda \in \Lambda$

$$\begin{aligned}\delta\left(A_\lambda, A_{\lambda_0}\right) &= \delta\left(U_\lambda(t)A_\lambda, A_{\lambda_0}\right)\\ &= \sup_{x\in A_\lambda}\inf_{y\in A_{\lambda_0}} d\left(U_\lambda(t)x, y\right)\\ &\leq \sup_{x\in B}\inf_{y\in A_{\lambda_0}} d\left(U_\lambda(t)x, y\right)\\ &\leq \sup_{x\in B}\inf_{y\in A_{\lambda_0}} d\left(U_\lambda(t)x, U_{\lambda_0}(t)x\right) + d\left(U_{\lambda_0}(t)x, y\right),\end{aligned}$$

so

$$\delta\left(A_\lambda, A_{\lambda_0}\right) \leq \sup_{x\in B} d\left(U_\lambda(t)x, U_{\lambda_0}(t)x\right) + \delta\left(U_{\lambda_0}(t)B, A_{\lambda_0}\right),$$

and the result follows. □

The idea of the following result is inspired by Theorem 23.14 p.41 in Sell and You [298]. It seems necessary to make some additional continuity assumptions (see assumption c)).

Proposition 2.93 *Let* (Λ, d_Λ) *be a metric space, let* $\lambda_0 \in \Lambda$ *be fixed, and let* $\{U_\lambda(t)\}_{t\in I}$ *be a family of continuous semiflow on* M *parametrized by* $\lambda \in \Lambda$. *We assume that for each* $\lambda \in \Lambda$, U_λ *is asymptotically smooth, and there exists* A_{λ_0} *a global attractor for* U_{λ_0}. *We assume in addition that there exists* B_0 *a bounded subset of* M, *such that:*
a) A_{λ_0} *attracts* B_0 *for* U_{λ_0}.
b) For each $\lambda \in \Lambda$, *and each* $x \in M$, *there exists* $t_0 = t_0(\lambda, x) \geq 0$, *such that*

$$U_\lambda(t)x \in B_0, \forall t \geq t_0.$$

c) For each bounded set $B \subset M$, *and each* $t^* \in I$, $U_\lambda(t)x \to U_{\lambda_0}(t)x$ *as* $\lambda \to \lambda_0$ *uniformly in* $(t,x) \in ([0,t^*] \cap I) \times B$.
Then there exists $\eta > 0$, *such that for all* $\lambda \in \overline{B}_\Lambda(\lambda_0, \eta)$, $U_\lambda(t)$ *has a global attractor* $A_\lambda \subset M$, *and the family* $\{A_\lambda\}_{\lambda\in\overline{B}_\Lambda(\lambda_0,\eta)}$ *is upper semi-continuous at* $\lambda_0 \in \Lambda$.

Proof For each $\lambda \in \Lambda$, we set

$$J_\lambda(B_0) := \{y \in B_0 : U_\lambda(t)(y) \in B_0, \forall t \in I\}.$$

Then $J_\lambda(B_0) \subset B_0$ is bounded, and positively invariant for U_λ, and for each $x \in M$, there exists $t \geq 0$, such that $U_\lambda(t)(x) \in J_\lambda(B_0)$. Since U_λ is asymptotically smooth, we deduce that $\omega_\lambda(J_\lambda(B_0)) = \bigcap_{s\geq 0}\overline{\bigcup_{t\geq s} U_\lambda(t)(J_\lambda(B_0))}$ is compact, invariant for U_λ, and attracts the points of M for U_λ. By applying Proposition 2.92 (with $A_\lambda = \omega_\lambda(J_\lambda(B_0)), \forall \lambda \in \Lambda \setminus \{\lambda_0\}, B = B_0$), we deduce that

$$\lim_{\lambda\to\lambda_0} \delta\left(\omega_\lambda(J_\lambda(B_0)), A_{\lambda_0}\right) = 0. \tag{2.19}$$

Let $\varepsilon > 0$ be fixed. Since A_{λ_0} is a global attractor for U_{λ_0}, it follows that A_{λ_0} is stable for U_{λ_0}, and A_{λ_0} attracts one of its neighborhood for U_{λ_0}. So there exists $\widehat{\varepsilon} \in (0, \varepsilon)$, such that A_{λ_0} attracts $\overline{N}\left(A_{\lambda_0}, \widehat{\varepsilon}\right)$ for U_{λ_0},

$$U_{\lambda_0}(t)\overline{N}\left(A_{\lambda_0}, \widehat{\varepsilon}\right) \subset \overline{N}\left(A_{\lambda_0}, \varepsilon/4\right), \forall t \geq 0,$$

and

$$U_{\lambda_0}(t^*)\overline{N}\left(A_{\lambda_0}, \widehat{\varepsilon}\right) \subset \overline{N}\left(A_{\lambda_0}, \widehat{\varepsilon}/4\right).$$

By using assumption c), we deduce that there exists $\eta_1 > 0$, such that for each $\lambda \in \Lambda$, with $d_\Lambda(\lambda_0, \lambda) \leq \eta_1$,

$$U_\lambda(t)\overline{N}\left(A_{\lambda_0}, \widehat{\varepsilon}\right) \subset \overline{N}\left(A_{\lambda_0}, \varepsilon/2\right), \forall t \in [0, t^*] \cap I,$$

and

$$U_\lambda(t^*)\overline{N}\left(A_{\lambda_0}, \widehat{\varepsilon}\right) \subset \overline{N}\left(A_{\lambda_0}, \widehat{\varepsilon}/2\right).$$

We set for each $\lambda \in \Lambda$, with $d_\Lambda(\lambda_0, \lambda) \leq \eta_1$,

$$B_\lambda = \overline{\bigcup_{t \in [0, t^*]} U_\lambda(t)\overline{N}\left(A_{\lambda_0}, \widehat{\varepsilon}\right)},$$

then B_λ is positively invariant for U_λ, and $\overline{N}\left(A_{\lambda_0}, \widehat{\varepsilon}\right) \subset B_\lambda \subset \overline{N}\left(A_{\lambda_0}, \varepsilon/2\right)$. Moreover by using (2.19) there exists $\eta \in (0, \eta_1)$, such that $\omega_\lambda\left(J_\lambda(B_0)\right) \subset \overline{N}\left(A_{\lambda_0}, \widehat{\varepsilon}/2\right)$, for each $\lambda \in \Lambda$, with $d_\Lambda(\lambda_0, \lambda) \leq \eta$. Then for each $\lambda \in \Lambda$, with $d_\Lambda(\lambda_0, \lambda) \leq \eta$, B_λ attracts the compact subset of M, and $A_\lambda = \omega_\lambda(B_\lambda)$ is a global attractor for $U_\lambda(t)$. Finally since $B_\lambda \subset \overline{N}\left(A_{\lambda_0}, \varepsilon/2\right)$, we deduce that

$$A_\lambda \subset \overline{N}\left(A_{\lambda_0}, \varepsilon/2\right),$$

and the result follows. □

In general the lower semi-continuity for attractors does not hold true. To see this consider the one dimensional equation

$$\begin{cases} x'(t) = (1 - x(t))(x(t)^2 - \lambda), \\ x(0) = x_0 \in \mathbb{R}, \end{cases}$$

where $\lambda \in [-1, 1]$ is a parameter. The attractor A_λ of this problem is given by

$$A_\lambda = \begin{cases} 1 \text{ if } \lambda < 0, \\ [0, 1] \text{ if } \lambda = 0 \\ [-\sqrt{\lambda}, 1] \text{ if } \lambda > 0. \end{cases}$$

For this example the family of the attractors is not lower semi-continuous at $\lambda = 0$ due to a saddle-node bifurcation for the stationary states. This situation also occurs when the system exhibits a backward bifurcation. Such a behavior arises quite often in epidemic systems. We refer the reader for instance to Dushoff, Huang and Castillo-

Chavez [98], Feng, Ruan, Teng and Wang [110], Hadeler and Van den Driessche [133], Ruan and Wang [290] and Van den Driessche and Watmough [86].

As mentioned above the global attractors can be non continuous at some point. To obtain a continuity result, we assume the following:

Assumption 2.94 There exists $t_0 \in I$ and a compact set $K \subset M$ such that

$$\bigcup_{\lambda \in \Lambda} A_\lambda \subset K,$$

and for all $\lambda_0 \in M$ and $x_0 \in K$ if $\lambda_k \to \lambda_0$ and $x_k \in A_{\lambda_k}$ with $x_k \to x_0$, then $U_{\lambda_k}(t_0)x_k \to U_{\lambda_0}(t_0)x_0$.

This "compactness" assumption is sufficient to ensure the family of global attractors is continuous at a large set of parameters.

Proposition 2.95 *Let Assumption* (2.94) *be satisfied. Then there exists a residual subset* $\Lambda_0 \subset \Lambda$ *such that the family of the global attractors* A_λ *is continuous at any point* $\lambda_0 \in \Lambda_0$. *Here recall* Λ_0 *residual in* Λ *means that it contains a countable intersection of open dense subsets of* Λ.

The proof of the above proposition can be found in [279], see also [143]. This result has been used by Babin and Pilyugin [15] to prove the continuity of the attractors for gradient systems.

Let us refer the reader to Smith and Waltam [315], Magal [236] and the references therein, where the authors studied the perturbation of a globally stable equilibrium. Such result can also be viewed as the continuity of an attractor.

Gradient systems

Recall that (X, d) denotes a metric space while the set I denotes either $[0, \infty)$ or $\mathbb{N}$.

Definition 2.96 (Gradient System) Let $\{U(t)\}_{t\in I}$ be a semiflow on X.

1) A function $\Phi \in C^0(X, \mathbb{R})$ is said to be a Lyapunov functional if

$$\Phi(S(t)x) \leq \Phi(x), \text{ for all } t \in I \text{ and } x \in X.$$

2) A Lyapunov functional Φ is said to be a strict Lyapunov functional if, furthermore, for $x \in X$,

$$\Phi(S(t)x) = \Phi(x)\ \forall t \in I \Rightarrow x \in X \text{ is an equilibrium point.}$$

3) A semiflow $\{U(t)\}_{t\in I}$ is a gradient system if it has a strict Lyapunov functional and if, either $I = \mathbb{N}$ or $[0, \infty)$ and U is a continuous semiflow.

In the following we denote by $\mathcal{E} \subset X$ the set of equilibrium points of U, that is

$$\mathcal{E} = \{z \in X | U(t)z = z, \forall t \geq 0\}.$$

Clearly, $\mathcal{E}$ is an invariant and closed set. If U is a discrete semiflow, $\mathcal{E}$ is simply the set of fixed points of $U(1)$. The following result, known as the *Invariance Principle of LaSalle*, is due to LaSalle [207]. It plays an important role in the theory of gradient systems. We set $I^- = \{-t | t \in I\}$.

Proposition 2.97 (Invariance Principle of LaSalle) *Let $\{U(t)\}_{t \in I}$ be a gradient system on X with a Lyapunov functional Φ. Then the following statements hold.*

1) If $z \in X$ is such that the positive orbit $\gamma^+(z)$ starting from z is relatively compact in X, then,

(i) The limit $\lim_{t \to \infty} \Phi(U(t)z) = \ell$ exists and $\Phi(v) = \ell$, for any $v \in \omega(z)$, the ω–limit set of z.

(ii) $\omega(z) \subset \mathcal{E}$; in particular, $\mathcal{E} \neq \emptyset$ and

$$\lim_{t \to \infty} d(U(t)z, \mathcal{E}) = 0.$$

2) Let $u_z \in C(I^-, X)$ be a negative orbit through some $z \in X$. If the negative orbit $u_z(I^-)$ is relatively compact, then the set $\alpha(z)$ is non-empty, compact, invariant, consists only of equilibrium points and $d(u_z(-t), \alpha(z)) \to 0$ as $t \to \infty$. Furthermore, the set $\alpha(z)$ is connected.

The existence of a Lyapunov functional can also be used to have an other descrition of the global attractors.

Theorem 2.98 *Let $\{U(t)\}_{t \in I}$ be an asymptotically smooth gradient system, which has the property that, for any bounded set $B \subset X$, there exists $\tau \in I$ such that the positives orbit of B from time τ, $\bigcup_{t \geq \tau} U(t)B$ is bounded. If the set of equilibrium points $\mathcal{E}$ is bounded, then the semiflow $\{U(t)\}_{t \in I}$ has a compact global attractor A and we have*

$$A = W^u(\mathcal{E}; U),$$

where $W^u(\mathcal{E}; U)$ is defined as the set of points $y \in X$ such that there exists a negative orbit u of U with

$$\begin{cases} u(0) = y, \\ d(u(t), \mathcal{E}) \to 0 \text{ as } t \to -\infty \end{cases}.$$

Furthermore, if $\mathcal{E}$ is a discrete set, then $\mathcal{E}$ is a finite set $\{e_1, \cdots, e_{n_0}\}$ and $A = \bigcup_{j=1,\cdots,n_0} W^u(e_j)$.

Gradient systems depending on a parameter may enjoy, under suitable assumptions, a continuity property for their global attractors. To investigate such a property, as above we consider a family of semiflows $\{U_\lambda(t)\}_{t \in I}$ on a metric space (X, d) and depending on the parameter $\lambda \in \Lambda$ with (Λ, d_Λ) a metric space. We assume that for each $\lambda \in \Lambda$, U_λ admits a global attractor A_λ. We fix $\lambda_0 \in \Lambda$ a non isolated parameter and we provide sufficient conditions for the continuity of the global attractor as $\lambda \to \lambda_0$.

To state a continuity result, we assume that following set of assumptions.

Assumption 2.99 We assume that the following properties are satisfied.

(i) There exists $\eta > 0$, $t_0 \in I$ and B_0 a bounded set (with bounded diameter) in (X, d) such that

$$\bigcup_{\lambda \in B_\Lambda(\lambda_0, \eta)} A_\lambda \subset B_0,$$

and, for any $\varepsilon > 0$, any $\tau \in I \cap |t_0, \infty)$, there exists $\theta \in (0, \eta)$ such that for all $\lambda \in B_\Lambda(\lambda_0, \theta)$ one has

$$d\left(S_\lambda(\tau)x_\lambda, S_{\lambda_0}(\tau)\right) \leq \varepsilon, \ \forall x_\lambda \in A_\lambda.$$

(ii) The set $\mathcal{E}^{\lambda_0} = \{z \in X | S_{\lambda_0}(t)z = z, \forall t \geq 0\}$ is finite, say

$$\mathcal{E}^{\lambda_0} = \left\{e_1^{\lambda_0}, \cdots, e_{n_0}^{\lambda_0}\right\}.$$

(iii) The global attractor A_{λ_0} is compact and given for any $t > 0$ as

$$A_{\lambda_0} = \bigcup_{j=1}^{n_0} W^u\left(e_j^{\lambda_0}; S_{\lambda_0}(t)\right).$$

Here recall that $W^u\left(e_j^{\lambda_0}; S_{\lambda_0}\right)$ is the set of points $y \in X$ such that there exists a negative orbit γ^- of S_{λ_0} with

$$\begin{cases} \gamma^-(0) = y, \\ \gamma^-(t) \to e_j^{\lambda_0} \text{ as } t \to -\infty. \end{cases}$$

(iv) For $j = 1, \cdots, n_0$ there exists a neighbourhood V_j of $e_j^{\lambda_0}$ in X such that

$$\lim_{\lambda \to \lambda_0} \delta\left(W^u\left(e_j^{\lambda_0}, S_{\lambda_0}; V_j\right), A_\lambda\right) = 0,$$

wherein $W^u\left(e_j^{\lambda_0}, S_{\lambda_0}; V_j\right)$ denotes the local unstable manifold of S_{λ_0} at $e_j^{\lambda_0}$ defined as the set of the points $y \in W^u\left(e_j^{\lambda_0}, S_{\lambda_0}\right)$ such that there exists a negative orbit γ^- of S_{λ_0} with

$$\begin{cases} \gamma^-(0) = y, \\ \gamma^-(t) \in V_j \ \forall t \leq 0. \end{cases}$$

Proposition 2.100 *Let Assumption 2.99 be satisfied. Then the global attractors A_λ is lower semi-continuous, in the sense that*

$$\lim_{\lambda \to \lambda_0} \delta\left(A_{\lambda_0}, A_\lambda\right) = 0.$$

We refer to [16], [143] and [138] for more general results about the lower semi-continuity of global attractors.

Gurtin-MacCamy population dynamics model

We now present an example of application to the so-called Gurtin-MacCamy model, taken from Ma and Magal [234]. The Gurtin-MacCamy model reads as the following age-structured equation

$$\begin{cases} \partial_t u(t,a) + \partial_a u(t,a) = -\mu u(t,a), & t > 0, \ a > 0, \\ u(t,0) = \alpha f\left(\int_0^\infty \beta(a) u(t,a) da\right), & t > 0, \end{cases}$$

with an initial data $u(0,a) = u_0(a)$ and $u_0 \in L^1_+((0,\infty),\mathbb{R})$. The parameters α, μ are positive while f takes the form

$$f(u) = ue^{-u},$$

and $\beta \in L^\infty_+(0,\infty)$ is such that

$$\int_0^\infty \beta(a) e^{-\mu a} da = 1.$$

This problem generates a strongly continuous semiflow $U = \{U(t)\}_{t\geq 0}$ in $L^1_+((0,\infty),\mathbb{R})$ given by

$$U(t)u_0 = u(t,\cdot) \text{ with } u(t,a) = \begin{cases} e^{-\mu t} u_0(a-t) \text{ if } a \geq t, \\ e^{-\mu a} b(t-a) \text{ if } a \leq t, \end{cases}$$

where $t \mapsto b(t)$ is the unique continuous solution of the following Volterra integral equation:

$$b(t) = \alpha f\left(\int_t^\infty \beta(a) e^{-\mu t} u_0(a-t) da + \int_0^t \beta(a) e^{-\mu a} b(t-a) da\right), \ t \geq 0.$$

This semiflow enjoys some dissipativity property in $L^1_+(0,\infty)$. Indeed for any $u_0 \in L^1_+(0,\infty)$ one has

$$\|U(t)u_0\|_{L^1(0,\infty)} \leq e^{-\mu t}\|u_0\|_{L^1(0,\infty)} + (1-e^{-\mu t})\Lambda, \ \forall t \geq 0,$$

so that U is bounded dissipative. Furthermore the semiflow U is asymptotically smooth, namely for any bounded set $B \subset L^1_+(0,\infty)$ one has

$$\lim_{t\to\infty} \kappa\left(U(t)B\right) = 0,$$

where κ denotes the Kuratowski measure of non-compactness in $L^1(0,\infty)$. These properties ensure that U admits a global compact attractor.

Now to discuss interior attractors and uniform persistence, we define the age $a^* \in (0,\infty]$ by

$$a^* = \sup\left\{a > 0 : \int_a^\infty \beta(\ell) e^{-\mu \ell} d\ell > 0\right\}.$$

This parameter a^* roughly corresponds to the maximal age at which individuals can reproduce. With this notation, we define the following partition of the state space $L^1_+(0,\infty)$:

$$M_0 = \left\{\varphi \in L^1_+(0,\infty) : \int_0^{a^*} \varphi(x)dx > 0\right\},$$

$$\partial M_0 = \left\{\varphi \in L^1_+(0,\infty) : \int_0^{a^*} \varphi(x)dx = 0\right\} = L^1_-(0,\infty) \setminus M_0.$$

The above partition of the state space is positively invariant by the semiflow $\{U(t)\}_{t\geq 0}$, in the sense that

$$U(t)M_0 \subset M_0 \text{ and } U(t)\partial M_0 \subset \partial M_0, \ \forall t \geq 0.$$

Moreover the following result holds.

Proposition 2.101 *Assume that $\alpha > 1$. Then the semiflow U is uniformly persistent in M_0, in the sense that there exists $\varepsilon > 0$ such that for any $u_0 \in M_0$ one has*

$$\liminf_{t\to\infty} \|U(t)u_0\|_{L^1(\mathbb{R})} \geq \varepsilon.$$

In addition the semiflow U has a compact global attractor in M_0, denoted by $\mathcal{A}_0$.

The condition $\alpha > 1$ corresponds to the instability of the trivial solution $u = 0$.

The existence of the interior attractor $\mathcal{A}_0$ can now be used to prove global stability results by constructing suitable lyapunov functional. We refer to [234] for more details and the proof of the following global stability result.

Theorem 2.102 *Assume that $1 < \alpha \leq e^2$. Then the unique positive equilibrium $a \mapsto \tilde{u}(a)$ is globally asymptotically stable for the semiflow U in M_0.*

One may notice that $\alpha = e^2$ is somehow sharp since one can find a sequence of function $\beta_n = \beta_n(a)$ and $\alpha_n \searrow e^2$ as $n \to \infty$ such that the unique positive equilibrium for the system with (α_n, β_n) is not globally stable and undergoes a Hopf bifurcation.

Bernoulli-Kermack-McKendrick epidemic model with age of infection

Global stability for age of infection structured epidemic systems have also been studied. A typical system reads as

$$\begin{cases} S'(t) = \Lambda - \gamma S(t) - S(t)\int_0^\infty \beta(a)i(t,a)da, \ t > 0, \\ (\partial_t + \partial_a)\, i(t,a) = -\mu(a)i(t,a), \ t > 0, \ a > 0, \\ i(t,0) = S(t)\int_0^\infty \beta(a)i(t,a)da, \ t > 0. \end{cases}$$

Here $S(t)$ denotes the density of suceptibles at time t while $i(t,a)$ denotes the density of infected individuals at time t, infected since a period of time $a > 0$.

Here the new infected come into the infected compartment at $a = 0$, that is at the boundary condition for the equation of the infected. This system has deeply studied in [238]. In particular the global stability of stationary states have been proved by using the construction of suitable Lyapunov functional. Note that the methodology developed in [238] is rather closed to the one recalled in the previous section including the construction of an interior attractor on which a specific Lyapunov function of Volterra type is well defined.

We also refer to [237] and the reference therin for the study of other epidemic systems with age of infection.

Chapter 3
Bifurcations

This chapter is concerned with bifurcation theory, that roughly consists in the study of the topological changes of the solutions of some equations as some parameters are varying. Along this chapter we shall mainly focus on bifurcation arising when a simple eigenvalue crosses zero, typically yielding the appearance (or the disappearance) of branches of solutions. The results presented below are mostly based on the implicit function theorem, recalled in Section 3.1. This theorem is used to study local branches of the solutions and is applies to state and prove the so-called Hopf-bifurcation theorem for systems of ordinary differential equations.

3.1 Implicit Function Theorem

Let $(X, \|\cdot\|)$ be a Banach space. Throughout this chapter the closed ball centered at $x \in X$ with radius $\delta \geq 0$, denoted by $\overline{B}_X(x, \delta)$, is defined by

$$\overline{B}_X(x, \delta) := \{y \in X : \ \|x - y\| \leq \delta\}$$

while the open ball centered at $x \in X$ with radius $\delta > 0$ is denoted by $B_X(x, \delta)$ and defined by

$$B_X(x, \delta) := \{y \in X : \ \|x - y\| < \delta\}.$$

We start this chapter by recalling the following result that will be used to prove the implicit function theorem below.

Theorem 3.1 *Let $(X, \|\cdot\|)$ be a Banach space. Let $0 \leq \kappa < 1$ and $\delta > 0$. Then the following properties hold*

(i) **(Global inverse)** *If $T : X \to X$ is a $\kappa-$ Lipschitz continuous map, then the map $I - T$ is a invertible and $(I - T)^{-1}$ is $(1 - \kappa)^{-1}$–Lipschitz continuous.*

(ii) **(Local inverse)** *If $T : \overline{B}_X(0, \delta) \to X$ is a κ–Lipschitz continuous map and $\|T(0)\| < \delta(1 - \kappa)$, then*

$$\overline{B}_X(0, \varrho) \subset (I - T)\left(\overline{B}_X(0, \delta)\right),$$

A. Ducrot et al., *Differential Equations and Population Dynamics II*,
Lecture Notes on Mathematical Modelling in the Life Sciences,
https://doi.org/10.1007/978-3-031-91409-6_3

when

$$\varrho := (1-\kappa)\delta - \|T(0)\|.$$

Moreover there exists a unique map $S : \overline{B}_X(0,\varrho) \to \overline{B}_X(0,\delta)$ *which is* $(1-\kappa)^{-1}$*–Lipschitz continuous and that satisfies*

$$(I-T)Sy = y, \ \forall y \in \overline{B}_X(0,\varrho). \tag{3.1}$$

Proof **Proof of** (i)**:** Let $y \in X$. Then we consider the fixed point problem

$$x = Tx + y. \tag{3.2}$$

Note that the Banach fixed point theorem applies to the κ–contraction map $\Phi : X \to X$ defined by $\Phi(x) = Tx+y$ (namely by using the iteration procedure $x_{n+1} = Tx_n+y$). Hence we deduce that (3.2) has a unique solution Sy satisfying

$$Sy = TSy + y \Leftrightarrow (I-T)\,Sy = y.$$

We conclude, $(I-T)$ is onto (i.e. $(I-T)(X) = X$), that is, for each $y \in X$, there exists a unique $x = S(y)$ satisfying $(I-T)Sy = y$.

Next we observe that, for each $x \in X$, we can fix $y = x - Tx$, by uniqueness of the fixed point problem (3.2), we deduce

$$Sy = x.$$

So the map $S : X \to X$ is onto (i.e. $S(X) = X$).

Moreover, by applying S on both sides of $(I-T)Sy = y$, and by setting $z = Sy$, we deduce that

$$S(I-T)z = z, \forall z \in X.$$

We conclude that $I-T : X \to X$ is invertible, and its inverse is S. Next by (3.2), we observe that we have, for any $y, \hat{y} \in X$,

$$Sy - S\hat{y} = TSy + y - (TS\hat{y} + \hat{y}),$$

thus

$$\|Sy - S\hat{y}\| \le \kappa\|Sy - S\hat{y}\| + \|y-\hat{y}\| \ \Leftrightarrow\ \|Sy - S\hat{y}\| \le (1-\kappa)^{-1}\|y-\hat{y}\|,$$

that completes the proof of (i).

Proof of (ii)**:** Let us fix $y \in \overline{B}_X(0,\varrho)$, and consider the map $\Phi : \overline{B}_X(0,\delta) \to X$ given by

$$\Phi(x) = Tx + y, \ \forall x \in \overline{B}_X(0,\delta).$$

Then for each $x \in \overline{B}_X(0,\delta)$, one has

$$\begin{aligned}\|\Phi(x)\| &\leq \|Tx\| + \|y\| \\ &\leq \|Tx - T(0)\| + \|T(0)\| + \varrho \\ &\leq \kappa\|x\| + \|T(0)\| + \varrho.\end{aligned}$$

Hence

$$\|\Phi(x)\| \leq \kappa\delta + \|T(0)\| + \varrho = \kappa\delta + \|T(0)\| + (1-\kappa)\delta - \|T(0)\| = \delta.$$

Therefore we deduce that

$$\Phi\left(\overline{B}_X(0,\delta)\right) \subset \overline{B}_X(0,\delta).$$

Moreover, since Φ is κ–Lipschitz continuous, there exists a unique $x \in \overline{B}_X(0,\delta)$ such that

$$\Phi(x) = x \Leftrightarrow x - T(x) = y.$$

Denoting by $S : \overline{B}_X(0,\varrho) \to \overline{B}_X(0,\delta)$ the map defined by the resolution of the above equation (i.e. $Sy = x$), one may observe that it satisfies (3.1). Moreover, by using the same argument as for the proof for (i), S is also $(1-\kappa)^{-1}$–Lipschitz continuous. □

We now turn to the statement of the implicit function theorem, that will be extensively used along this chapter.

Theorem 3.2 (Implicit function theorem) *Let X, Y and Z be three given Banach spaces. Let $U \subset X$ and $V \subset Y$ be two open sets. We assume that $F : U \times V \to Z$ is continuous function, which is differentiable with respect to the second variable y and the map $(x,y) \to \partial_y F(x,y) \in \mathcal{L}(Y,Z)$ is continuous. We assume that there exists $(x_0,y_0) \in U \times V$ such that*

$$F(x_0,y_0) = 0_Z.$$

We further assume that $\partial_y F(x_0,y_0) \in \mathcal{L}(Y,Z)$ is one to one and onto and its inverse is a bounded linear operator, that is

$$\partial_y F(x_0,y_0)^{-1} \in \mathcal{L}(Z,Y).$$

Then there exist $r > 0$, $\delta > 0$ with $\overline{B}_X(x_0,r) \subset U$ and $\overline{B}_Y(y_0,\delta) \subset V$ and there exists a unique continuous map $S : \overline{B}_X(x_0,r) \to \overline{B}_Y(y_0,\delta)$ with $Sx_0 = y_0$ such that

$$F(x,Sx) = 0_Z,\ \forall x \in \overline{B}_X(x_0,r).$$

Proof Without loss of generality we can assume that $x_0 = 0_X$ and $y_0 = 0_Y$. We set $L = \partial_y F(0,0) \in \mathcal{L}(Y,Z)$. We observe that

$$F(x,y) = 0_Z \Leftrightarrow Ly - F(x,y) = Ly \Leftrightarrow y - L^{-1}F(x,y) = y \Leftrightarrow y - (y - L^{-1}F(x,y)) = 0_Y,$$

and we set

$$\Lambda(x,y) := y - L^{-1}F(x,y).$$

So we will apply the local invertibility result Theorem 3.1-(ii) to find the zeros of

$$y - \Lambda(x, y) = 0_Y.$$

So, let us prove that Λ satisfies the assumptions of (ii) in Theorem 3.1. We observe that

$$\partial_y \Lambda(x, y) = I - L^{-1} \partial_y F(x, y).$$

One deduces that

$$\partial_y \Lambda(0_X, 0_Y) = I - L^{-1} \partial_y F(0_X, 0_Y) = 0_{\mathcal{L}(Y)}.$$

Let $\kappa \in (0, 1)$ be fixed. Since the map $(x, y) \to \partial_y F(x, y)$ is continuous, we can find $\delta > 0$ and $r_0 > 0$ such that

$$\|\partial_y \Lambda(x, y)\|_{\mathcal{L}(Y)} \le \kappa, \ \forall y \in \overline{B}_Y(0_Y, \delta), \ \forall x \in \overline{B}_X(0_X, r_0).$$

By using the formula (obtained by taking the derivative of the function $s \to \Lambda(x, y + s(\hat{y} - y))$)

$$\Lambda(x, \hat{y}) - \Lambda(x, y) = \int_0^1 \partial_y \Lambda(x, y + s(\hat{y} - y))(\hat{y} - y) \mathrm{d}s,$$

we deduce that

$$\|\Lambda(x, \hat{y}) - \Lambda(x, y)\| \le \kappa \|\hat{y} - y\|, \ \forall y, \hat{y} \in \overline{B}_Y(0_Y, \delta), \ \forall x \in \overline{B}_X(0_X, r_0).$$

Moreover one has

$$\Lambda(0_X, 0_Y) = 0_Y - L^{-1} F(0_X, 0_Y) = 0_Y,$$

so that due to the continuity of F, one can find $r \in (0, r_0)$ such that

$$\|\Lambda(x, 0_Y)\| \le (1 - \kappa)\delta, \ \forall x \in \overline{B}_X(0_X, r).$$

Now let $x \in \overline{B}_X(0_X, r)$ be given. By applying Theorem 3.1-(ii) to the map $T : \overline{B}_Y(0_Y, \delta) \to Y$, given by $Ty = \Lambda(x, y)$, it follows that for each $x \in \overline{B}_X(0_X, r)$, there exists a unique $y \in B_Y(0_Y, \delta)$, such that

$$y - \Lambda(x, y) = 0_Y.$$

Set

$$Sx = y.$$

By using the uniqueness result in Theorem 3.1-(ii), we deduce that $Sx_0 = y_0$. It remains to prove the continuity of S. To that aim we fix $x \in \overline{B}_X(0_X, r)$ and we vary $\widehat{x} \in \overline{B}_X(0_X, r)$, and we observe that

$$S(\widehat{x}) - S(x) = \Lambda(\widehat{x}, S(\widehat{x})) - \Lambda(\widehat{x}, S(x)) + \Lambda(\widehat{x}, S(x)) - \Lambda(x, S(x)),$$

hence

$$\|S(\widehat{x}) - S(x)\| \leq \kappa \|S(\widehat{x}) - S(x)\| + \|\Lambda(\widehat{x}, S(x)) - \Lambda(x, S(x))\|.$$

Therefore

$$\|S(\widehat{x}) - S(x)\| \leq (1-\kappa)^{-1} \|\Lambda(\widehat{x}, S(x)) - \Lambda(x, S(x))\|,$$

and by using the continuity of the map $\widehat{x} \longrightarrow \Lambda(\widehat{x}, S(x))$, we deduce that $S(\widehat{x}) \longrightarrow S(x)$ as $\widehat{x} \longrightarrow x$. □

3.2 Bifurcating Branch of Equilibria: Finite Dimensional Case

In this section, we discuss some results to describe local branches of solutions for some finite dimensional equations. We start by discussing two simple examples.

Example 3.3 (Logistic equation) The equilibria of the logistic equation satisfy a quadratic equation of the form

$$u(\lambda \kappa - u) = 0$$

for some given and fixed parameter $\kappa \neq 0$ while $\lambda \in \mathbb{R}$ is a varying parameter. Then the solutions of the above equation are given by the two branches of solutions

$$u = 0 \text{ and } u = \lambda \kappa.$$

The two branches of solutions cross each other at $\lambda = 0$. This special point is called a bifurcation point for the branch of trivial solution $u = 0$ in the plane (λ, u).

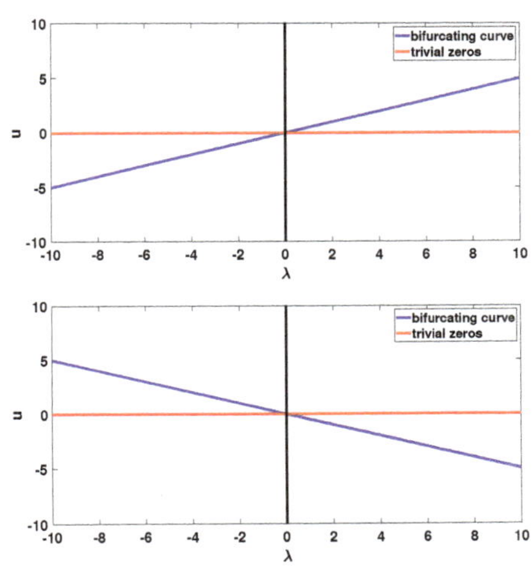

Fig. 3.1: *The figure on the top and the figure on the bottom correspond respectively to* $\kappa = 0.5$ *and* $\kappa = -0.5$.

Example 3.4 (Bernoulli-Verhulst equation) We consider now the equation of the form

$$U(\lambda\,\kappa - U^2) = 0$$

with $\kappa \neq 0$. Then the solution are given by

$$U = 0 \text{ and } \lambda\,\kappa = U^2.$$

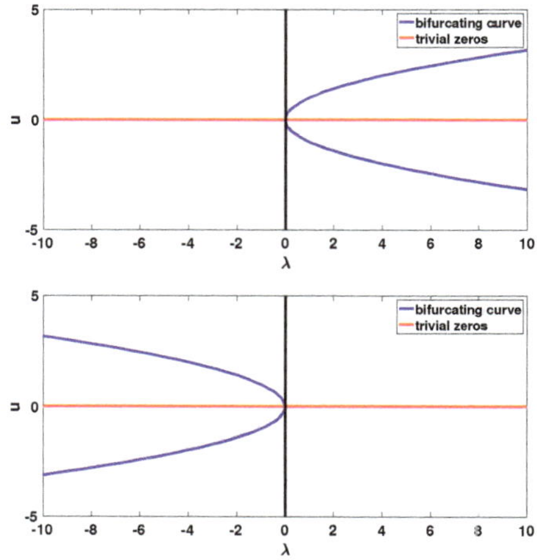

Fig. 3.2: *The figure on the top and the figure on the bottom correspond respectively to* $\kappa = 1$ *and* $\kappa = -1$.

Theorem 3.5 (Bifurcating Branch of Equilibria) *Let* $e \in \mathbb{R}^n$ *with* $e \neq 0_{\mathbb{R}^n}$. *Let* I *be an open interval in* $\mathbb{R}$ *and let* $B_{\mathbb{R}^n}(0, r)$ *be the open ball of* $\mathbb{R}^n$ *centered on* $0_{\mathbb{R}^n}$ *with radius* $r > 0$. *Let* $\lambda_0 \in I$ *and* $F : I \times B_{\mathbb{R}^n}(0_{\mathbb{R}^n}, r) \to \mathbb{R}^n$ *a two times continuously differentiable map satisfying*

(i) $F(\lambda, 0_{\mathbb{R}^n}) = 0_{\mathbb{R}^n}, \ \forall \lambda \in I$;

(ii) $\mathrm{N}(\partial_x F(\lambda_0, 0_{\mathbb{R}^n})) = \mathbb{R}e$, *where* $\mathbb{R}e := \{\lambda e : \lambda \in \mathbb{R}\}$;

(iii) $\partial_\lambda \partial_x F(\lambda_0, 0_{\mathbb{R}^n})e \notin \mathrm{R}(\partial_x F(\lambda_0, 0_{\mathbb{R}^n}))$.

Then there exist a constant $\delta > 0$ *and a continuous function* $(\lambda^*, x^*) : (-\delta, \delta) \subset \mathbb{R} \to I \times \mathbb{R}^n$ *satisfying*

$$\lambda^*(0_{\mathbb{R}}) = \lambda_0, x^*(0_{\mathbb{R}}) = 0_{\mathbb{R}^n} \text{ and } x^*(s) \neq 0_{\mathbb{R}^n}, \ \forall s \neq 0_{\mathbb{R}},$$

and

$$F\left(\lambda^*(s), x^*(s)\right) = 0_{\mathbb{R}^n}, \ \forall s \in (-\delta, \delta).$$

Moreover the curve $s \to x^*(s)$ *is tangent to* $\mathbb{R}e$ *at* $s = 0$, *that is to say that*

$$\lim_{s\to 0}\frac{x^*(s)}{s}=e.$$

Furthermore, there is a neighborhood of $(\lambda_0,0)$ *such that any zero of* F *either belongs to this curve or has the form* $(\lambda,0)$.

Proof Let Z denotes a complement space of $\mathbb{R}e$ in $\mathbb{R}^n$, namely

$$\mathbb{R}e\oplus Z=\mathbb{R}^n.$$

Fix $\delta>0$ and $\varrho>0$ such that $\delta(\|e\|+\varrho)\le r$ and let $G:(-\delta,\delta)\times I\times B_Z(0,\varrho)\to\mathbb{R}^n$ be the map defined by

$$G(s,\lambda,z)=\begin{cases}\dfrac{F(\lambda,s(e+z))}{s}, & \text{if } s\neq 0,\\ \partial_x F(\lambda,0)(e+z), & \text{if } s=0,\end{cases}$$

whenever $(s,\lambda,z)\in(-\delta,\delta)\times I\times B_Z(0,\varrho)$.

Since F is of class C^2, it readily follows that G is of class C^1. Next note that one has $G(0,\lambda_0,0)=0$ and

$$\partial_{(\lambda,z)}G(0,\lambda_0,0)(\lambda,z)=\partial_\lambda G(0,\lambda_0,0)\lambda+\partial_z G(0,\lambda_0,0)z,$$

hence

$$\partial_{(\lambda,z)}G(0,\lambda_0,0)(\lambda,z)=\lambda\,\partial_\lambda\partial_x F(\lambda_0,0)e+\partial_x F(\lambda_0,0)z.$$

Now let us prove that the operator $L:(\lambda,z)\to\partial_{(\lambda,z)}G(0,\lambda_0,0)(\lambda,z)$ is invertible from $\mathbb{R}\times Z$ into $\mathbb{R}^n$. Note that since $\dim(Z)=n-1$, it is sufficient to prove that $\mathrm{N}(L)=0$. To that aim we argue by contradiction by assuming that $\mathrm{N}(L)\neq 0$. Let $(\lambda,z)\in\mathrm{N}(L)\setminus\{0\}$ be given. Then one has

$$\lambda\,\partial_\lambda\partial_x F(\lambda_0,0)e=-\partial_x F(\lambda_0,0)z\in\mathrm{R}(\partial_x F(\lambda_0,0)).$$

Due to (iii) this implies that $(\lambda,z)=(0,0)$, a contradiction that ensures the invertibility of L.

Now since

$$G(0,\lambda_0,0)=\partial_x F(\lambda_0,0)e=0,$$

the implicit function theorem applies to the function G (with $x=s$ and $y=(\lambda,z)$) and we deduce that there exist $\delta_0>0$ sufficiently small and a continuous map $(\lambda^*,z^*):(-\delta_0,\delta_0)\to I\times Z$ with $\lambda^*(0)=\lambda_0$ and $z^*(0)=0$ and satisfying

$$F(\lambda^*(s),s(e+z^*(s)))=0,\ \forall s\in(-\delta_0,\delta_0).$$

The result follows by setting $x^*(s)=s(e+z^*(s))$. □

3.3 Application to the Scalar Generalized Logistic Equation

Consider a one dimensional logistic equation

$$U'(t) = \lambda \kappa U(t) - U(t)^m, \text{ for } t \geq 0, \; U(0) = U_0 \geq 0,$$

where $\kappa \in \mathbb{R} \setminus \{0\}$ is a fixed positive parameter, while $\lambda \in \mathbb{R}$ is a varying parameter and $m \geq 2$ is an integer. Then the equilibria correspond to the zeros of the function

$$F(\lambda, U) = \lambda \kappa U - U^m.$$

First we have

$$F(\lambda, 0) = 0, \; \forall \lambda \in \mathbb{R},$$

while

$$\partial_U F(\lambda, U) = \lambda \kappa - m U^{m-1},$$

hence

$$\partial_U F(\lambda, 0) = \lambda \kappa.$$

Set $\lambda_0 = 0$ so that we get

$$\partial_U F(0, 0) = 0.$$

If we regard the partial derivative as a linear map $\partial_U F(0,0) : U \to \partial_U F(0,0) U$ from $\mathbb{R}$ into itself, one has

$$\dim(\mathrm{N}(\partial_U F(0,0))) = \dim(\mathbb{R}) = 1 \text{ and } \mathrm{R}\,(\partial_U F(0,0)) = \{0_{\mathbb{R}}\}.$$

Furthermore one also has

$$\partial_\lambda \partial_U F(0,0) = \kappa,$$

and the linear map $e \to \partial_\lambda \partial_U F(0,0) e \in \mathcal{L}(\mathbb{R}, \mathbb{R})$ satisfies for each $e \neq 0$,

$$\partial_\lambda \partial_U F(0,0) e \notin \mathrm{R}(\partial_U F(0,0)).$$

So we can apply Theorem 3.5 and we find a branch of equilibrium. Actually, this branch of equilibrium is obtained as

$$\lambda(s) = s^{n-1}/\kappa^{-1} \text{ and } U(s) = s.$$

Remark 3.6 From this example, the condition (iii) of Theorem 3.5 corresponds to $\kappa \neq 0$. We can see that this assumption is needed.

Remark 3.7 The cases $m = 2$ and $m = 3$ provide different types of bifurcating branches. We may observe that Theorem 3.5 is not informing us about the shape of the bifurcating branches.

3.4 Application to the $n-$Dimensional Logistic Equation

Consider the $n-$dimensional logistic equation

$$\begin{cases} U_i'(t) = \sum_{j=1}^n A_{ij}(\lambda)U_j(t) - \sum_{j=1}^n \kappa_{ij}U_j(t)U_i(t),\ t \geq 0, \\ U_i(0) = U_{0i} \geq 0. \end{cases} \tag{3.3}$$

We assume that

$$\kappa_{ij} > 0,\ \text{for all } i, j = 1, \dots, n,$$

and

$$\lambda \in (\lambda_0 - \delta_0, \lambda_0 + \delta_0),\ \text{for some } \delta_0 > 0 \text{ and } \lambda_0 \in \mathbb{R}.$$

Assumption 3.8 We assume that the function $\lambda \to A(\lambda) = \left(A_{ij}(\lambda)\right)_{i,j=1,\dots,n}$ is twice continuously differentiable and satisfies the following properties

(i) There exists a vector $e \in \mathbb{R}^n$ such that $e \gg 0$ and $\mathrm{N}(A(\lambda_0)) = \mathbb{R}e$;

(ii) $\partial_\lambda A(\lambda_0)e \notin \mathrm{R}(A(\lambda_0))$.

Now the equilibria of the $n-$dimensional logistic equation above correspond to the zeros of the map $F : I \times \mathbb{R}^n \to \mathbb{R}^n$, with $I = (\lambda_0 - \delta_0, \lambda_0 + \delta_0)$, given by

$$F_i(\lambda, U)_i = \sum_{j=1}^n A_{ij}(\lambda)U_j - \sum_{j=1}^n \kappa_{ij}U_jU_i,\ i = 1, \dots, n. \tag{3.4}$$

Next the following result holds.

Theorem 3.9 *Let Assumption 3.8 be satisfied. Then there exist some constant $\delta > 0$ and a continuous function $(\lambda^*, U^*) : (-\delta, \delta) \to I \times \mathbb{R}^n$ satisfying*

$$\lambda^*(0) = \lambda_0 \text{ and } U^*(0) = 0,$$
$$U^*(s) \neq 0, \forall s \in (-\delta, \delta) \setminus \{0\},$$

and

$$F(\lambda^*(s), U^*(s)) = 0, \forall s \in (-\delta, \delta).$$

Moreover for $|s|$ small enough

$$\begin{cases} U^*(s) \gg 0,\ \text{if } s > 0, \\ U^*(s) \ll 0,\ \text{if } s < 0. \end{cases} \tag{3.5}$$

Furthermore, there is a neighbourhood of $(\lambda_0, 0)$ such that any zero of F either belongs to this curve or belongs to the curve $(\lambda, 0)$.

Proof In order to apply the Theorem 3.5 it is sufficient to observe that

$$\partial_x F(\lambda_0, 0) = A(\lambda_0).$$

Moreover since the curve $s \to U^*(s)$ given by Theorem 3.5 is tangent to $\mathbb{R}e$ at $s = 0$ and since $e \gg 0$ we deduce (3.5). □

Example 3.10 If we consider for example the matrix

$$A(\lambda) = \begin{pmatrix} -\pi + d_1\lambda & \beta \\ \pi & -\beta + d_2\lambda \end{pmatrix}$$

where $\pi > 0$, $d_1 > 0$, $d_2 > 0$ and $\beta > 0$. Then for $\lambda = 0$, we have $\mathrm{N}(A(0)) = \mathbb{R}e$ with

$$e = \begin{pmatrix} \beta \\ \pi \end{pmatrix}.$$

Moreover

$$A(\lambda)e = \lambda \begin{pmatrix} d_1\beta \\ d_2\pi \end{pmatrix} = \lambda De,$$

where $D := \mathrm{Diag}(d_1, d_2)$ and

$$\partial_\lambda A(\lambda)e = De.$$

Moreover

$$A(0)\begin{pmatrix} \alpha_1 \\ \alpha_2 \end{pmatrix} = \alpha_1\pi \begin{pmatrix} -1 \\ 1 \end{pmatrix} - \alpha_2\beta \begin{pmatrix} -1 \\ 1 \end{pmatrix} = (\alpha_1\pi - \alpha_2\beta)\begin{pmatrix} -1 \\ 1 \end{pmatrix}.$$

Since $d_1\beta > 0$ and $d_2\pi > 0$, we deduce that

$$De \neq \gamma \begin{pmatrix} -1 \\ 1 \end{pmatrix}, \forall \gamma \in \mathbb{R},$$

which implies that

$$\partial_\lambda A(0)e \notin \mathrm{R}(A(0)).$$

For this two dimensional matrix example, the zeros of the corresponding two dimensional logistic system reads as the following algebraic system of equations

$$\begin{cases} 0 = (-\pi + d_1\lambda)U_1 + \beta U_2 - (\kappa_{11}U_1 + \kappa_{12}U_2)U_1 \\ 0 = \pi U_1 + (-\beta + d_2\lambda)U_2 - (\kappa_{21}U_1 + \kappa_{22}U_2)U_2 \end{cases}$$

for which we can no longer compute the equilibria explicitly. Therefore Theorem 3.9 is particularly useful when no simplification arises.

3.5 Bifurcating Branch of Equilibria: Infinite Dimensional Case

In order to prove an infinite dimensional version of Theorem 3.5 we will need the following result (see Brezis [37, Corollary 2.7 p. 35]).

Theorem 3.11 *Let X and Y be two Banach spaces and let L be a continuous linear operator from X into Y that is bijective, namely one-to-one and onto. Then L^{-1} is also continuous from Y into X.*

The infinite dimensional version of Theorem 3.5 reads as follows.

Theorem 3.12 *[Bifurcating Branch of Equilibria] Let I be an open interval in $\mathbb{R}$ and let X and Y be two Banach spaces. Let $\lambda_0 \in I$, $r > 0$ and $F : I \times B_X(0, r) \to Y$ a twice continuously differentiable map satisfying*

(i) $F(\lambda, 0_X) = 0_Y$, $\forall \lambda \in I$;

(ii) $\dim(\mathrm{N}(\partial_x F(\lambda_0, 0_X))) = 1$;

(iii) $\partial_\lambda \partial_x F(\lambda_0, 0_X)e \notin \mathrm{R}(\partial_x F(\lambda_0, 0_X))$ *for some* $e \in \mathrm{N}(\partial_x F(\lambda_0, 0_X)) \setminus \{0\}$.

(iv) $\operatorname{codim}\big(\mathrm{R}(\partial_x F(\lambda_0, 0_X))\big) = 1$ *which is equivalent to say that there exists $v \in Y$ such that*

$$Y = \mathbb{R}v \oplus \mathrm{R}(\partial_x F(\lambda_0, 0_X))$$

whenever $v \notin \mathrm{R}(\partial_x F(\lambda_0, 0_X))$ (which implies that $v \neq 0$).

Then the conclusions of Theorem 3.5 hold.

Proof Let Z be a closed complement space of $\mathbb{R}e$ in X (such a subspace exists due to the Analytic form of the Hahn-Banach theorem, see [37, Example p. 38])). That is to say that

$$X = \mathbb{R}e \oplus Z.$$

Consider the map $B : \mathbb{R} \times Z \to Y$ given by

$$B(\lambda, z) = \lambda \hat{e} + L_0 z,$$

wherein we have set $\hat{e} := \partial_\lambda \partial_x F(\lambda_0, 0)e$ and $L_0 = \partial_x F(\lambda_0, 0)$. Then in order to prove the theorem, it is sufficient to prove that B is a bijection from $\hat{X} := \mathbb{R} \times Z$ onto Y. To do so we first observe that since $\operatorname{codim} \mathrm{R}(L_0) = 1$ and $\hat{e} \notin \mathrm{R}(L_0)$ we have

$$Y = \mathbb{R}\hat{e} \oplus \mathrm{R}(L_0).$$

Now since $X = \mathbb{R}e \oplus Z$ one deduces that $\mathrm{R}(L_0) = L_0(Z)$ and B is surjective. Next by using again the fact that $\hat{e} \notin \mathrm{R}(L_0)$ and $z \in Z$ we deduce that

$$B(\lambda, z) = 0 \iff \lambda \hat{e} + L_0 z = 0 \iff \lambda = 0_{\mathbb{R}} \text{ and } z = 0_X.$$

The invertiblity is B follows from Theorem 3.11. Finally the result follows by using the same arguments as in the proof of Theorem 3.5. □

3.6 Bifurcating Branch of periodic orbit: Hopf bifurcation Theorem

3.6.1 Example of the Poincaré normal form and a first Hopf bifurcation Theorem

In order to understand the idea of the Hopf bifurcation theorem (see Hopf [162]), we first consider the so called Poincaré normal form. This normal form was introduced by Poincaré in his memoir on celestial mechanics [274] in 1893. This system is the following

$$\begin{pmatrix} x'(t) \\ y'(t) \end{pmatrix} = \begin{pmatrix} \alpha & -\omega \\ \omega & \alpha \end{pmatrix} \begin{pmatrix} x(t) \\ y(t) \end{pmatrix} + \kappa \left(x(t)^2 + y(t)^2 \right) \begin{pmatrix} x(t) \\ y(t) \end{pmatrix}, \tag{3.6}$$

where the bifurcation parameter α varies from negative values to positive values, while the parameters

$$\omega \neq 0 \text{ and } \kappa \neq 0$$

are fixed.

Complex number reformulation: Set

$$z(t) = x(t) + i\, y(t).$$

Then

$$z'(t) = x'(t) + i\, y'(t) = \alpha\,(x + i\, y) - \omega y + \omega\, i\, x + \kappa\, |z|^2 z = \alpha\,(x + i\, y) + i\omega\,(x + i\, y) + \kappa\, |z|^2 z$$

and we obtain the Poincaré normal form

$$z'(t) = \lambda\, z + \kappa\, |z|^2 z, \text{ and } z(0) = z_0 \in \mathbb{C}, \tag{3.7}$$

where

$$\lambda = \alpha + i\omega.$$

The equation (3.7) is nothing but a complex-valued Bernoulli-Verhulst equation. In order to derive an explicit formula for the solution we first set

$$\widehat{z}(t) = e^{-i\,\omega t} z(t).$$

We obtain

$$\widehat{z}'(t) = -i\,\omega \widehat{z} + e^{-i\,\omega t} z'(t),$$

it follows that

$$\widehat{z}'(t) = \alpha \widehat{z} + \kappa\, |\widehat{z}|^2 \widehat{z}, \text{ and } z(0) = z_0 \in \mathbb{C}. \tag{3.8}$$

We are now in position to use the explicit formula for the Bernoulli-Verhulst equation, and we deduce that the solution of equation (3.7) is given by the explicit formula

$$z(t) = \frac{e^{\lambda t} z_0}{\left(1 - 2\kappa \int_0^t e^{2\,\mathrm{Re}(\lambda)\,\sigma} |z_0|^2 d\sigma\right)^{\frac{1}{2}}}. \tag{3.9}$$

Moreover if we set $\rho(t) = |z|^2$, then we obtain

$$\rho(t)' = (z\overline{z})' = z'\overline{z} + z\overline{z}' = 2\,\mathrm{Re}(\lambda)|z|^2 + 2\,\kappa\,|z|^4$$

which is a logistic equation

$$\rho' = 2\,\mathrm{Re}(\lambda)\,\rho + 2\,\kappa\,\rho^2. \tag{3.10}$$

The equilibrium for equation (3.10)

$$\overline{\rho} = -\frac{\mathrm{Re}(\lambda)}{\kappa}.$$

So the equilibrium $\overline{\rho}$ is strictly positive if and only if $\mathrm{Re}(\lambda)\,\kappa < 0$. Moreover in order to understand the stability of the equilibrium $\overline{\rho}$, we can make the change of variable $x = \rho - \overline{\rho}$ and we obtain

$$x' = 2\kappa(x + \overline{\rho})x.$$

Now the equilibrium 0 of the above equation is stable if and only if $\kappa < 0$ and unstable if $\kappa > 0$. We summarize this in the following theorem.

Theorem 3.13 *The system* (3.6) *has a periodic orbit if and only if*

$$\mathrm{Re}(\lambda)\,\kappa < 0. \tag{3.11}$$

Moreover the periodic orbit (when it exists) is unique. Furthermore, the periodic orbit (when it exists) is locally asymptotically stable only if $\kappa < 0$ *and unstable if* $\kappa > 0$.

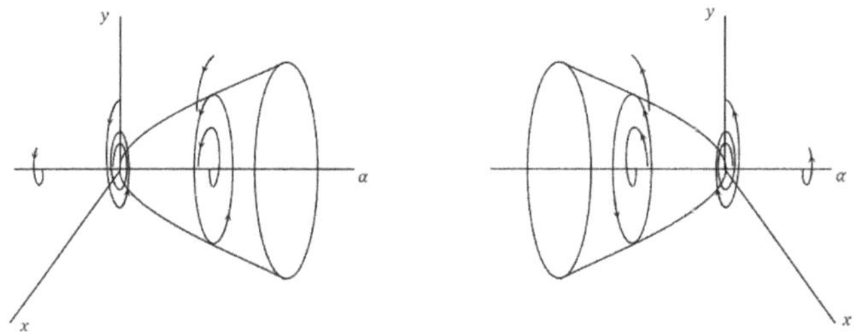

Fig. 3.3: *When $\kappa < 0$, we use $\alpha = \mathrm{Re}(\lambda)$ as a bifurcation parameter, and when α passes from negative to positive through 0, a stable periodic orbit is appearing. Then we say that the system has undergone a supercritical Hopf bifurcation (see the figure on the left). When $\kappa > 0$, we use $\alpha = \mathrm{Re}(\lambda)$ as a bifurcation parameter, when α passes from positive to negative through 0, an unstable periodic orbit is appearing. Then we say that the system has undergone a subcritical Hopf bifurcation (see the figure on the right).*

3.6.2 Hopf Bifurcation Theorem

Let I be an open interval in $\mathbb{R}$ containing 0 and $r > 0$ be fixed. Let $F : I \times B_{\mathbb{R}^n}(0, r) \to \mathbb{R}^n$ be a continuously differential map. Consider the ordinary differential equation

$$U'(t)x = F(\mu, U(t)x), \ \forall t \geq 0, \ U(0)x = x \in \mathbb{R}^n. \tag{3.12}$$

Before stating the main result of this section, we first describe the assumptions the will be needed.

Assumption 3.14 We assume that $(\mu, x) \to F(\mu, x)$ is twice continuously differentiable on $I \times B_{\mathbb{R}^n}(0, r)$ into $\mathbb{R}^n$ and satisfies the following properties

(i) $F(\mu, 0_{\mathbb{R}^n}) = 0_{\mathbb{R}^n}, \ \forall \mu \in I$;

(ii) **(Transversality condition)** For each $\mu \in I$, there exists a pair of conjugated simple eigenvalues of $\partial_x F(\mu, 0)$ denoted by $\lambda(\mu)$ and $\overline{\lambda(\mu)}$ written as

$$\lambda(\mu) = \alpha(\mu) + i\omega(\mu), \text{ with } \alpha(\mu), \ \omega(\mu) \in \mathbb{R}$$

such that the map $\mu \to \lambda(\mu)$ is continuously differentiable,

$$\omega(0) > 0, \ \alpha(0) = 0 \text{ and } \frac{\mathrm{d}\alpha(0)}{\mathrm{d}\mu} \neq 0.$$

(iii) The only eigenvalues of $\partial_x F(0, 0)$ which are equal to $i\omega(0)k$ or $-i\omega(0)k$ for some integer $k \in \mathbb{N}$ are $\lambda(0), \overline{\lambda(0)}$. In other words, one has

$$\sigma(\partial_x F(0, 0)) \cap i\omega(0)\mathbb{Z} = \left\{\lambda(0), \overline{\lambda(0)}\right\}. \tag{3.13}$$

Remark 3.15 The condition (3.13) implies $0 \notin \sigma(\partial_x F(0,0))$.

Theorem 3.16 (Hopf bifurcation) *Let Assumption 3.14 be satisfied. Then there exist* $\varepsilon^* > 0$ *and continuous maps,* $\varepsilon \to \mu(\varepsilon)$ *from* $[0, \varepsilon^*)$ *to* $\mathbb{R}$, $\varepsilon \to x(\varepsilon)$ *from* $[0, \varepsilon^*)$ *to* $\mathbb{R}^n$, $\varepsilon \to \gamma(\varepsilon)$ *from* $[0, \varepsilon^*)$ *to* $\mathbb{R}$, *such that for each* $\varepsilon \in [0, \varepsilon^*)$ *there exists a* $\gamma(\varepsilon)$*-periodic function* u_ε *of class* C^1 *which is a solution of* (3.12) *with the parameter* $\mu = \mu(\varepsilon)$ *and with initial condition* $u_\varepsilon(0) = x(\varepsilon)$. *Moreover the branch of periodic orbit is bifurcating from* 0 *at* $\mu = 0$, *that is to say that*

$$\mu(0) = 0, \ x(0) = 0 \text{ and } x(\varepsilon) \neq 0, \ \forall \varepsilon \in (0, \varepsilon^*).$$

Furthermore one also has

$$\gamma(0) = 2\pi\omega(0).$$

3.7 Proof of the Hopf Bifurcation Theorem

Throughout this section we assume that Assumption 3.14 is satisfied and we aim at proving Theorem 3.16. Up to time rescaling we assume without loss of generality that $\omega(0) = 1$. Now to prove the Hopf bifurcation theorem above, we first rewrite the system under a more convenient form. Using a change of basis we can rewrite (3.12) under the following form

$$\begin{cases} \dot{x}_1(t) = B(\mu)x_1(t) + G_1(\mu, x_1(t), x_2(t)) \in \mathbb{R}^2, \\ \dot{x}_2(t) = C(\mu)x_2(t) + G_2(\mu, x_1(t), x_2(t)) \in \mathbb{R}^{n-2}, \end{cases} \tag{3.14}$$

with some initial data

$$(x_1(0), x_2(0)) \in \mathbb{R}^2 \times \mathbb{R}^{n-2}.$$

Moreover the following properties are satisfied:

(i) The functions $\mu \to B(\mu)$, $\mu \to C(\mu)$ are of class C^1;

(ii) The matrix $B(\mu)$ takes the following form $B(\mu) = \begin{pmatrix} \alpha(\mu) & \omega(\mu) \\ -\omega(\mu) & \alpha(\mu) \end{pmatrix}$, with $\alpha(0) = 0$, $\omega(0) = 1$ as well as the transversality condition $\alpha'(0) \neq 0$;

(iii) The matrix $I - e^{2\pi C(0)}$ is invertible from $\mathbb{R}^{n-2}$ into itself;

(iv) The functions G_i for $i = 1, 2$ are of class C^2;

(v) $G_i(\mu, 0, 0) = 0$, for $i = 1, 2$ and for all μ in a small neighboorhood of 0;

(vi) $\partial_{x_i} G_i(\mu, 0, 0) = 0$, for $i = 1, 2$ and for μ in a small neighboorhood of 0.

Remark 3.17 The property (iii) is a consequence of the assumption (3.13). Indeed, let P an invertible matrix such that

$$J = P^{-1}C(0)P = \begin{pmatrix} J_{\lambda_1} & 0 & \dots & 0 \\ 0 & J_{\lambda_2} & \ddots & \vdots \\ \vdots & \ddots & \ddots & 0 \\ 0 & \dots & 0 & J_{\lambda_m} \end{pmatrix}$$

is a Jordan decomposition of $C(0)$. We have

$$e^{2\pi J} = \begin{pmatrix} e^{2\pi J_{\lambda_1}} & 0 & \dots & 0 \\ 0 & e^{2\pi J_{\lambda_2}} & \ddots & \vdots \\ \vdots & \ddots & \ddots & 0 \\ 0 & \dots & 0 & e^{2\pi J_{\lambda_m}} \end{pmatrix},$$

and

$$x - e^{2\pi C(0)}x = 0 \Leftrightarrow x = e^{2\pi C(0)}x \Leftrightarrow P^{-1}x = e^{2\pi J}P^{-1}x.$$

Therefore

$$N\left(I - e^{2\pi C(0)}\right) = \{0\} \Leftrightarrow N\left(I - e^{2\pi J_{\lambda_i}}\right) = \{0\}, \forall i = 1, \dots, m \Leftrightarrow e^{2\pi\lambda_i} \neq 1, \forall i = 1, \dots, m.$$

The system (3.14) is equivalent to

$$\frac{dx(t)}{dt} = A(\mu)x + G(\mu, x), t > 0, x \in \mathbb{R}^n,$$

with some initial value $x(0) \in \mathbb{R}^n$ and

$$A(\mu) = \begin{pmatrix} B(\mu) & 0 \\ 0 & C(\mu) \end{pmatrix}, \ G(\mu, x) = \begin{pmatrix} G_1(\mu, x_1, x_2) & 0 \\ 0 & G_2(\mu, x_1, x_2) \end{pmatrix}.$$

In order to look for nontrivial $2\pi\gamma-$periodic solutions, we consider the rescaled function

$$\hat{x}(t) := x(\gamma t),$$

so that, omitting the hat for the sake of simplicity, the system becomes

$$x'(t) = \gamma A(\mu)x + \gamma G(\mu, x), \tag{3.15}$$

and we look for $2\pi-$periodic solution of the above system. A solution $t \to x(t)$ is on some time interval $[0, 2\pi]$ satisfies

$$x(t) = e^{\gamma A(\mu)t}x(0) + \gamma \int_0^t e^{\gamma A(\mu)(t-s)}G(\mu, x(s))ds, \ \forall t \in [0, 2\pi].$$

In order to investigate the existence of $2\pi-$periodic solution of the above integral integral equation, let us define the map $F : \mathbb{R}^2 \times C_{2\pi}([0, 2\pi]; \mathbb{R}^n) \to C_0([0, 2\pi]; \mathbb{R}^n)$ by

$$F(\gamma,\mu,x)(t) = x(t) - e^{\gamma A(\mu)t}x(0) - \gamma \int_0^t e^{\gamma A(\mu)(t-s)} G(\mu, x(s))\mathrm{d}s,$$

whenever $(\gamma,\mu,x) \in \mathbb{R}^2 \times C_{2\pi}([0,2\pi];\mathbb{R}^n)$.

The space $C_{2\pi}([0,2\pi];\mathbb{R}^n)$ which we define here, is the space continuous functions h satisfying $h(0) = h(2\pi)$, and the space $C_0([0,2\pi];\mathbb{R}^n)$ is the space of continuous functions h satisfying $h(0) = 0$. Both spaces $C_{2\pi}([0,2\pi];\mathbb{R}^n)$ and $C_0([0,2\pi];\mathbb{R}^n)$ are understood as Banach spaces endowed with the supremum norm on $[0,2\pi]$.

We now aim at investigating the zeros of the equation

$$F(\gamma,\mu,x) = 0,$$

for (γ,μ,x) close to $(1,0,0_{\mathbb{R}^n})$, using the implicit function theorem.

Due to the finite dimensional setting, one can calculate the following derivatives directly

$$\begin{aligned} F_x(1,0,0)(x)(t) &= x(t) - e^{A(0)t}x(0), \\ F_{\mu,x}(1,0,0)(x)(t) &= -A'(0)te^{A(0)t}x(0), \\ F_{\gamma,x}(1,0,0)(x)(t) &= -A(0)te^{A(0)t}x(0). \end{aligned} \tag{3.16}$$

Now using (iii), it is readily checked that $\mathrm{N}(F_x(1,0,0)) = \mathrm{span}\{u_0, u_1\}$, wherein the functions u_0 and u_1 are given by

$$u_0(t) = \begin{pmatrix} \sin(t) \\ \cos(t) \\ 0 \\ \vdots \\ 0 \end{pmatrix}, \quad u_1(t) = \begin{pmatrix} \cos(t) \\ -\sin(t) \\ 0 \\ \vdots \\ 0 \end{pmatrix}. \tag{3.17}$$

Define $X_1 := \mathrm{span}\{u_0, u_1\}$ and fix X_2 the complement space of X_1 as given by

$$X_2 = \{z \in C_{2\pi}([0,2\pi];\mathbb{R}^n) : \int_0^{2\pi} z(t)u_i(t)\mathrm{d}t = 0_{\mathbb{R}^n} \text{ for } i = 0,1\},$$

where the multiplication under the above integral is understood componentwise.

By using the fact that $L^2((0,2\pi);\mathbb{R}^n)$ is a Hilbert space (see Brezis [37, Remark 5 p. 137]) we know that

$$L^2((0,2\pi);\mathbb{R}^n) = X_1 \oplus X_1^{\perp},$$

where

$$X_1^{\perp} = \{z \in L^2((0,2\pi);\mathbb{R}^n) : \int_0^{2\pi} z(t)u_i(t)\mathrm{d}t = 0_{\mathbb{R}^n} \text{ for } i = 0,1\}.$$

It follows that $C_{2\pi}([0,2\pi];\mathbb{R}^n)$ inherits the properties of $L^2((0,2\pi);\mathbb{R}^n)$. Indeed, $v \in C_{2\pi}([0,2\pi];\mathbb{R}^n)$ implies $v \in L^2((0,2\pi);\mathbb{R}^n)$ and therefore

$$v = v_1 + w_1$$

where $v_1 \in X_1$ and $w_1 \in X_1^{\perp}$.

Moreover $X_1 \subset C_{2\pi}([0, 2\pi]; \mathbb{R}^n)$ so we deduce that

$$w_1 = v - v_1 \in C_{2\pi}([0, 2\pi]; \mathbb{R}^n).$$

We conclude that

$$C_{2\pi}([0, 2\pi]; \mathbb{R}^n) = X_1 \oplus X_2.$$

Let us now define the map $h : \mathbb{R}^3 \times X_2 \to C_0([0, 2\pi], \mathbb{R}^n)$ by

$$h(s, \gamma, \mu, z) = \begin{cases} \dfrac{F(\gamma, \mu, s(u_0 + z))}{s}, & \text{if } s \neq 0, \\ F_x(\gamma, \mu, 0)(u_0 + z), & \text{if } s = 0. \end{cases}$$

Then one can prove that h is of class C^1. One has $h(0, 1, 0, 0) = 0$ while the partial derivative with respect to (γ, μ, z) is given by

$$\partial_{(\gamma,\mu,z)} h(0, 1, 0, 0)(\hat{\gamma}, \hat{\mu}, \hat{z}) = F_x(1, 0, 0)\hat{z} + \hat{\gamma} F_{\gamma,x}(1, 0, 0)u_0 + \hat{\mu} F_{\mu,x}(1, 0, 0)u_0,$$

that more explicitly rewrites using (3.16) as

$$\partial_{(\gamma,\mu,z)} h(0, 1, 0, 0)(\hat{\gamma}, \hat{\mu}, \hat{z})(t) = \hat{z}(t) - e^{A(0)t}\hat{z}(0) - \hat{\gamma} t A(0) e^{A(0)t} u_0(0) - \hat{\mu} t A'(0) e^{A(0)t} u_0(0). \tag{3.18}$$

We now claim that

Claim 3.18 *The bounded linear operator*

$$\partial_{(\gamma,\mu,v)} h(0, 1, 0, 0) \in \mathcal{L}\left(\mathbb{R}^2 \times X_2, C_0([0, 2\pi]; \mathbb{R}^n)\right)$$

is invertible.

Proof (Proof of Claim 3.18) To prove this claim, let $y \in C_0([0, 2\pi]; \mathbb{R}^n)$ be given. Set

$$M := \partial_{(\gamma,\mu,v)} h(0, 1, 0, 0)$$

and let us investigate the equation

$$M(\hat{\gamma}, \hat{\mu}, \hat{z}) = y. \tag{3.19}$$

To do so, for any $z = (z_1, z_2)^T \in \mathbb{R}^n = \mathbb{R}^2 \times \mathbb{R}^{n-2}$ we define the projection maps $\pi_1 : \mathbb{R}^n \to \mathbb{R}^2$ and $\pi_2 : \mathbb{R}^n \to \mathbb{R}^{n-2}$ by $\pi_1 z = z_1$ and $\pi_2 z = z_2$.

We set $\hat{z}_i := \pi_i \hat{z}$ and $y_i = \pi_i y$ for $i = 1, 2$.

From (3.17) it follows that $\pi_2 u_i(0) = 0$, $i = 1, 2$. Therefore by applying π_1 and π_2 on both side of (3.19), the system becomes for all $t \in [0, 2\pi]$,

$$\begin{cases} \hat{z}_1(t) - e^{B(0)t}\hat{z}_1(0) - \hat{\gamma} t B(0) e^{B(0)t}\pi_1 u_0(0) - \hat{\mu} t B'(0) e^{B(0)t}\pi_1 u_0(0) = y_1(t), \\ \hat{z}_2(t) - e^{C(0)t}\hat{z}_2(0) = y_2(t). \end{cases} \tag{3.20}$$

Since $\hat{z}_2$ is 2π–periodic and $I - e^{2\pi C(0)}$ is invertible, one obtains

$$\hat{z}_2(t) = e^{tC(0)}\left(I - e^{2\pi C(0)}\right)^{-1} y_2(2\pi) + y_2(t),$$

so that $\hat{z}_2 \in C_{2\pi}\left([0, 2\pi], \mathbb{R}^{n-2}\right)$.

We now turn to the resolution of the first equation is (3.20). To do so, note that one has

$$tB(0)e^{tB(0)}\pi_1 u_0(0) = t(\cos t, -\sin t)^T,$$

while

$$tB'(0)e^{tB(0)}\pi_1 u_0(0) = t\alpha'(0)(\sin t, \cos t)^T + t\beta'(0)(\cos t, -\sin t)^T.$$

As a consequence, the first equation in (3.20) rewrites as finding $\hat{z}_1 \in \pi_1 X_2$, $(\hat{\gamma}, \hat{\mu}) \in \mathbb{R}^2$ such that

$$\hat{z}_1(t) - e^{B(0)t}\hat{z}_1(0) + c_1 t \begin{pmatrix} \sin t \\ \cos t \end{pmatrix} + c_2 t \begin{pmatrix} \cos t \\ -\sin t \end{pmatrix} = y_1(t). \tag{3.21}$$

with

$$c_1 = -\alpha'(0)\hat{\mu}, \quad c_2 = -\hat{\gamma} - \beta'(0)\hat{\mu}.$$

Since $\tilde{z}_1$ is 2π–periodic, taking $t = 2\pi$ in the above equation yields, since $e^{B(0)2\pi} = I$,

$$\tilde{z}_1(2\pi) - e^{B(0)2\pi}\tilde{z}_1(0) = \tilde{z}_1(0) - e^{B(0)2\pi}\tilde{z}_1(0) = 0.$$

and (3.21) becomes

$$c_1 2\pi \begin{pmatrix} 0 \\ 1 \end{pmatrix} + 2\pi c_2 \begin{pmatrix} 1 \\ 0 \end{pmatrix} = y_1(2\pi) =: \begin{pmatrix} y_1^1(2\pi) \\ y_1^2(2\pi) \end{pmatrix}.$$

Therefore we obtain $c_1 = y_1^2(2\pi)/2\pi$ and $c_2 = y_1^1(2\pi)/2\pi$. Since $\alpha'(0) \neq 0$, this allows to recover $\hat{\mu}$ and $\hat{\gamma}$ that are given by the following expressions:

$$\hat{\mu} = -\frac{y_1^2(2\pi)}{2\pi\alpha'(0)} \text{ and } \hat{\gamma} = -\frac{y_1^1(2\pi)}{2\pi} + \frac{\beta'(0)y_1^2(2\pi)}{2\pi\alpha'(0)}.$$

Next we focus on $\hat{z}_1$. To do so we set

$$Y_1(t) = y_1(t) - c_1 t \begin{pmatrix} \sin t \\ \cos t \end{pmatrix} - c_2 t \begin{pmatrix} \cos t \\ -\sin t \end{pmatrix},$$

so that $\hat{z}_1 \in \pi_1 X_2$ satisfies

$$\hat{z}_1(t) - e^{B(0)t}\hat{z}_1(0) = Y_1(t), \ \forall t \in [0, 2\pi].$$

Here note that $Y_1 \in C_0([0,2\pi],\mathbb{R}^2) \cap C_{2\pi}([0,2\pi],\mathbb{R}^2)$. Consider now the linear operator $Q : C_{2\pi}([0,2\pi];\mathbb{R}^2) \to C_{2\pi}([0,2\pi];\mathbb{R}^2)$ defined by

$$Q(z)(t) = e^{B(0)t}z(0),\ \forall t \in [0,2\pi].$$

so that the above equation rewrites as

$$\hat{z}_1 \in \pi_1 X_2 \text{ and } (I-Q)\hat{z}_1 = Y_1 \in C_0([0,2\pi],\mathbb{R}^2) \cap C_{2\pi}([0,2\pi],\mathbb{R}^2).$$

Now note that Q is a projector (i.e. $Q^2 = Q$) and since $t \to e^{B(0)t}$ is 2π–periodic, one has

$$\mathrm{R}(I-Q) = C_0([0,2\pi],\mathbb{R}^2) \cap C_{2\pi}([0,2\pi],\mathbb{R}^2),$$

while

$$N(I-Q) = \mathrm{R}(Q) = \operatorname{span}\{\pi_1 u_0, \pi_1 u_1\}.$$

Since $C_{2\pi}([0,2\pi],\mathbb{R}^2) = \pi_1 X_2 \oplus \operatorname{span}\{\pi_1 u_0, \pi_1 u_1\}$, we obtain that the linear bounded operator $(I-Q)_{|\pi_1 X_2}$ is bijective from $\pi_1 X_2$ onto $C_0([0,2\pi],\mathbb{R}^2) \cap C_{2\pi}([0,2\pi],\mathbb{R}^2)$ and this invertible on these spaces (see Theorem 3.11). Hence we end-up with

$$\hat{z}_1 = (I-Q)^{-1}_{|\pi_1 X_2} Y_1.$$

To sum-up the above analysis, we have obtained for each $y \in C_0([0,2\pi],\mathbb{R}^n)$ the exists a unique $(\hat{\gamma},\hat{\mu},\hat{z}) \in \mathbb{R}^2 \times C_{2\pi}([0,2\pi],\mathbb{R}^n)$ satisfying (3.19) and this completes the proof of Claim 3.18. □

To conclude the proof of the Hopf bifurcation theorem, we apply the implicit function theorem to the function $h : \mathbb{R}^3 \times X_2 \to C_0([0,2\pi],\mathbb{R}^n)$ and we deduce that there exists a C^1–mapping $(\gamma,\mu,z) : (-\delta,\delta) \to \mathbb{R}^2 \times X_2$, for some $\delta > 0$ small enough, such that

$$h(s,\gamma(s),\mu(s),z(s)) = 0,\ \forall s \in (-\delta,\delta).$$

By the definition of h, this is equivalent to say that

$$F(\gamma(s),\mu(s),s(u_0 + z(s))) = 0,$$

when $s \neq 0$ with $(\gamma(0),\mu(0),z(0)) = (1,0,0)$.

3.8 Remarks and notes

3.8.1 Classification of bifurcations for systems in dimension 2

In this part, we will show some bifurcations which are related to the changes in the number or stability of equilibria and closed orbits, or related to the changes in saddle connections as a parameter is varied. For two-dimensional system, the eigenvalues of the Jacobian are the key to the stability of the equilibrium. If the equilibrium is stable, the eigenvalues λ_1, λ_2 must both lie in the left half-plane and either the eigenvalues are both real and negative or they are complex conjugates with negative real parts. To destabilize the equilibrium, we need one or both of the eigenvalues to cross into the right half-plane as the parameter varies. Hopf bifurcation presented in section 3.6 is the scenario in which two complex conjugate eigenvalues simultaneously cross the imaginary axis into the right half-plane. For the cases in which a real eigenvalue passes through $\lambda = 0$, we explore saddle-node, transcritical, and pitchfork bifurcations in the following. The Hopf bifurcation, Saddle-node coalescence of cycles, and Homoclinic bifurcation are common ways in which limit cycles are created or destroyed. We describe the definition of each bifurcation by using a simple prototypical example. This section is taken from [318] and we refer to the books [54], [200], [318] for more details.

3.8.1.1 Saddle-node bifurcation

The saddle-node bifurcation is the basic mechanism by which equilibria are created and destroyed (i.e., where on one side of a parameter value there are two equilibria and on the other side there are no equilibria). As a parameter is varied, two equilibria known as saddle and node move toward each other, collide, and mutually annihilate. The prototypical example in two dimension of a saddle-node bifurcation is given by

$$\begin{aligned} x'(t) &= \mu - x^2 \\ y'(t) &= -y \end{aligned} \tag{3.22}$$

where $x, y \in \mathbb{R}^1$. For $\mu > 0$, there are two equilibria, a stable node at $(\sqrt{\mu}, 0)$ and a saddle at $(-\sqrt{\mu}, 0)$. The saddle and node approach each other as μ decreases, then collide when $\mu = 0$, and finally disappear when $\mu < 0$. Even after the equilibria have annihilated each other, they continue to influence the flow. They leave a ghost, a bottleneck region that sucks trajectories in and delays them before allowing passage out the other side. Figure 3.4 shows the phase portrait as μ varies.

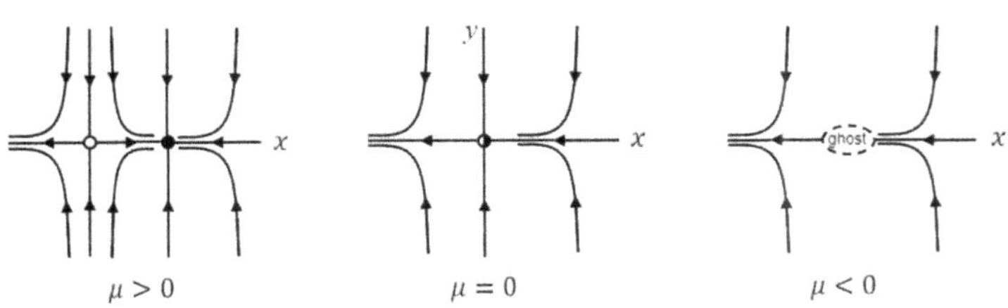

Fig. 3.4: *Saddle-node bifurcation*

3.8.1.2 Transcritical bifurcation

There are certain situations where a equilibrium must exist for all values of a parameter and can never be destroyed. However, such a equilibrium may change its stability as the parameter is varied. The transcritical bifurcation is the mechanism for such changes in stability of equilibrium.

Consider the transcritical bifurcation system

$$\begin{aligned} x'(t) &= \mu x - x^2 \\ y'(t) &= -y \end{aligned} \tag{3.23}$$

where $x, y, \mu \in \mathbb{R}^1$. Figure 3.5 shows the vector field as μ varies. For $\mu < 0$, there is a stable equilibrium at $(0, 0)$ and an unstable equilibrium at $(\mu, 0)$. As μ increases, the unstable equilibrium at $(\mu, 0)$ approaches the origin, and coalesces with it when $\mu = 0$. When $\mu > 0$, the origin loses stability and become unstable and $(\mu, 0)$ becomes stable.

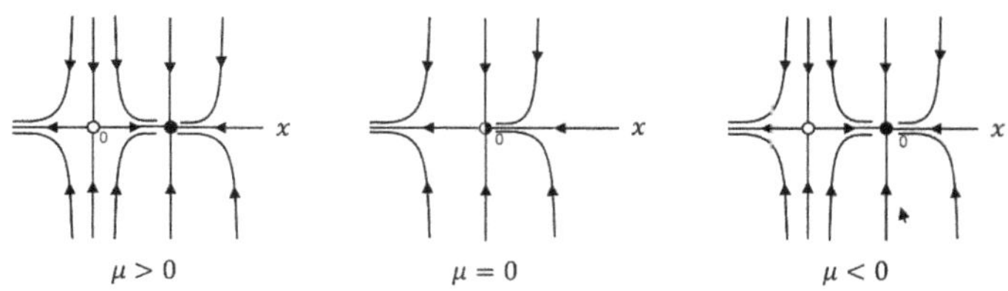

Fig. 3.5: *Transcritical bifurcation*

3.8.1.3 Subcritical and supercritical pitchfork bifurcation

Like Hopf bifurcations, pitchfork bifurcations come in both super- and subcritical varieties. The pitchfork bifurcation is common in physical problems that have a symmetry. For example, many problems have a spatial symmetry between left and right. Thus, equilibria tend to appear and disappear in symmetrical pairs. There are two very different types of pitchfork bifurcation. The simpler type is called supercritical which will be shown first.

Consider the supercritical pitchfork system

$$\begin{aligned} x'(t) &= \mu x - x^3 \\ y'(t) &= -y \end{aligned} \tag{3.24}$$

where $x, y, \mu \in \mathbb{R}^1$. For $\mu < 0$, the only equilibrium is a stable node at the origin. For $\mu = 0$, the origin is still stable, but now we have very slow decay along the x-direction instead of exponential decay. For $\mu > 0$,the origin loses stability and gives birth to two new stable equilibria symmetrically located at $(\sqrt{\mu}, 0)$ and $(-\sqrt{\mu}, 0)$. You can check that the origin is a saddle and the other two equilibria are stable nodes by computing the Jacobian at each point. Figure 3.6 shows the phase portrait as μ varies.

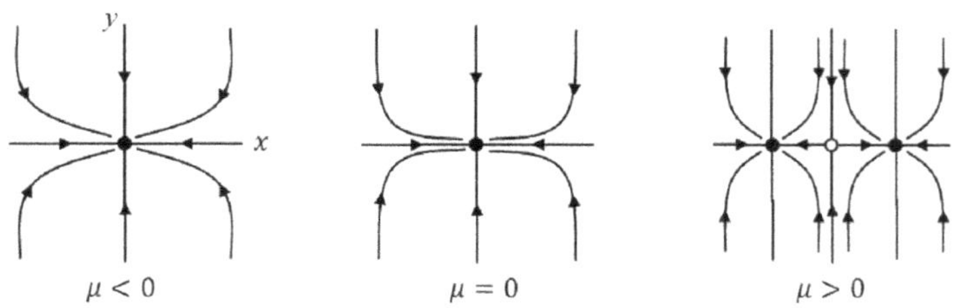

Fig. 3.6: *Supercritical pitchfork bifurcation*

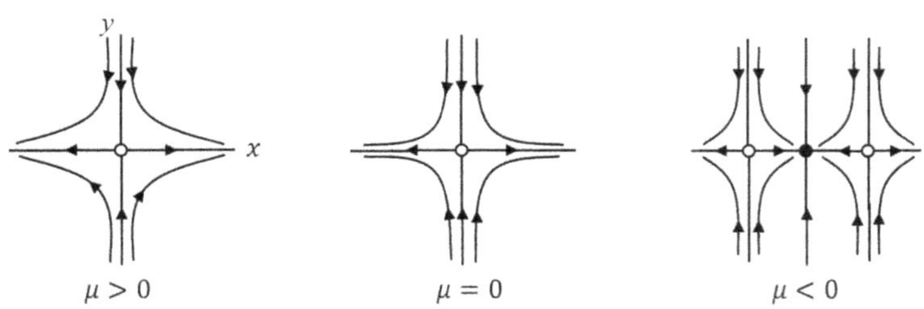

Fig. 3.7: *Subcritical pitchfork bifurcation*

Consider the subcritical pitchfork system

$$\begin{aligned} x'(t) &= \mu x + x^3 \\ y'(t) &= -y \end{aligned} \tag{3.25}$$

where $x, y, \mu \in \mathbb{R}^1$. Compared to the supercritical pitchfork system, the pitchfork is inverted. The nonzero equilibria $(\sqrt{\mu}, 0)$ and $(-\sqrt{\mu}, 0)$ are unstable, and exist for $\mu < 0$ which motivates the term "subcritical". The origin is stable for $\mu < 0$ and unstable for $\mu > 0$, as in the supercritical case. Figure 3.7 shows the phase portrait as μ varies.

3.8.1.4 Saddle-node coalescence of cycles

A bifurcation in which two limit cycles coalesce and annihilate is called a saddle-node bifurcation of cycles. Consider the following example:

$$\begin{aligned} r'(t) &= \mu r + r^3 - r^5 \\ \theta'(t) &= \omega + br^2 \end{aligned} \tag{3.26}$$

The subcritical Hopf bifurcation occurs at $\mu = 0$ and we leave this case as an exercise. Now we concentrate on the dynamics for $\mu < 0$.

Regard the radial equation $r'(t) = \mu r + r^3 - r^5$ as a one-dimensional system. This system undergoes a saddle-node bifurcation of equilibria at $\mu_c = -1/4$. Returning to the two-dimensional system, these equilibria correspond to circular limit cycles. Figure 3.8 plots the radial phase portraits and the corresponding behavior in the phase plane. A stable and unstable cycle collide and disappear as μ decreases through μ_c. The origin remains stable throughout which does not participate in this bifurcation.

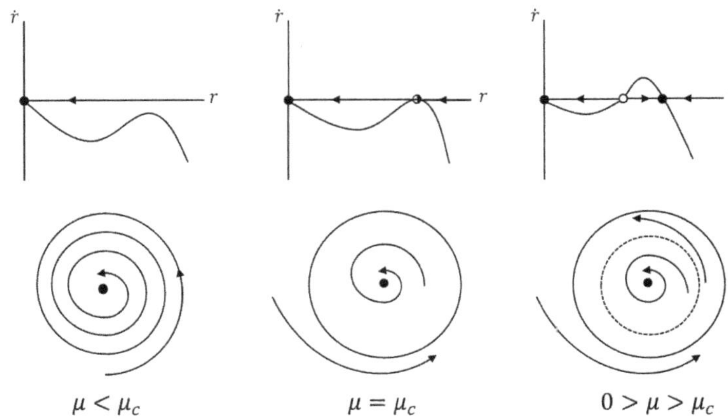

Fig. 3.8: *Saddle-node coalescence of cycles*

3.8.1.5 Homoclinic bifurcation or saddle-loop bifurcation

In this scenario, part of a limit cycle moves closer and closer to a saddle point. At the bifurcation the cycle touches the saddle point and becomes a homoclinic orbit. This is called homoclinic bifurcation. Consider the system:

$$\begin{aligned} x' &= y \\ y' &= \mu y + x - x^2 + xy \end{aligned} \tag{3.27}$$

Figure 3.9 plots the phase portraits as μ varies. Numerically, the bifurcation is found to occur at $\mu_c \approx -0.8645$. For $\mu < \mu_c$, a stable limit cycle passes close to a saddle point at the origin (see Figure 3.9 a). As μ increases to μ_c, the limit cycle swells and bangs into the saddle, creating a homoclinic orbit (see Figure 3.9C). Once $\mu > \mu_c$, the saddle connection breaks and the loop is destroyed (see Figure 3.9d).

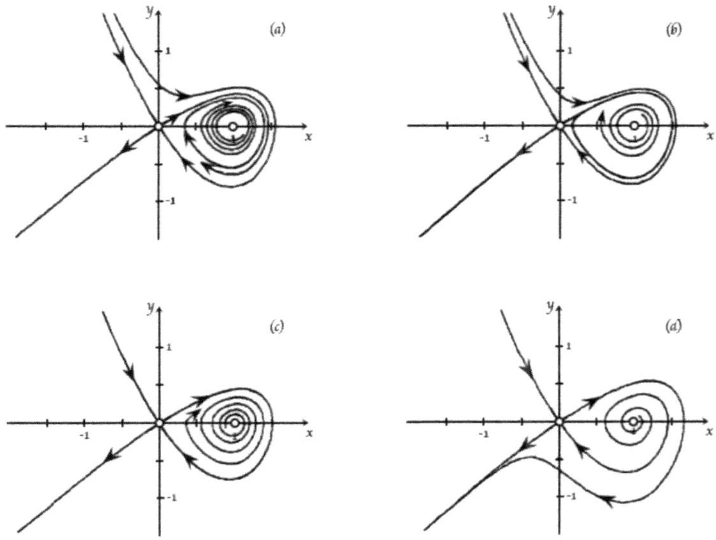

Fig. 3.9: *Homoclinic bifurcation*

3.8.2 More about implicit function Theorem

An analytic version of the implicit function theorem can be found in Theorem 15.3 p.151 in the book of Deimling [75]. Such a result provides an analytic dependency of the solution with respect to the parameters. Furthermore, an implicit function Theorem in scale of Banach spaces is also presented in Theorem 15.8 p. 162 in [75].

3.8.3 More about Hopf bifurcation

We need extend Theorem 3.13 in order to determine the direction of the Hopf bifurcation and the stability of the bifurcating periodic solutions described in Theorem 3.16. The following result is taken from the book by Chow et al. [54].

Consider a C^∞ system defined in a neighborhood of the origin in $\mathbb{R}^2$

$$\begin{aligned} x'(t) &= f(x, y, \mu) \\ y'(t) &= g(x, y, \mu) \end{aligned} \tag{3.28}$$

where $x, y, \mu \in \mathbb{R}^1$. We first describe the following assumption that will be needed.

Assumption 3.19 We suppose that the origin is an equilibrium of (3.28), and, for μ near 0, there exists a pair of conjugated eigenvalues of the linear part of (3.28) around the origin denoted by $\lambda(\mu)$ and $\overline{\lambda(\mu)}$ written as

$$\lambda(\mu) = \alpha(\mu) + i\omega(\mu), \text{ with } \alpha(\mu),\ \omega(\mu) \in \mathbb{R},$$

such that the map $\mu \longrightarrow \lambda(\mu)$ is continuously differentiable,

$$\omega(0) = \omega_0 > 0,\ \alpha(0) = 0 \text{ and } \frac{d\alpha(0)}{d\mu} \neq 0.$$

Before we state the main result about the property of the Hopf bifurcation, we derive a normal form for (3.28) for $\mu = 0$. By making a suitable linear change of coordinates $z = x + iy$, equation (3.28) for $\mu = 0$ becomes

$$\begin{aligned} z'(t) &= i\omega_0 z + F(z, \overline{z}) \\ \overline{z}'(t) &= -i\omega_0 \overline{z} + \overline{F(z, \overline{z})}. \end{aligned} \tag{3.29}$$

Since the eigenvalues of the linear part of (3.28) for $\mu = 0$ are $\lambda_{1,2} = \pm i\omega_0$, by a polynomial change of variables

$$z = w + \sum_{2 \le k+l \le m} b_{kl} w^k \overline{w}^l,$$

and System (3.29) takes the form

$$w' = i\omega_0 w + C_1 w^2 \overline{w} + \cdots + O(|w|^{2k+3}).$$

Theorem 3.20 *Let Assumption 3.14 be satisfied and* $\mathrm{Re}(C_1) \neq 0$. *Then there are* $\sigma > 0$ *and a neighborhood* U *of* $(x, y) = (0, 0)$ *such that*

(i) *if* $|\mu| < \sigma$ *and* $\mu \,\mathrm{Re}(C_1)\dfrac{d\alpha(0)}{d\mu} < 0$, *the system (3.28) has exactly one limit cycle inside* U;

(ii) *if* $|\mu| < \sigma$ *and* $\mu \,\mathrm{Re}(C_1)\dfrac{d\alpha(0)}{d\mu} \geq 0$, *the system (3.28) has no periodic orbit inside* U.

Moreover, the limit cycle is stable (unstable) if $\mathrm{Re}(C_1) < 0$ $(\mathrm{Re}(C_1) > 0)$, *and it tends to the equilibrium* $(0, 0)$ *as* $\mu \to 0$.

3.8.4 Bifurcations for finite dimensional systems

We refer to Golubitsky, Stewart and Schaeffer [126], Chicone and Latushkin [50], Chow and Hale [53], Chow, Li and Wang [54], Hassard, Kazarinoff and Wan [151], Kuznetsov [201], Guckenheimer, and Holmes [128], Hale and Koçak [141], Guo and Wu [130], looss [184], looss and Joseph [185], Kolmogorov [193], Marsden and McCracken [247], Wiggins [350], Wiggins [352], Perko [270], Luo [232], Roussarie [287], Seydel [299], Rasmussen [278], Han and Yu [147] and Demazure [77], Luo [232] for more results on bifurcations for finite dimensional systems.

Here we will show the backward bifurcation in SI epidemic model with vaccination. In compartmental models for the transmission of communicable diseases there is usually a basic reproductive number R_0, representing the average number of cases produced by a single infective introduced into a pool of susceptible population during its infective period. There is a disease-free equilibrium which is asymptotically stable (the infection dies out) for $R_0 < 1$ and unstable for $R_0 > 1$. At $R_0 = 1$, the usual situation is that an endemic equilibrium which is asymptotically stable bifurcates, and the infection persists. Even if the endemic equilibrium is unstable, the instability commonly arises from a Hopf bifurcation and the infection still persists but in an oscillatory manner. More precisely, as R_0 increases through 1 there is an exchange of stability between the disease-free equilibrium, which exists for all values of R_0 and the endemic equilibrium which only exists on one side of the bifurcation point $R_0 > 1$ (which is negative as well as unstable and thus biologically meaningless if $R_0 < 1$). The bifurcation leading from the disease free state to the infected state is forward or supercritical and we call the type of bifurcation a forward bifurcation or transcritical bifurcation, in which the endemic equilibrium exists only for $R_0 > 1$, so that there is no possibility of an endemic state when $R_0 < 1$. The equilibrium infective population size depends continuously on R_0 and prevalence is an increasing function of R_0 at the bifurcation point.

A phenomenon of considerable interest in theoretical epidemiology is that of the existence a backward or subcritical bifurcation at $R_0 = 1$. In systems exhibiting a backward bifurcation, the endemic equilibrium exists for $R_0 < 1$, so that under certain initial conditions it is possible for an invasion to succeed, or for an established endemic state to persist, with $R_0 < 1$.

The behavior at a bifurcation may be described graphically by the bifurcation curve, which is the graph of equilibrium infective population size I^* as a function of the basic reproductive number R_0. The forward bifurcation and backward bifurcation curves have the form shown in Figure 3.10.

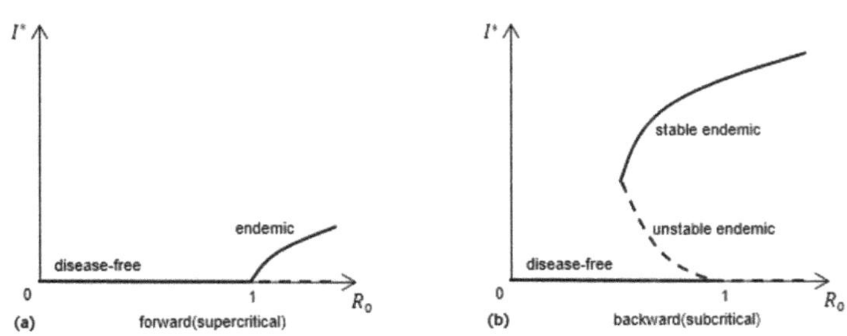

Fig. 3.10: *The forward bifurcation and backward bifurcation curves*

Brauer [36] have demonstrated the existence of backward bifurcations in the following simple vaccination SIS model

$$\begin{aligned} S' &= \Lambda(N) - \beta(N)SI - (\mu + \phi)S + \gamma I + \theta V, \\ I' &= \beta(N)SI + \sigma\beta(N)VI - (\mu + \gamma)I, \\ V' &= \phi S - \sigma\beta(N)VI - (\mu + \theta)V, \end{aligned} \tag{3.30}$$

where $N = S + I + V$. $N' = \Lambda(N) - \mu N$ and $\lim_{t\to\infty} N(t) = K$ for every choice of initial values and by the theory of asymptotically autonomous systems, N could be replaced by K and S by $K - I - V$ to give the qualitatively equivalent system which is the basic vaccination model

$$\begin{aligned} I' &= \beta(K)[K - I - (1-\sigma)V]I - (\mu+\gamma)I, \\ V' &= \phi[K - I] - \sigma\beta(K)VI - (\mu+\theta+\phi)V. \end{aligned} \tag{3.31}$$

The system (3.31) has a backward bifurcation at $R(\phi) = 1$ if and only if $B < 0$ when β is chosen to make $C = 0$, where $B = (\mu + \theta + \sigma\phi) + \sigma(\mu+\gamma) - \sigma\beta K$, $C = \frac{(\mu+\gamma)(\mu+\theta+\phi)}{\beta} - (\mu+\theta+\sigma\phi)K$ and $R(\phi) = \frac{\beta K}{\mu+\gamma}\frac{\mu+\theta+\sigma\phi}{\mu+\theta-\phi}$.

We refer to the papers of Brauer [36], Hadeler et al. [133], Kribs-Zaleta et al. [196] and Dushoff et al.[98], and references therein for more references and results on this subject.

3.8.5 Bifurcations for infinite dimensional systems

The theory developed in this chapter can also be extended and applied to infinite dimensional systems. The results presented in this chapter are based on the original three papers by Crandall and Rabinowitz [67, 68, 69] which were considering bifurcation for infinite dimensional system (parabolic equations).

In the 1970s, several Hopf bifurcation theorems were obtained for infinite dimensional systems in order to establish the bifurcation of periodic solutions of the Navier–Stokes equations. Such results are usually based on the so-called Liapunov–Schmidt or center manifold reduction approach. We refer to Iudovich [186], Sattinger [293], Iooss [183], Joseph and Sattinger [189], Marsden [246], Marsden and McCraken [247], Crandall and Rabinowitz [69], Henry [153], Da Prato and Lunardi [72], and Kielhöfer [191] for results on the subject. The Hopf bifurcation theorem has also been extended to functional differential equations, see Hale [140], Hale and Verduyn Lunel [144], Diekmann et al. [83] and the references cited therein. We also refer to a sketch of the Hopf bifurcation theorem by Golubitsky and Rabinowitz in [163] for a nice commentary on Hopf bifurcation theorem and more references.

It has been observed that Hopf bifurcation also occurs in age-structured models (see Prüss [275], Cushing [71], Swart [320], Kostova and Li [195], Bertoni [28]). A recent extension of Hopf bifurcation theorem to non-densely defined Cauchy problems has been obtained in Liu, Magal and Ruan [221] by using the center manifolds theory for semilinear equations with non- dense domain given in Magal and Ruan [239]. Center-unstable manifold theorem for non-densely defined Cauchy problems, and the stability of bifurcation periodic orbits by Hopf bifurcation are provided in Magal Liu and Ruan [222]. Several examples of differential equations, such as delay differential equations, parabolic equations with nonlinear and non-local boundary conditions can be put into the abstract non-densely defined Cauchy problems. More results and examples of abstract non-densely defined Cauchy problems can be found in Magal and Ruan [239, 240]. Thus, Hopf bifurcation theorem for abstract non-densely defined Cauchy problems obtained in [221] have a wide range of applicability. In particular, they provide an application to age structured population dynamics models (see the book of Webb [349] and Magal and Ruan [239, 240] for more results on this topic). A Hopf bifurcation theorem for second order semi-linear Gurtin-MacCamy equation are presented in Ducrot, Kang and Magal [90].

We refer to Henry [153], Magal and Ruan [240] and Ducrot, Kang and Magal [91], Kielhöfer [191], Robinson [283], Kielhöfer [191], Haragus and Iooss [148] for more results on bifurcations for infinite dimensional systems.

3.8.6 Example of application to the pattern formation to predator prey system

Spatial pattern formation with reaction diffusion system is a nice class of problem. For the study on pattern formation mechanisms which generate pattern in a variety of diverse areas, we refer to the books of Murray [263] [264] and [265].

Spatial, temporal and spatiotemporal patterns could occur in the reaction–diffusion system via Turing instability, Hopf bifurcation, Turing–Hopf bifurcation, and so on. Turing instability could lead to non-uniform spatial patterns which could generate biological patterns by gene activation. We refer to Levin and Segel [212], Malchow et al. [245] and references cited therein for related work on Turing instability and Turing pattern formation of reaction–diffusion systems. Temporal periodic patterns could be generated via Hopf bifurcation in the reaction-diffusion system, such as diffusive predator–prey models with suitable functional response. In a given reaction–diffusion system, when the Hopf bifurcation curve and Turing bifurcation curve intersect at the codimension-2 bifurcation point, a new bifurcation, called Turing–Hopf bifurcation, occurs which generates spatiotemporal patterns. The interaction of Turing and Hopf bifurcations in chemical, physical and predator–prey systems has been observed and analysed (see, e.g., the papers Rovinsky and Menzinger [288], Meixner [251], Wit et al.[74], Baurmann et al. [24], Shi and Ruan [304]). Ducrot et al. [87] study Turing and Turing–Hopf bifurcations for a one-dimensional nonlocal reaction–diffusion–advection equationa.

To study the spatiotemporal dynamics of the predator–prey model, Shi and Ruan [304] consider the following partial differential equation model under homogeneous Neumann boundary conditions:

$$\begin{cases} \frac{\partial u}{\partial t} - d_1 \Delta u = ru\left(1 - \frac{u}{K}\right) - \frac{muv}{(1+au)(1+bv)}, x \in \Omega, t > 0, \\ \frac{\partial v}{\partial t} - d_2 \Delta v = sv\left(1 - \frac{v}{hu}\right), x \in \Omega, t > 0, \\ \frac{\partial u}{\partial \nu} = \frac{\partial v}{\partial \nu} = 0, x \in \partial\Omega, t > 0, \\ u(x,0) = u_0(x) \geq 0, v(x,0) = v_0(x) \geq 0, x \in \Omega. \end{cases}$$

Here, $u(x,t)$ and $v(x,t)$ stand for the densities of the prey and predators at location $x \in \Omega$ and time t, respectively; $\Omega \subset R^N (N \leq 3)$ is a bounded domain with smooth boundary $\partial\Omega$; ν is the outward unit normal vector of the boundary $\partial\Omega$.

The homogeneous Neumann boundary conditions indicate that the predator–prey system is self-contained with zero population flux across the boundary. The positive constants d_1 and d_2 are diffusion coefficients, and the initial data $u_0(x)$ and $v_0(x)$ are non-negative continuous functions. They show the evolution of the spatial pattern of the prey at 0, 2000 and 10000 iterations, with random perturbation of the steady states to a predator prey system in the following figure. We see from this figure that the spotted and labyrinth patterns prevail over the whole domain.

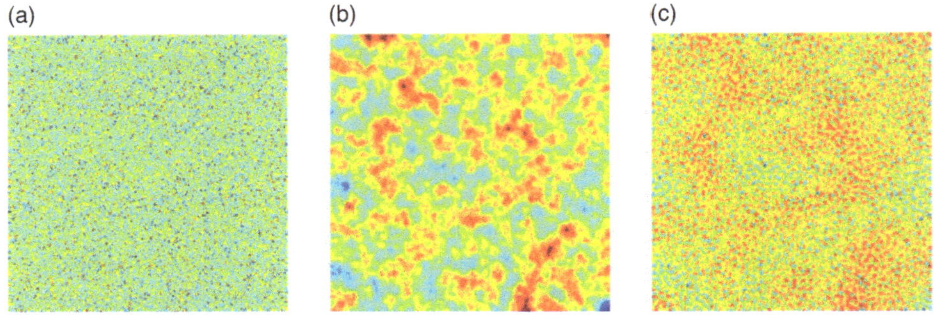

Fig. 3.11: *Snapshots of contour pictures of the time evolution of the prey population density at different instants for the reaction– diffusion system in a 2D spatial domain*

Chapter 4
Center Manifold and Center Unstable Manifold Theory

4.1 Introduction

In this section we consider differential equations of the form

$$u'(t) = Au(t) + F(u(t)), \forall t \in \mathbb{R}, \text{ and } u(0) = u_0 \in \mathbb{R}^n, \tag{4.1}$$

where $A \in M_n(\mathbb{R})$ is a given n by n matrix while $F : \mathbb{R}^n \to \mathbb{R}^n$ is a of class $C^k(\mathbb{R}^n)$ with $k \geq 1$.

Assumption 4.1 We assume that there exists an equilibrium $\overline{u} \in \mathbb{R}^n$ point for system (4.1), that is satisfying the identity

$$A\overline{u} + F(\overline{u}) = 0_{\mathbb{R}^n}.$$

Replacing u by $v = u - \overline{u} \Leftrightarrow u = v + \overline{u}$ in (4.1), we obtain

$$v'(t) = Av(t) + F(v(t) + \overline{u}) + A\overline{u}, \forall t \in \mathbb{R}, \text{ and } v(0) = u_0 - \overline{u} \in \mathbb{R}^n.$$

Hence by defining

$$\widehat{A} := A + DF(\overline{u}) \in M_n(\mathbb{R}) \text{ and } \widehat{F}(x) := F(x + \overline{u}) + A\overline{u} - DF(\overline{u})x,$$

and dropping the hat for the simplicity of the notations, the point $0_{\mathbb{R}^n}$ becomes an equilibrium point. Therefore without loss of generality we shall assume in the sequel that the following assumption is satisfied.

Assumption 4.2 We assume that

$$F(0) = 0_{\mathbb{R}^n} \text{ and } DF(0) = 0_{M_n(\mathbb{R})}. \tag{4.2}$$

A. Ducrot et al., *Differential Equations and Population Dynamics II*,
Lecture Notes on Mathematical Modelling in the Life Sciences,
https://doi.org/10.1007/978-3-031-91409-6_4

4.2 State Space Decomposition of a Linear Equation

Consider first the linear system

$$u'(t) = Au(t), \forall t \in \mathbb{R}, \text{ and } u(0) = u_0 \in \mathbb{R}^n. \tag{4.3}$$

We denote by $\sigma(A) \subset \mathbb{C}$ the set of the eigenvalues of A. The spectrum of A consists in the disjoint union of the **stable spectrum** $\sigma_s(A)$, the **central spectrum** $\sigma_c(A)$ and the **unstable spectrum** $\sigma_u(A)$, given by

$$\begin{aligned} \sigma_s(A) &= \{\lambda \in \sigma(A) : \text{Re}\,(\lambda) < 0\}, \\ \sigma_c(A) &= \{\lambda \in \sigma(A) : \text{Re}\,(\lambda) = 0\}, \\ \sigma_u(A) &= \{\lambda \in \sigma(A) : \text{Re}\,(\lambda) > 0\}. \end{aligned} \tag{4.4}$$

Assumption 4.3 We assume that $\sigma_c(A) = \sigma(A) \cap i\mathbb{R} \neq \emptyset$.

The generalized eigenspaces of A corresponding to the spectral decomposition $\sigma_s(A)$, $\sigma_c(A)$, and $\sigma_u(A)$ provide a state space decomposition

$$\mathbb{C}^n = X_s^{\mathbb{C}} \oplus X_c^{\mathbb{C}} \oplus X_u^{\mathbb{C}}, \tag{4.5}$$

with

$$\begin{aligned} X_s^{\mathbb{C}} &= \bigoplus_{\lambda \in \sigma_s(A)} N((\lambda I - A)^n), \\ X_c^{\mathbb{C}} &= \bigoplus_{\lambda \in \sigma_c(A)} N((\lambda I - A)^n), \\ X_u^{\mathbb{C}} &= \bigoplus_{\lambda \in \sigma_u(A)} N((\lambda I - A)^n), \end{aligned} \tag{4.6}$$

and A is regarded as a linear operator on $\mathbb{C}^n$.

Let $k = s, c, u$ be given. Let us observe that since A is a real valued matrix, we have

$$\overline{(\lambda I - A)^n x} = \left(\overline{\lambda} I - A\right)^n \overline{x}, \forall \lambda \in \mathbb{C},$$

therefore one deduces that

$$\lambda \in \sigma_k(A) \Leftrightarrow \overline{\lambda} \in \sigma_k(A),$$

and

$$x \in X_k^{\mathbb{C}} \Leftrightarrow \overline{x} \in X_k^{\mathbb{C}}.$$

Hence it follows that

$$x \in X_k^{\mathbb{C}} \Rightarrow \text{Re}(x) = \frac{x + \overline{x}}{2} \in X_k^{\mathbb{C}} \text{ and } \text{Im}(x) = \frac{x - \overline{x}}{2i} \in X_k^{\mathbb{C}}.$$

Moreover, since one has

$$x \in X_k^{\mathbb{C}} \Leftrightarrow \pm ix \in X_k^{\mathbb{C}},$$

it follows that

$$X_k^{\mathbb{C}} = X_k \oplus i X_k,$$

wherein we have set

$$X_k = \mathrm{Re}\left(X_k^{\mathbb{C}}\right) = \left\{\mathrm{Re}(x) : x \in X_k^{\mathbb{C}}\right\} \subset \mathbb{R}^n. \tag{4.7}$$

Hence the space X_k $(k = s, c, u)$ is the real part of the subspace of $X_k^{\mathbb{C}}$ spanned by the generalized eigenvectors of A corresponding to the eigenvalues in $\sigma_k(A)$.

Now, we obtain the following **state space decomposition**

$$\mathbb{R}^n = X_s \oplus X_c \oplus X_u, \tag{4.8}$$

which satisfies

$$A(X_s) \subset X_s, A(X_c) \subset X_c \text{ and } A(X_u) \subset X_u. \tag{4.9}$$

For $k = s, c, u$, we define $A_k : X_k \to X_k$ the bounded linear operator as

$$A_k x = Ax, \forall x \in X_k.$$

Then A_k becomes a bounded linear operator on X_k, which we write for short $A_k \in \mathcal{L}(X_k)$. The linear operator A_k is also called **the part of** A **in** X_k.

By construction, we have the following lemma.

Lemma 4.4 *The spectrum of A_s, A_c and A_u satisfy the following identities*

$$\sigma(A_s) = \sigma_s(A), \sigma(A_c) = \sigma_c(A) \textit{ and } \sigma(A_u) = \sigma_u(A). \tag{4.10}$$

Define $\Pi_s, \Pi_c, \Pi_u \in \mathcal{L}(\mathbb{R}^n)$ the projectors (i.e. $\Pi^2 = \Pi$) satisfying the following properties

$$\mathrm{N}(I - \Pi_s) = X_s \text{ and } \mathrm{N}(\Pi_s) = X_c \oplus X_u,$$

$$\mathrm{N}(I - \Pi_c) = X_c \text{ and } \mathrm{N}(\Pi_c) = X_s \oplus X_u,$$

$$\mathrm{N}(I - \Pi_u) = X_u \text{ and } \mathrm{N}(\Pi_u) = X_s \oplus X_c,$$

wherein $\mathrm{N}(\Pi)$ denotes the null space of Π.

Finally we combine the above projectors to define

$$\Pi_h := \Pi_s + \Pi_u$$

the projector on the **hyperbolic space** $X_h = X_s \oplus X_u$, and

$$\Pi_{cs} := \Pi_c + \Pi_s$$

the projector on the **center-stable space** $X_{cs} = X_c \oplus X_s$, and

$$\Pi_{cu} := \Pi_c + \Pi_u$$

the projector on the **center-unstable space** $X_{cu} = X_c \oplus X_u$.

Each of these projections commutes with the matrix A, and therefore the corresponding subspaces are invariant under the flow of (4.3). That is to say that

$$\Pi_k A = A \Pi_k, \forall k = s, c, u, h, cs, cu. \tag{4.11}$$

Note that the invariance under the flow of (4.3) directly follows from the following formula for the exponential of the matrix A, namely,

$$e^{At} = I + (At) + \frac{(At)^2}{2!} + \frac{(At)^3}{3!} + \ldots = \sum_{k=0}^{+\infty} \frac{(At)^k}{k!},$$

from which we deduce that

$$\Pi_k e^{At} = e^{At} \Pi_k, \forall k = s, c, u, h, cs, cu. \tag{4.12}$$

Let us now define the two quantities

$$\beta_+ = \min \{\text{Re}(\lambda) : \lambda \in \sigma_u(A)\} > 0,$$

and

$$\beta_- = \min \{-\text{Re}(\lambda) : \lambda \in \sigma_s(A)\} > 0.$$

And define also the number β by

$$\beta = \min(\beta_-, \beta_+)/2 > 0, \tag{4.13}$$

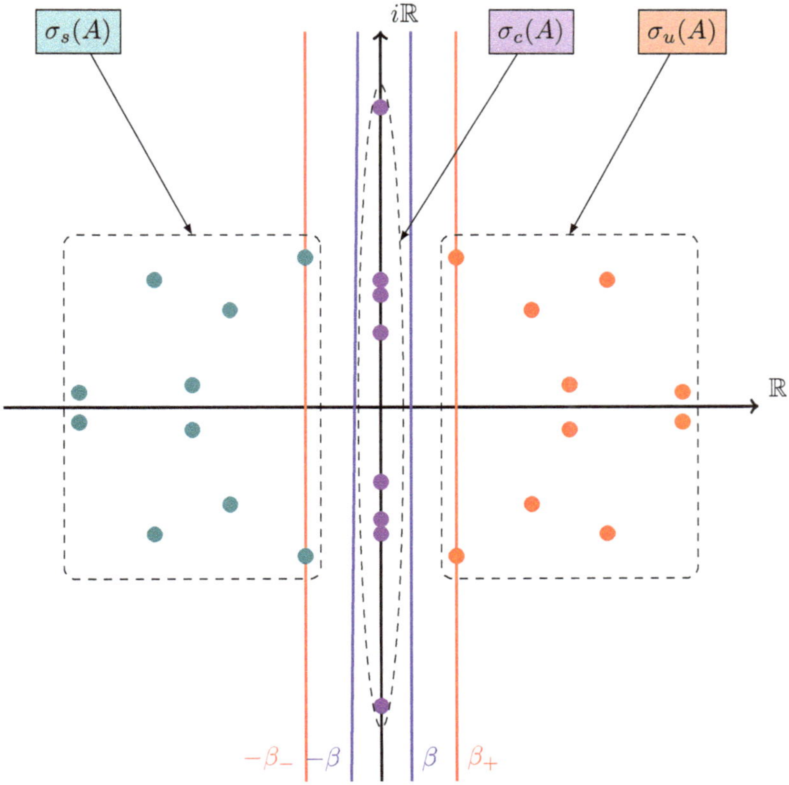

Fig. 4.1: *In this Figure, we illustrate the different parts of the spectrum $\sigma_s(A)$ (green dots), $\sigma_c(A)$ (purple dots), and $\sigma_u(A)$ (red dots) as well as the constants β, β_-, and β_+. The blue vertical lines separate the purely imaginary spectrum of A (purple dots) and the hyperbolic spectrum of A (red and green dots).*

In the following we collect some important information on the growth of the flow generated by (4.3) on various invariant sub-spaces.

Lemma 4.5 *There exists a non increasing map $\varepsilon \mapsto M_c(\varepsilon)$ from $(0, +\infty)$ to itself such that*

$$\|e^{At}\Pi_c\|_{\mathcal{L}(\mathbb{R}^n)} \leq M_c(\varepsilon)e^{\varepsilon|t|}, \forall t \in \mathbb{R}, \forall \varepsilon > 0. \tag{4.14}$$

There exist two constant number $M_s > 0$ and $M_u > 0$ such that

$$\begin{aligned} \|e^{At}\Pi_u\|_{\mathcal{L}(\mathbb{R}^n)} &\leq M_u e^{(\beta-\varepsilon)t}, \forall t \leq 0, \forall \varepsilon \geq 0, \\ \|e^{At}\Pi_s\|_{\mathcal{L}(\mathbb{R}^n)} &\leq M_s e^{(-\beta+\varepsilon)t}, \forall t \geq 0, \forall \varepsilon \geq 0. \end{aligned} \tag{4.15}$$

Remark 4.6 In general we may have $M_c(\varepsilon) \to +\infty$ as $\varepsilon \to 0$.

Proof The result is based on the stability theorem for linear system (see [89, Theorem 3.17 p.101] in the first volume of this book). We have for $t \geq 0$

$$e^{-\varepsilon|t|}e^{At}\Pi_c = e^{(A_c-\varepsilon I)t}\Pi_c$$

an the spectrum $\sigma(A_c - \varepsilon I)$ contains only complex number with strictly negative real part.

So by using the stability theorem for linear systems, we deduce that for each $\varepsilon > 0$

$$M_c^+(\varepsilon) := \sup_{t\geq 0} e^{-\varepsilon|t|}\|e^{At}\Pi_c\|_{\mathcal{L}(\mathbb{R}^n)} < +\infty.$$

Similarly, for $t \leq 0$

$$e^{-\varepsilon|t|}e^{At}\Pi_c = e^{-(A_c+\varepsilon I)|t|}\Pi_c$$

an the spectrum $\sigma(-A_c - \varepsilon I)$ contains only complex number with strictly negative real part. So by applying again the stability theorem for linear systems, we obtain for $\varepsilon > 0$

$$M_c^-(\varepsilon) := \sup_{t\leq 0} e^{-\varepsilon|t|}\|e^{At}\Pi_c\|_{\mathcal{L}(\mathbb{R}^n)} < +\infty.$$

We conclude that we can define $M_c(\varepsilon) = \max\left(M_c^+(\varepsilon), M_c^-(\varepsilon)\right)$ and we obtain (4.14).

To prove the last part of the lemma, we observe that (4.13) implies that

$$\beta < \beta_- \text{ and } \beta < \beta_+,$$

and by using again the stability theorem we deduce that for each $\varepsilon \geq 0$

$$\widehat{M}_u(\varepsilon) := \sup_{t\leq 0} e^{-(\beta-\varepsilon)t}\|e^{At}\Pi_u\|_{\mathcal{L}(\mathbb{R}^n)} = \sup_{t\geq 0} e^{-\varepsilon t}\|e^{(\beta I-A)t}\Pi_u\|_{\mathcal{L}(\mathbb{R}^n)} < +\infty,$$

$$\widehat{M}_s(\varepsilon) := \sup_{t\geq 0} e^{-(-\beta+\varepsilon)t}\|e^{At}\Pi_s\|_{\mathcal{L}(\mathbb{R}^n)} = \sup_{t\geq 0} e^{-\varepsilon t}\|e^{(\beta I+A)t}\Pi_s\|_{\mathcal{L}(\mathbb{R}^n)} < +\infty,$$

and by choosing $M_s = \widehat{M}_s(0)$ and $M_u = \widehat{M}_u(0)$ the proof is completed. □

Example 4.7 The above lemma is sharp. To see this, one can consider the matrix

$$A := \begin{pmatrix} 0 & 1 \\ 0 & 0 \end{pmatrix} \in M_2(\mathbb{R}),$$

which is a 2 by 2 Jordan block.

The spectrum of A is reduced to 0, that is

$$\sigma(A) = \{0\},$$

while the exponential of the matrix is given by

$$e^{At} := \begin{pmatrix} 1 & t \\ 0 & 1 \end{pmatrix} \in M_2(\mathbb{R}).$$

Now, by using the norm

$$\left\| \begin{pmatrix} x \\ y \end{pmatrix} \right\|_1 = |x| + |y|, \quad \forall \begin{pmatrix} x \\ y \end{pmatrix} \in \mathbb{R}^2,$$

one deduces that

$$1+t = \|e^{At}e_2\|_1 \le \|e^{At}\|_{\mathcal{L}(\mathbb{R}^n)} = \sup_{\|x\|_1 \le 1} \|e^{At}x\|_1 \le \sup_{p\in[0,1]} \|e^{At}\left(pe_1 + (1-p)e_2\right)\|_1 \le 2+t.$$

So, we obtain for this example

$$M_c(\varepsilon) \le \sup_{t\in\mathbb{R}} e^{-\varepsilon|t|}\,(2+t) < +\infty$$

and

$$M_c(\varepsilon) \ge \sup_{t\in\mathbb{R}} e^{-\varepsilon|t|}\,(1+t) \to +\infty, \text{ as } \varepsilon \to 0^+.$$

The following lemma characterizes the space X_c. These characterizations will be used to define the center manifold.

Lemma 4.8 (Characterization of center space) *We have the following*

$$X_c = \left\{ x \in \mathbb{R}^n : \sup_{t\in\mathbb{R}} \|\Pi_h e^{At}x\| < +\infty \right\}, \tag{4.16}$$

and for each $\eta \in (0, \beta)$,

$$X_c = \left\{ x \in \mathbb{R}^n : \sup_{t\in\mathbb{R}} e^{-\eta|t|}\|e^{At}x\| < +\infty \right\}. \tag{4.17}$$

Proof If $x \in X_c$ then $\Pi_h x = 0$. We deduce that the map $t \to \Pi_h e^{At}x = e^{At}\Pi_h x = 0$ is bounded function. Next assume that there exists a constant $C \in (0, +\infty)$ such that

$$\sup_{t\in\mathbb{R}} \|\Pi_h e^{At}x\| \le C.$$

Then by Lemma 4.5, we have for each $\eta \in (0, \beta)$:

$$\|e^{At}x\| = \|e^{At}(\Pi_c + \Pi_h)x\| \le \|e^{At}\Pi_c x\| + \|\Pi_h e^{At}x\| \le M(\eta)e^{\eta|t|} + C,$$

and we deduce that

$$\sup_{t\in\mathbb{R}} e^{-\eta|t|}\|e^{At}x\| \le M(\eta) + C.$$

Finally assume that

$$\|e^{At}x\| \le e^{\eta|t|}C, \forall t \in \mathbb{R},$$

for some $C > 0$, and some $\eta \in (0, \beta)$.

Then by using (4.15), we deduce that for all $t \le 0$, and $\varepsilon > 0$

$$\|\Pi_u x\| = \|e^{At}\Pi_u e^{-At}x\| \le Me^{(\beta-\varepsilon)t}\|e^{-At}x\| \le Me^{(\beta-\varepsilon)t}e^{-\eta t}C,$$

and by choosing $\varepsilon < \beta - \eta \le \beta_+ - \eta$, when t goes to $-\infty$ we deduce that

$$\Pi_u x = 0.$$

Similar arguments also provide $\Pi_s x = 0$. The proof is completed. □

The following lemma characterizes X_{cu}. These characterizations will be used as a definition for the center-unstable manifold.

Lemma 4.9 (Characterization of center-unstable space) *We have the following*

$$X_{cu} = \left\{ x \in \mathbb{R}^n : \sup_{t \le 0} \|\Pi_s e^{At} x\| < +\infty \right\}, \tag{4.18}$$

and for each $\eta \in (0, \beta)$,

$$X_{cu} = \left\{ x \in \mathbb{R}^n : \sup_{t \le 0} e^{-\eta |t|} \|e^{At} x\| < +\infty \right\}. \tag{4.19}$$

Proof Assume that

$$\sup_{t \le 0} \|\Pi_s e^{At} x\| \le C,$$

for some $C > 0$. Then by using (4.15), we have

$$\|\Pi_s x\| = \|e^{-At} \Pi_s e^{At} x\| \le M_s e^{(-\beta+\varepsilon)(-t)} \|\Pi_s e^{At} x\| \le M_s e^{(-\beta+\varepsilon)(-t)} C, \forall t \le 0,$$

and when $t \to -\infty$ we obtain $\Pi_s x = 0$, and we deduce that $x \in X_{cu}$.

Conversely, if $x \in X_{cu}$, then

$$\Pi_s e^{At} x = e^{At} \Pi_s x = 0, \forall t \le 0.$$

Next, assume that

$$\sup_{t \le 0} e^{-\eta |t|} \|e^{At} x\| \le C,$$

for some $C > 0$. Then by using again (4.15), we have

$$\|\Pi_s x\| = \|e^{A|t|} \Pi_s e^{At} x\| \le M_s e^{(-\beta+\varepsilon)|t|} \|\Pi_s e^{At} x\| \le M_s e^{(-\beta+\varepsilon)|t|} e^{\eta |t|} C, \forall t \le 0,$$

by choosing again $\varepsilon < \beta - \eta$, and letting $t \to -\infty$ yields $\Pi_s x = 0$.

Conversely assume that $\Pi_s x = 0$, by Lemma 4.5, we have for all $t \le 0$,

$$e^{-\eta |t|} \|e^{At} x\| \le e^{-\eta |t|} \left[\|e^{At} \Pi_c x\| + \|e^{At} \Pi_u x\| \right] \le e^{-\eta |t|} \left[M_c(\varepsilon) e^{\varepsilon |t|} + M_u e^{-(\beta-\varepsilon)|t|} \right]$$

so by choosing $0 < \varepsilon < \eta$ and $0 < \varepsilon < \beta_+$ we deduce

$$e^{-\eta |t|} \|e^{At} x\| \le M_c(\varepsilon) + M_u, \forall t \le 0.$$

The proof is completed. □

To conclude this section, consider the norm

$$|x| = \|\Pi_s x\| + \|\Pi_c x\| + \|\Pi_u x\|.$$

Since $I = \Pi_s + \Pi_c + \Pi_u$, by using the triangle inequality one obtains

$$\|x\| = \|\Pi_s x + \Pi_c x + \Pi_u x\| \leq |x|.$$

Next since Π_s Π_s and Π_u are continuous linear maps we get

$$|x| \leq \left(\|\Pi_s\|_{\mathcal{L}(\mathbb{R}^n)} + \|\Pi_c\|_{\mathcal{L}(\mathbb{R}^n)} + \|\Pi_u\|_{\mathcal{L}(\mathbb{R}^n)}\right) \|x\|,$$

therefore the two norms $|.|$ and $\|.\|$ are equivalent (this property remains true for infinite dimensional spaces).

Next let us observe that

$$|\Pi_k x| = \|\Pi_k^2 x\| = \|\Pi_k x\| \leq |x|, \forall k = s, u, c.$$

Hence without loss of generality we can make the following assumption.

Assumption 4.10 We assume that

$$\|\Pi_k\|_{\mathcal{L}(\mathbb{R}^n)} \leq 1, \forall k = s, u, c,$$

wherein the norm of operator is defined by

$$\|\Pi_k\|_{\mathcal{L}(\mathbb{R}^n)} := \sup_{\|x\| \leq 1} \|\Pi_k x\|.$$

4.3 Center Manifold Theory

Let $F : \mathbb{R}^n \to \mathbb{R}^n$ be a Lipschitz continuous map. In this section, we investigate the existence and smoothness of the center manifold for a nonlinear semiflow $\{U(t)\}_{t \geq 0}$ on $\mathbb{R}^n$ generated by solutions of the Cauchy problem

$$\frac{du(t)}{dt} = Au(t) + F(u(t)), \text{ for } t \geq 0, \text{ with } u(0) = x \in \mathbb{R}^n. \tag{4.20}$$

A solution of (4.20) is a continuous map $t \in [0, +\infty) \to U(t)x$ satisfying the fixed point problem

$$U(t)x = x + A \int_0^t U(s)x ds + \int_0^t F(U(s)x) ds, \forall t \geq 0.$$

By using the variation of constant formula [89, Proposition 2.21 p. 64], we deduce that the above fixed point problem is equivalent to

$$U(t)x = e^{At}x + \int_0^t e^{A(t-s)} F(U(s)x) \mathrm{d}s, \forall t \geq 0. \tag{4.21}$$

The variation of constant formula can be used to study the local stability of equilibrium for semi-linear Cauchy problems (see [89, Chapter 6] the first volume of this book). The section shows that it is also a crucial tool to understand the center manifold theory.

The variation of constant formula (4.21) can be rewritten into the following more condensed form

$$U(t)x = e^{At}x + \left(e^{A\cdot} * F(U(.)x)\right)(t), \forall t \geq 0,$$

where the convolution is defined as

$$\left(e^{A\cdot} * f(.)\right)(t) := \int_0^t e^{A(t-s)} f(s)\, ds, \forall t \geq 0. \tag{4.22}$$

By [89, Theorem 8.11 p. 232], and since F is Lipschitz continuous, we know that for each $x \in \mathbb{R}^n$, (4.21) has a unique solution $t \to U(t)x$ from $[0, +\infty)$ into $\mathbb{R}^n$. Moreover, the family $\{U(t)\}_{t\geq 0}$ defines a continuous semiflow. That is

(i) $U(0) = I$ and $U(t)U(s) = U(t+s), \forall t, s \geq 0$;
(ii) The map $(t, x) \to U(t)x$ is continuous from $[0, +\infty) \times \mathbb{R}^n$ into $\mathbb{R}^n$.

Furthermore,

$$\|U(t)x - U(t)y\| \leq e^{\|F\|_{\text{Lip}} t} \|x - y\| \text{ for all } t \geq 0,$$

where $\|F\|_{\text{Lip}}$ denotes the Lipschitz semi-norm, giben by

$$\|F\|_{\text{Lip}} := \sup_{x,y\in\mathbb{R}^n : x\neq y} \frac{\|F(x) - F(y)\|}{\|x - y\|}.$$

4.3.1 Weighted spaces of exponential bounded continuous functions

Let $(Y, \|.\|_Y)$ be a Banach space. Let $\eta \in \mathbb{R}$ be a constant and $I \subset \mathbb{R}$ be an interval. Define

$$BC^{\eta}(I, Y) = \left\{ f \in C(I, Y) : \sup_{t\in I} e^{-\eta|t|} \|f(t)\|_Y < +\infty \right\}.$$

It is well known (see Chapter 2 of the first volume of this book [89]) that $BC^{\eta}(I, Y)$ is a Banach space when it is endowed with the norm

$$\|f\|_{BC^{\eta}(I,Y)} = \sup_{t\in I} e^{-\eta|t|} \|f(t)\|_Y .$$

We observe that

$$\|f\|_{BC^{\eta}(I,Y)} \leq M \Leftrightarrow \|f(t)\|_Y \leq M e^{\eta |t|}, \forall t \in I.$$

Therefore such spaces characterize the exponential growth speed of the function at plus or minus infinity.

In this chapter, the examples of set I used are the following

$$I = [0, +\infty), \ I = (-\infty, 0], \text{ and } I = (-\infty, +\infty),$$

The weighted spaces $BC^{\eta}(\mathbb{R}, Y)$ are invariant by a shift.

Lemma 4.11 (Shift-invariance) *Let $\eta > 0$ be given. Then for each $f \in BC^{\eta}(\mathbb{R}, Y)$ the map $t \to f(t + t^{\star})$ belongs to $BC^{\eta}(\mathbb{R}, Y)$, and we have the following estimate holds*

$$e^{-\eta|t^{\star}|} \|f\|_{BC^{\eta}(\mathbb{R},Y)} \leq \left\|f(. + t^{\star})\right\|_{BC^{\eta}(\mathbb{R},Y)} \leq e^{\eta|t^{\star}|} \|f\|_{BC^{\eta}(\mathbb{R},Y)} .$$

Proof By using the triangle inequality, we have

$$|t + t^{\star}| \leq |t| + |t^{\star}| \Leftrightarrow |t + t^{\star}| - |t| \leq |t^{\star}| \Leftrightarrow -\left[|t| - |t + t^{\star}|\right] \leq |t^{\star}|,$$

and

$$|t| = |t + t^{\star} - t^{\star}| \leq |t + t^{\star}| + |-t^{\star}| \Leftrightarrow |t| - |t + t^{\star}| \leq |t^{\star}|,$$

hence we obtain

$$-|t^{\star}| \leq -\left[|t| - |t + t^{\star}|\right] \leq |t^{\star}|.$$

We deduce that

$$e^{-\eta|t^{\star}|} \leq e^{-\eta\left[|t|-|t+t^{\star}|\right]} \leq e^{\eta|t^{\star}|},$$

and by using

$$\sup_{t\in\mathbb{R}} e^{-\eta|t|} \|f(t + t^{\star})\| = \sup_{t\in\mathbb{R}} e^{-\eta\left[|t|-|t+t^{\star}|\right]} e^{-\eta|t+t^{\star}|} \|f(t + t^{\star})\|$$

we obtain by setting $\sigma = t + t^{\star}$,

$$e^{-\eta|t^{\star}|} \sup_{\sigma\in\mathbb{R}} e^{-\eta|\sigma|} \|f(\sigma)\| \leq \sup_{t\in\mathbb{R}} e^{-\eta|t|} \|f(t + t^{\star})\| \leq e^{\eta|t^{\star}|} \sup_{\sigma\in\mathbb{R}} e^{-\eta|\sigma|} \|f(\sigma)\|$$

and the proof is completed. □

The family $\left\{\left(BC^{\eta}(I, Y), \|.\|_{BC^{\eta}(I,Y)}\right)\right\}_{\eta>0}$ forms a **scale of Banach spaces**, that is, if $0 < \zeta < \eta$ then $BC^{\zeta}(I, Y) \subset BC^{\eta}(I, Y)$ and the embedding is continuous; more precisely, we have

$$\|f\|_{BC^{\eta}(I,Y)} \leq \|f\|_{BC^{\zeta}(I,Y)}, \quad \forall f \in BC^{\zeta}(I, Y).$$

Let $(Z, \|.\|_Z)$ be a Banach spaces. From now on, we denote by $\mathrm{Lip}(Y, Z)$ (respectively $\mathrm{Lip}_B(Y, Z)$) the space of Lipschitz (respectively Lipschitz and bounded) maps from Y into Z. Define the semi-norm

$$\|F\|_{\mathrm{Lip}(Y,Z)} := \sup_{x,y\in Y: x\neq y} \frac{\|F(x) - F(y)\|_Z}{\|x - y\|_Y}.$$

The above Lipschitz norm is an extended notion of the supremum norm of the first derivative. So $\|.\|_{\mathrm{Lip}(Y,Z)}$ is only a semi-norm and not a norm, because

$$\|F\|_{\mathrm{Lip}(Y,Z)} = 0$$

whenever F is a constant function.

Definition 4.12 (Center manifold) Let $\eta \in (0, \beta)$ be given. We define the η−**center manifold** or simply **center manifold** $\mathcal{M}_c^\eta$ as the set of the points x in $\mathbb{R}^n$ satisfying the two following properties

(i) There exists $t \in \mathbb{R} \mapsto u(t)$ a complete orbit of the semiflow U passing through x at $t = 0$. That is

$$U(t-s)u(s) = u(t), \forall t \geq s, \text{ and } u(0) = x.$$

(ii) The exponential growth of u is bounded by η when t goes to $\pm\infty$, which means that

$$u \in BC^\eta\left(\mathbb{R}, \mathbb{R}^n\right) \Leftrightarrow \|u(t)\| \leq Me^{\eta|t|}, \forall t \in \mathbb{R} \text{ (for some constant } M > 0).$$

For short the center manifold can be rewritten as

$$\mathcal{M}_c^\eta = \left\{x \in \mathbb{R}^n : \exists u \in BC^\eta\left(\mathbb{R}, \mathbb{R}^n\right) \text{ a complete orbit of } \{U(t)\}_{t\geq 0}, \text{ with } u(0) = x\right\}.$$

Lemma 4.13 (Invariance of the center manifold by the semiflow) *Let $\eta \in (0, \beta)$ be given. Then $\mathcal{M}_c^\eta$ is invariant by the semiflow U. That is $U(t)\mathcal{M}_c^\eta = \mathcal{M}_c^\eta, \forall t \geq 0$.*

Proof Let $t^\star > 0$ be given and let us prove that $U(t^\star)\mathcal{M}_c^\eta \subset \mathcal{M}_c^\eta$. Let $x \in \mathcal{M}_c^\eta$ be given and let $u \in BC^\eta\left(\mathbb{R}, \mathbb{R}^n\right)$ be a complete orbit of U with $u(0) = x$. Then by construction we have $y = u(t^\star) = U(t^\star)x$, so

$$y \in U(t^\star)\mathcal{M}_c^\eta.$$

Moreover $t \mapsto u_{t^\star}(t) = u\left(t + t^\star\right)$ is also a complete orbit, and by Lemma 4.11 we know that the shifted orbit satisfies

$$t \mapsto u_{t^\star}(t) = u\left(t + t^\star\right) \in BC^\eta\left(\mathbb{R}, \mathbb{R}^n\right).$$

One concludes that $y \in \mathcal{M}_c^\eta$.

Conversely, let us prove that $\mathcal{M}_c^\eta \subset U(t^\star)\mathcal{M}_c^\eta$. Let $y \in \mathcal{M}_c^\eta$ be given. Then there exists $u \in BC^\eta\left(\mathbb{R}, \mathbb{R}^n\right)$ a complete orbit of U with $u(0) = y$. Let $x = u(-t^\star)$ and $t \mapsto u_{-t^\star}(t) = u(t - t^\star)$. Then $U(t^\star)x = y$ because $t \mapsto u(t)$ is a complete orbit of U, and x belongs to $\mathcal{M}_c^\eta$ because $t \mapsto u_{-t^\star}(t)$ is a complete orbit of U, and by Lemma 4.15 we deduce that

$$u_{-t^\star} \in BC^\eta\left(\mathbb{R}, \mathbb{R}^n\right).$$

The proof is completed. □

4.3.2 Reduced equation

In this chapter, we will see that, if $\|F\|_{\text{Lip}(\mathbb{R}^n,\mathbb{R}^n)}$ is small enough, then we can find a map $\Psi : X_c \to X_h$ such that

$$\mathcal{M}_c^\eta = \{x_c + \Psi(x_c) : x_c \in X_c\}.$$

Assume for a moment that this property is true. Then we can reformulate the definition of the center manifold by saying that

$$\Pi_c \mathcal{M}_c^\eta = X_c,$$

and

$$x \in \mathcal{M}_c^\eta \Leftrightarrow \Pi_h x = \Psi(\Pi_c x).$$

As consequence of Lemma 4.13, the set $\mathcal{M}_c^\eta$ is invariant by $U(t)$, so that we get

$$U(t)x \in \mathcal{M}_c^\eta, \forall t \geq 0, \forall x \in \mathcal{M}_c^\eta.$$

Therefore we obtain the following property

$$\Pi_h U(t)x = \Psi(\Pi_c U(t)x), \forall t \geq 0, \forall x \in \mathcal{M}_c^\eta.$$

We deduce that on the manifold $\mathcal{M}_c^\eta$, we can reduce the dimension of the system (4.20), by projecting (4.20) on X_c. This leads us to the following equation for $u_c(t) = \Pi_c U(t)x$

$$u_c'(t) = A_c\, u_c(t) + \Pi_c F\big(u_c(t) + \Psi(u_c(t))\big). \quad \textbf{(Reduced Equation)}$$

The advantage of the center manifold is that the dimension of reduced system has the same dimension than the space X_c, which is much smaller than n in general. The center manifold contains all the bounded orbited (for example periodic orbits, homoclinic orbit of an equilibrium, and heteroclinic orbits) of the original system (4.20).

The disadvantage of the center manifold is that the function Ψ is not explicit in genaral. In this chapter, we will see in particular that if all the derivatives up to order m satisfy

$$\Pi_h D^k F(0) = 0, \forall k = 1, \ldots, m, \tag{4.23}$$

then all the derivatives of the center manifold Ψ satisfy the same properties, that is,

$$D^k \Psi(0) = 0, \forall k = 1, \ldots, m.$$

In that case, by using a Taylor expansion at 0 of the center manifold, we can obtain some n^{th}-order approximation locally around 0 for the reduce system. Furthermore, in the next Chapter 5, we will prove that it is always possible to make some changes of variable in order to satisfy the condition (4.23) at any order.

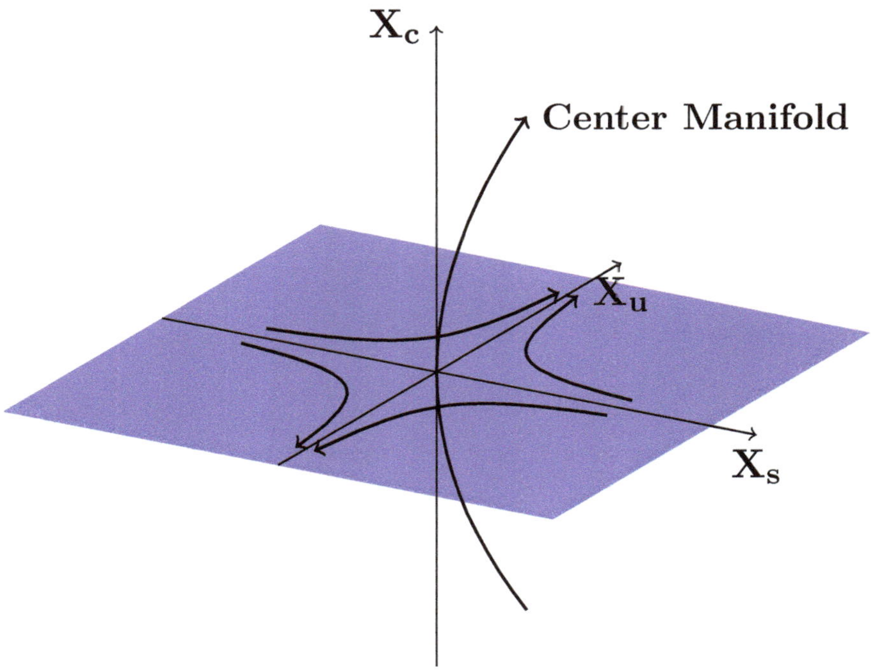

Fig. 4.2: *Schematic representation of the center manifold.*

4.3.3 Existence of the center manifold

In this subsection, we investigate the existence of the center manifold.

Definition 4.14 Let us recall that a function $u : \mathbb{R} \to \mathbb{R}$ is a **complete orbit** for the semiflow $\{U(t)\}_{t\geq 0}$ if and only if the function $t \mapsto u(t)$ on $\mathbb{R}$ satisfies

$$u(t) = U(t-s)u(s), \forall t, s \in \mathbb{R} \text{ with } t \geq s,$$

where $\{U(t)\}_{t\geq 0}$ denotes the continuous semiflow generated by (4.21).

This is equivalent to say that for each $s \in \mathbb{R}$ (fixed) the function $t \mapsto u(t)$ from $[s, \infty)$, satisfies

$$u(t) = u(s) + A \int_0^{t-s} u(s+r)dr + \int_0^{t-s} F\left(u(s+r)\right) dr, \forall t \geq s,$$

or equivalently by using the variation of constant formula

$$u(t) = e^{A(t-s)}u(s) + \int_s^t e^{A(t-\sigma)}F(u(\sigma))d\sigma, \forall t \geq s,$$

or (for short) by using the convolutions

$$u(t) = e^{A(t-s)}u(s) + \left(e^{A\cdot} * F\left(u(s+.)\right)\right)(t-s), \forall t \geq s. \quad (4.24)$$

In the above intergal equation s and $u(s)$ are fixed, and the map $t \in [s, +\infty) \mapsto u(t)$ is a fixed solution of the above integral equation.

Recall that for $0 < \eta < \beta$, the $\eta-$center manifold $\mathcal{M}_c^\eta$ in defined in Definition 4.12. Recall also that from Lemma 4.11, for all $u \in BC^\eta(\mathbb{R}, \mathbb{R}^n)$ and each $\tau \in \mathbb{R}$ one has

$$e^{-\eta|\tau|}\|u\|_{BC^\eta(\mathbb{R},\mathbb{R}^n)} \leq \|u(.+\tau)\|_{BC^\eta(\mathbb{R},\mathbb{R}^n)} \leq e^{\eta|\tau|}\|u\|_{BC^\eta(\mathbb{R},\mathbb{R}^n)}.$$

By Lemma 4.13, for each $\eta > 0$, $\mathcal{M}_c^\eta$ is invariant under the semiflow $\{U(t)\}_{t\geq 0}$, that is,

$$U(t)\mathcal{M}_c^\eta = \mathcal{M}_c^\eta, \quad \forall t \geq 0.$$

Moreover, we say that $\{U(t)\}_{t\geq 0}$ is **reduced on** $\mathcal{M}_c^\eta$ if there exists a map $\Psi : X_c \to X_h$ such that

$$\mathcal{M}_c^\eta = \text{Graph}(\Psi) = \{x_c + \Psi(x_c) : x_c \in X_c\}.$$

Before proving the main results of this section, we first provide some important preliminary lemmas.

Lemma 4.15 *Let Assumption 4.3 be satisfied. Let $\tau > 0$ be fixed. Then for each $f \in C([0,\tau], \mathbb{R}^n)$, each $t \in [0,\tau]$ and $k = s, c, u$, we have*

$$\Pi_k\left(e^{A\cdot} * f(.)\right)(t) = \left(e^{A\cdot} * \Pi_k f(.)\right)(t) = \left(e^{A_k\cdot} * \Pi_k f(.)\right)(t), \forall t \geq 0. \quad (4.25)$$

Furthermore, for each $\gamma > -\beta$, there exists $\widehat{C}_{s,\gamma} := \dfrac{2M_s}{\beta+\gamma} > 0$ such that for each $f \in C([0,\tau], \mathbb{R}^n)$ and each $t \in [0,\tau]$, we have

$$e^{-\gamma t}\|\Pi_s\left(e^{A\cdot} * f\right)(t)\| \leq \widehat{C}_{s,\gamma} \sup_{s\in[0,t]} e^{-\gamma s}\|\Pi_s f(s)\|. \quad (4.26)$$

Proof From Lemma 4.5 we have for each $\varepsilon > 0$

$$\|e^{At}\Pi_s\|_{\mathcal{L}(\mathbb{R}^n)} \leq M_s e^{(-\beta+\varepsilon)t}, \forall t \geq 0.$$

Therefore we obtain

$$\begin{aligned} e^{-\gamma t}\left\|\Pi_s\left(e^{A\cdot} * f\right)(t)\right\| &\leq e^{-\gamma t}\int_0^t \|e^{A(t-s)}\Pi_s f(s)\|\,\mathrm{d}s \\ &\leq M_s \int_0^t e^{(-\beta+\varepsilon-\gamma)(t-s)}\mathrm{d}s \,\sup_{s\in[0,t]} e^{-\gamma s}\|\Pi_s f(s)\|, \end{aligned}$$

and

$$\int_0^t e^{(-\beta+\varepsilon-\gamma)(t-s)}ds = \int_0^t e^{-(\beta+\gamma-\varepsilon)s}ds \leq \frac{1}{\beta+\gamma-\varepsilon}.$$

The result follows by choosing $\varepsilon = (\beta+\gamma)/2$. □

Convolution's formula: By using the convolution formula (4.22) we deduce that

$$\left(e^{A\cdot} * f\right)(t) = e^{A(t-s)}\left(e^{A\cdot} * f\right)(s) + \left(e^{A\cdot} * f\left(s+.\right)\right)(t-s), \tag{4.27}$$

whenever $t, s \in [0,\tau]$ with $s \leq t$, and $f \in C\left([0,\tau], \mathbb{R}^n\right)$ (for some $\tau > 0$).

By using the above formula, namely (4.27), together with the property of $e^{A_s t}$ we obtain the following lemma.

Lemma 4.16 *Let Assumption 4.3 be satisfied. Then the following properties hold true:*

(i) *For each $\eta \in [0,\beta)$, each $f \in BC^\eta(\mathbb{R},\mathbb{R}^n)$, and each $t \in \mathbb{R}$,*

$$K_s(f)(t) := \lim_{\sigma\to-\infty} \Pi_s\left(e^{A\cdot} * f(\sigma+.)\right)(t-\sigma) \text{ exists.}$$

(ii) *For each $\eta \in [0,\beta)$, K_s is a bounded linear operator from $BC^\eta(\mathbb{R},\mathbb{R}^n)$ into itself. More precisely, for each $\nu \in (-\beta,0)$, we have*

$$\|K_s\|_{\mathcal{L}(BC^\eta(\mathbb{R},\mathbb{R}^n),BC^\eta(\mathbb{R},X_s))} \leq \widehat{C}_{s,\nu}\, \|\Pi_s\|_{\mathcal{L}(\mathbb{R}^n)}, \forall\eta \in [0,-\nu],$$

where $\widehat{C}_{s,\nu} = \dfrac{2M_s}{\beta+\nu} > 0$ is the constant introduced in (4.26).

(iii) *For each $\eta \in [0,\beta)$, each $f \in BC^\eta(\mathbb{R},\mathbb{R}^n)$, and each $t, s \in \mathbb{R}$ with $t \geq s$,*

$$K_s(f)(t) - e^{A_s(t-s)}K_s(f)(s) = \Pi_s\left(e^{A\cdot} * f(s+.)\right)(t-s).$$

Remark 4.17 The definition of K_s, using a limit as above, can be extended to infinite dimensional systems (see Magal and Ruan [240]). Nevertheless for ordinary differential equation we have the following equivalent reformulation

$$\begin{aligned} K_s(f)(t) &= \lim_{\sigma\to-\infty}\int_0^{t-\sigma} e^{A_s(t-\sigma-l)}\Pi_s f(\sigma+l)dl \\ &= \lim_{\sigma\to-\infty}\int_\sigma^t e^{A_s(t-s)}\Pi_s f(s)ds \\ &= \lim_{\sigma\to-\infty}\int_0^{t-\sigma} e^{A_s\theta}\Pi_s f(t-\theta)d\theta. \end{aligned}$$

Hence this leads us to the following explicit formula

$$K_s(f)(t) = \int_0^{+\infty} e^{A_s\theta}\Pi_s f(t-\theta)d\theta.$$

Proof Let us start by proving (i) and (iii). Let $\eta \in (0, \beta)$ be fixed. By using (4.27), we have for each $t, s, r \in \mathbb{R}$ with $r \le s \le t$, and each $f \in BC^{\eta}(\mathbb{R}, \mathbb{R}^n)$ that

$$\left(e^{A\cdot} * f(r+.)\right)(t-r) = e^{A(t-s)}\left(e^{A\cdot} * f(r+.)\right)(s-r) + \left(e^{A\cdot} * f(s+.)\right)(t-s).$$

Projecting the above equation on X_s yields, for all $t, s, r \in \mathbb{R}$ with $r \le s \le t$,

$$\Pi_s\left(e^{A\cdot} * f(r+.)\right)(t-r) = e^{A_s(t-s)}\Pi_s\left(e^{A\cdot} * f(r+.)\right)(s-r) + \Pi_s\left(e^{A\cdot} * f(s+.)\right)(t-s). \tag{4.28}$$

Let $\nu \in (-\beta, -\eta)$ be fixed. Then by using (4.26) and (4.28), we have for all $t, s, r \in \mathbb{R}$ with $r \le s \le t$

$$\begin{aligned}
&\left\|\Pi_s\left(e^{A\cdot} * f(r+.)\right)(t-r) - \Pi_s\left(e^{A\cdot} * f(s+.)\right)(t-s)\right\| \\
&= \left\|e^{A_s(t-s)}\Pi_s\left(e^{A\cdot} * f(r+.)\right)(s-r)\right\| \\
&\le M_s e^{-\beta(t-s)}\widehat{C}_{s,\nu}e^{\nu(s-r)} \sup_{l\in[0,s-r]} e^{-\nu l}\|\Pi_s f(r+l)\| \\
&= M_s\widehat{C}_{s,\nu}e^{-\beta(t-s)}e^{\nu(s-r)} \sup_{\sigma\in[r,s]} e^{-\nu(\sigma-r)}\|\Pi_s f(\sigma)\| \\
&= M_s\widehat{C}_{s,\nu}e^{-\beta(t-s)}e^{\nu s} \sup_{\sigma\in[r,s]} e^{-\nu\sigma}e^{\eta|\sigma|}e^{-\eta|\sigma|}\|\Pi_s f(\sigma)\| \\
&\le \|\Pi_s f\|_{BC^{\eta}(\mathbb{R},\mathbb{R}^n)} M_s\widehat{C}_{s,\nu}e^{-\beta(t-s)}e^{\nu s} \sup_{\sigma\in[r,s]} e^{-\nu\sigma}e^{\eta|\sigma|}.
\end{aligned}$$

Hence, for all $s, r \in \mathbb{R}_-$ with $s \ge r$, we obtain

$$\begin{aligned}
&\left\|\Pi_s\left(e^{A\cdot} * f(r+.)\right)(t-r) - \Pi_s\left(e^{A\cdot} * f(s+.)\right)(t-s)\right\| \\
&\le \|\Pi_s f\|_{BC^{\eta}(\mathbb{R},\mathbb{R}^n)} M_s\widehat{C}_{s,\nu}e^{-\beta(t-s)}e^{\nu s} \sup_{\sigma\in[r,s]} e^{-(\nu+\eta)\sigma}.
\end{aligned}$$

Because $-(\nu+\eta) > 0$, we have

$$\begin{aligned}
&\left\|\Pi_s\left(e^{A\cdot} * f(r+.)\right)(t-r) - \Pi_s\left(e^{A\cdot} * f(s+.)\right)(t-s)\right\| \\
&\le \|\Pi_s f\|_{BC^{\eta}(\mathbb{R},\mathbb{R}^n)} M_s\widehat{C}_{s,\nu}e^{-\beta(t-s)}e^{\nu s}e^{-(\nu+\eta)s} \\
&= \|\Pi_s f\|_{BC^{\eta}(\mathbb{R},\mathbb{R}^n)} M_s\widehat{C}_{s,\nu}e^{-\beta t}e^{(\beta-\eta)s}.
\end{aligned}$$

Since $\beta - \eta > 0$, by using Cauchy sequences, we deduce that

$$K_s(f)(t) = \lim_{s\to-\infty}\Pi_s\left(e^{A\cdot} * f(s+.)\right)(t-s) \text{ exists.}$$

Taking the limit as r goes to $-\infty$ in (4.28), we obtain (iii).

Let us prove the property (ii). To that aim, let $\nu \in (-\beta, 0)$ and $\eta \in [0, -\nu]$ be fixed. Then for each $f \in BC^{\eta}(\mathbb{R}, \mathbb{R}^n)$ and each $t \in \mathbb{R}$, we have

$$\begin{aligned}
\|K_s(f)(t)\| &= \lim_{r\to-\infty}\left\|\Pi_s\left(e^{A\cdot} * f(r+.)\right)(t-r)\right\| \\
&\le \widehat{C}_{s,\nu}\limsup_{r\to-\infty} e^{\nu(t-r)}\sup_{l\in[0,t-r]} e^{-\nu l}\|\Pi_s f(r+l)\| \\
&= \widehat{C}_{s,\nu}\limsup_{r\to-\infty} e^{\nu(t-r)}\sup_{\sigma\in[r,t]} e^{-\nu(\sigma-r)}\|\Pi_s f(\sigma)\| \\
&= \widehat{C}_{s,\nu}\limsup_{r\to-\infty} e^{\nu t}\sup_{\sigma\in[r,t]} e^{-\nu\sigma}e^{\eta|\sigma|}e^{-\eta|\sigma|}\|\Pi_s f(\sigma)\| \\
&= \widehat{C}_{s,\nu}e^{\nu t}\|\Pi_s f\|_\eta \sup_{\sigma\in(-\infty,t]} e^{-\nu\sigma}e^{\eta|\sigma|}.
\end{aligned}$$

Since $(\nu+\eta)\le 0$, we deduce that if $t\le 0$,

$$\begin{aligned}
e^{-\eta|t|}\|K_s(f)(t)\| &\le \widehat{C}_{s,\nu}e^{(\nu+\eta)t}\|\Pi_s f\|_\eta \sup_{\sigma\in(-\infty,t]} e^{-(\nu+\eta)\sigma} = \widehat{C}_{s,\nu}e^{(\nu+\eta)t}\|\Pi_s f\|_\eta e^{-(\nu+\eta)t} \\
&= \widehat{C}_{s,\nu}\|\Pi_s f\|_\eta,
\end{aligned}$$

and since $(\eta-\nu)>0$, it follows that if $t\ge 0$,

$$\begin{aligned}
e^{-\eta|t|}\|K_s(f)(t)\| &\le \widehat{C}_{s,\nu}e^{(\nu-\eta)t}\|\Pi_s f\|_\eta \sup_{\sigma\in(-\infty,t]} e^{-\nu\sigma}e^{\eta|\sigma|} \\
&\le \widehat{C}_{s,\nu}\|\Pi_s f\|_\eta e^{(\nu-\eta)t}\max\Big(\sup_{\sigma\in(-\infty,0]} e^{-(\nu+\eta)\sigma}, \sup_{\sigma\in[0,t]} e^{(\eta-\nu)\sigma}\Big) \\
&= \widehat{C}_{s,\nu}\|\Pi_s f\|_\eta e^{(\nu-\eta)t}e^{(\eta-\nu)t} = \widehat{C}_{s,\nu}\|\Pi_s f\|_\eta .
\end{aligned}$$

This completes the proof. □

Lemma 4.18 *Let Assumption 4.3 be satisfied. Let $\eta\in[0,\beta)$ be fixed. Then we have the following:*

(i) *For each $f\in BC^\eta(\mathbb{R},\mathbb{R}^n)$ and each $t\in\mathbb{R}$,*

$$K_u(f)(t) := -\int_t^{+\infty} e^{-A_u(l-t)}\Pi_u f(l)dl := -\lim_{r\to+\infty}\int_t^r e^{-A_u(l-t)}\Pi_u f(l)dl$$

exists.

(ii) *K_u is a bounded linear operator from $BC^\eta(\mathbb{R},\mathbb{R}^n)$ into $BC^\eta(\mathbb{R},X_u)$ and*

$$\|K_u\|_{\mathcal{L}(BC^\eta(\mathbb{R},\mathbb{R}^n))} \le \frac{M_u\|\Pi_u\|_{\mathcal{L}(\mathbb{R}^n)}}{(\beta-\eta)}.$$

(iii) *For each $f\in BC^\eta(\mathbb{R},\mathbb{R}^n)$ and each $t,s\in\mathbb{R}$ with $t\ge s$,*

$$K_u(f)(t) - e^{A_u(t-s)}K_u(f)(s) = \Pi_u\left(e^{A\cdot} * f(s+.)\right)(t-s).$$

Proof By Lemma 4.5, we have

$$\left\|e^{-A_u t}\right\|_{\mathcal{L}(X_u)} \le M_u e^{-\beta t}, \forall t\ge 0. \tag{4.29}$$

By using (4.29) and the same argument as in the proof of Lemma 4.16, we obtain (i) and (ii). Moreover, for each $s, t, r \in \mathbb{R}$ with $s \leq t \leq r$, we have

$$\begin{aligned}\int_s^r e^{A_u(s-l)}\Pi_u f(l)dl &= \int_s^t e^{A_u(s-l)}\Pi_u f(l)dl + \int_t^r e^{A_u(s-l)}\Pi_u f(l)dl \\ &= \int_s^t e^{A_u(s-l)}\Pi_u f(l)dl + e^{A_u(s-t)}\int_t^r e^{A_u(t-l)}\Pi_u f(l)dl.\end{aligned}$$

It follows that

$$e^{A_u(t-s)}\int_s^r e^{A_u(s-l)}\Pi_u f(l)dl = \int_s^t e^{A_u(t-l)}\Pi_u f(l)dl + \int_t^r e^{A_u(t-l)}\Pi_u f(l)dl.$$

When $r \to +\infty$, we obtain for all $s, t \in \mathbb{R}$ with $s \leq t$ that

$$\begin{aligned}-e^{A_u(t-s)}K_{u,\eta}(f)(s) &= \int_0^{t-s} e^{A_u(t-s-r)}\Pi_u f(s+r)dr - K_{u,\eta}(f)(t) \\ &= \Pi_u\left(e^{A.} * f(s+.)\right)(t-s) - K_{u,\eta}(f)(t).\end{aligned}$$

This gives (iii). □

Lemma 4.19 *Let Assumption 4.3 be satisfied. Let $\eta \in (0, \beta)$ be fixed. For each $x_c \in X_c$, each $f \in BC^\eta(\mathbb{R}, \mathbb{R}^n)$, and each $t \in \mathbb{R}$, denote*

$$K_1(x_c)(t) := e^{A_c t}x_c, \quad K_c(f)(t) := \int_0^t e^{A_c(t-s)}\Pi_c f(s)ds.$$

Then K_1 (respectively K_c) is a bounded linear operator from X_c into $BC^\eta(\mathbb{R}, X_c)$ (respectively from $BC^\eta(\mathbb{R}, \mathbb{R}^n)$ into $BC^\eta(\mathbb{R}, X_c)$, and

$$\|K_1\|_{\mathcal{L}(X_c, BC^\eta(\mathbb{R},\mathbb{R}^n))} \leq \max\left(\sup_{t\geq 0}\left\|e^{(A_c-\eta I)t}\right\|, \sup_{t\geq 0}\left\|e^{-(A_c+\eta I)t}\right\|\right),$$

$$\|K_c\|_{\mathcal{L}(BC^\eta(\mathbb{R},\mathbb{R}^n))} \leq \|\Pi_c\|_{\mathcal{L}(\mathbb{R}^n)} \max\left(\int_0^{+\infty}\left\|e^{(A_c-\eta I)l}\right\|dl, \int_0^{+\infty}\left\|e^{-(A_c+\eta I)l}\right\|dl\right).$$

Proof The proof is straightforward. □

Lemma 4.20 *Let Assumption 4.3 be satisfied. Let $\eta \in (0, \beta)$ be fixed. Then a function*

$$u \in BC^\eta(\mathbb{R}, \mathbb{R}^n),$$

is a complete orbit for the semiflow $\{U(t)\}_{t\geq 0}$ if and only if

$$u(t) = K_1(\Pi_c u(0))(t) + K_c(F(u(.)))(t) + K_u(F(u(.)))(t) + K_s(F(u(.)))(t), \forall t \in \mathbb{R}. \tag{4.30}$$

Proof Let $u \in BC^\eta(\mathbb{R}, \mathbb{R}^n)$ be given. Assume first that u is a complete orbit of $\{U(t)\}_{t\geq 0}$. Then for $k \in \{c, u, s\}$ we have, for all $l, r \in \mathbb{R}$ with $r \leq l$,

$$\Pi_k u(l) = e^{A_k(l-r)}\Pi_k u(r) + \int_r^l e^{A_k(l-s)}\Pi_k F(u(s))ds.$$

Thus this rewrites as

$$\frac{d\Pi_k u(t)}{dt} = A_k\Pi_k u(t) + \Pi_k F(u(t)), \quad \forall t \in \mathbb{R}.$$

From this ordinary differential equation and $k = c$, we first deduce that

$$\Pi_c u(t) = e^{A_c t}\Pi_c u(0) + \int_0^t e^{A_c(t-s)}\Pi_c F(u(s))ds, \forall t \in \mathbb{R}. \tag{4.31}$$

Next, for each $t, l \in \mathbb{R}$, we have

$$\Pi_u u(t) = e^{A_u(t-l)}\Pi_u u(l) + \int_l^t e^{A_u(t-s)}\Pi_u F(u(s))ds, \forall t, l \in \mathbb{R}.$$

It follows that for each $l \geq 0$,

$$\left\| e^{-A_u(l-t)}\Pi_u u(l)\right\| \leq M_u \left\|\Pi_u\right\|_{\mathcal{L}(\mathbb{R}^n)} e^{-\beta(l-t)} e^{\eta l} \left\|u\right\|_{BC^\eta(\mathbb{R},\mathbb{R}^n)} .$$

So when l goes to $+\infty$, we obtain

$$\Pi_u u(t) = -\int_t^{+\infty} e^{A_u(t-s)}\Pi_u F(u(s))ds, \quad \forall t \in \mathbb{R}. \tag{4.32}$$

Furthermore, we have for all $t, l \in \mathbb{R}$ with $t \geq l$ that

$$\Pi_s u(t) = e^{A_s(t-l)}\Pi_s u(l) + \Pi_s\left(e^{A\cdot} * F(u(l+.))\right)(t-l),$$

and for each $l \leq 0$ that

$$\left\| e^{A_s(t-l)}\Pi_s u(l)\right\| \leq e^{-\beta t} M_s \left\|\Pi_s u\right\|_\eta e^{(\beta-\eta)l}.$$

Taking $l \to -\infty$, we obtain

$$\Pi_s u(t) = K_s\left(F(u(.))\right)(t), \quad \forall t \in \mathbb{R}. \tag{4.33}$$

Finally, by summing up (4.31), (4.32), and (4.33), we obtain (4.30).

Conversely, assume that u is a solution of (4.30). Then

$$\Pi_c u(t) = e^{A_c t}\Pi_c u(0) + \int_0^t e^{A_c(t-s)}\Pi_c F(u(s))ds, \quad \forall t \in \mathbb{R}.$$

It follows that

$$\frac{d\Pi_c u(t)}{dt} = A_c\Pi_c u(t) + \Pi_c F(u(t)), \quad \forall t \in \mathbb{R}.$$

Thus, for $l, r \in \mathbb{R}_-$ with $r \leq l$, we have

$$\Pi_c u(l) = e^{A(t-s)}\Pi_c u(r) + \Pi_c \left(e^{A.} * F(u(s+.))\right)(t-s).$$

Moreover, by using Lemma 4.16 (iii) and Lemma 4.18 (iii), we deduce that for all $t, s \in \mathbb{R}$ with $t \geq s$

$$\Pi_s u(t) - e^{A(t-s)}\Pi_s u(s) = \Pi_s \left(e^{A.} * F(u(s+.))\right)(t-s),$$
$$\Pi_u u(t) - e^{A(t-s)}\Pi_u u(s) = \Pi_u \left(e^{A.} * F(u(s+.))\right)(t-s).$$

Therefore, u satisfies (4.24) and is a complete orbit of $\{U(t)\}_{t\geq 0}$. □

Let $\eta \in (0, \beta)$ be fixed. We rewrite (4.30) as the following fixed point problem: To find $u \in BC^{\eta}(\mathbb{R}, \mathbb{R}^n)$ such that

$$u = K_1(\Pi_c u(0)) + K_2\Phi_F(u), \tag{4.34}$$

where K_1 has been defined in Lemma 4.19, Φ_F is a *Nemytskii operator* which is defined by

$$\Phi_F(u)(t) = F(u(t)), \quad \forall t \in \mathbb{R}.$$

and

$$K_2 \in \mathcal{L}\left(BC^{\eta}(\mathbb{R}, \mathbb{R}^n), BC^{\eta}(\mathbb{R}, \mathbb{R}^n)\right),$$

is the linear operator defined by

$$K_2 = K_c + K_s + K_u.$$

Moreover, since $F \in \text{Lip}(\mathbb{R}^n, \mathbb{R}^n)$, one has

$$\Phi_F \in \text{Lip}\left(BC^{\eta}(\mathbb{R}, \mathbb{R}^n), BC^{\eta}(\mathbb{R}, \mathbb{R}^n)\right), \text{ and } \|\Phi_F\|_{\text{Lip}} \leq \|F\|_{\text{Lip}}.$$

In addition, the linear operators K_1 and K_2 enjoy the following estimates:

$$\|K_1\|_{\mathcal{L}(X_c, BC^{\eta}(\mathbb{R},\mathbb{R}^n))} \leq \max\left(\sup_{t\geq 0}\left\|e^{(A_c-\eta Id)t}\right\|, \sup_{t\geq 0}\left\|e^{-(A_c+\eta Id)t}\right\|\right),$$

and for each $\nu \in (-\beta, 0)$, we have

$$\|K_2\|_{\mathcal{L}(BC^{\eta}(\mathbb{R},\mathbb{R}^n))} \leq \gamma(\nu, \eta), \forall \eta \in (0, -\nu],$$

where

$$\begin{aligned}\gamma(\nu, \eta) := {} & \frac{2M_s}{\beta + \nu}\|\Pi_s\|_{\mathcal{L}(\mathbb{R}^n)} + \frac{M_u\|\Pi_u\|_{\mathcal{L}(\mathbb{R}^n)}}{(\beta - \eta)} \\ & + \|\Pi_c\|_{\mathcal{L}(\mathbb{R}^n)} \max\left(\int_0^{+\infty}\left\|e^{(A_c-\eta I)l}\right\| dl, \int_0^{+\infty}\left\|e^{-(A_c+\eta I)l}\right\| dl\right).\end{aligned} \tag{4.35}$$

Recall also that, due to Lemma 4.20, the η-center manifold is given by

$$\mathcal{M}_c^{\eta} = \{x \in \mathbb{R}^n : \exists u \in BC^{\eta}(\mathbb{R}, \mathbb{R}^n) \text{ a solution of (4.34) and } u(0) = x\}. \tag{4.36}$$

We are now in position to state and prove the existence of a center manifold for (4.24). We refer the readers to Vanderbauwhede and Iooss [340, Theorem 1, p.129] and to Magal and Ruan [239] for extension to semilinear equations with non-densely defined operators.

Theorem 4.21 *Let Assumption 4.3 be satisfied. Let $\eta \in (0,\beta)$ be fixed. Assume that $F \in \text{Lip}(\mathbb{R}^n,\mathbb{R}^n)$ with*

$$\|F\|_{\text{Lip}(\mathbb{R}^n,\mathbb{R}^n)}\ \|K_2\|_{\mathcal{L}(BC^\eta(\mathbb{R},\mathbb{R}^n))} < 1.$$

Then there exists a Lipschitz continuous map $\Psi : X_c \to X_h$ such that

$$\mathcal{M}_c^\eta = \{x_c + \Psi(x_c) \in \mathbb{R}^n : x_c \in X_c\},$$

or equivalently

$$\mathcal{M}_c^\eta = \{x \in \mathbb{R}^n : \Pi_c x = x_c, \Pi_h x = \Psi(x_c),\ and\ x_c \in X_c\}.$$

Moreover, we have the following properties:

(i)

$$\sup_{x_c \in X_c} \|\Psi(x_c)\| \le \|K_s + K_u\|_{\mathcal{L}(BC^\eta(\mathbb{R},\mathbb{R}^n))} \sup_{x\in\mathbb{R}^n} \|\Pi_h F(x)\|.$$

(ii)

$$\|\Psi\|_{\text{Lip}(X_c,X_h)} \le \frac{\|K_s + K_u\|_{\mathcal{L}(BC^\eta(\mathbb{R},\mathbb{R}^n))}\ \|F\|_{\text{Lip}(\mathbb{R}^n,\mathbb{R}^n)}\ \|K_1\|_{\mathcal{L}(X_c,BC^\eta(\mathbb{R},\mathbb{R}^n))}}{1 - \|K_2\|_{\mathcal{L}(BC^\eta(\mathbb{R},\mathbb{R}^n))}\ \|F\|_{\text{Lip}(\mathbb{R}^n,\mathbb{R}^n)}}. \tag{4.37}$$

(iii) *For each $u \in C(\mathbb{R},\mathbb{R}^n)$, the following statement are equivalent*

(a) *$u \in BC^\eta(\mathbb{R},\mathbb{R}^n)$ is a complete orbit of $\{U(t)\}_{t\ge0}$.*

(b) *The map $u_c : \mathbb{R} \to X_c$ is a solution of the reduced equation*

$$\frac{du_c(t)}{dt} = A_c u_c(t) + \Pi_c F\left[u_c(t) + \Psi\left(u_c(t)\right)\right], \tag{4.38}$$

and

$$u(t) = u_c(t) + \Psi(u_c(t)), \forall t \in \mathbb{R}.$$

Proof (i) Since $\|F\|_{\text{Lip}}\ \|K_2\|_{\mathcal{L}(BC^\eta(\mathbb{R},\mathbb{R}^n))} < 1$ we can use the global inverse theorem (see Theorem 3.1 in Chapter 3), and we deduce that the map $(Id - K_2\Phi_F)$ is invertible, the inverse $(Id - K_2\Phi_F)^{-1}$ is Lipschitz continuous, and

$$\left\|(Id - K_2\Phi_F)^{-1}\right\|_{\text{Lip}(BC^\eta(\mathbb{R},\mathbb{R}^n))} \le \frac{1}{1 - \|K_2\|_{\mathcal{L}(BC^\eta(\mathbb{R},\mathbb{R}^n))}\ \|F\|_{\text{Lip}(\mathbb{R}^n,\mathbb{R}^n)}}. \tag{4.39}$$

Let $x \in \mathbb{R}^n$ be fixed. By Lemma 4.20, we know that $x \in V_\eta$ if and only if there exists $u_{\Pi_c x} \in BC^\eta(\mathbb{R},\mathbb{R}^n)$, such that $u_{\Pi_c x}(0) = x$ and

$$u_{\Pi_c x}(t) = K_1(\Pi_c x)(t) + K_2\Phi_F\left(u_{\Pi_c x}\right)(t), \forall t \in \mathbb{R}.$$

So

$$\mathcal{M}_c^\eta = \left\{ (Id - K_2\Phi_F)^{-1} K_1(x_c)\,(0) : x_c \in X_c \right\}.$$

We define $\Psi : X_c \to X_h$ by

$$\Psi(x_c) = \Pi_h \left[(Id - K_2\Phi_F)^{-1} K_1(x_c) \right](0), \forall x_c \in X_c.$$

Then we have

$$\mathcal{M}_c^\eta = \{ x_c + \Psi(x_c) : x_c \in X_c \}.$$

For each $x_c \in X_c$, set

$$u_{x_c} = (Id - K_2\Phi_F)^{-1} K_1(x_c),$$

which is equivalent to

$$u_{x_c} = K_1(x_c) + K_2\Phi_F\left(u_{x_c}\right).$$

By projecting this last equation on X_h, we obtain

$$\Pi_h u_{x_c} = [K_s + K_u]\,\Phi_{\Pi_h F}\left(u_{x_c}\right),$$

and we obtain

$$\Psi(x_c) = \Pi_h u_{x_c}(0) = [K_s + K_u]\,\Phi_{\Pi_h F}\left(u_{x_c}\right)(0) \tag{4.40}$$

and assertion (i) follows.

Assertion (ii) follows from (4.39) and (4.40).

It remains to prove assertion (iii). Assume first that $u \in BC^\eta(\mathbb{R}, \mathbb{R}^n)$ is a complete orbit of $\{U(t)\}_{t \geq 0}$. By using the invariance property of the center manifold (see Lemma 4.13), we deduce that

$$u(t) \in \mathcal{M}_c^\eta, \forall t \in \mathbb{R},$$

and we deduce that

$$\Pi_h u(t) = \Psi(\Pi_c u(t)), \quad \forall t \in \mathbb{R}.$$

Moreover, by projecting (4.24) on X_c, we have for each $t, s \in \mathbb{R}$ with $t \geq s$ that

$$\Pi_c u\,(t) = e^{A_c(t-s)}\Pi_c u(s) + \int_0^{t-s} e^{A_c(t-s-l)}\Pi_c F\,(u\,(s+l))\,dl.$$

Thus, the map $t \mapsto \Pi_c u\,(t)$ is a solution of the reduced equation (4.38).

Conversely, assume that $u \in C(\mathbb{R}, \mathbb{R}^n)$ satisfies (iii)-(b). Then we have

$$\Pi_h u(t) = \Psi(\Pi_c u(t)), \quad \forall t \in \mathbb{R},$$

and $\Pi_c u(.) : \mathbb{R} \to X_c$ becomes a solution of (4.38). Set $x = u(0)$. We know that $x \in \mathcal{M}_c^\eta$ (since $\Pi_h x = \Psi(\Pi_c x)$), therefore by using the definition of the center manifold $\mathcal{M}_c^\eta$, we deduce that there exists a function $v \in BC^\eta(\mathbb{R}, \mathbb{R}^n)$, which is a complete orbit of $\{U(t)\}_{t\geq 0}$, and $v(0) = x$. But since $\mathcal{M}_c^{\eta}$ is invariant under the semiflow, we deduce that

$$\Pi_h v(t) = \Psi(\Pi_c v(t)), \quad \forall t \in \mathbb{R},$$

and $\Pi_c v(.) : \mathbb{R} \to X_c$ is a solution of (4.38). Finally, since $\Pi_c v(0) = \Pi_c u(0)$, and since F and Ψ are Lipschitz continuous, we deduce that (4.38) has at most one solution. It follows that

$$\Pi_c v(t) = \Pi_c u(t), \forall t \in \mathbb{R},$$

and by construction

$$\Pi_h v(t) = \Psi(\Pi_c v(t)) = \Psi(\Pi_c u(t)) = \Pi_h u(t), \quad \forall t \in \mathbb{R}.$$

Thus, we get

$$u(t) = v(t), \ \forall t \in \mathbb{R}.$$

Therefore, $u \in BC^{\eta}(\mathbb{R}, \mathbb{R}^n)$ is a complete orbit of $\{U(t)\}_{t\geq 0}$. □

We defined $\text{Lip}_B(\mathbb{R}^n, \mathbb{R}^n)$, as the space of the maps $F : \mathbb{R}^n \to \mathbb{R}^n$ which are Lipschitz continuous and bounded, that is satisfying

$$\|F\|_{\text{Lip}} = \sup_{x,y \in \mathbb{R}^n : x \neq y} \frac{\|F(x) - F(y)\|}{\|x - y\|} < +\infty.$$

and

$$\|F\|_{\infty} = \sup_{x \in \mathbb{R}^n} \|F(x)\| < +\infty.$$

Lemma 4.22 *Let Assumption 4.3 be satisfied. Assume that $F \in \text{Lip}_B(\mathbb{R}^n, \mathbb{R}^n)$. Then one has*

$$\mathcal{M}_c^{\eta} = \mathcal{M}_c^{\xi}, \quad \forall \eta, \xi \in (0, \beta).$$

Proof Let $\eta, \xi \in (0, \beta)$ be such that $\xi < \eta$. Let $x \in \mathcal{M}_c^{\xi}$ be given. By the definition of $\mathcal{M}_c^{\xi}$ there exists $u \in BC^{\xi}(\mathbb{R}, \mathbb{R}^n)$, a complete orbit of $\{U(t)\}_{t\geq 0}$, such that $u(0) = x$. But $BC^{\xi}(\mathbb{R}, \mathbb{R}^n) \subset BC^{\eta}(\mathbb{R}, \mathbb{R}^n)$, so $u \in BC^{\eta}(\mathbb{R}, \mathbb{R}^n)$, and we deduce that $x \in \mathcal{M}_c^{\eta}$.

Conversely, let $x \in \mathcal{M}_c^{\eta}$ be fixed. By the definition of $\mathcal{M}_c^{\eta}$ there exists $u \in BC^{\eta}(\mathbb{R}, \mathbb{R}^n)$, a complete orbit of $\{U(t)\}_{t\geq 0}$, such that $u(0) = x$. By Lemma 4.20 we deduce that u is a solution of

$$u = K_1(\Pi_c u(0)) + K_2 \Phi_F(u).$$

But $K_1(\Pi_c u(0)) \in BC^{\xi}(\mathbb{R}, \mathbb{R}^n)$ and F is bounded, so we have $\Phi_F(u) \in BC^0(\mathbb{R}, \mathbb{R}^n) \subset BC^{\xi}(\mathbb{R}, \mathbb{R}^n)$ and

$$K_2 \Phi_F(u) \in BC^{\xi}(\mathbb{R}, \mathbb{R}^n).$$

Hence, $u \in BC^{\xi}(\mathbb{R}, \mathbb{R}^n)$ and

$$u = K_1(\Pi_c u(0)) + K_2 \Phi_F(u).$$

Using again Lemma 4.20 once more, we obtain that $x \in \mathcal{M}_c^{\xi}$. □

4.3.4 Smoothness of the Center Manifold

In the sequel, we will use the following notations. Let $k \geq 1$ be an integer, let $Y_1, Y_2, .., Y_k$, Y and Z be Banach spaces. Denote $\mathcal{L}^{(k)}(Y_1, Y_2, .., Y_k, Z)$ (resp. $\mathcal{L}^{(k)}(Y, Z)$) the space of bounded k-linear maps (or multilinear map) from $Y_1 \times \ldots \times Y_k$ into Z (resp. from Y^k into Z). We recall that if L is a k-linear map from Y^k into Z, then its norm is defined as

$$\|L\|_{\mathcal{L}^{(k)}(Y,Z)} = \sup_{\|u_1\| \leq 1, \ldots, \|u_k\| \leq 1} \|L(u_1, \ldots, u_k)\|.$$

Let V be an open subset of Y, and let $W \in C^k(V, Z)$ be fixed. We choose the convention that if $l = 1, \ldots, k-1$ and $u, \widehat{u} \in V$ with $u \neq \widehat{u}$, the quantity

$$\sup_{u_1, \ldots, u_l \in B_Y(0,1)} \frac{\left\| \left[D^l W(u) - D^l W(\widehat{u})\right](u_1, \ldots, u_l) - D^{l+1} W(\widehat{u})(u - \widehat{u}, u_1, \ldots, u_l) \right\|}{\|u - \widehat{u}\|}$$

goes to 0 as $\|u - \widehat{u}\| \to 0$.

Set

$$C_b^k(V, Z) := \left\{ W \in C^k(V, Z) : |W|_{j,V} := \sup_{u \in V} \left\| D^j W(u) \right\|_{\mathcal{L}^{(j)}(Y,Z)} < +\infty, \ 0 \leq j \leq k \right\}.$$

For each $\eta \in [0, \beta)$, consider $K_h : BC^{\eta}(\mathbb{R}, \mathbb{R}^n) \to BC^{\eta}(\mathbb{R}, X_h)$, the bounded linear operator defined by

$$K_h = K_s + K_u,$$

where K_s and K_u are the bounded linear operators defined, respectively, in Lemma 4.16 and Lemma 4.18.

For each $\varrho > 0$ we define some open and closed neighbourhood of X_c as

$$\textbf{(Open neighbourhood of } X_c\textbf{)} \quad V_{\varrho} := \{x \in \mathbb{R}^n : \|\Pi_h x\| < \varrho\},$$

and

$$\textbf{(Closed neighbourhood of } X_c\textbf{)} \quad \overline{V}_{\varrho} := \{x \in \mathbb{R}^n : \|\Pi_h x\| \leq \varrho\}.$$

Moreover for each $\eta \geq 0$, we define

$$\overline{V}_{\varrho}^{\eta} := \left\{ u \in BC^{\eta}(\mathbb{R}, \mathbb{R}^n) : u(t) \in \overline{V}_{\varrho}, \forall t \in \mathbb{R} \right\}.$$

Note that since $\overline{V}_{\varrho}$ is a closed and convex subset of $\mathbb{R}^n$, the subset $\overline{V}_{\varrho}^{\eta}$ is also a closed convex subset of $BC^{\eta}(\mathbb{R}, \mathbb{R}^n)$ for each $\eta \geq 0$.

We make the following assumption.

Assumption 4.23 Let $k \geq 1$ be an integer and let $\eta, \widehat{\eta} \in (0, \beta)$ be such that

$$k\eta < \widehat{\eta} < \beta.$$

Let $\varrho > 0$ be given and assume the following.

(i) The function F satisfies

$$F \in \text{Lip}\left(\mathbb{R}^n, \mathbb{R}^n\right) \cap C_b^k\left(V_\varrho, \mathbb{R}^n\right).$$

(ii)

$$\varrho_0 := \|K_h\|_{\mathcal{L}\left(BC^0(\mathbb{R},\mathbb{R}^n)\right)} \|\Pi_h F\|_\infty < \varrho,$$

where

$$\|F\|_\infty = \sup_{x \in \mathbb{R}^n} \|F(x)\|.$$

(iii)

$$\sup_{\theta \in [\eta, \widehat{\eta}]} \|K_2\|_{\mathcal{L}\left(BC^\theta(\mathbb{R},\mathbb{R}^n)\right)} \|F\|_{\text{Lip}(\mathbb{R}^n,\mathbb{R}^n)} < 1.$$

Note that by using (4.35) we deduce that

$$\sup_{\theta \in [\eta, \widehat{\eta}]} \|K_2\|_{\mathcal{L}\left(BC^\theta(\mathbb{R},\mathbb{R}^n)\right)} < +\infty.$$

Thus, Assumption 4.23 makes sense.

Following the approach of Vanderbauwhede [338, Corollary 3.6] and Vanderbauwhede and Iooss [340, Theorem 2], we obtain the following result on the smoothness of the center manifold.

Theorem 4.24 (Smoothness of the Center Manifold) *Let Assumptions 4.3 and 4.23 be satisfied. Then the map Ψ obtained in Theorem 4.21 belongs to the space $C_b^k(X_c, X_h)$.*

Definition 4.25 Let (M, d) be a metric space and $H : M \to M$ be a map. A fixed point $\overline{x} \in M$ of H is said to be **attracting** if

$$\lim_{n \to +\infty} H^n(x) = \overline{x} \quad \text{for each } x \in M.$$

The following lemma is an extension of the Fibre contraction theorem (which corresponds to the case $k = 1$). This result is taken from [338, Corollary3.6].

Theorem 4.26 (Fibre contraction) *Let $k \geq 1$ be an integer and let $(M_0, d_0), \ldots, (M_k, d_k)$ be complete metric spaces, and consider a map $H : M_0 \times \ldots \times M_k \to M_0 \times \cdots \times M_k$ having the following the form*

$$H(x_0, x_1, \ldots, x_k) = \left(H_0(x_0), H_1(x_0, x_1), \ldots, H_k(x_0, x_1, \ldots, x_k)\right),$$

where each map $H_l : M_0 \times M_1 \times \ldots \times M_l \to M_l$ (for each $l = 0, \ldots, k$) is a uniform contraction with respect to the last variable. That is, H_0 is a contraction, and for each $l = 1, \ldots, k$, there exists a constant $0 \leq \tau_l < 1$ such that

$$d_l\left(H_l\left(x_0,x_1,\ldots,x_{l-1},x_l\right),H_l\left(x_0,x_1,\ldots,x_{l-1},\widehat{x}_l\right)\right)\leq\tau_l\ d_l\left(x_l,\widehat{x}_l\right),$$

whenever $(x_0,x_1,\ldots,x_{l-1})\in M_0\times M_1\times\ldots\times M_{l-1}$, *and* $x_l,\widehat{x}_l\in M_l$.

Then F *has a unique fixed point* $(\overline{x}_0,\overline{x}_1,\ldots,\overline{x}_k)$, *satisfying*

$$\begin{aligned}\overline{x}_0&=H_0\left(\overline{x}_0\right),\\ \overline{x}_1&=H_1\left(\overline{x}_0,\overline{x}_1\right),\\ &\vdots\\ \overline{x}_k&=H_k\left(\overline{x}_0,\overline{x}_1,\ldots,\overline{x}_k\right).\end{aligned}$$

If we assume assume in addition that each map (for each $l=1,\ldots,k$*)*

$$H_l\left(.,\overline{x}_l\right):M_0\times M_1\times\ldots\times M_{l-1}\to M_l$$

is continuous, then $(\overline{x}_0,\overline{x}_1,\ldots,\overline{x}_k)$ *is an attracting fixed point of* H.

Proof We prove the lemma for $k=1$. The proof for any integer $k\geq 1$ can be easily derived from this case. By the Banach fixed point theorem, H_0 has a unique fixed point $\overline{x}_0\in M_0$, and the map $x_1\to H_1(\overline{x}_0,x_1)$ also has a unique fixed point $\overline{x}_1\in M_1$. It is clear that $(\overline{x}_0,\overline{x}_1)$ is the unique fixed point of H, so we only need to prove its attractivity.

Let $(x_0,x_1)\in M_0\times M_1$. Consider the sequence $(x_0(n),x_1(n))$ defined by

$$(x_0(0),x_1(0)):=(x_0,x_1)$$

and

$$(x_0(n+1),x_1(n+1)):=(H_0(x_0(n)),H_1(x_0(n),x_1(n))),\ \forall n\geq 0.$$

Since H_0 is a contraction, it is clear that $\lim_{n\to+\infty}x_0(n)=\overline{x}_0$. It remains to show that $\lim_{n\to+\infty}x_1(n)=\overline{x}_1$. We observe first that

$$\begin{aligned}d(x_1(n+1),\overline{x}_1)&=d(H_1(x_0(n),x_1(n)),H_1(\overline{x}_0,\overline{x}_1))\\ &\leq d(H_1(x_0(n),x_1(n)),H_1(x_0(n),\overline{x}_1))+d\left(H_1(x_0(n),\overline{x}_1),H_1(\overline{x}_0,\overline{x}_1)\right)\\ &\leq\tau_1 d(x_1(n),\overline{x}_1)+\alpha_n,\end{aligned}$$

where

$$\alpha_n:=d\left(H_1(x_0(n),\overline{x}_1),H_1(\overline{x}_0,\overline{x}_1)\right)\to 0\text{ as }n\to+\infty.$$

Setting $\delta_n:=d(x_1(n),\overline{x}_1)$, we obtain

$$\delta_{n+1}\leq\tau_1\delta_n+\alpha_n,\ \forall n\geq 0.$$

Since $\tau_1<1$, it is not difficult to prove that $\{\delta_n\}$ is bounded sequence and

$$\limsup_{n\to+\infty}\delta_n\leq\tau_1\limsup_{n\to+\infty}\delta_n.$$

Hence, $\limsup_{n\to+\infty}\delta_n=0$, and the proof is completed. □

We recall that the map $\Psi : X_c \to X_h$ is defined by

$$\Psi(x_c) = \Pi_h (I - K_2\Phi_F)^{-1} (K_1 x_c)(0), \quad \forall x_c \in X_c.$$

We define the map $\Gamma_0 : BC^\eta(\mathbb{R}, X_c) \to BC^\eta(\mathbb{R}, \mathbb{R}^n)$ by

$$\Gamma_0(u) = (I - K_2\Phi_F)^{-1}(u), \quad \forall u \in BC^\eta(\mathbb{R}, X_c). \tag{4.41}$$

For each $\delta \geq 0$, we define the bounded linear operator $L : BC^\delta(\mathbb{R}, \mathbb{R}^n) \to X_h$ by

$$L(u) = \Pi_h u(0), \quad \forall u \in BC^\delta(\mathbb{R}, X_c).$$

Then we have

$$\Psi(x_c) = L\Gamma_0(K_1 x_c), \quad \forall x_c \in X_c,$$

and we deduce that $\Gamma_0 : BC^\eta(\mathbb{R}, X_c) \to BC^\eta(\mathbb{R}, \mathbb{R}^n)$ is the unique solution of the fixed problem

$$\Gamma_0 = J + K_2 \circ \Phi_F \circ (\Gamma_0),$$

where $J(u) = u$ is the continuous imbedding from $BC^\eta(\mathbb{R}, X_c)$ into $BC^\eta(\mathbb{R}, \mathbb{R}^n)$.

From (4.41), we deduce that for each $u \in BC^\eta(\mathbb{R}, X_c)$,

$$\|\Gamma_0(u) - u\|_{BC^\eta(\mathbb{R},\mathbb{R}^n)} \leq \|K_2\|_{\mathcal{L}(BC^\eta(\mathbb{R},\mathbb{R}^n), BC^\eta(\mathbb{R},\mathbb{R}^n))} \|F\|_\infty,$$

$$\|\Pi_h \Gamma_0(u)(t)\|_{BC^0(\mathbb{R},\mathbb{R}^n)} \leq \|K_h\|_{\mathcal{L}(BC^0(\mathbb{R},\mathbb{R}^n))} \|\Pi_h F\|_\infty = \varrho_0, \quad \forall t \in \mathbb{R}.$$

For each subset $E \subset BC^\eta(\mathbb{R}, X_c)$, denote

$$M_{0,E} = \left\{ \Theta \in C\left(E, \overrightarrow{V}^0_{\varrho_0}\right) : \sup_{u \in E} \|\Theta(u) - u\|_{BC^\eta(\mathbb{R},\mathbb{R}^n)} < +\infty \right\}$$

and set

$$M_0 = M_{0,BC^\eta(\mathbb{R},X_c)}.$$

From the above remarks, it follows that Γ_0 (respectively $\Gamma_0 \mid_E$) must be an element of M_0 (respectively $M_{0,E}$). Since $\overrightarrow{V}^0_{\varrho_0}$ is a closed subset of $BC^\eta(\mathbb{R}, \mathbb{R}^n)$, we know that for each subset $E \subset BC^\eta(\mathbb{R}, X_c)$, $M_{0,E}$ is a complete metric space endowed with the metric

$$d_{0,E}\left(\Theta, \widetilde{\Theta}\right) = \sup_{u \in E} \left\| \Theta(u) - \widetilde{\Theta}(u) \right\|_{BC^\eta(\mathbb{R},\mathbb{R}^n)}.$$

Set

$$d_0 = d_{0,BC^\eta(\mathbb{R},X_c)}.$$

Lemma 4.27 (Regularity of the Nemytskii operator) *Let E be a Banach space and $W \in C^1_b(V_\varrho, E)$. Let $\xi \geq \delta \geq 0$ be fixed. Define $\Phi_W : V^\eta_\varrho \to BC^\xi(\mathbb{R}, E)$, $\Phi_{DW} : V^\eta_\varrho \to BC^\xi(\mathbb{R}, \mathcal{L}(\mathbb{R}^n, E))$, and $\Phi^{(1)}_W : V^\eta_\varrho \to \mathcal{L}\left(BC^\delta(\mathbb{R}, \mathbb{R}^n), BC^\xi(\mathbb{R}, E)\right)$ for each $t \in \mathbb{R}$, each $u \in V^\eta_\varrho$, and each $v \in BC^\delta(\mathbb{R}, \mathbb{R}^n)$ by*

$$\Phi_W(u)(t) := W(u(t)) \in E,$$
$$\Phi_{DW}(u)(t) := DW(u(t)) \in \mathcal{L}(\mathbb{R}^n, E),$$
$$\Phi_W^{(1)}(u)(v)(t) := DW(u(t))(v(t)) \in E,$$

respectively. Then we have the following:

(a) *If* $\xi > 0$, *then* Φ_W *and* Φ_{DW} *are continuous.*
(b) *For each* $u, v \in V_\varrho^\eta$, $\Phi_W^{(1)}(u) \in \mathcal{L}\left(BC^\delta(\mathbb{R}, \mathbb{R}^n), BC^\xi(\mathbb{R}, E)\right)$,

$$\begin{aligned}
&\left\|\Phi_W^{(1)}(u) - \Phi_W^{(1)}(v)\right\|_{\mathcal{L}\left(BC^\delta(\mathbb{R},\mathbb{R}^n), BC^\xi(\mathbb{R},E)\right)} \\
&= \sup_{w \in BC^\delta(\mathbb{R},\mathbb{R}^n)} \left\|\Phi_W^{(1)}(u)(w) - \Phi_W^{(1)}(v)(w)\right\|_{BC^\xi(\mathbb{R},E)} \\
&\le \sup_{t\in\mathbb{R}} e^{-(\xi-\delta)|t|} \|DW(u(t)) - DW(v(t))\|_{\mathcal{L}(\mathbb{R}^n,E)} \\
&= \|\Phi_{DW}(u) - \Phi_{DW}(v)\|_{BC^{\xi-\delta}(\mathbb{R},\mathcal{L}(\mathbb{R}^n,E))}
\end{aligned}$$

and

$$\left\|\Phi_W^{(1)}(u)\right\|_{\mathcal{L}\left(BC^\delta(\mathbb{R},\mathbb{R}^n), BC^\xi(\mathbb{R},E)\right)} \le \|\Phi_{DW}(u)\|_{BC^{\xi-\delta}(\mathbb{R},\mathcal{L}(\mathbb{R}^n,E))} \le |W|_{1,V_\varrho}.$$

(c) *If* $\xi > \delta$, *then* $\Phi_W^{(1)}$ *is continuous.*
(d) *If* $\xi \ge \delta \ge \eta$, *we have for each* $u, \widehat{u} \in V_\varrho^\eta$ *that*

$$\left\|\Phi_W(u) - \Phi_W(\widehat{u}) - \Phi_W^{(1)}(\widehat{u})(u - \widehat{u})\right\|_{BC^\xi(\mathbb{R},E)} \le \|u - \widehat{u}\|_{BC^\delta(\mathbb{R},\mathbb{R}^n)} \varkappa_{\xi-\delta}(u, \widehat{u})$$

where

$$\varkappa_{\xi-\delta}(u, \widehat{u}) = \sup_{s\in[0,1]} \|\Phi_{DW}(su + (1-s)\widehat{u}) - \Phi_{DW}(\widehat{u})\|_{BC^{\xi-\delta}(\mathbb{R},\mathcal{L}(\mathbb{R}^n,E))},$$

and if $\xi > \delta \ge \eta$, *we have (by continuity of* Φ_{DW}*)*

$$\varkappa_{\xi-\delta}(u, \widehat{u}) \to 0 \quad as \quad \|u - \widehat{u}\|_{BC^\eta(\mathbb{R},\mathbb{R}^n)} \to 0.$$

Proof We first prove that $\Phi_W \in C_b^0\left(V_\varrho^\eta, BC^\xi(\mathbb{R}, E)\right)$. For each $u, \widehat{u} \in V_\varrho^\eta$ and each $R > 0$, we have

$$\begin{aligned}
\|\Phi_W(u) - \Phi_W(\widehat{u})\|_{BC^\xi(\mathbb{R},E)} &= \sup_{t\in\mathbb{R}} e^{-\xi|t|} \|W(u(t)) - W(\widehat{u}(t))\|_E \\
&\le \max\left(\sup_{|t|\le R} e^{-\xi|t|} \|W(u(t)) - W(\widehat{u}(t))\|_E, 2\|W\|_0 e^{-\xi R}\right).
\end{aligned} \tag{4.42}$$

Fix some arbitrary $\varepsilon > 0$. Let $R > 0$ be such that $2\|W\|_0 e^{-\xi R} < \varepsilon$ and denote $\Omega = \{\widehat{u}(t) : |t| \le R\}$. Since W is continuous and Ω is compact, we can find $\delta_1 > 0$ such that

$$\|W(x) - W(\widehat{x})\| \le \varepsilon \text{ if } \widehat{x} \in \Omega, \text{ and } \|x - \widehat{x}\| \le \delta_1.$$

Let $\delta_2 = e^{-\eta R}\delta_1$. If $\|u - \widehat{u}\|_{BC^{\eta}(\mathbb{R},\mathbb{R}^n)} \leq \delta_2$, then

$$\|u(t) - \widehat{u}(t)\| \leq \delta_1, \forall t \in [-R, R],$$

and (4.42) implies

$$\|\Phi_W(u) - \Phi_W(\widehat{u})\|_{BC^{\xi}(\mathbb{R},E)} \leq \varepsilon.$$

Remark that here we only used the fact that $W \in C_b^0(V_\varrho, E)$. Therefore, replacing E by $\mathcal{L}(E, \mathbb{R}^n)$, we may apply the same argument to $DW \in C_b^0(V_\varrho, \mathcal{L}(\mathbb{R}^n, E))$ to find that $\Phi_{DW} \in C_b^0(V_\varrho^\eta, BC^{\xi}(\mathbb{R}, \mathcal{L}(\mathbb{R}^n, E)))$.

We now prove that $\Phi_W^{(1)} \in C\left(V_\varrho^\eta, \mathcal{L}\left(BC^\delta(\mathbb{R},\mathbb{R}^n), BC^{\xi}(\mathbb{R},E)\right)\right)$. For each $u, \widehat{u} \in V_\varrho^\eta$ and each $v \in BC^\delta(\mathbb{R},\mathbb{R}^n)$, we have

$$\begin{aligned}\left\|\Phi_W^{(1)}(u)(v)\right\|_{BC^{\xi}(\mathbb{R},E)} &= \sup_{t\in\mathbb{R}} e^{-\xi|t|}\|DW(u(t))(v(t))\|_E \\ &\leq \sup_{t\in\mathbb{R}} e^{-(\xi-\delta)|t|}\|DW(u(t))\|_{\mathcal{L}(\mathbb{R}^n,E)}\, e^{-\delta|t|}\|v(t)\|_{\mathbb{R}^n} \\ &\leq \|\Phi_{DW}(u)\|_{BC^{\xi-\delta}(\mathbb{R},\mathcal{L}(\mathbb{R}^n,E))}\|v\|_{BC^\delta(\mathbb{R},\mathbb{R}^n)},\end{aligned}$$

and

$$\begin{aligned}&\left\|\left[\Phi_W^{(1)}(u) - \Phi_W^{(1)}(\widehat{u})\right](v)\right\|_{BC^{\xi}(\mathbb{R},E)} \\ &= \sup_{t\in\mathbb{R}} e^{-\xi|t|}\|DW(u(t))(v(t)) - DW(\widehat{u}(t))(v(t))\|_E \\ &\leq \|\Phi_{DW}(u) - \Phi_{DW}(\widehat{u})\|_{BC^{\xi-\delta}(\mathbb{R},\mathcal{L}(\mathbb{R}^n,E))}\|v\|_{BC^\delta(\mathbb{R},\mathbb{R}^n)}.\end{aligned}$$

Thus, if $\xi \geq \delta$, we have for each $u \in V_\varrho^\eta$ that

$$\Phi_W^{(1)}(u) \in \mathcal{L}\left(BC^\delta(\mathbb{R},\mathbb{R}^n), BC^{\xi}(\mathbb{R},E)\right), \quad \forall u \in V_\varrho^\eta,$$

and if $\xi > \delta$,

$$\Phi_W^{(1)} \in C\left(V_\varrho^\eta, \mathcal{L}\left(BC^\delta(\mathbb{R},\mathbb{R}^n), BC^{\xi}(\mathbb{R},E)\right)\right).$$

Since V_ϱ is an open and convex subset of $\mathbb{R}^n$, we have the following classical formula

$$W(x) - W(y) = \int_0^1 DW(sx + (1-s)y)(x-y)\,ds, \quad \forall x, y \in V_\varrho.$$

Therefore, for each $u, \widehat{u} \in V_\varrho^\eta$,

$$\begin{aligned}
&\left\|\Phi_W(u)-\Phi_W(\widehat{u})-\Phi_W^{(1)}(\widehat{u})(u-\widehat{u})\right\|_{BC^{\xi}(\mathbb{R},E)}\\
&=\sup_{t\in\mathbb{R}} e^{-\xi|t|}\left\|W(u(t))-W(\widehat{u}(t))-DW(\widehat{u}(t))(u(t)-\widehat{u}(t))\right\|\\
&\leq \sup_{t\in\mathbb{R}}\sup_{s\in[0,1]} e^{-\xi|t|}\left\|[DW(su(t)+(1-s)\widehat{u}(t))-DW(\widehat{u}(t))](u(t)-\widehat{u}(t))\right\|\\
&\leq \|u-\widehat{u}\|_{BC^{\delta}(\mathbb{R},\mathbb{R}^n)}\sup_{s\in[0,1]}\|\Phi_{DW}(su+(1-s)\widehat{u})-\Phi_{DW}(\widehat{u})\|_{BC^{\xi-\delta}(\mathbb{R},\mathcal{L}(\mathbb{R}^n,E))}.
\end{aligned}$$

The proof is complete. □

The following Lemma is taken from Vanderbauwhede and Iooss [340, Lemma 3].

Lemma 4.28 *Let E be a Banach space and $W\in C_b^1\left(V_\varrho,E\right)$. Let Φ_W and $\Phi_W^{(1)}$ be defined as in Lemma 4.27. Let $\Theta\in C\left(BC^\eta(\mathbb{R},X_c),V_\varrho^\eta\right)$ be such that*

(a) *Θ is of class C^1 from $BC^\eta(\mathbb{R},X_c)$ into $BC^{\eta+\mu}(\mathbb{R},\mathbb{R}^n)$ for each $\mu>0$.*
(b) *The derivative of Θ takes the form*

$$D\Theta(u)(v)=\Theta^{(1)}(u)(v),\quad \forall u,v\in BC^\eta(\mathbb{R},X_c),$$

for some globally bounded map

$$\Theta^{(1)}:BC^\eta(\mathbb{R},X_c)\to\mathcal{L}\left(BC^\eta(\mathbb{R},X_c),BC^\eta(\mathbb{R},\mathbb{R}^n)\right).$$

Then

$$\Phi_W\circ\Theta\in C_b^0\left(BC^\eta(\mathbb{R},X_c),BC^\eta(\mathbb{R},E)\right)\cap C^1\left(BC^\eta(\mathbb{R},X_c),BC^{\eta+\mu}(\mathbb{R},E)\right)$$

for each $\mu>0$, and

$$D\left(\Phi_W\circ\Theta\right)(u)(v)=\Phi_W^{(1)}\left(\Theta(u)\right)\left(\Theta^{(1)}(u)(v)\right),$$

for each $u,v\in BC^\eta(\mathbb{R},X_c)$.

Proof By using Lemma 4.27, it follows that

$$\Phi_W\circ\Theta\in C_b^0\left(BC^\eta(\mathbb{R},X_c),BC^\eta(\mathbb{R},E)\right)$$

and

$$u\to\Phi_W^{(1)}\left(\Theta(u)\right)\Theta^{(1)}(u)\in C\left(BC^\eta(\mathbb{R},X_c),\mathcal{L}\left(BC^\eta(\mathbb{R},X_c),BC^{\eta+\mu}(\mathbb{R},E)\right)\right).$$

Let $u,\widehat{u}\in BC^\eta(\mathbb{R},X_c)$. By Lemma 4.27, we also have

$$\begin{aligned}
&\left\|\Phi_W(\Theta(u)) - \Phi_W(\Theta(\widehat{u})) - \Phi_W^{(1)}(\Theta(\widehat{u}))\left(\Theta^{(1)}(\widehat{u})(u-\widehat{u})\right)\right\|_{BC^{\eta+\mu}(\mathbb{R},E)} \\
&\quad \le \left\|\Phi_W(\Theta(u)) - \Phi_W(\Theta(\widehat{u})) - \Phi_W^{(1)}(\Theta(\widehat{u}))(\Theta(u)-\Theta(\widehat{u}))\right\|_{BC^{\eta+\mu}(\mathbb{R},E)} \\
&\qquad + \left\|\Phi_W^{(1)}(\Theta(\widehat{u}))\left[\Theta(u)-\Theta(\widehat{u})-\Theta^{(1)}(\widehat{u})(u-\widehat{u})\right]\right\|_{BC^{\eta+\mu}(\mathbb{R},E)} \\
&\quad \le \|\Theta(u)-\Theta(\widehat{u})\|_{BC^{\eta+\mu/2}(\mathbb{R},\mathbb{R}^n)}\,\varkappa_{\mu/2}(\Theta(u),\Theta(\widehat{u})) \\
&\qquad + \|\Phi_{DW}(\Theta(\widehat{u}))\|_{BC^{\mu/2}(\mathbb{R},\mathcal{L}(\mathbb{R}^n,E))}\left\|\Theta(u)-\Theta(\widehat{u})-\Theta^{(1)}(\widehat{u})(u-\widehat{u})\right\|_{BC^{\eta+\mu/2}(\mathbb{R},\mathbb{R}^n)}
\end{aligned}$$

and the result follows. □

One may extend the previous lemma to any order $k > 1$.

Lemma 4.29 *Let E be a Banach space and let $W \in C_b^k(V_\varrho, E)$ (for some integer $k \ge 1$). Let $l \in \{1,\ldots,k\}$ be an integer. Suppose $\xi \ge 0$, $\mu \ge 0$ are two real numbers and $\delta_1, \delta_2, \ldots, \delta_l \ge 0$ such that $\xi = \mu + \delta_1 + \delta_2 + \ldots + \delta_l$. Define*

$$\Phi_{D^{(l)}W}(u)(t) := D^{(l)}W(u(t)), \forall t \in \mathbb{R}, \forall u \in V_\varrho^\eta,$$

$$\Phi_W^{(l)}(u)(u_1,u_2,\ldots,u_l)(t) := D^{(l)}W(u(t))(u_1(t),u_2(t),\ldots,u_l(t)),$$
$$\forall t \in \mathbb{R}, \forall u \in V_\varrho^\eta, \forall u_i \in BC^{\delta_i}(\mathbb{R},\mathbb{R}^n),\ \text{for } i = 1,\ldots,l.$$

Then we have the following:

(a) *If $\xi > 0$, then $\Phi_{D^{(l)}W} : V_\varrho^\eta \to BC^\xi\left(\mathbb{R}, \mathcal{L}^{(l)}(\mathbb{R}^n, E)\right)$ is continuous.*
(b) *For each $u, v \in V_\varrho^\eta$, $\Phi_W^{(l)}(u) \in \mathcal{L}^{(l)}\left(BC^{\delta_1}(\mathbb{R},\mathbb{R}^n),\ldots,BC^{\delta_l}(\mathbb{R},\mathbb{R}^n); BC^\xi(\mathbb{R},E)\right)$,*

$$\begin{aligned}
&\left\|\Phi_W^{(l)}(u) - \Phi_W^{(l)}(v)\right\|_{\mathcal{L}^{(l)}\left(BC^{\delta_1}(\mathbb{R},\mathbb{R}^n),\ldots,BC^{\delta_l}(\mathbb{R},\mathbb{R}^n);BC^\xi(\mathbb{R},E)\right)} \\
&\quad \le \left\|\Phi_{D^{(l)}W}(u) - \Phi_{D^{(l)}W}(v)\right\|_{BC^\mu\left(\mathbb{R},\mathcal{L}^{(l)}(\mathbb{R}^n,E)\right)}
\end{aligned}$$

and

$$\begin{aligned}
&\left\|\Phi_W^{(l)}(u)\right\|_{\mathcal{L}^{(l)}\left(BC^{\delta_1}(\mathbb{R},\mathbb{R}^n),\ldots,BC^{\delta_l}(\mathbb{R},\mathbb{R}^n);BC^\xi(\mathbb{R},E)\right)} \\
&\quad \le \left\|\Phi_{D^{(l)}W}(u)\right\|_{BC^\mu\left(\mathbb{R},\mathcal{L}^{(l)}(\mathbb{R}^n,E)\right)} \le |W|_{l,V_\varrho}.
\end{aligned}$$

(c) *If $\mu > 0$, then $\Phi_W^{(l)}$ is continuous.*
(d) *If $\delta_1 \ge \eta$, we have for each $u, \widehat{u} \in V_\varrho^\eta$ (with the convention for the derivatives defined at the beginning of this section) that*

$$\begin{aligned}
&\left\|\Phi_W^{(l-1)}(u) - \Phi_W^{(l-1)}(\widehat{u}) - \Phi_W^{(l)}(\widehat{u})(u-\widehat{u})\right\|_{\mathcal{L}^{(l-1)}\left(BC^{\delta_2}(\mathbb{R},\mathbb{R}^n),\ldots,BC^{\delta_l}(\mathbb{R},\mathbb{R}^n);BC^\xi(\mathbb{R},E)\right)} \\
&\quad \le \|u-\widehat{u}\|_{BC^{\delta_1}(\mathbb{R},\mathbb{R}^n)}\,\varkappa_\mu^{(l)}(u,\widehat{u}),
\end{aligned}$$

where

$$\varkappa_\mu^{(l)}(u,\widehat{u}) = \sup_{s\in[0,1]}\left\|\Phi_{D^{(l)}W}(su+(1-s)\widehat{u}) - \Phi_{D^{(l)}W}(\widehat{u})\right\|_{BC^\mu\left(\mathbb{R},\mathcal{L}^{(l)}(\mathbb{R}^n,E)\right)},$$

and if $\mu > 0$, we have by continuity of $\Phi_{D^{(l)}W}$ that

$$\varkappa_{\mu}^{(l)}(u,\widehat{u}) \to 0 \text{ as } \|u-\widehat{u}\|_{BC^{\eta}(\mathbb{R},\mathbb{R}^n)} \to 0.$$

Proof This proof is similar to that of Lemma 4.27. □

In the following lemma we use a formula for the kth differential of the composition of two maps. This formula is taken from Avez [14, p. 38]. This formula corrects the one used in Vanderbauwhede [338, Proof of Lemma 3.11].

Lemma 4.30 *Let E be a Banach space and let $W \in C_b^k(V_\varrho, E)$. Let Φ_W and $W^{(k)}$ be defined as above. Let $\Theta \in C(BC^{\eta}(\mathbb{R}, X_c), V_\varrho^{\eta})$ be such that*

(a) *Θ is of class C^k from $BC^{\eta}(\mathbb{R}, X_c)$ into $BC^{k\eta+\mu}(\mathbb{R}, \mathbb{R}^n)$ for each $\mu > 0$;*
(b) *for each $l = 1, \ldots, k$, its derivative takes the form*

$$D^l\Theta(u)(v_1, v_2, \ldots, v_l) = \Theta^{(l)}(u)(v_1, v_2, \ldots, v_l), \forall u, v_1, v_2, \ldots, v_l \in BC^{\eta}(\mathbb{R}, X_c),$$

for some globally bounded $\Theta^{(l)} : BC^{\eta}(\mathbb{R}, X_c) \to \mathcal{L}^{(l)}(BC^{\eta}(\mathbb{R}, X_c); BC^{\eta}(\mathbb{R}, \mathbb{R}^n))$.

Then $\Phi_W \circ \Theta \in C_b^0(BC^{\eta}(\mathbb{R}, X_c), BC^{\eta}(\mathbb{R}, E)) \cap C^k(BC^{\eta}(\mathbb{R}, X_c), BC^{k\eta+\mu}(\mathbb{R}, E))$ for each $\mu > 0$. Moreover, for each $l = 1, \ldots, k$ and each $u, v_1, v_2, \ldots, v_l \in BC^{\eta}(\mathbb{R}, X_c)$,

$$D^l(\Phi_W \circ \Theta)(u)(v) = (\Phi_W \circ \Theta)^{(l)}(u)(v_1, v_2, \ldots, v_l)$$

for some globally bounded $(\Phi_W \circ \Theta)^{(l)} : BC^{\eta}(\mathbb{R}, X_c) \to \mathcal{L}^{(l)}(BC^{\eta}(\mathbb{R}, X_c); BC^{\eta}(\mathbb{R}, E))$. More precisely, we have for $j = 1, \ldots, k$ that

(i) $(\Phi_W \circ \Theta)^{(j)}(u) = \Phi_W^{(1)}(\Theta(u)) D^{(j)}\Theta(u) + \widetilde{\Phi}_{W,j}(u)$;
(ii) $\widetilde{\Phi}_{W,1}(u) = 0$;
(iii) *for $j > 1$, the map $\widetilde{\Phi}_{W,j}(u)$ is a finite sum $\sum_{\lambda \in \Lambda_j} \widetilde{\Phi}_{W,\lambda,j}(u)$, where for each $\lambda \in \Lambda_j$ the map $\widetilde{\Phi}_{W,\lambda,j}(u) : BC^{\eta}(\mathbb{R}, X_c) \to \mathcal{L}^{(j)}(BC^{\eta}(\mathbb{R}, X_c), BC^{\eta}(\mathbb{R}, E))$ has the following form*

$$\widetilde{\Phi}_{W,\lambda,j}(u)(u_1, u_2, \ldots, u_j) = \Phi_W^{(l)}(\Theta(u)) \begin{pmatrix} D^{(r_1)}\Theta(u)\left(u_{i_1^{r_1}}, u_{i_2^{r_1}}, \ldots, u_{i_{r_1}^{r_1}}\right), \ldots, \\ D^{(r_l)}\Theta(u)\left(u_{i_1^{r_l}}, \ldots, u_{i_{r_l}^{r_l}}\right) \end{pmatrix}$$

with $2 \le l \le j$, $1 \le r_i \le j-1$ for $1 \le i \le l$,

$$r_1 + r_2 + \ldots + r_l = j,$$

$$\left\{i_1^{r_m}, \ldots, i_{r_m}^{r_m}\right\} \subset \{1, \ldots, j\}, \forall m = 1, \ldots, l$$

$$\left\{i_1^{r_m}, \ldots, i_{r_m}^{r_m}\right\} \cap \left\{i_1^{r_n}, \ldots, i_{r_n}^{r_n}\right\} = \emptyset, \text{ if } m \neq n,$$

$$i_1^{r_m} \le i_2^{r_m} \le \ldots \le i_{r_m}^{r_m}, \forall m = 1, \ldots, l,$$

and each $\lambda \in \Lambda_j$ corresponds to each particular such a choice.

Proof This proof is similar to that of Lemma 4.28. □

Proof (Proof of Theorem 4.24) :

Step 1 (Existence of a fixed point): Let k, η, and $\widehat{\eta}$ be the numbers introduced in Assumption 4.23. Let $\mu > 0$ be such that $k\eta + (2k-1)\mu = \widehat{\eta}$. We first apply the fibre contraction Theorem 4.26. For each subset $E \subset BC^{\eta}(\mathbb{R}, X_c)$, let

$$M_{0,E} := \left\{ \Theta_0 \in C^0(E, BC^{\eta}(\mathbb{R}, X_c)) \, : \, \sup_{u \in E} \|\Theta_0(u) - u\| < +\infty \right\},$$

equipped with the metric $d_{0,E}(\Theta_0, \widehat{\Theta}_0) := \sup_{u \in E} \|\Theta_0(u) - \widehat{\Theta}_0(u)\|_{BC^{\eta}(\mathbb{R}, X_c)}$, which turns it into a complete metric space. For each $j = 1, \ldots, k$, define $M_{j,E}$ as the Banach space of all bounded continuous maps

$$\Theta_j : E \to \mathcal{L}^{(j)} \left(BC^{\eta}(\mathbb{R}, X_c), BC^{j\eta + (2j-1)\mu}(\mathbb{R}, \mathbb{R}^n) \right)$$

such that

$$\left| \Theta_j \right|_j = \sup_{u \in E} \left\| \Theta_j(u) \right\|_{\mathcal{L}^{(j)}\left(BC^{\eta}(\mathbb{R}, X_c), BC^{j\eta+(2j-1)\mu}(\mathbb{R}, \mathbb{R}^n)\right)} < +\infty.$$

For $j = 0, \ldots, k$, define the map $H_{j,E} : M_{0,E} \times M_{1,E} \times \ldots \times M_{j,E} \to M_{j,E}$ as follows: If $j = 0$, set for each $u \in E$ that

$$H_{0,E}(\Theta_0)(u) = u + K_2 \circ \Phi_F \circ \Theta_0(u).$$

If $j = 1$, set for each $u \in E$ that

$$H_{1,E}(\Theta_0, \Theta_1)(u)(.) = J^1 + K_2 \circ \Phi_F^{(1)}(\Theta_0(u)) \circ \Theta_1(u), \tag{4.43}$$

where J^1 is the continuous imbedding from $BC^{\eta}(\mathbb{R}, X_c)$ into $BC^{\eta+\mu}(\mathbb{R}, \mathbb{R}^n)$.
If $k \geq 2$, set for each $j = 2, \ldots, k$ and each $u \in E$ that

$$\begin{aligned} & H_{j,E}(\Theta_0, \Theta_1, \ldots, \Theta_j)(u) \\ & \quad = K_2 \circ \Phi_F^{(1)}(\Theta_0(u)) \circ \Theta_j(u) + \widehat{H}_{j,E}(\Theta_0, \Theta_1, \ldots, \Theta_{j-1})(u), \end{aligned} \tag{4.44}$$

where

$$\widehat{H}_{j,E}(\Theta_0, \Theta_1, \ldots, \Theta_{j-1})(u) = \sum_{\lambda \in \Lambda_j} \widehat{H}_{\lambda,j,E}(\Theta_0, \Theta_1, \ldots, \Theta_{j-1})(u)$$

and

$$\begin{aligned} & \widehat{H}_{\lambda,j,E}(\Theta_0, \Theta_1, \ldots, \Theta_{j-1})(u)(u_0, u_1, \ldots, u_j) \\ & \quad = K_2 \circ \Phi_F^{(l)}(\Theta_0(u)) \left(\Theta_{r_1}(u) \left(u_{i_1^{r_1}}, u_{i_2^{r_1}}, \ldots, u_{i_{r_1}^{r_1}} \right), \ldots, \Theta_{r_l}(u) \left(u_{i_1^{r_l}}, \ldots, u_{i_{r_l}^{r_l}} \right) \right) \end{aligned}$$

with the same constraints as in Lemma 4.30 for λ, r_j, l, and $i_k^{r_j}$.

Define

$$H_j = H_{j, BC^{\eta}(\mathbb{R}, X_c)} \quad \text{for each } j = 0, \ldots, k.$$

In the above definition one has to consider K_2 as a linear operator from $BC^{j\eta+(2j-1)\mu}(\mathbb{R},\mathbb{R}^n)$ into $BC^{j\eta+(2j-1)\mu}(\mathbb{R},\mathbb{R}^n)$, and $\Phi_F^{(l)}(\Theta_0(u))$ as an element of

$$\mathcal{L}^{(j)}\left(BC^{r_1\eta+(2r_1-1)\mu}(\mathbb{R},\mathbb{R}^n),\ldots,BC^{r_l\eta+(2r_l-1)\mu}(\mathbb{R},\mathbb{R}^n);BC^{j\eta+(2j-1)\mu}(\mathbb{R},\mathbb{R}^n)\right).$$

Notice that

$$j\eta+(2j-1)\mu>\sum_{k=1}^{l}r_k\eta+(2r_k-1)\mu \qquad (\Leftrightarrow j\eta+(2j-1)\mu>j\eta+(2j-l)\mu)$$

since $2\leq l\leq j$ and

$$r_1+r_2+\ldots+r_l=j.$$

Finally, define $H:M_0\times M_1\times\ldots\times M_k\to M_0\times M_1\times\ldots\times M_k$ by

$$H(\Theta_0,\Theta_1,\ldots,\Theta_k)=(H_0(\Theta_0),H_1(\Theta_0,\Theta_1),\ldots,H_k(\Theta_0,\Theta_1,\ldots,\Theta_k)).$$

We now check that the conditions of the fibre contraction Theorem 4.26 are satisfied. We have already shown that H_0 is a contraction on $M_{0,E}$ (from the existence part of the center manifold). It follows from (4.43) and (4.44) that H_j $(1\leq j\leq k)$ is a contraction on X_j. More precisely, from Assumption 4.23-(iii), we have for each $j=1,\ldots,k$ that

$$\begin{aligned}
&\sup_{u\in V_\varrho^\eta}\left\|K_2\circ\Phi_F^{(1)}(u)\right\|_{\mathcal{L}\left(BC^{j\eta+(2j-1)\mu}(\mathbb{R},\mathbb{R}^n),BC^{j\eta+(2j-1)\mu}(\mathbb{R},\mathbb{R}^n)\right)}\\
&\leq\|K_2\|_{\mathcal{L}\left(BC^{j\eta+(2j-1)\mu}(\mathbb{R},\mathbb{R}^n)\right)}\sup_{u\in V_\varrho^\eta}\left\|\Phi_F^{(1)}(u)\right\|_{\mathcal{L}\left(BC^{j\eta+(2j-1)\mu}(\mathbb{R},\mathbb{R}^n),BC^{j\eta+(2j-1)\mu}(\mathbb{R},\mathbb{R}^n)\right)}\\
&\leq\sup\nolimits_{\theta\in[\eta,\widehat{\eta}]}\|K_2\|_{\mathcal{L}\left(BC^\theta(\mathbb{R},\mathbb{R}^n)\right)}|F|_{1,V_\varrho}\\
&\leq\sup\nolimits_{\theta\in[\eta,\widehat{\eta}]}\|K_2\|_{\mathcal{L}\left(BC^\theta(\mathbb{R},\mathbb{R}^n)\right)}\|F\|_{\mathrm{Lip}(\mathbb{R}^n,\mathbb{R}^n)}<1.
\end{aligned}$$

Thus, each H_j is a contraction with respect to Θ_j. The fixed point of H_0 is Γ_0, and we denote by $\Gamma=(\Gamma_0,\Gamma_1,\ldots,\Gamma_k)$ the fixed point of H. Moreover, for $\mu=0$, each H_j is still a contraction so we have for each $j=1,\ldots,k$ that

$$\sup_{u\in BC^\eta(\mathbb{R},X_c)}\left\|\Gamma_j(u)\right\|_{\mathcal{L}^{(j)}\left(BC^\eta(\mathbb{R},\mathbb{R}^n),BC^{j\eta}(\mathbb{R},\mathbb{R}^n)\right)}<+\infty.$$

Step 2 (Attractivity of the fixed point): In this part, we apply the fibre contraction Theorem 4.26 to prove that for each compact subset C of $BC^\eta(\mathbb{R},X_c)$ and each $\Theta\in M_0\times M_1\times\ldots\times M_k$,

$$\lim_{m\to+\infty}H_C^m(\Theta\mid_C)=\Gamma\mid_C. \tag{4.45}$$

Let C be a compact subset of $BC^\eta(\mathbb{R},X_c)$. From the definition of H_C, it is clear that

$$\Gamma\mid_C=H_C(\Gamma\mid_C)$$

and from the step 1, it is not difficult to see that for each $j = 0, \ldots, k$, $H_{j,C}$ is a contraction. In order to apply the fibre contraction Theorem 4.26, it remains to prove that for each $j = 1, \ldots, k$, $H_{j,C}\left(\Theta_{0,C}, \Theta_{1,C}, \ldots, \Theta_{j-1,C}, \Gamma_j \mid_C\right) \in M_j$ dependents continuously on $\left(\Theta_{0,C}, \Theta_{1,C}, \ldots, \Theta_{j-1,C}\right) \in M_{0,C} \times M_{1,C} \times \ldots \times M_{j-1,C}$.

We have

$$\begin{aligned}&H_j\left(\Theta_{0,C}, \Theta_{1,C}, \ldots, \Theta_{j-1,C}, \Gamma^{(j)} \mid_C\right)(u)\\&\quad = K_2 \circ \Phi_F^{(1)}\left(\Theta_{0,C}(u)\right) \circ \Gamma^{(j)}(u) + \widehat{H}_j\left(\Theta_{0,C}, \Theta_{1,C}, \ldots, \Theta_{j-1,C}\right)(u).\end{aligned}$$

Since $\Gamma^{(j)}(u) \in \mathcal{L}^{(j)}\left(BC^{\eta}(\mathbb{R}, \mathbb{R}^n), BC^{j\eta}(\mathbb{R}, \mathbb{R}^n)\right)$ and $\Phi(u) \in V_{\varrho}^{\eta}$, we can consider $\Phi_F^{(1)}$ as a map from V_{ϱ}^{η} into $\mathcal{L}\left(BC^{j\eta}(\mathbb{R}, \mathbb{R}^n), BC^{j\eta+(2j-1)\mu}(\mathbb{R}, \mathbb{R}^n)\right)$, and by Lemma 4.30 (c) this map is continuous.

Indeed, let $\Theta_0, \widehat{\Theta}_0 \in M_0$ be two maps. Then we have

$$\begin{aligned}&\sup_{u\in C}\left\|K_2 \circ \left[\Phi_F^{(1)}\left(\Theta_0(u)\right) - \Phi_F^{(1)}\left(\widehat{\Theta}_0(u)\right)\right] \circ \Gamma^{(j)}(u)\right\|_{\mathcal{L}^{(j)}\left(BC^{\eta}(\mathbb{R},X_c), BC^{j\eta+(2j-1)\mu}(\mathbb{R},\mathbb{R}^n)\right)}\\&\quad \leq \|K_2\|_{\mathcal{L}\left(BC^{j\eta+(2j-1)\mu}(\mathbb{R},\mathbb{R}^n)\right)}\\&\qquad \cdot \sup_{u\in C}\left\|\left[\Phi_F^{(1)}\left(\Theta_0(u)\right) - \Phi_F^{(1)}\left(\widehat{\Theta}_0(u)\right)\right] \circ \Gamma^{(j)}(u)\right\|_{\mathcal{L}^{(j)}\left(BC^{\eta}(\mathbb{R},X_c), BC^{j\eta+(2j-1)\mu}(\mathbb{R},\mathbb{R}^n)\right)}\\&\quad \leq \|K_2\|_{\mathcal{L}\left(BC^{j\eta+(2j-1)\mu}(\mathbb{R},\mathbb{R}^n)\right)} \sup_{u\in C}\left\|\Gamma^{(j)}(u)\right\|_{\mathcal{L}^{(j)}\left(BC^{\eta}(\mathbb{R},X_c), BC^{j\eta}(\mathbb{R},\mathbb{R}^n)\right)}\\&\qquad \cdot \sup_{u\in C}\left\|\Phi_F^{(1)}\left(\Theta_0(u)\right) - \Phi_F^{(1)}\left(\widehat{\Theta}_0(u)\right)\right\|_{\mathcal{L}^{(j)}\left(BC^{j\eta}(\mathbb{R},\mathbb{R}^n), BC^{j\eta+(2j-1)\mu}(\mathbb{R},\mathbb{R}^n)\right)}\end{aligned}$$

and by Lemma 4.27 we have

$$\begin{aligned}&\sup_{u\in C}\left\|\Phi_F^{(1)}\left(\Theta_0(u)\right) - \Phi_F^{(1)}\left(\widehat{\Theta}_0(u)\right)\right\|_{\mathcal{L}^{(j)}\left(BC^{j\eta}(\mathbb{R},\mathbb{R}^n), BC^{j\eta+(2j-1)\mu}(\mathbb{R},\mathbb{R}^n)\right)}\\&\quad \leq \sup_{u\in C}\left\|\Phi_{DF}\left(\Theta_0(u)\right) - \Phi_{DF}\left(\widehat{\Theta}_0(u)\right)\right\|_{BC^{(2j-1)\mu}(\mathbb{R},\mathcal{L}(\mathbb{R}^n,\mathbb{R}^n))}\\&\quad \leq \max\begin{pmatrix}\sup_{|t|\geq R} e^{-(2j-1)\mu|t|}\left\|DF\left(\Theta_0(u)(t)\right) - DF\left(\widehat{\Theta}_0(u)(t)\right)\right\|_{\mathcal{L}(\mathbb{R}^n,\mathbb{R}^n)},\\ \sup_{|t|\leq R} e^{-(2j-1)\mu|t|}\left\|DF\left(\Theta_0(u)(t)\right) - DF\left(\widehat{\Theta}_0(u)(t)\right)\right\|_{\mathcal{L}(\mathbb{R}^n,\mathbb{R}^n)}\end{pmatrix}.\end{aligned}$$

Since $\widehat{\Theta}_0$ is continuous, C is compact, it follows that $\widehat{\Theta}_0(C)$ is compact, and by Ascoli's theorem (see for example Lang [202]), the set $\widehat{C} = \overline{\bigcup_{|t|\leq R, u\in C}\left\{\widehat{\Theta}_0(u)(t)\right\}}$ is compact. But since $DF(.)$ is continuous, we have that for each $\varepsilon > 0$, there exists $\eta > 0$, such that for each $x, y \in \mathbb{R}^n$,

$$d\left(x, \widehat{C}\right) \leq \eta, \quad d\left(y, \widehat{C}\right) \leq \eta, \quad \text{and } \|x - y\| \leq \eta \Rightarrow \|DF(x) - DF(y)\| \leq \varepsilon.$$

Hence, the map $\Theta_{0,C} \to K_2 \circ \Phi_F^{(1)}\left(\Theta_{0,C}(.)\right) \circ \Gamma^{(j)}(.)$ is continuous.

It remains to consider $1 \leq r_i \leq j-1$, $r_1 + r_2 + \ldots + r_l = j$. We have

$$\left\|K_2 \circ \left[\Phi_F^{(l)}\left(\Theta_0(u)\right) - \Phi_F^{(l)}\left(\widehat{\Theta}_0(u)\right)\right]\left(\widetilde{\Theta}_{r_1}(u), \ldots, \widetilde{\Theta}_{r_l}(u)\right)\right\|_{\mathcal{L}^{(j)}\left(BC^{\eta}(\mathbb{R},X_c), BC^{j\eta+(2j-1)\mu}(\mathbb{R},\mathbb{R}^n)\right)}$$
$$\leq \|K_2\|_{\mathcal{L}\left(BC^{j\eta+(2j-1)\mu}(\mathbb{R},\mathbb{R}^n), BC^{j\eta+(2j-1)\mu}(\mathbb{R},\mathbb{R}^n)\right)}$$
$$\sup_{u\in C}\left\|\left[\Phi_F^{(l)}\left(\Theta_0(u)\right) - \Phi_F^{(l)}\left(\widehat{\Theta}_0(u)\right)\right]\left(\widetilde{\Theta}_{r_1}(u), \ldots, \widetilde{\Theta}_{r_l}(u)\right)\right\|_{\mathcal{L}^{(j)}\left(BC^{\eta}(\mathbb{R},X_c), BC^{j\eta+(2j-1)\mu}(\mathbb{R},\mathbb{R}^n)\right)}$$
$$\leq \|K_2\|_{\mathcal{L}\left(BC^{j\eta+(2j-1)\mu}(\mathbb{R},\mathbb{R}^n), BC^{j\eta+(2j-1)\mu}(\mathbb{R},\mathbb{R}^n)\right)}$$
$$\cdot\left\|\Phi_F^{(l)}\left(\Theta_0(u)\right) - \Phi_F^{(l)}\left(\widehat{\Theta}_0(u)\right)\right\|_{\mathcal{L}^{(l)}\left(\prod_{p=1,\ldots,l} BC^{r_p\eta+(2r_p-1)\mu}(\mathbb{R},\mathbb{R}^n); BC^{j\eta+(2j-1)\mu}(\mathbb{R},\mathbb{R}^n)\right)}$$
$$\cdot\prod_{p=1,\ldots,l}\left\|\widetilde{\Theta}_{r_p}(u)\right\|_{\mathcal{L}^{(j)}\left(BC^{\eta}(\mathbb{R},X_c), BC^{r_p\eta+(2r_p-1)\mu}(\mathbb{R},\mathbb{R}^n)\right)}$$

and by Lemma 4.29 we have

$$\sup_{u\in C}\left\|\Phi_F^{(l)}\left(\Theta_0(u)\right) - \Phi_F^{(l)}\left(\widehat{\Theta}_0(u)\right)\right\|_{\mathcal{L}^{(l)}\left(\prod_{p=1,\ldots,l} BC^{r_p\eta+(2r_p-1)\mu}(\mathbb{R},\mathbb{R}^n); BC^{j\eta+(2j-1)\mu}(\mathbb{R},\mathbb{R}^n)\right)}$$
$$\leq \sup_{u\in C}\left\|\Phi_{D^{(l)}F}\left(\Theta_0(u)\right) - \Phi_{D^{(l)}F}\left(\widehat{\Theta}_0(u)\right)\right\|_{BC^{\delta}\left(\mathbb{R},\mathcal{L}^{(l)}(\mathbb{R}^n,\mathbb{R}^n)\right)}$$

with $\delta = (j\eta + (2j-1)\mu) - \sum_{k=1}^{l} r_k\eta + (2r_k - 1)\mu > 0$. By using the same compactness arguments as previously, we deduce that

$$\sup_{u\in C}\left\|\Phi_{D^{(l)}F}\left(\Theta_0(u)\right) - \Phi_{D^{(l)}F}\left(\widehat{\Theta}_0(u)\right)\right\|_{BC^{\delta}\left(\mathbb{R},\mathcal{L}^{(l)}(\mathbb{R}^n,\mathbb{R}^n)\right)} \to 0$$

as $d_{0,C}(\Theta_0, \widehat{\Theta}_0) \to 0$. We conclude that the continuity condition of the fibre contraction Theorem 4.26 is satisfied for each $H_{j,C}$ and (4.45) follows.

Step 3: In order to prove Theorem 4.24 it now remains to prove that for each $u, v \in BC^{\eta}(\mathbb{R}, X_c), \forall j = 1, \ldots, k$,

$$\Gamma_{j-1}(u) - \Gamma_{j-1}(v) = \int_0^1 \Gamma_j(s(u-v)+v)(u-v)\,ds, \tag{4.46}$$

where the last integral is a Riemann integral. As assumed that (4.46) is satisfied, we deduce that $\Gamma_0 : BC^{\eta}(\mathbb{R}, X_c) \to BC^{k\eta+(2k-1)\mu}(\mathbb{R}, \mathbb{R}^n)$ is k-times continuously differentiable, and since

$$\Psi(x_c) = L \circ \Gamma_0 \circ K_1(x_c) = \Pi_h\Gamma_0(K_1(x_c))(0),$$

and $Lu = \Pi_h u(0)$ is a bounded linear operator from $BC^{k\eta+(2k-1)\mu}(\mathbb{R}, \mathbb{R}^n)$ into X_h, it follows that $\Psi : X_c \to X_h$ is k-times continuously differentiable.

We now prove (4.46). Set

$$\Theta^0 = \left(\Theta_0^0, \Theta_1^0, \ldots, \Theta_k^0\right)$$

with

$$\Theta_0^0(u) = u, \Theta_1^0(u) = J, \text{ and } \Theta_j^0 = 0, \forall j = 2, \ldots, k$$

and set

$$\Theta^m = \left(\Theta_0^m, \Theta_1^m, \ldots, \Theta_k^m\right) = H^m\left(\Theta^0\right), \forall m \geq 1.$$

Then from Lemma 4.30, we know that $\Theta_0^m : BC^\eta(\mathbb{R}, X_c) \to BC^{k\eta+(2k-1)\mu}(\mathbb{R}, \mathbb{R}^n)$ is a C^k-map and

$$D^j\Theta_0^m(u) = \Theta_j^m(u), \quad \forall j = 1, \ldots, k, \quad \forall u \in BC^\eta(\mathbb{R}, X_c).$$

For each $u, v \in BC^\eta(\mathbb{R}, X_c)$ and each $\forall j = 1, \ldots, k, \forall m \geq 1$,

$$\Theta_{j-1}^m(u) - \Theta_{j-1}^m(v) = \int_0^1 \Theta_j^m(s(u-v)+v)(u-v)\, ds.$$

Let $u, v \in BC^\eta(\mathbb{R}, X_c)$ be fixed. Denote

$$C = \{s(u-v)+v : s \in [0,1]\}.$$

Then clearly C is a compact set, and from step 2, we have for each $j = 0, \ldots, k$ that

$$\sup_{w \in C} \left\| \Theta_j^m(w) - \Gamma_j(w) \right\|_{BC^{j\eta+(2j-1)\mu}(\mathbb{R},\mathbb{R}^n)} \to 0, \quad \text{as } m \to +\infty.$$

Thus, (4.46) follows. □

It follows from the foregoing treatment that we can obtain the derivatives of $\Gamma_0(u)$ at $u = 0$. Assume that $F(0) = 0$ and $DF(0) = 0$, we have

$$\begin{aligned}
&D\Gamma_0(0) = J, \\
&D^{(2)}\Gamma_0(0)(u_1, u_2) = K_2 \circ \Phi_F^{(2)}(0)\left(D\Gamma_0(0)(u_1), D\Gamma_0(0)(u_2)\right), \\
&D^{(3)}\Gamma_0(0)(u_1, u_2, u_3) = K_2 \circ \Phi_F^{(2)}(0)\left(D^{(2)}\Gamma_0(0)(u_1, u_3), D\Gamma_0(0)(u_2)\right) \\
&\qquad + K_2 \circ \Phi_F^{(2)}(0)\left(D\Gamma_0(0)(u_1), D^{(2)}\Gamma_0(0)(u_2, u_3)\right) \\
&\qquad + K_2 \circ \Phi_F^{(3)}(0)\left(D\Gamma_0(0)(u_1), D\Gamma_0(0)(u_2), D\Gamma_0(0)(u_3)\right), \\
&\quad \vdots \\
&D^{(l)}\Gamma_0(0) = \sum_{\lambda \in \Lambda_j} K_2 \circ \Phi_F^{(l)}(0)\left(D^{(r_1)}\Gamma(0), \ldots, D\Gamma^{(r_l)}(0)\right).
\end{aligned} \tag{4.47}$$

We have the following Lemma.

Lemma 4.31 *Let Assumptions 4.3 and 4.23 be satisfied. Assume also that $F(0) = 0$ and $DF(0) = 0$. Then*

$$\Psi(0) = 0, \text{ and } D\Psi(0) = 0,$$

and if $k > 1$,

$$D^j\Psi(0)(x_1, \ldots, x_n) = \Pi_h D^{(l)}\Gamma_0(0)(K_1x_1, \ldots, K_1x_n)(0),$$

where $D^{(l)}\Gamma_0(0)$ is given by (4.47). *In particular, if $k > 1$ and*

$$\Pi_h D^j F(0)\mid_{X_c \times \ldots \times X_c} = 0, \text{ for } 2 \leq j \leq k,$$

then

$$D^j\Psi(0) = 0, \quad \textit{for } 1 \le j \le k.$$

In the context of Hopf bifurcation, we need an explicit formula for $D^2\Psi(0)$. Since $D\Gamma_0(0) = J$, we obtain from the above formula that $\forall x_1, x_2 \in X_c$,

$$D^2\Psi(0)\,(x_1, x_2) = \Pi_h K_h \left[D^{(2)}F(0)\,(K_1x_1, K_1x_2)\right](0),$$

where

$$K_h = K_s + K_u, \quad K_1(x_c)(t) := e^{A_c t}x_c,$$

$$K_u(f)(t) := -\int_0^{+\infty} e^{-A_u l}\Pi_u f(t+l)dl,$$

and

$$K_s(f)(t) := \int_0^{+\infty} e^{A_s l}\Pi_s f(t-l)\mathrm{d}l.$$

Hence,

$$\begin{aligned}
&D^2\Psi(0)\,(x_1, x_2)\\
&= -\int_0^{+\infty} e^{-A_u l}\Pi_u D^{(2)}F(0)\left(e^{A_c l}x_1, e^{A_c l}x_2\right)dl\\
&\quad + \int_0^{+\infty} e^{A_s l}\Pi_s D^{(2)}F(0)\left(e^{-A_c l}x_1, e^{-A_c l}x_2\right)dl.
\end{aligned}$$

Assume that $\mathbb{R}^n$ is a complex Banach space and F is twice continuously differentiable in $\mathbb{R}^n$ considered as a $\mathbb{C}$-Banach space. We assume in addition that A_c is diagonalisable, and denote by $\{v_1, \ldots, v_n\}$ a basis of X_c such that for each $i = 1, \ldots, n$, $A_c v_i = \lambda_i v_i$. Then by Assumption 4.3 , we must have $\lambda_i \in i\mathbb{R}, \forall i = 1, \ldots, n$. Moreover, for each $i, j = 1, \ldots, n$,

$$\begin{aligned}
D^2\Psi(0)\left(v_i, v_j\right) &= -\int_0^{+\infty} e^{(\lambda_i+\lambda_j)l}e^{-A_u l}\Pi_u D^{(2)}F(0)\left(v_i, v_j\right)dl\\
&\quad + \int_0^{+\infty} e^{A_s l}\Pi_s D^{(2)}F(0)\left(e^{-\lambda_i l}v_i, e^{-\lambda_j l}v_j\right)dl\\
&= \left(\left(\lambda_i + \lambda_j\right)I - A_u\right)^{-1}\Pi_u D^{(2)}F(0)\left(v_i, v_j\right)\\
&\quad + \int_0^{+\infty} e^{-(\lambda_i+\lambda_j)l}e^{A_s l}\Pi_s D^{(2)}F(0)\left(v_i, v_j\right)dl\\
&= \left(\left(\lambda_i + \lambda_j\right)I - A_u\right)^{-1}\Pi_u D^{(2)}F(0)\left(v_i, v_j\right)\\
&\quad - \left(-\left(\lambda_i + \lambda_j\right)I + A_s\right)^{-1}\Pi_s D^{(2)}F(0)\left(v_i, v_j\right).
\end{aligned}$$

Thus,

$$\begin{aligned}
D^2\Psi(0)\left(v_i, v_j\right) &= \left(\left(\lambda_i + \lambda_j\right)I - A_u\right)^{-1}\Pi_u D^{(2)}F(0)\left(v_i, v_j\right)\\
&\quad + \left(\left(\lambda_i + \lambda_j\right)I - A_s\right)^{-1}\Pi_s D^{(2)}F(0)\left(v_i, v_j\right).
\end{aligned} \tag{4.48}$$

Note that by Assumption 4.3 $i\mathbb{R} \subset \rho(A_s)$ and $i\mathbb{R} \subset \rho(A_u)$, so the above formula is well defined.

As in Vanderbauwhede and Iooss [340, Theorem 3], we have the following theorem about the existence of the local center manifold.

Theorem 4.32 *Let Assumption 4.3 be satisfied. Let $F : \mathbb{R}^n \to \mathbb{R}^n$ be a map. Assume there exists an integer $k \geq 1$ such that F is k-times continuously differentiable in some neighbourhood of 0 with $F(0) = 0$ and $DF(0) = 0$. Then there exist a neighbourhood Ω of the origin in $\mathbb{R}^n$ and a map $\Psi \in C_b^k(X_c, X_h)$, with $\Psi(0) = 0$ and $D\Psi(0) = 0$, such that the following properties hold:*

(i) *If I is an interval of $\mathbb{R}$ and $x_c : I \to X_c$ is a solution of*

$$\frac{dx_c(t)}{dt} = A_c x_c(t) + \Pi_c F\left[x_c(t) + \Psi(x_c(t))\right] \tag{4.49}$$

such that

$$u(t) := x_c(t) + \Psi(x_c(t)) \in \Omega, \forall t \in I,$$

then for each $t, s \in I$ with $t \geq s$,

$$u(t) = u(s) + A\int_s^t u(l)dl + \int_s^t F(u(l))\, dl.$$

(ii) *If $u : \mathbb{R} \to \mathbb{R}^n$ is a map such that for each $t, s \in \mathbb{R}$ with $t \geq s$,*

$$u(t) = u(s) + A\int_s^t u(l)dl + \int_s^t F(u(l))\, dl$$

and

$$u(t) \in \Omega, \quad \forall t \in \mathbb{R},$$

then

$$\Pi_h u(t) = \Psi(\Pi_c u(t)), \forall t \in \mathbb{R},$$

and $\Pi_c u : \mathbb{R} \to X_c$ is a solution of (4.49).

(iii) *If $k \geq 2$, then for each $x_1, x_2 \in X_c$,*

$$\begin{aligned} &D^2\Psi(0)(x_1, x_2) \\ &= -\int_0^{+\infty} e^{-A_u l}\Pi_u D^{(2)}F(0)\left(e^{A_c l}x_1, e^{A_c l}x_2\right) dl \\ &\quad + \int_0^{+\infty} e^{A_s l}\Pi_s D^{(2)}F(0)\left(e^{-A_c l}x_1, e^{-A_c l}x_2\right) dl. \end{aligned}$$

Moreover, $\mathbb{C}^n$ is a $\mathbb{C}$-Banach space, and if $\{v_1, \ldots, v_n\}$ is a basis of X_c such that for each $i = 1, \ldots, n$, $A_c v_i = \lambda_i v_i$, with $\lambda_i \in i\mathbb{R}$, then for each $i, j = 1, \ldots, n$,

$$\begin{aligned} D^2\Psi(0)(v_i, v_j) &= \left((\lambda_i + \lambda_j) I - A_u\right)^{-1} \Pi_u D^{(2)}F(0)(v_i, v_j) \\ &\quad + \left((\lambda_i + \lambda_j) I - A_s\right)^{-1} \Pi_s D^{(2)}F(0)(v_i, v_j). \end{aligned}$$

Proof Set for each $r > 0$ that

$$F_r(x) = F(x)\chi_c\left(r^{-1}\Pi_c(x)\right)\chi_h\left(r^{-1}\|\Pi_h(x)\|\right), \forall x \in \mathbb{R}^n,$$

where $\chi_c : X_c \to [0, +\infty)$ is a C^∞ map with $\chi_c(x) = 1$ if $\|x\| \le 1$, $\chi_c(x) = 0$ if $\|x\| \ge 2$, and $\chi_h : [0, +\infty) \to [0, +\infty)$ is a C^∞ map with $\chi_h(x) = 1$ if $|x| \le 1$, $\chi_h(x) = 0$ if $|x| \ge 4.4$. Then by using the same arguments as in the proof of Theorem 3 in [340], we deduce that there exists $r_0 > 0$, such that for each $r \in (0, r_0]$, F_r satisfies Assumption 4.23. By applying Theorem 4.24 to

$$\frac{du(t)}{dt} = Au(t) + F_r(u(t)), \quad t \ge 0, \text{ and } u(0) = x$$

for $r > 0$ small enough, the result follows. □

4.4 Existence and Stability of the Center Unstable Manifold

In this section we will assume that the center-unstable spectrum is not empty.

Assumption 4.33 We assume that $\sigma_{cu}(A) = \{\lambda \in \sigma(A) : \mathrm{Re}\,\lambda \ge 0\} \ne \emptyset$.

Let

$$\beta_- = \min\{-\mathrm{Re}(\lambda) : \lambda \in \sigma_s(A)\} > 0,$$

or equivalently

$$\beta_- = -\lim_{t\to+\infty} \frac{\ln\left(\|e^{A_s t)}\|_{\mathcal{L}(X_s)}\right)}{t} > 0.$$

From now on, we fix $\beta \in (0, \beta_-)$. From the proof of Lemma 4.5, we deduce that we can find a constant number $M_s > 0$ such that

$$\|e^{At}\Pi_s\|_{\mathcal{L}(\mathbb{R}^n)} \le M_s e^{(-\beta+\varepsilon)t}, \forall t \ge 0, \forall \varepsilon \ge 0. \tag{4.50}$$

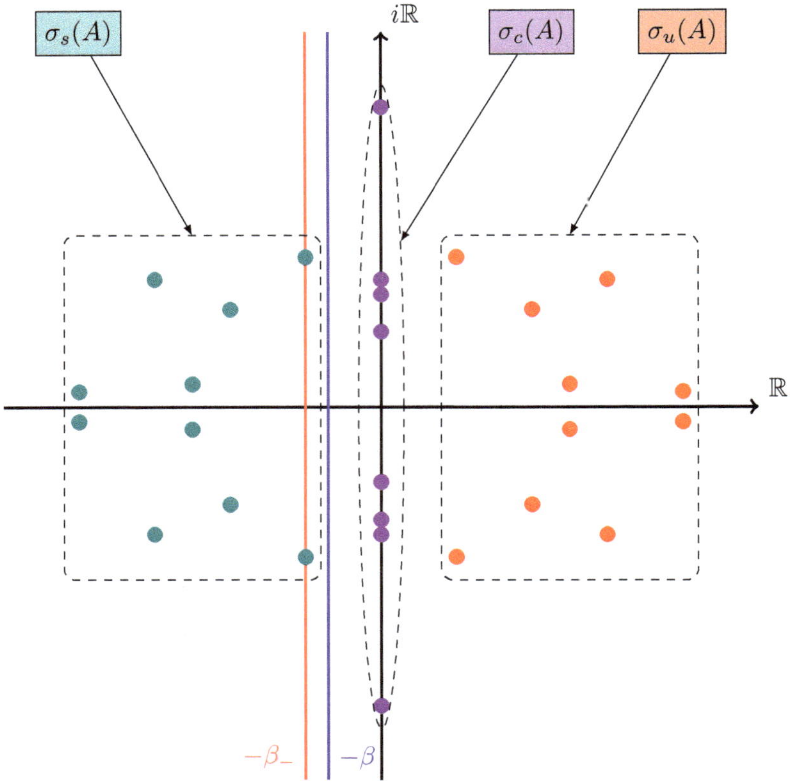

Fig. 4.3: *In this Figure, we illustrate the different parts of the spectrum* $\sigma_s(A)$, $\sigma_c(A)$ *and* $\sigma_u(A)$ *as well as* β_-.

Let us recall that a function $u : \mathbb{R} \to \mathbb{R}$ is a **negative orbit** for the semiflow $\{U(t)\}_{t\geq 0}$ if and only if the function $t \to u(t)$ on $\mathbb{R}$ satisfies

$$u(t) = U(t-s)u(s), \forall t, s \in (-\infty, 0], \text{ with } t \geq s,$$

where $\{U(t)\}_{t\geq 0}$ is a continuous semiflow generated by (4.20).

That is equivalent to say that for each $s \in (-\infty, 0]$ (fixed) the function $t \to u(t)$ from $[s, 0]$ satisfies

$$u(t) = u(s) + A\int_0^{t-s} u(s+r)dr + \int_0^{t-s} F\left(u(s-r)\right)dr, \forall t \in [s, 0],$$

or equivalently by using the variation of constant formula

$$u(t) = e^{A(t-s)}u(s) + \int_s^t e^{A(t-\sigma)}F(u(\sigma))d\sigma, \forall t \in [s, 0],$$

or (for short) by using the convolutions

$$u(t) = e^{A(t-s)}u(s) + \left(e^{A\cdot} * F\left(u(s+.)\right)\right)(t-s), \forall t \in [s,0]. \tag{4.51}$$

Definition 4.34 Let $\eta \in (0, \beta_-)$. The η-**center-unstable manifold** of (4.20), denoted by $\mathcal{M}_{cu}^{\eta}$, is the set of all points $x \in \mathbb{R}^n$ such that there exists $u \in BC^{\eta}(\mathbb{R}_-, \mathbb{R}^n)$, a negative orbit of $\{U(t)\}_{t\geq 0}$, such that $u(0) = x$. That is to say that

$$\mathcal{M}_{cu}^{\eta} = \left\{x \in \mathbb{R}^n : \exists u \in BC^{\eta}(\mathbb{R}_-, \mathbb{R}^n),\ \text{a negative orbit of}\ \{U(t)\}_{t\geq 0},\ \text{such that}\ u(0) = x\right\}. \tag{4.52}$$

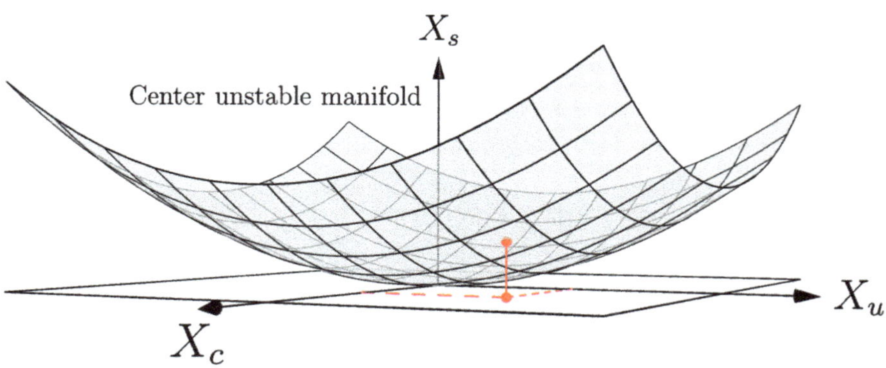

Fig. 4.4: *Schematic representation of the center-unstable manifold. In this figure we plot the linear center-unstable manifold and the center-unstable manifold which corresponds to a surface tangent at* 0 *to the linear centerunstable manifold. The surface representing the center-unstable manifold is delimited by the dashed curves.*

For each $\eta > 0$, $\mathcal{M}_{cu}^{\eta}$ is invariant under the semiflow $\{U(t)\}_{t\geq 0}$, that is,

$$U(t)\mathcal{M}_{cu}^{\eta} = \mathcal{M}_{cu}^{\eta}, \quad \forall t \geq 0.$$

Moreover, we say that $\{U(t)\}_{t\geq 0}$ is *reduced on* $\mathcal{M}_{cu}^{\eta}$ if there exists a map $\Psi_{cu} : X_{cu} \to X_s$ such that

$$\mathcal{M}_{cu}^{\eta} = \text{Graph}\left(\Psi_{cu}\right) = \left\{x_{cu} + \Psi_{cu}\left(x_{cu}\right) : x_{cu} \in X_{cu}\right\}.$$

The proof of the following lemma is similar to the proof of Lemma 4.16.

Lemma 4.35 *Let Assumption 4.33 be satisfied. Then*

(i) *For each* $\eta \in [0, \beta_-)$, *each* $f \in BC^{\eta}(\mathbb{R}_-, \mathbb{R}^n)$, *and each* $t \in \mathbb{R}_-$,

$$K_s(f)(t) := \lim_{r\to-\infty} \Pi_s\left(e^{A\cdot} * f(r+.)\right)(t-r)\ \textit{exists.}$$

(ii) *For each* $\eta \in [0, \beta_-)$, K_s *is a bounded linear operator from* $BC^{\eta}(\mathbb{R}_-, \mathbb{R}^n)$ *into* $BC^{\eta}(\mathbb{R}_-, X_s)$. *More precisely, for each* $\nu \in (-\beta_-, 0)$, *there exists a constant*

$\widehat{C}_{s,\nu} > 0$ *such that*

$$\|K_s\|_{\mathcal{L}(BC^\eta(\mathbb{R}_-,\mathbb{R}^n),BC^\eta(\mathbb{R}_-,X_s))} \leq \widehat{C}_{s,\nu}, \forall \eta \in [0,-\nu].$$

(iii) *For each* $\eta \in [0,\beta_-)$, *each* $f \in BC^\eta(\mathbb{R}_-,\mathbb{R}^n)$, *and each* $t, s \in \mathbb{R}$ *with* $t \geq s$,

$$K_s(f)(t) - e^{A(t-s)}K_s(f)(s) = \Pi_s\left(e^{A.} * f(s+.)\right)(t-s).$$

Remark 4.36 As explained before in the chapter we can expressed $K_s(f)(t)$ as

$$K_s(f)(t) = \int_0^{+\infty} e^{A_s\theta}\Pi_s f(t-\theta)d\theta, \forall t \leq 0.$$

The following lemmas can be proved by using similar argument as for Lemma 4.19.

Lemma 4.37 *Let Assumption 4.33 be satisfied. Let* $\eta \in (0,\beta_-)$ *be fixed. For each* $x_{cu} \in X_{cu}$, *each* $f \in BC^\eta(\mathbb{R}_-,\mathbb{R}^n)$, *and each* $t \in (-\infty, 0]$, *denote*

$$K_1(x_{cu})(t) := e^{A_{cu}t}x_{cu}, \quad K_{cu}(f)(t) := \int_0^t e^{A_{cu}(t-s)}\Pi_{cu}f(s)ds,$$

where $\Pi_{cu} = \Pi_c + \Pi_u$. *Then* K_1 *is a bounded linear operator from* X_{cu} *into* $BC^\eta(\mathbb{R}_-,X_{cu})$ *and*

$$\|K_1\|_{\mathcal{L}(X_{cu},BC^\eta(\mathbb{R}_-,\mathbb{R}^n))} \leq \sup_{t\geq 0}\left\|e^{-(A_{cu}+\eta I)t}\right\| < +\infty,$$

$$\|K_{cu}\|_{\mathcal{L}(BC^\eta(\mathbb{R}_-,\mathbb{R}^n))} \leq \|\Pi_{cu}\|_{\mathcal{L}(\mathbb{R}^n)}\int_0^{+\infty}\left\|e^{-(A_{cu}+\eta I)l}\right\|dl < +\infty.$$

Lemma 4.38 *Let Assumption 4.33 be satisfied. Let* $\eta \in (0,\beta_-)$ *and* $u \in BC^\eta(\mathbb{R}_-,\mathbb{R}^n)$ *be fixed. Then* u *is a negative complete orbit of* $\{U(t)\}_{t\geq 0}$ *if and only if for each* $t \in \mathbb{R}_-$,

$$u(t) = K_1(\Pi_{cu}u(0))(t) + K_{cu}(F(u(.)))(t) + K_s(F(u(.)))(t), \tag{4.53}$$

where $\Pi_{cu} = \Pi_c + \Pi_u$.

Proof This proof is similar to the proof of Lemma 4.20. □

Let $\eta \in (0,\beta_-)$ be fixed. Rewrite equation (4.53) as the following fixed point problem: To find $u \in BC^\eta(\mathbb{R}_-,\mathbb{R}^n)$ such that

$$u = K_1(\Pi_{cu}u(0)) + K_2\Phi_F(u), \tag{4.54}$$

where the nonlinear operator $\Phi_F \in \text{Lip}\left(BC^\eta(\mathbb{R}_-,\mathbb{R}^n), BC^\eta(\mathbb{R}_-,\mathbb{R}^n)\right)$ is defined by

$$\Phi_F(u)(t) = F(u(t)), \quad \forall t \in \mathbb{R}_-,$$

and the linear operator $K_2 \in \mathcal{L}\left(BC^\eta(\mathbb{R}_-,\mathbb{R}^n), BC^\eta(\mathbb{R}_-,\mathbb{R}^n)\right)$ is defined by

$$K_2 = K_{cu} + K_s.$$

Moreover, we have the following estimates

$$\|K_1\|_{\mathcal{L}(X_{cu},BC^{\eta}(\mathbb{R}_-,\mathbb{R}^n))} \leq \sup_{t\geq 0}\left\|e^{-(A_{cu}+\eta I)t}\right\|,$$

$$\|\Phi_F\|_{\text{Lip}} \leq \|F\|_{\text{Lip}},$$

and for each $\nu \in (-\beta_-, 0)$, we have

$$\|K_2\|_{\mathcal{L}(BC^{\eta}(\mathbb{R},\mathbb{R}^n))} \leq \gamma(\nu,\eta), \forall \eta \in (0, -\nu],$$

where

$$\gamma(\nu,\eta) := \widehat{C}_{s,\nu} + \|\Pi_{cu}\|_{\mathcal{L}(\mathbb{R}^n)} \int_0^{+\infty} \left\|e^{-(A_{cu}+\eta I)l}\right\| dl. \tag{4.55}$$

Furthermore, by Lemma 4.38, the η-center-unstable manifold is given by

$$\mathcal{M}_{cu}^{\eta} = \{x \in \mathbb{R}^n : \exists u \in BC^{\eta}(\mathbb{R}_-,\mathbb{R}^n) \text{ a solution of (4.54) and } u(0) = x\}. \tag{4.56}$$

We state the existence of center-unstable manifolds for the Cauchy problem (4.20) which can be proved similarly as Theorem 4.21.

Theorem 4.39 (Global center-unstable manifold) *Let Assumption 4.33 be satisfied. Let $\eta \in (0, \beta_-)$ be fixed and $\delta_0 = \delta_0(\eta) > 0$ be such that*

$$\delta_0 \|K_2\|_{\mathcal{L}(BC^{\eta}(\mathbb{R}_-,\mathbb{R}^n))} < 1.$$

Then for each $F \in \text{Lip}(\mathbb{R}^n, \mathbb{R}^n)$ with $\|F\|_{\text{Lip}(\mathbb{R}^n,\mathbb{R}^n)} \leq \delta_0$, there exists a Lipschitz continuous map $\Psi_{cu} : X_{cu} \to X_s$ such that

$$\mathcal{M}_{cu}^{\eta} = \{x_{cu} + \Psi_{cu}(x_{cu}) : x_{cu} \in X_{cu}\}.$$

Moreover, we have the following properties:

(i)

$$\sup_{x_{cu}\in X_{cu}} \|\Psi_{cu}(x_{cu})\| \leq \|K_s\|_{\mathcal{L}(BC^{\eta}(\mathbb{R}_-,\mathbb{R}^n))} \sup_{x\in\mathbb{R}^n} \|\Pi_s F(x)\|.$$

(ii) *We have*

$$\|\Psi_{cu}\|_{\text{Lip}(X_{cu},X_s)} \leq \frac{\|K_s\|_{\mathcal{L}(BC^{\eta}(\mathbb{R}_-,\mathbb{R}^n))} \|F\|_{\text{Lip}(\mathbb{R}^n,\mathbb{R}^n)} \|K_1\|_{\mathcal{L}(X_{cu},BC^{\eta}(\mathbb{R}_-,\mathbb{R}^n))}}{1 - \|K_2\|_{\mathcal{L}(BC^{\eta}(\mathbb{R}_-,\mathbb{R}^n))} \|F\|_{\text{Lip}(\mathbb{R}^n,\mathbb{R}^n)}}. \tag{4.57}$$

We now state and prove the existence of local center-unstable manifolds.

Theorem 4.40 (Local center-unstable manifold) *Let Assumption 4.33 be satisfied. Let $r > 0$ and $F : B_{\mathbb{R}^n}(0,r) \to \mathbb{R}^n$ be a map. Assume that there exists an integer $k \geq 1$ such that F is k-time continuously differentiable in $B_{\mathbb{R}^n}(0,r)$ with*

$$F(0) = 0 \text{ and } DF(0) = 0.$$

Then there exists a neighbourhood Ω *of the origin in* $\mathbb{R}^n$ *and a map* $\Psi_{cu} \in C_b^k(X_{cu}, X_s)$ *with*

$$\Psi_{cu}(0) = 0 \text{ and } D\Psi_{cu}(0) = 0,$$

such that

$$M_{cu} = \{x_{cu} + \Psi_{cu}(x_{cu}) : x_{cu} \in X_{cu}\}$$

is a locally invariant manifold by the semiflow generated by (4.20) around 0.

More precisely, the following properties hold:

(i) *If* I *is an interval of* $\mathbb{R}$ *and* $x_{cu} : I \to X_{cu}$ *is a solution of*

$$\frac{dx_{cu}(t)}{dt} = A_{cu}x_{cu}(t) + \Pi_{cu}F(x_{cu}(t) + \Psi_{cu}(x_{cu}(t))) \text{ **(reduced equation)**} \tag{4.58}$$

such that

$$u(t) := x_{uc}(t) + \Psi_{cu}(x_{uc}(t)) \in \Omega, \forall t \in I,$$

then for each $t, s \in I$ *with* $t \geq s$,

$$u(t) = u(s) + A\int_s^t u(l)dl + \int_s^t F(u(l))\,dl.$$

(ii) *If* $u : (-\infty, 0] \to \mathbb{R}^n$ *is a map such that for each* $t, s \in (-\infty, 0]$ *with* $t \geq s$,

$$u(t) = u(s) + A\int_s^t u(l)dl + \int_s^t F(u(l))\,dl$$

and

$$u(t) \in \Omega, \quad \forall t \in (-\infty, 0],$$

then

$$\Pi_s u(t) = \Psi_{cu}(\Pi_{cu}u(t)), \forall t \in (-\infty, 0],$$

and $\Pi_{cu}u : (-\infty, 0] \to X_{cu}$ *is a solution of* (4.58).

Proof In order to prove the local center-unstable manifold theorem, we apply Theorem 4.39 to the Cauchy problem

$$\frac{du}{dt} = Au(t) + F_r(u(t)), t \geq 0, u(0) = x \in \mathbb{R}^n,$$

where $F_r : \mathbb{R}^n \to \mathbb{R}^n$ is the following truncated function

$$F_r(x) = F(x)\chi_{cu}\left(r^{-1}\Pi_{cu}(x)\right)\chi_s\left(r^{-1}\|\Pi_s(x)\|\right), \forall x \in \mathbb{R}^n,$$

$\chi_{cu} : X_{cu} \to [0, +\infty)$ is a C^∞ map with $\chi_{cu}(x) \leq 1$ and

$$\chi_{cu}(x) = \begin{cases} 1, \text{ if } \|x\| \leq 1, \\ 0, \text{ if } \|x\| \geq 2, \end{cases}$$

and $\chi_s : [0, +\infty) \to [0, +\infty)$ is a C^∞ map with $\chi_s(y) \leq 1, \forall y \geq 0$, and

$$\chi_s(y) = \begin{cases} 1, & \text{if } |y| \le 1, \\ 0, & \text{if } |y| \ge 2. \end{cases}$$

The smoothness of Ψ_{cu} is obtained by applying the same arguments as in Magal and Ruan [239] to the above truncated system, and the result follows. □

The following theorem is the main result of this section. This result is proved for discrete time systems with bounded Lipschitz map F in Vanderbauwhede [337] and for ordinary differential equations in Vanderbauwhede [338] and Chow, Li and Wang [54].

Theorem 4.41 (Stability of the center-unstable manifold) *Let Assumption 4.33 be satisfied. Let $\eta \in (0, \beta_-)$ be fixed. Then there exists $\delta_1(\eta) \in (0, \delta_0)$ (where $\delta_0 > 0$ is the constant introduced in Theorem 4.39), such that for each $F \in \text{Lip}(\mathbb{R}^n, \mathbb{R}^n)$ with $\|F\|_{\text{Lip}(\mathbb{R}^n,\mathbb{R}^n)} \le \delta_1(\eta)$, there exists a continuous map $H_{cu} : \mathbb{R}^n \to \mathcal{M}_{cu}^\eta$ such that for each $x \in \mathbb{R}^n$,*

$$\mathcal{M}_{cu}^\eta \cap \widetilde{V}_\eta(x) = \{H_{cu}(x)\},$$

where

$$\widetilde{V}_\eta(x) = \left\{ y \in \mathbb{R}^n : \sup_{t \ge 0} e^{\eta t} \|U(t)y - U(t)x\| < +\infty \right\}.$$

More precisely, for each $x \in \mathbb{R}^n$, there is a constant $M_\eta = M_\eta(x) > 0$ such that

$$\|U(t)H_{cu}(x) - U(t)x\| \le e^{-\eta t} M_\eta \|H_{cu}(x) - x\|, \forall t \ge 0.$$

Before proving the theorem we give some preliminary lemmas. Recall that

$$BC^{-\eta}(\mathbb{R}_+, \mathbb{R}^n) = \left\{ w \in C(\mathbb{R}_+, \mathbb{R}^n) : \|w\|_\eta = \sup_{t \in \mathbb{R}_+} e^{\eta t} \|w(t)\| < +\infty \right\}.$$

In order to determine $\widetilde{V}_\eta(x)$, we have to find all $w \in BC^{-\eta}(\mathbb{R}_+, \mathbb{R}^n)$ such that $t \to U(t)x + w(t)$ is a solution of

$$u(t) = e^{At}u(0) + \left(e^{A.} * F(u(.))\right)(t), \ \forall t \ge 0. \tag{4.59}$$

Lemma 4.42 *Let Assumption 4.33 be satisfied. Let $\eta \in (0, \beta_-)$ be fixed and $w \in BC^{-\eta}(\mathbb{R}_+, \mathbb{R}^n)$. Then the map $t \to U(t)x + w(t)$ is a solution of (4.59) if and only if for each $t \ge 0$,*

$$\begin{aligned} w(t) =& e^{A_s t}\Pi_s w(0) + \left(e^{A_s} * \Pi_s \left[F(U(.)x + w(.)) - F(U(.)x)\right]\right)(t) \\ &- \int_t^{+\infty} e^{A_{cu}(t-s)}\Pi_{cu} \left[F(U(s)x + w(s)) - F(U(s)x)\right] ds. \end{aligned} \tag{4.60}$$

Proof Let $w \in BC^{-\eta}(\mathbb{R}_+, \mathbb{R}^n)$ be fixed. Assume first that $t \to U(t)x + w(t)$ is a solution of (4.59). Then we have for each $t, s \in [0, +\infty)$ with $t \ge s$, that

$$\begin{aligned} U(t)x + w(t) =& e^{A(t-s)} \left(U(s)x + w(s)\right) \\ &+ \left(e^{A.} * F(U(s+.)x + w(s+.))\right)(t-s) \end{aligned}$$

and

$$U(t)x = e^{A(t-s)}U(s)x + \left(e^{A.} * F(U(s+.)x)\right)(t-s).$$

Then

$$\begin{aligned} w(t) =& e^{A(t-s)}w(s) \\ &+ \left(e^{A.} * [F(U(s+.)x + w(s+.)) - F(U(s+.)x)]\right)(t-s). \end{aligned} \tag{4.61}$$

By projecting the above equation on X_{cu}, we obtain for each $t, s \in [0, +\infty)$ with $t \geq s$ that

$$\Pi_{cu}w(t) = e^{A_{cu}(t-s)}\Pi_{cu}w(s) + \int_s^t e^{A_{cu}(t-l)}\Pi_{cu}\left[F(U(l)x + w(l)) - F(U(l)x)\right] dl.$$

Then

$$\Pi_{cu}w(s) = e^{-A_{cu}(t-s)}\Pi_{cu}w(t) - \int_s^t e^{A_{cu}(s-l)}\Pi_{cu}\left[F(U(l)x + w(l)) - F(U(l)x)\right] dl.$$

We have $\left\|e^{-A_{cu}(t-s)}\right\|_{\mathcal{L}(X_{cu})} \leq \min\left\{e^{\frac{\eta}{2}|t-s|}M_{c,\frac{\eta}{2}},\ e^{-\eta_1(t-s)}M_u\right\}$, $\eta_1 > 0$, $\forall t \geq s$, here η, $M_{c,\frac{\eta}{2}}$ and M_u are constants (see Magal and Ruan [239] for details). Since $w \in BC^{-\eta}(\mathbb{R}_+, \mathbb{R}^n)$, we obtain for each $t, s \in [0, +\infty)$ with $t \geq s$ that

$$\left\|e^{-A_{cu}(t-s)}\Pi_{cu}w(t)\right\| \leq \min\left\{e^{\frac{\eta}{2}|t-s|}M_{c,\frac{\eta}{2}},\ e^{-\eta_1(t-s)}M_u\right\} \|\Pi_{cu}\|_{\mathcal{L}(\mathbb{R}^n)} \|w\|_\eta e^{-\eta t}.$$

Then

$$\left\|e^{-A_{cu}(t-s)}\Pi_{cu}w(t)\right\| \to 0 \text{ as } t \to +\infty.$$

Thus

$$\Pi_{cu}w(t) = -\int_t^{+\infty} e^{A_{cu}(t-l)}\Pi_{cu}\left[F(U(l)x + w(l)) - F(U(l)x)\right] dl, \forall t \geq 0. \tag{4.62}$$

By projecting (4.61) on X_s we obtain for each $t \geq 0$ that

$$\Pi_s w(t) = e^{A_s t}\Pi_s w(0) + \left(e^{A_s} * \Pi_s\left[F(U(.)x + w(.)) - F(U(.)x)\right]\right)(t). \tag{4.63}$$

So by summing (4.62) and (4.63), we obtain (4.60). Conversely, assume that w satisfies (4.60). Then by projecting (4.60) on X_s we obtain for each $t \geq 0$ that

$$\Pi_s w(t) = \Pi_s e^{At}w(0) + \Pi_s\left(e^{A.} * [F(U(.)x + w(.)) - F(U(.)x)]\right)(t).$$

Then

$$\Pi_s(U(t)x + w(t)) = \Pi_s e^{At}(w(0) + x) + \Pi_s \left(e^{A\cdot} * F(U(.)x + w(.))\right)(t). \quad (4.64)$$

Furthermore, by projecting (4.60) on X_{cu} we obtain for each $t \geq 0$ that

$$\Pi_{cu} w(t) = -\int_t^{+\infty} e^{A(t-s)} \Pi_{cu} \left[F(U(s)x + w(s)) - F(U(s)x)\right] ds.$$

Thus

$$\begin{aligned}
\Pi_{cu} w(t) - e^{A_{cu}t}\Pi_{cu} w(0) &= -\int_t^{+\infty} e^{A(t-s)} \Pi_{cu} \left[F(U(s)x + w(s)) - F(U(s)x)\right] ds \\
&\quad + \int_0^{+\infty} e^{A(t-s)} \Pi_{cu} \left[F(U(s)x + w(s)) - F(U(s)x)\right] ds \\
&= \int_0^{t} e^{A(t-s)} \Pi_{cu} \left[F(U(s)x + w(s)) - F(U(s)x)\right] ds.
\end{aligned}$$

Hence

$$\Pi_{cu}(U(t)x + w(t)) = e^{A_{cu}t}\Pi_{cu}(x + w(0)) + \int_0^t e^{A(t-s)} \Pi_{cu} F(U(s)x + w(s)) ds. \quad (4.65)$$

By summing up (4.64) and (4.65), we deduce that $t \to U(t)x + w(t)$ is a solution of (4.59). □

Rewrite (4.60) in the following abstract form

$$w = \widetilde{K}_1(w_s) + \widetilde{K}_2 \widetilde{\Phi}(x, w),$$

where $\widetilde{K}_1 : X_s \to BC^{-\eta}(\mathbb{R}_+, X_s)$, $\widetilde{K}_2 : BC^{-\eta}(\mathbb{R}_+, \mathbb{R}^n) \to BC^{-\eta}(\mathbb{R}_+, \mathbb{R}^n)$, and $\widetilde{\Phi} : \mathbb{R}^n \times BC^{-\eta}(\mathbb{R}_+, \mathbb{R}^n) \to BC^{-\eta}(\mathbb{R}_+, \mathbb{R}^n)$ are defined as follows

$$\begin{aligned}
&\widetilde{K}_1(x_s)(t) = e^{A_s t} x_s, t \in \mathbb{R}_+, \\
&\widetilde{K}_2(f)(t) = \left(e^{A_s \cdot} * \Pi_s f\right)(t) - \textstyle\int_t^{+\infty} e^{A_{cu}(t-s)} \Pi_{cu} f(s) ds, \forall t \in \mathbb{R}_+, \\
&\widetilde{\Phi}(x, f)(t) = F(U(t)x + f(t)) - F(U(t)x), \forall t \in \mathbb{R}_+.
\end{aligned}$$

One has

$$\left\|\widetilde{\Phi}(x, f)(t)\right\| = \|F(U(t)x + f(t)) - F(U(t)x)\| \leq e^{-\eta t} \|F\|_{\text{Lip}} \|f\|_\eta. \quad (4.66)$$

Lemma 4.43 *Let Assumption 4.33 be satisfied. Let $\eta \in (0, \beta_-)$ be fixed. Then*

$$\widetilde{K}_1 \in \mathcal{L}(X_s, BC^{-\eta}(\mathbb{R}_+, \mathbb{R}^n)) \text{ and } \widetilde{K}_2 \in \mathcal{L}(BC^{-\eta}(\mathbb{R}_+, \mathbb{R}^n), BC^{-\eta}(\mathbb{R}_+, \mathbb{R}^n))$$

with

$$\left\|\widetilde{K}_2\right\|_{\mathcal{L}(BC^{-\eta}(\mathbb{R}_+, \mathbb{R}^n), BC^{-\eta}(\mathbb{R}_+, \mathbb{R}^n))} \leq \gamma(\eta) := \widehat{C}_{s,-\eta} + \|\Pi_{cu}\|_{\mathcal{L}(\mathbb{R}^n)} \int_0^{+\infty} \left\|e^{-(A_{cu} + \eta I)l}\right\| dl$$

where $\widehat{C}_{s,-\eta} > 0$ is a constant, and

$$\widetilde{\Phi}(x,0)=0,$$

$$\widetilde{\Phi}(x,.)\in \text{Lip}\left(BC^{-\eta}\left(\mathbb{R}_+,\mathbb{R}^n\right),BC^{-\eta}\left(\mathbb{R}_+,\mathbb{R}^n\right)\right),\forall x\in\mathbb{R}^n,$$

with

$$\left\|\widetilde{\Phi}(x,.)\right\|_{\text{Lip}}\leq\|F\|_{\text{Lip}}.$$

Proof This proof is straightforward. □

Proof (Proof of Theorem 4.41) Let $\eta\in(0,\beta_-)$ and $x\in\mathbb{R}^n$ be fixed. Let $\delta_0>0$ be the constant introduced in Theorem 4.39. Let $\delta_1^*\in(0,\delta_0)$ be such that

$$\delta_1^*\gamma(\eta)<1. \tag{4.67}$$

Then for each $F\in\text{Lip}\left(\mathbb{R}^n,\mathbb{R}^n\right)$ with $\|F\|_{\text{Lip}}\leq\delta_1^*$, we obtain that for each $(x,w_s)\in\mathbb{R}^n\times X_s$, there exists a unique solution $w=\widetilde{w}(x,w_s)\in BC^{-\eta}\left(\mathbb{R}_+,\mathbb{R}^n\right)$ such that

$$w=\widetilde{K}_1(w_s)+\widetilde{K}_2\widetilde{\Phi}(x,w)$$

and

$$w=(Id-\widetilde{K}_2\widetilde{\Phi}(x,.))^{-1}\widetilde{K}_1(w_s).$$

We have

$$\|\widetilde{w}(x,w_s)-\widetilde{w}(x,\widetilde{w}_s)\|_{\eta}\leq l\,\|w_s-\widetilde{w}_s\|,\forall x\in\mathbb{R}^n,\forall w_s,\widetilde{w}_s\in X_s,$$

where l depends on η and $\|F\|_{\text{Lip}}$ but stays bounded as $\|F\|_{\text{Lip}}\to 0$. To see the continuous dependence of $\widetilde{w}(x,w_s)$ on $x\in\mathbb{R}^n$, we remark that (4.67) and the continuity of $\gamma(\eta)$ imply that $\gamma(\zeta)\delta_1^*<1$ for some $\zeta\equiv(\eta,\beta)$. Replacing η by ζ in the above argument, we conclude that $\widetilde{w}(x,w_s)$ belongs in fact to the space $BC^{-\zeta}\left(\mathbb{R}_+,\mathbb{R}^n\right)$, which is continuously imbedded in $BC^{-\eta}(\mathbb{R}_+,\mathbb{R}^n)$. More precisely, we have

$$\begin{aligned}\|\widetilde{w}(x,w_s)\|_{\zeta}&\leq\left\|\widetilde{K}_1\right\|_{\mathcal{L}(X_s,BC^{-\zeta}(\mathbb{R}_+,\mathbb{R}^n))}\|w_s\|+\left\|\widetilde{K}_2\right\|_{\mathcal{L}(BC^{-\zeta}(\mathbb{R}_+,\mathbb{R}^n))}\left\|\widetilde{\Phi}(x+x_0,w)\right\|_{\zeta}\\&\leq\left\|\widetilde{K}_1\right\|_{\mathcal{L}(X_s,BC^{-\zeta}(\mathbb{R}_+,\mathbb{R}^n))}\|w_s\|+\left\|\widetilde{K}_2\right\|_{\mathcal{L}(BC^{-\zeta}(\mathbb{R}_+,\mathbb{R}^n))}\|F\|_{\text{Lip}}\|\widetilde{w}(x,w_s)\|_{\zeta}.\end{aligned}$$

Therefore, we obtain an estimate independent of x,

$$\|\widetilde{w}(x,w_s)\|_{\zeta}\leq\frac{\left\|\widetilde{K}_1\right\|_{\mathcal{L}(X_s,BC^{-\zeta}(\mathbb{R}_+,\mathbb{R}^n))}\|w_s\|}{1-\left\|\widetilde{K}_2\right\|_{\mathcal{L}(BC^{-\zeta}(\mathbb{R}_+,\mathbb{R}^n))}\|F\|_{\text{Lip}}}<+\infty.$$

Moreover, we have

$$\begin{aligned}
&\widetilde{w}(x+x_0,w_s)-\widetilde{w}(x_0,w_s)\\
&\quad=\widetilde{K}_1(w_s)+\widetilde{K}_2\widetilde{\Phi}(x+x_0,\widetilde{w}(x+x_0,w_s))-\left[\widetilde{K}_1(w_s)+\widetilde{K}_2\widetilde{\Phi}(x_0,\widetilde{w}(x_0,w_s))\right]\\
&\quad=\widetilde{K}_2\left[\widetilde{\Phi}(x+x_0,\widetilde{w}(x+x_0,w_s))-\widetilde{\Phi}(x_0,\widetilde{w}(x_0,w_s))\right]\\
&\quad=\widetilde{K}_2\left[\widetilde{\Phi}(x+x_0,\widetilde{w}(x+x_0,w_s))-\widetilde{\Phi}(x+x_0,\widetilde{w}(x_0,w_s))\right]\\
&\qquad+\widetilde{K}_2\left[\widetilde{\Phi}(x+x_0,\widetilde{w}(x_0,w_s))-\widetilde{\Phi}(x_0,\widetilde{w}(x_0,w_s))\right].
\end{aligned}$$

Then

$$\begin{aligned}
&\|\widetilde{w}(x+x_0,w_s)-\widetilde{w}(x_0,w_s)\|_\eta\\
&\quad\le\left\|\widetilde{K}_2\right\|_{\mathcal{L}(BC^{-\eta}(\mathbb{R}_+,\mathbb{R}^n))}\|F\|_{\mathrm{Lip}}\|\widetilde{w}(x+x_0,w_s)-\widetilde{w}(x_0,w_s)\|_\eta\\
&\qquad+\left\|\widetilde{K}_2\right\|_{\mathcal{L}(BC^{-\eta}(\mathbb{R}_+,\mathbb{R}^n))}\left\|\widetilde{\Phi}(x+x_0,\widetilde{w}(x_0,w_s))-\widetilde{\Phi}(x_0,\widetilde{w}(x_0,w_s))\right\|_\eta.
\end{aligned}$$

Thus

$$\begin{aligned}
&\|\widetilde{w}(x+x_0,w_s)-\widetilde{w}(x_0,w_s)\|_\eta\\
&\le\frac{\left\|\widetilde{K}_2\right\|_{\mathcal{L}(BC^{-\eta}(\mathbb{R}_+,\mathbb{R}^n))}}{1-\left\|\widetilde{K}_2\right\|_{\mathcal{L}(BC^{-\eta}(\mathbb{R}_+,\mathbb{R}^n))}\|F\|_{\mathrm{Lip}}}\left\|\widetilde{\Phi}(x+x_0,\widetilde{w}(x_0,w_s))-\widetilde{\Phi}(x_0,\widetilde{w}(x_0,w_s))\right\|_\eta.
\end{aligned}$$

For fixed $w\in BC^{-\zeta}(\mathbb{R}_+,\mathbb{R}^n)$ we claim that the mapping $x\to\widetilde{\Phi}(x,w)$ is continuous from $\mathbb{R}^n$ into $BC^{-\eta}(\mathbb{R}_+,\mathbb{R}^n)$. In fact, by using (*4.67*), we have

$$\left\|\widetilde{\Phi}(x+x_0,w)-\widetilde{\Phi}(x_0,w)\right\|_\eta=\sup_{t\in\mathbb{R}_+}e^{\eta t}\|\mathcal{H}(t)\|,$$

where

$$\begin{aligned}
\mathcal{H}(t):=&\left[F(U(t)(x+x_0)+w(t))-F(U(t)x_0+w(t))\right]\\
&-\left[F(U(t)(x+x_0))-F(U(t)x_0)\right].
\end{aligned}$$

Thus

$$\left\|\widetilde{\Phi}(x+x_0,w)-\widetilde{\Phi}(x_0,w)\right\|_\eta=\max\left(\sup_{0\le t\le T}e^{\eta T}\|\mathcal{H}(t)\|,\ 2e^{(\eta-\zeta)T}\|F\|_{\mathrm{Lip}}\|w\|_\zeta\right).$$

By the continuity of $x\to U(t)(x)$ uniformly with respect to $t\in[0,T]$, we obtain

$$\limsup_{x\to0}\left\|\widetilde{\Phi}(x+x_0,w)-\widetilde{\Phi}(x_0,w)\right\|_\eta\le2e^{(\eta-\zeta)T}\|F\|_{\mathrm{Lip}}\|w\|_\zeta,\ T\ge0.$$

So when T goes to $+\infty$, we obtain

$$\lim_{x\to0}\left\|\widetilde{\Phi}(x+x_0,w)-\widetilde{\Phi}(x_0,w)\right\|_\eta=0.$$

From this and the fact that $\widetilde{w}(x, w_s) \in BC^{-\zeta}(\mathbb{R}_+, \mathbb{R}^n)$, it follows that $\widetilde{w} : \mathbb{R}^n \times X_s \to BC^{-\eta}(\mathbb{R}_+, \mathbb{R}^n)$ is continuous.

Define a map $\Gamma : \mathbb{R}^n \times X_s \to X_{cu}$ by

$$\Gamma(x, w_s) = \Pi_{cu}(Id - \widetilde{K}_2\widetilde{\Phi}(x, .))^{-1}\widetilde{K}_1(w_s)(0), \ \forall x \in \mathbb{R}^n, \ w_s \in X_s.$$

Notice that $\Gamma : \mathbb{R}^n \times X_s \to X_{cu}$ is continuous and Γ is Lipschitz continuous with respect to w_s with

$$\|\Gamma(x, .)\|_{\text{Lip}} \leq \|\Pi_{0cu}\|_{\mathcal{L}(\mathbb{R}^n)} \frac{\left\|\widetilde{K}_1\right\|_{\mathcal{L}(X_s, BC^{-\eta}(\mathbb{R}_+, \mathbb{R}^n))}}{1 - \left\|\widetilde{K}_2\right\|_{\mathcal{L}(BC^{-\eta}(\mathbb{R}_+, \mathbb{R}^n))} \|F\|_{\text{Lip}}}.$$

We have by construction that

$$y \in \widetilde{V}_\eta(x) \Leftrightarrow y = x + w \text{ with } \Pi_{cu} w = \Gamma(x, \Pi_s w).$$

Then

$$\begin{aligned}\widetilde{V}_\eta(x) &= \{x_s + w_s + x_{cu} + \Gamma(x, w_s) : w_s \in X_s\} \\ &= \{z + x_{cu} + \Gamma(x, z - x_s) : z \in X_s\}.\end{aligned}$$

Consider the map $\Theta : \mathbb{R}^n \times X_s \to X_{cu}$ defined by

$$\Theta(x, z) = x_c + \Gamma(x, z - x_s), \forall x \in \mathbb{R}^n, \ z \in X_s,$$

we have $\widetilde{V}_\eta(x) = \{z + \Theta(x, z) : z \in X_s\}$. Since $\Gamma : \mathbb{R}^n \times X_s \to X_{cu}$ is continuous and $\Gamma(x, w_s)$ is Lipschitz continuous with respect to w_s, so is Θ, and $\|\Theta\|_{\text{Lip}} \leq \|\Gamma\|_{\text{Lip}}$. Finally, we look for $y \in \mathbb{R}^n$, such that

$$\Pi_s y = \Psi_{cu}(\Pi_{cu} y) \text{ and } \Pi_{cu} y =: \Theta(x, \Pi_s y).$$

But by (4.57), we deduce that $\|\Psi_{cu}\|_{\text{Lip}} \to 0$ as $\|F\|_{\text{Lip}} \to 0$. So (4.57) and (4.67) imply that there exists $\delta_1 \in (0, \delta_1^*)$ such that for each $F \in \text{Lip}(\mathbb{R}^n, \mathbb{R}^n)$ with $\|F\|_{\text{Lip}} \leq \delta_1$,

$$\|\Theta(x, .)\|_{\text{Lip}} \|\Psi_{cu}\|_{\text{Lip}} < 1.$$

Thus, there exists for each $x \in \mathbb{R}^n$ a unique $\widetilde{y}_{cu}(x) \in X_{cu}$ such that

$$\Theta(x, \Psi_{cu}(\widetilde{y}_{cu}(x))) = \widetilde{y}_{cu}(x)$$

and the map $\widetilde{y}_c : \mathbb{R}^n \to X_{cu}$ is continuous. By setting $H_{cu}(x) = \widetilde{y}_{cu}(x) + \Psi_{cu}(\widetilde{y}_{cu}(x))$ the result follows. □

Theorem 4.44 (Local uniform convergence) *Let Assumption 4.33 be satisfied. Let $\eta \in (0, \beta_-)$ be fixed. Then there exists $\delta_1(\eta) \in (0, \delta_0)$ (where $\delta_0 > 0$ is the constant introduced in Theorem 4.39), such that for each $F \in \text{Lip}(\mathbb{R}^n, \mathbb{R}^n)$ with $\|F\|_{\text{Lip}(\mathbb{R}^n, \mathbb{R}^n)} \leq \delta_1(\eta)$, the following holds: for each $\widetilde{x} \in \mathcal{M}_{cu}^\eta$ and for each $\varepsilon > 0$, there exists some $\delta > 0$ such that*

$$\|U(t)x - U(t)H_{cu}(x)\| \leq \varepsilon e^{-\eta t}, \ \forall t \geq 0, \tag{4.68}$$

for all $x \in \mathbb{R}^n$ *with* $\|x - \widetilde{x}\| < \delta$.

Proof Let $\widetilde{x} \in \mathcal{M}_{cu}^{\eta}$ be fixed. The proof of Theorem 4.43 implies that

$$\widetilde{w}(x, \Psi_{cu}(\widetilde{y}_c(x)) - \Pi_{0s}x)(t) = U(t)(H_{cu}(x)) - U(t)(x), \forall t \geq 0, \ \ \forall x \in \mathbb{R}^n,$$

where

$$H_{cu}(x) = \widetilde{y}_c(x) + \Psi_{cu}(\widetilde{y}_c(x)) \in \mathcal{M}_{cu}^{\eta}.$$

It is clear that $H_{cu}(\widetilde{x}) = \widetilde{x}$ if $\widetilde{x} \in \mathcal{M}_{cu}^{\eta}$ and hence

$$\widetilde{w}(\widetilde{x}, \Psi_{cu}(\widetilde{y}_c(\widetilde{x})) - \Pi_s\widetilde{x}) = U(t)(H_{cu}(\widetilde{x})) - U(t)(\widetilde{x}) = 0, \forall t \geq 0, \forall \widetilde{x} \in \mathcal{M}_{cu}^{\eta}.$$

Let $\widetilde{x} \in \mathcal{M}_{cu}^{\eta}$ and $\varepsilon > 0$. By the continuity of $\widetilde{w} : \mathbb{R}^n \times X_s \to BC^{-\eta}(\mathbb{R}_+, \mathbb{R}^n)$ and $\widetilde{y}_c : \mathbb{R}^n \to X_{cu}$, we can find some $\delta > 0$ such that

$$||\widetilde{w}(x, \Psi_{cu}(\widetilde{y}_c(x)) - \Pi_{0s}x) - \widetilde{w}(\widetilde{x}, \Psi_{cu}(\widetilde{y}_c(\widetilde{x})) - \Pi_s\widetilde{x})||_{\eta} \leq \varepsilon$$

whenever $x \in \mathbb{R}^n$ and $||x - \widetilde{x}|| < \delta$. Therefore,

$$\begin{aligned}
&\sup_{t \in \mathbb{R}_+} e^{\eta t}||U(t)(H_{cu}(x)) - U(t)(x)|| \\
&= \sup_{t \in \mathbb{R}_+} e^{\eta t}||\widetilde{w}(x, \Psi_{cu}(\widetilde{y}_c(x)) - \Pi_s x)(t)|| \\
&= \sup_{t \in \mathbb{R}_+} e^{\eta t}||\widetilde{w}(x, \Psi_{cu}(\widetilde{y}_c(x)) - \Pi_s x)(t) - \widetilde{w}(\widetilde{x}, \Psi_{cu}(\widetilde{y}_c(\widetilde{x})) - \Pi_s\widetilde{x})(t)|| \\
&\leq \varepsilon \text{ if } x \in \mathbb{R}^n \text{ and } ||x - \widetilde{x}|| < \delta.
\end{aligned}$$

The proof is complete. □

Remark 4.45 Our presentations focused on center-unstable manifolds. However, similar results can be established for center-stable manifolds.

4.5 Remarks and Notes

The chapter is devoted to the center manifold theory that was presented by Vanderbauwhede in [338] and Vanderbauwhede and Iooss [340]. Our presentation is inspired by Magal and Ruan [239] where Vanderbauwhede's center manifold's results was extended to an abstract class of Cauchy problems on Banach spaces. The second part of this chapter about the center-unstable manifold is inspired by Liu, Magal and Ruan [222]. More results and references on the center manifold theory will be given in the sequel.

Center manifold

The classical center manifold theory was initially established by Pliss [273] and Kelley [190], and later developed and refined by Carr [44], Sijbrand [310], Vanderbauwhede [338], among others. In the case of a single equilibrium, the center manifold theorem asserts that if a finite-dimensional system has a non-hyperbolic equilibrium, there exists a center manifold in the vicinity of this equilibrium. This manifold is tangent to the generalized eigenspace associated with eigenvalues that have zero real parts, allowing the analysis of the system near the non-hyperbolic equilibrium to be reduced to an ordinary differential equation on this lower-dimensional invariant manifold. This often leads to a significant reduction in dimensionality, leading to simpler calculations and providing clearer geometric insights.

The center manifold theory has numerous important applications in the study of dynamical systems, such as in the analysis of bifurcations, stability, and perturbations. It is also widely employed in applied fields like biology, engineering, and physics, with notable references including Carr [44] and Hassard et al. [151].

There are two classical methods for proving the existence of center manifolds. The first is Hadamard's method (Hadamard [132]), also known as the graph transformation method, which is a geometric approach based on constructing graphs over linearized spaces (see Hirsch et al. [155] and Chow et al. [56, pp. 266–291 and 355–385]). The second method is the Lyapunov-Perron method (Lyapunov [233], Perron [271]), also called the variation of constants method, which is more analytical and finds the manifold as the fixed point of a particular integral equation.The technique was originally introduced by Krylov and Bogoliubov [198], and further developed by Hale [134, 136], with additional contributions from Ball [17], Chow and Lu [58], Yi [360], and others. The smoothness of center manifolds can be established through various methods, including the use of contraction mappings in a scale of Banach spaces (Vanderbauwhede and van Gils [339]), the fiber contraction mapping technique (Hirsch et al. [155]), and the Henry lemma (Henry [153], Chow and Lu [57]), among other techniques (Chow et al. [55]). For more comprehensive results and references on center manifold theory, see the monographs by Carr [44], Chow and Hale [53], Chow et al. [54], Sell and You [298], Wiggins [351], as well as survey papers by Bates and Jones [21], Vanderbauwhede [337], and Vanderbauwhede and Iooss [340].

In recent years, significant attention has been given to the study of center manifolds in infinite-dimensional systems, leading to the development of center manifold theory for various such systems. These include partial differential equations (Bates and Jones [21], Da Prato and Lunardi [72], Henry [153], Scheel [295]), semiflows in Banach spaces (Bates et al. [22], Chow and Lu [58], Gallay [116], Scarpellini [294], Vanderbauwhede [337], Vanderbauwhede and van Gils [339]), and delay differential equations (Hale [137], Hale and Verduyn Lunel [144], Diekmann and van Gils [81, 82], Diekmann et al. [83], Hupkes and Verduyn Lunel [176]). Other areas of development include infinite-dimensional non-autonomous differential equations (Mielke [256, 257], Chicone and Latushkin [49]) and partial functional differential equations (Lin et al. [217], Faria et al. [107], Krisztin [197], Nguyen and Wu [336], Wu [356]).

Infinite-dimensional systems often lack several of the favourable properties found in finite-dimensional systems. For instance, the initial value problem may not be well-posed, solutions might not extend backward in time, the regularity of solutions may be lacking, and the domain of operators might not be dense in the state space. As a result, center manifold reduction becomes crucial in the study of infinite-dimensional systems, as it allows for the analysis of ordinary differential equations on finite-dimensional center manifolds. Vanderbauwhede and Iooss [340] outlined minimal conditions under which the approach developed by Vanderbauwhede [338] can be extended to infinite-dimensional systems.
In Magal and Ruan [239, 240] the authors investigate a center manifold for non densely Cauchy problems of the form

$$u'(t) = Au(t) + F(u(t)), \text{ for } t \geq 0, \text{ and } u(0) = x \in \overline{D(A)},$$

where $A : D(A) \subset X \to X$ is a, possibly non densely defined, linear operator on a Banach space while $F : \overline{D(A)} \to X$ is a locally Lipschitz continuous function.
More recently, Ducrot and Magal [94] extended such an idea to a class of abstract second order semi-linear differential equations on the real line and the authors prove the existence of periodic wave train arising from a Hopf bifurcation.

Let us also mention that center manifold can also be used to study some singularly perturbed problem. An example will be given in Chapter 8 in this book.

Center-unstable manifold

Given a non-hyperbolic equilibrium, the center-unstable manifold is a locally invariant manifold for the semiflow, tangent to the generalized eigenspace associated with eigenvalues that have non-negative real parts (Kelley [190]). The local center-unstable manifold plays a significant role in applications due to its stability properties. Compared to the center manifold, it is often easier to use in practice because a point (in a local neighbourhood around the equilibrium) belongs to the center-unstable manifold if and only if there exists a negative orbit, remaining in a small neighbourhood of the equilibrium, that passes through the point at time $t = 0$. For the center manifold, by contrast, a complete orbit is required. This property is leveraged in the following section to establish the uniqueness of traveling waves for the Fisher-KPP problem.

Center-unstable manifolds in infinite-dimensional dynamical systems have been widely studied. For example, Armbruster et al. [8] investigated center-unstable manifolds for the Kuramoto-Sivashinsky equation. Chow and Lu [57] discussed the existence and smoothness of global center-unstable manifolds for both semilinear and fully nonlinear differential evolution equations. Dell'Antonio and D'Onofrio [76] studied center-unstable manifolds for the Navier-Stokes equation, while Nakanishi and Schlag [266] established properties of center-unstable and center-stable manifolds around soliton manifolds for the nonlinear Klein-Gordon equation. Turyn [334] obtained a center-unstable manifold theorem for parametrically excited surface waves, and Stumpf [319] discussed center-unstable manifolds for differential equations with state-dependent delays.

Let us mention that Liu, Magal and Ruan [222] construct and study a center-unstable manifold for non densely defined abstract Cauchy problems

$$u'(t) = Au(t) + F(u(t)), \text{ for } t \geq 0, \text{ and } u(0) = x \in \overline{D(A)},$$

where $A : D(A) \subset X \to X$ is a linear operator, possibly with a non dense domain, acting on a Banach space and $F : \overline{D(A)} \to X$ is a locally Lipschitz continuous function. In particular, taking advantage of the stability property of the center-unstable manifold, the authors prove in this paper a stability result for the periodic orbits arsing from a Hopf bifurcation.

Existence and the Uniqueness of Traveling Waves for Fisher-KPP Equation

In this section we illustrate some consequence of the global attractors theory and the center-unstable manifold to give a short proof for the existence and uniqueness of traveling waves for the so-called Fisher-KPP equation. This section is taken from Ducrot, Langlais and Magal [92].

Fisher-Kolmogorov-Petrovski-Piskunov's traveling waves problem

Let us consider a logistic reaction-diffusion equation

$$\partial_t N = \partial_{xx} N + \lambda N \left[1 - \frac{N}{\kappa} \right]. \tag{4.69}$$

with $\kappa > 0$ and $\lambda > 0$. The above equation corresponds to a logistic equation with diffusion, which is also called Fisher-Kolmogorov-Petrovski-Piskunov (or for short Fisher-KPP) equation [113, 194]. Classically, λ corresponds to the growth rate of the population while $\kappa > 0$ denotes the carrying capacity of the environment. Then if we look for traveling wave solutions, that is special solution of the form

$$N(t, x) = U(x - ct),$$

where $c \in \mathbb{R}$ denotes the wave speed, it is known (see for instance [12]) that for each $c \geq c^* = 2\sqrt{\lambda}$ this logistic equation has a unique (up to translation) traveling wave solution connecting $N = 0$ to $N = \kappa$. This means that for each $c \geq c^*$, there exists a non-increasing function $U \equiv U_c(x)$ such that

$$\begin{cases} U''(x) + cU'(x) + \lambda U(x) \left[1 - \dfrac{U(x)}{\kappa} \right] = 0, \quad x \in \mathbb{R}, \\ U(-\infty) = \kappa, \quad U(\infty) = 0. \end{cases} \tag{4.70}$$

The spatial dynamics of the Fisher-KPP equation (4.69) is well known and, in many cases, strongly related to traveling wave solutions. The literature about this topic is very wide. We only quote some of them, see for instance [12, 35, 146, 204, 208, 286, 335, 342] as well as references therein.

In this section, we discuss the existence and uniqueness of solutions for (4.70) by using invariant manifold techniques. To that aim, fix $c \geq 2$ and let us notice that (4.70) can be re-written as

$$\left(\frac{d}{dx} + \frac{c}{2}\right)^2 U - \frac{c^2}{4} U + U(1 - U) = 0.$$

Next setting

$$\begin{cases} U_1 := U \\ U_2 := \left(\frac{d}{dx} + \frac{c}{2}\right) U. \end{cases}$$

we obtain the following first order system of ordinary differential equations

$$\begin{cases} \left(\frac{d}{dx} + \frac{c}{2}\right) U_1 = U_2 \\ \left(\frac{d}{dx} + \frac{c}{2}\right) U_2 = \frac{c^2}{4} U_1 - U_1(1 - U_1). \end{cases}$$

Set

$$\alpha := \frac{c}{2}$$

and the above system becomes

$$\begin{cases} \frac{dU_1}{dx} = -\alpha U_1 + U_2 \\ \frac{dU_2}{dx} = -\alpha U_2 + \left(\alpha^2 - 1\right) U_1 + U_1^2. \end{cases} \tag{4.71}$$

Note that this system is monotone increasing on $[0, +\infty)^2$ whenever

$$\alpha \geq 1. \tag{4.72}$$

Moreover one has

$$\frac{d\left(\alpha U_1 + U_2\right)}{dt} = -U_1(x) + U_1(x)^2 = -U_1(x)(1 - U_1(x)) \tag{4.73}$$

and the points

$$\overline{U}^0 := (0, 0) \text{ and } \overline{U}^1 := (1, \alpha)$$

are the only equilibria of the system in $[0, +\infty)^2$.

Existence of traveling waves

Since $[0,1]\times[0,\alpha]$ is invariant by the semiflow $\{T(t)\}_{t\geq 0}$ generated by the system (4.71), there exists a connected subset $A\subset[0,1]\times[0,\alpha]$, which is the global attractor of the semiflow T on $[0,1]\times[0,\alpha]$. This global attractor A contains both equilibria $(0,0)$ and $(1,\alpha)$, and by considering the linear functional $P:\mathbb{R}^2\to\mathbb{R}$

$$P(U_1,U_2)=U_1$$

we deduce that $P(A)$ is compact and connected and contains $P(0,0)=0$ and $P(1,\alpha)=1$. Hence one concludes that

$$P(A)=[0,1].$$

Moreover we have

$$T(t)A=A,\forall t\geq 0.$$

Therefore $\{T(t)\}_{t\in\mathbb{R}}$ is a flow on A, and it follows that there exists a complete orbit $(U_1,U_2)\in C^1(\mathbb{R},\mathbb{R}^2)$ of system (4.71) such that

$$(U_1(t),U_2(t))\in A,\forall t\geq 0,$$

and passing at $t=0$ through the point

$$(U_1,U_2)\ \text{with}\ U_1=1/2\ \text{and}\ U_2\in[0,\alpha].$$

By using (4.73) we deduce that

$$\lim_{t\to+\infty}(U_1(t),U_2(t))=(0,0)\ \text{and}\ \lim_{t\to-\infty}(U_1(t),U_2(t))=(1,\alpha).$$

Therefore A contains the equilibria, and all the traveling wave solutions going from $(1,\alpha)$ to $(0,0)$.

4.5.1 Uniqueness of the traveling waves

In order to prove the uniqueness of the heteroclinic orbit going from $(1,\alpha)$ to $(0,0)$, we will study the center-unstable manifold around the equilibrium $(1,\alpha)$.

Linearized equation at $\overline{U}^1=(1,\alpha)$: The matrix of the linearized equation of system (4.71) at $(1,\alpha)$ reads as

$$L_U=\begin{bmatrix}-\alpha & 1\\ \alpha^2+1 & -\alpha\end{bmatrix},$$

the characteristic equation is given by

$$\begin{aligned}(\alpha+\lambda)^2-\alpha^2-1=0 &\Leftrightarrow \lambda^2+2\alpha\lambda+\alpha^2-\alpha^2-1=0\\ &\Leftrightarrow \lambda^2+2\alpha\lambda-1=0,\end{aligned}$$

hence the spectrum of L_U is given by

$$\sigma(L_U) = \left\{\lambda_U^-, \lambda_U^+\right\},$$

with

$$\lambda_U^- := -\alpha - \sqrt{\alpha^2+1} < 0 < \lambda_U^+ := -\alpha + \sqrt{\alpha^2+1}.$$

It follows from Theorem 4.40 that the center-unstable manifold at $(1, \alpha)$ is a one dimensional locally invariant manifold. Since the center-unstable manifold at $(1, \alpha)$ contains the point of any negative orbit staying in some neighbourhood (small enough) of $(1, \alpha)$. It follows that the traveling wave (or the complete orbit) going from $(1, \alpha)$ (at $t = -\infty$) to $(0, 0)$ (to $t = +\infty$) is unique (we refer for instance to [335] for an other proof).

The precise result reads as follows.

Theorem 4.46 *Assume that $\alpha \geq 1$ (that reads $c \geq 2$). Then there exists at most one traveling wave going from $(1, \alpha)$ to $(0, 0)$ for* (4.70). *More precisely, there exists a unique solution $U^*(x) = \left(U_1^*(x), U_2^*(x)\right)$ of system* (4.71) *satisfying*

$$\lim_{x\to-\infty} U^*(x) = (1, \alpha) \text{ and } \lim_{x\to+\infty} U^*(x) = (0, 0).$$

Note that another proof can be found in Aronson and Weinberger [12] in 1978. We also refer to Ducrot, Langlais, and Magal [92] for an example with non-uniqueness for traveling wave profile in some epidemic problem. Note also that a similar transformation of the traveling wave system to a monotone system (of ordinary differential equations) was also observed in the early paper by Wu and Zou [357].

Chapter 5
Normal Forms

In studying nonlinear dynamical problems, we usually restrict our analysis to the flow on the centre manifold near a equilibrium solution. The centre manifold theory of chapter 4 gives a reduction in the dimension of the system. The aim of the normal form theory is to find a nonlinear analytic transformation of coordinates to change the nonlinear system restricted to the center manifold into a "simplest" form, called *normal form*. Both are useful for the study of bifurcation problems. In this chapter, we concentrate our efforts to introduce the methods to compute the normal form for reduced differential equations on the center manifold. Our presentation is inspired by Liu, Magal and Ruan [223] where the normal form theory for an abstract class of Cauchy problems on Banach spaces is presented.

5.1 Introduction

In this chapter, we will consider the computation of normal form associated with the flow on the center manifold for the differential equation of the form

$$u'(t) = Au(t) + F(u(t)), \forall t \in \mathbb{R}, \tag{5.1}$$

where $A \in M_n(\mathbb{R})$ is a n by n matrix, $u \in \mathbb{R}^n$ (or $\mathbb{C}^n$) and $F : \mathbb{R}^n \to \mathbb{R}^n$ is C^k with $F(0) = 0$ and $DF(0) = 0$, $k > 1$.The notations X_i, Π_i, A_i with $i = c, s, u, h$, which we will use in the following, are defined in chapter 4.

Assumption 5.1 We assume that $\sigma_c(A) = \sigma(A) \cap i\mathbb{R} \neq \emptyset$.

The center manifold theorem of Chapter 4 tells us that there exists a C^k-smooth function $\Psi : X_c \to X_h$ with $\Psi(0) = 0$, $D\Psi(0) = 0$, such that

$$M = \{x_c + \Psi(x_c) : x_c \in X_c\} \quad \textbf{(center manifold)}$$

is a locally invariant center manifold of (5.1). Moreover, as Π_c is defined on $\mathbb{R}^n$, we can project (5.1) on X_c and obtain the following **reduced system**

A. Ducrot et al., *Differential Equations and Population Dynamics II*,
Lecture Notes on Mathematical Modelling in the Life Sciences,
https://doi.org/10.1007/978-3-031-91409-6_5

$$\frac{du_c(t)}{dt} = A_c u_c(t) + \Pi_c F\left[u_c(t) + \Psi\left(u_c(t)\right)\right]. \tag{5.2}$$

The qualitative behaviour of the flow of (5.1) near the equilibrium solution $u = 0$ is determined by its behaviour on the center manifold.

We consider the computation of the normal form of the reduced system (5.2) in the center manifold. This requires computing the center manifold before evaluating the normal forms for the reduced system (5.2). One needs to realize that the center manifold Ψ is known only through an implicit fixed point procedure. In general, the only information available to compute the Taylor's expansion and normal form of the reduced system is the following result in chapter 4.

Lemma 5.2 *Let Assumptions 5.1 be satisfied. Let $r > 0$ and $F : B_{\mathbb{R}^n}(0, r) \to \mathbb{R}^n$ be a k-time continuously differentiable map ($k > 1$) with $F(0) = 0$, $DF(0) = 0$ and*

$$\Pi_h D^j F(0) \mid_{X_c \times X_c \times \ldots \times X_c} = 0 \textit{ for each } j = 2, \ldots, k.$$

Then

$$D^j \Psi(0) = 0 \textit{ for each } j = 1, \ldots, k.$$

The Taylor's expansion of the reduced system (5.2) is simply given by

$$\frac{du_c(t)}{dt} = A_c u_c(t) + \sum_{l=1}^{k} \frac{1}{l!} \Pi_c D^l F(0)\left(u_c(t), \ldots, u_c(t)\right) + h.o.t.$$

when

$$D^l \Psi(0) = 0 \text{ for each } l = 1, \ldots, k.$$

In this chapter we give two other possible approaches based on lemma 5.2. We compute the Taylor's expansion of the reduced system (5.2) which is achieved by considering a sequence of change of variables, and after this first step, the usual method of normal forms in Chow and Li and Wang [54], Chow and Hale [53], Bibikov [30], Vanderbauwhede [338] and Wiggins [352] is applied to compute a normal form for (5.2). Another idea is that at each step j, a different type of change of variables is used to project the original equation (5.1) on the centre manifold and, simultaneously, to eliminate the non-resonant terms of order j from the equation (5.2).

Before proceeding further it will be useful to first establish some basic terminology, definitions and results that we will frequently use throughout this chapter.

5.1.1 Notations and basic results

Let $m \geq 1$ be given. Since we will mainly be interested along this chapter in higher order derivative, let us recall that if $F : X \to Y$ is a smooth map between vector spaces X and Y, the differential of order of m at some point $x \in X$, $DF(x)$, is a bounded $m-$linear and symmetric map, acting from X^m into Y. We denote by $\mathcal{L}_s\left(X^m, Y\right)$ the set of the bounded m-linear and symmetric maps from $X^m = X \times X \times \ldots \times X$ into Y.

In the sequel, if Y denotes a closed subspace of $\mathbb{R}^n$, we shall more particularly make use of the space $\mathcal{L}_s\left((\mathbb{R}^n)^m, Y\right)$ and $\mathcal{L}_s\left(X_c^m, Y\right)$, with $Y = \mathbb{R}^n$ or $Y = X_h$ as special cases.

In what follows we will use the fact that m-linear symmetric maps can be represented by combinations of homogeneous polynomials of degree m. Let us recall the reason. We define $V^m(X_c, Y)$ the linear space of vector homogeneous polynomials of degree m. More precisely, let $l = \dim(X_c)$, $\Lambda = \left\{b_j\right\}_{j=1,\ldots,l}$ be a basis of X_c and $x = (x_1, ..., x_l) \in \mathbb{R}^l$ be the coordinates with respect to this basis, $V^m(X_c, Y)$ is the space of finite linear combinations of maps of the form

$$x_c = \sum_{j=1}^{l} x_j b_j \in X_c \mapsto x_1^{n_1} x_2^{n_2} ... x_l^{n_l} V$$

where $n_1 + n_2 + ... + n_l = m$, $n_j \geq 0$ are integers and $V \in Y$.
Define now the map $\mathcal{G} : \mathcal{L}_s\left(X_c^m, Y\right) \to V^m(X_c, Y)$ by

$$\mathcal{G}(L)(x_c) = L(x_c, \cdots, x_c), \ \forall L \in \mathcal{L}_s\left(X_c^m, Y\right).$$

Let $G \in V^m(X_c, Y)$, we have $G(x_c) = \frac{1}{m!} D^m G(0)(x_c, \cdots, x_c)$. So

$$\mathcal{G}^{-1}(G) = \frac{1}{m!} D^m G(0).$$

In other words, we have

$$L = \frac{1}{m!} D^m G(0) \Leftrightarrow G(x_c) = L\left(x_c, ..., x_c\right), \ \forall x_c \in X_c.$$

It follows that $\mathcal{G}$ is a bijection from $\mathcal{L}_s\left(X_c^m, Y\right)$ into $V^m(X_c, Y)$.

In order to use the usual formalism in the context of normal form theory, we now define the Lie bracket.

Definition 5.3 For each $G_m \in V^m(X_c, Y)$, we define the *Lie bracket*

$$[A, G_m](x_c) := DG_m(x_c)\left(Ax_c\right) - AG_m(x_c), \ \forall x_c \in X_c.$$

Let $A_c \in \mathcal{L}\left(X_c\right)$ be the part of A in X_c, then we obtain

$$[A, G_m](x_c) = DG_m(x_c)\left(A_c x_c\right) - AG_m(x_c), \forall x_c \in X_c.$$

Setting $L := \frac{1}{m!} D^m G(0) \in \mathcal{L}_s\left(X_c^m, X_h\right)$, we have

$$DG(x_c)(y) = mL\left(y, x_c, ..., x_c\right), \quad DG(x_c) A_c x_c = mL\left(A_c x_c, x_c, ..., x_c\right),$$

and

$$[A, G](x_c) = \frac{d}{dt}\left[L(e^{A_c t} x_c, \ldots, e^{A_c t} x_c)\right](0) - AL(x_c, \ldots, x_c). \tag{5.3}$$

Define $\Theta_m^c : V^m(X_c, X_c) \to V^m(X_c, X_c)$ by

$$\Theta_m^c\left(G_m^c\right) := [A_c, G_m^c], \forall G_m^c \in V^m(X_c, X_c) \tag{5.4}$$

and $\Theta_m^h : V^m(X_c, X_h) \to V^m(X_c, X_h)$ by

$$\Theta_m^h\left(G_m^h\right) := [A, G_m^h], \forall G_m^h \in V^m(X_c, X_h).$$

Let $p := \dim(X_h)$ and let us identify X_c and X_h with $\mathbb{R}^l$ and $\mathbb{R}^p$. We shall denote by x and y the elements of $\mathbb{R}^l$ and $\mathbb{R}^p$, respectively. Let $H_m^l(\mathbb{R}^q) = V^m(\mathbb{R}^l, \mathbb{R})^q$, that is the linear space of all q-dimensional vector valued functions with components in $V^m(\mathbb{R}^l, \mathbb{R})$. Then there is a $l \times l$ matrix $\mathcal{A}$ and a $p \times p$ matrix $\mathcal{B}$ such that the operators Θ_m^c and Θ_m^h become

$$\Theta_m^c\left(\widetilde{G}_m^c\right) := D\left(\widetilde{G}_m^c\right)\mathcal{A} - \mathcal{A}\left(\widetilde{G}_m^c\right), \forall \widetilde{G}_m^c \in H_m^l(\mathbb{R}^l)$$

and

$$\Theta_m^h\left(\widetilde{G}_m^h\right) := D\left(\widetilde{G}_m^h\right)\mathcal{A} - \mathcal{B}\left(\widetilde{G}_m^h\right), \forall G_m^h \in H_m^l(\mathbb{R}^p),$$

respectively. Note also if $e_1, \cdots, e_q$ denotes a basis of $\mathbb{R}^q$, then a basis for $H_m^l(\mathbb{R}^q)$ can be chosen with the form

$$\Phi_m = \left\{ x_1^{n_1} x_2^{n_2} \ldots x_l^{n_l} e_j, j = 1, \cdots, q, \sum_{j=1}^{l} n_j = m, n_j \geq 0 \text{ are integers} \right\}.$$

We also endow this set with the reverse lexicographic ordering on $(j, n_1, \ldots, n_l)$. To do so, let us recall that the reverse lexicographic ordering on $\mathbb{N}^{l+1}$ is a partial order defined by saying that $(i_1, \cdots, i_{l+1})$ precedes $(j_1, \cdots, j_{l+1})$ if $i_k = j_k$ for all $k = 1, .., l+1$ or

$$i_1 > j_1 \text{ or } \begin{cases} i_1 = j_1 \\ i_2 > j_2 \end{cases} \quad \text{or} \quad \begin{cases} i_1 = j_1 \\ \ldots \\ i_l = j_l \\ i_{l+1} > j_{l+1} \end{cases}.$$

Hence to order now the elements in the basis Φ_m of $H_m^l(\mathbb{R}^q)$, we say $x^\xi e_i \in \Phi_m$ with $\xi = (\xi_1, ..., \xi_l)$ precedes $x^\chi e_j \in \Phi_m$ with $\chi = (\chi_1, ..., \chi_l)$ if and only if $(i, \xi_1, \cdots, \xi_l)$ precedes $(j, \chi_1, \cdots, \chi_l)$ for the reverse lexicographic ordering in $\mathbb{N}^{l+1}$ or equivalently, if and only if the first non-zero difference $i - j, \xi_1 - \chi_1, ..., \xi_l - \chi_l$ is positive.

Example 5.4 Let $\Psi = \{e_1, e_2\}$ be the standard basis of $\mathbb{R}^2$, that is, $e_1 = \begin{pmatrix} 1 \\ 0 \end{pmatrix}$ and $e_2 = \begin{pmatrix} 0 \\ 1 \end{pmatrix}$. Then

$$H_2^2(\mathbb{R}^2) = \left\{ x_1^{d_1} x_2^{d_2} \varphi \middle| \varphi \in \mathbb{R}^2, \ d_1 + d_2 = 2, \ d_j \geq 0, \ j = 1, 2, \text{ are integers} \right\}$$

and

$$H_3^2(\mathbb{R}^2) = \left\{ x_1^{d_1} x_2^{d_2} \varphi \middle| \varphi \in \mathbb{R}^2, d_1 + d_2 = 3,\ d_j \geq 0,\ \ j = 1, 2, \text{ are integers} \right\}.$$

The bases Φ_2 and Φ_3 with the elements in the reverse lexicographic ordering are

$$\Phi_2 = \left\{ \begin{pmatrix} 0 \\ x_1^2 \end{pmatrix}, \begin{pmatrix} 0 \\ x_1 x_2 \end{pmatrix}, \begin{pmatrix} 0 \\ x_2^2 \end{pmatrix}, \begin{pmatrix} x_1^2 \\ 0 \end{pmatrix}, \begin{pmatrix} x_1 x_2 \\ 0 \end{pmatrix}, \begin{pmatrix} x_2^2 \\ 0 \end{pmatrix} \right\},$$

and

$$\Phi_3 = \left\{ \begin{pmatrix} 0 \\ x_1^3 \end{pmatrix}, \begin{pmatrix} 0 \\ x_1^2 x_2 \end{pmatrix}, \begin{pmatrix} 0 \\ x_1 x_2^2 \end{pmatrix}, \begin{pmatrix} 0 \\ x_2^3 \end{pmatrix}, \begin{pmatrix} x_1^3 \\ 0 \end{pmatrix}, \begin{pmatrix} x_1^2 x_2 \\ 0 \end{pmatrix}, \right.$$
$$\left. \begin{pmatrix} x_1 x_2^2 \\ 0 \end{pmatrix}, \begin{pmatrix} x_2^3 \\ 0 \end{pmatrix} \right\}$$

respectively.

If $\sigma(\mathcal{A}) = \{\lambda_1, \cdots, \lambda_l\}$, $\sigma(\mathcal{B}) = \{\mu_1, \cdots, \mu_p\}$, we have the following result (see Lemma 2.5 of the book [53] by S.-N. Chow and J. K. Hale)

Lemma 5.5 *The spectra of Θ_m^c and Θ_m^h are the sets*

$$\sigma\left(\Theta_m^c\right) = \left\{ \lambda \cdot d - \lambda_j : j = 1, \cdots, l,\ \sum_{j=1}^{l} d_j = m \right\} \tag{5.5}$$

and

$$\sigma\left(\Theta_m^h\right) = \left\{ \lambda \cdot d - \mu_j : j = 1, \cdots, p,\ \sum_{j=1}^{l} d_j = m \right\}, \tag{5.6}$$

respectively, where $\lambda \cdot d = \lambda_1 d_1 + \cdots + \lambda_l d_l$.

Definition 5.6 (Resonance conditions) The following relations are called resonant conditions:

$$\lambda \cdot d - \lambda_j = 0, \text{ for all } \lambda_j \in \sigma(\mathcal{A}), \text{ all } d = (d_1, \cdots, d_l) \text{ with} \sum_{j=1}^{l} d_j \geq 2. \tag{5.7}$$

Let $x_1, \cdots, x_l$ be coordinates with respect to the standard basis $e_1, \cdots, e_l$ of $\mathbb{R}^l$ in which the matrix $\mathcal{A}$ is in Jordan normal form with diagonal elements $\lambda_1, \cdots, \lambda_l$. Then a monomial $x_1^{d_1} x_2^{d_2} \ldots x_l^{d_l} e_j (\sum_{j=1}^{l} d_j = m \geq 2$, and $1 \leq j \leq l, d_j \geq 0$ are integers) is called a resonant monomial of order m if and only if (5.7) holds for j and d.

Definition 5.7 (Non-resonance conditions) We say that equation (5.1) satisfies the nonresonance conditions of order k if

$$\lambda \cdot d - \mu_j \neq 0, \text{ for all } \mu_j \in \sigma(\mathcal{B}), \text{ all } d = (d_1, \cdots, d_l) \text{ with } \sum_{j=1}^{l} d_j = k.$$

5.1.2 An Outline on the Methods

Description of the method for computing the normal form up to terms of order 3: Assume first that

$$\Pi_h D^2 F(0) \mid_{X_c \times X_c} \neq 0.$$

Let $G_2^h \in V^2(X_c, X_h)$ be a vector-valued homogeneous polynomial of degree 2. Consider the following global change of variable

$$v := u - G_2^h(\Pi_c u) \Leftrightarrow \begin{cases} \Pi_c v = \Pi_c u \\ \Pi_h v = \Pi_h u - G_2^h(\Pi_c u) \end{cases} \Leftrightarrow u = v + G_2^h(\Pi_c v). \tag{5.8}$$

Then we obtain

$$\begin{aligned} \frac{dv(t)}{dt} &= \frac{du(t)}{dt} - DG_2^h(\Pi_c u)(\Pi_c \frac{du(t)}{dt}) \\ &= Au + F(u) - DG_2^h(\Pi_c u)(\Pi_c [Au + F(u)]) \\ &= A\left[v + G_2^h(\Pi_c v)\right] + F(v + G_2^h(\Pi_c v)) \\ &\quad - DG_2^h(\Pi_c v)\left(\Pi_c\left[Av + F\left(v + G_2^h(\Pi_c v)\right)\right]\right). \end{aligned}$$

Thus we obtain the following system after the change of variable

$$\frac{dv(t)}{dt} = Av(t) + H(v(t)) \text{ for } t \geq 0 \text{ and } v(0) = x \in \overline{D(A)}, \tag{5.9}$$

where

$$H(v) = F\left(v + G_2^h(\Pi_c v)\right) - \left[A, G_2^h\right](\Pi_c v) - DG_2^h(\Pi_c v)\left(\Pi_c F(v + G_2^h(\Pi_c v))\right)$$

and $[\cdot, \cdot]$ is the Lie bracket

$$\left[A, G_2^h\right](v_c) = DG_2^h(v_c)(A_c v_c) - A_h G_2^h(v_c), \ \forall v_c \in X_c.$$

We can rewrite H as

$$\begin{aligned} H(v) = F(v) - \left[A, G_2^h\right](\Pi_c v) + \left[F\left(v + G_2^h(\Pi_c v)\right) - F(v)\right] \\ - DG_2^h(\Pi_c v)\left(\Pi_c F(v + G_2^h(\Pi_c v))\right). \end{aligned}$$

Since $DF(0) = 0$, we obtain

$$\frac{1}{2!}\Pi_h D^2 H(0)(v_c, v_c) = \frac{1}{2!}\Pi_h D^2 F(0)(v_c, v_c) - \left[A, G_2^h\right](v_c).$$

Therefore, in order to cancel out the second order term we need to solve

$$\left[A, G_2^h\right](v_c) = \frac{1}{2!}\Pi_h D^2 F(0)(v_c, v_c) \text{ with } G_2^h \in V^2(X_c, X_h) \tag{5.10}$$

which is the key point to compute the normal form for reduced system. This question will be solved in the next section.

If (5.10) is solved, then by applying the center manifold theorem of chapter 4 and Lemma 5.2 to system (5.9), we deduce that the reduced system of (5.9) has the following form

$$\frac{dv_c}{dt} = A_c v_c + \Pi_c F\left(v_c + G_2^h(v_c)\right) + R(v_c), \tag{5.11}$$

where

$$R(v_c) = \Pi_c F\left(v_c + G_2^h(v_c) + \Psi(v_c)\right) - \Pi_c F\left(v_c + G_2^h(v_c)\right)$$

and $\Psi : X_c \to X_h$ is a local center manifold of the new system (5.9) satisfying

$$\Psi(0) = 0, \; D\Psi(0) = 0, \text{ and } D^2\Psi(0) = 0.$$

Now assume that F is C^4-smooth locally around 0 (so is Ψ). Then we see that $R(v_c)$ is of order 4 and the Taylor's expansion of the reduced system (5.11) at the order 3 is given by

$$\begin{aligned}\frac{dv_c}{dt} &= A_c v_c + \frac{1}{2!}\Pi_c D^2 F(0)(v_c, v_c) \\ &+ \frac{1}{2!}\left\{\Pi_c D^2 F(0)\left(G_2^h(v_c), v_c\right) + \Pi_c D^2 F(0)\left(v_c, G_2^h(v_c)\right)\right\} \\ &+ \frac{1}{3!}\Pi_c D^3 F(0)[v_c, v_c, v_c] + h.o.t.\end{aligned}$$

Therefore, in order to compute the Taylor's expansion of the reduced system at the order 3, we only need to compute G_2^h. Then we can apply the normal form theory in Chow and Li and Wang [54], Chow and Hale [53], Bibikov [30], Vanderbauwhede [338] and Wiggins [352] to the Taylor's expansion of the reduced system.

An alternative approach, to compute both the Taylor's expansion and the normal form of the reduced system would be to use the following change of variables

$$u := v + G_2(\Pi_c v)$$

wherein

$$G_2 \in V^2(X_c, \mathbb{R}^n).$$

In this case

$$u := v + G_2(\Pi_c v) \Leftrightarrow \begin{cases} \Pi_c u = \Pi_c v + \Pi_c G_2(\Pi_c v) \\ \Pi_h u = \Pi_h v + \Pi_h G_2(\Pi_c v). \end{cases}$$

Then the map $\xi_c(v_c) = v_c + \Pi_c G_2(v_c)$ from X_c into itself is only locally invertible around 0. This type of change of variables leads to the normal form theory for reduced system on the center manifold directly.

5.2 Normal Form Theory - Nonresonant Type Results

Corresponding to the two methods mentioned earlier, we consider two types of change of variables, namely, $G_m^h \in V^m(X_c, X_h)$ and $G_m \in V^m(X_c, \mathbb{R}^n)$, respectively in the following.

5.2.0.1 $G_m^h \in V^m(X_c, X_h)$

The aim of using this type of change of variables is to get the Taylor's expansion of the reduced system (5.2). Consider the following change of variables

$$u = v + G_m^h(\Pi_c v), \; G_m^h \in V^m(X_c, X_h). \tag{5.12}$$

Then

$$G_m^h(v_c) := L_m(v_c, v_c, ..., v_c), \forall\, \Pi_c v := v_c \in X_c,$$

where $L_m = \frac{1}{m!} D^m G_m^h(0)$. The map $v_c \to A G_m^h(v_c)$ is differentiable and

$$D\left(A G_m^h\right)(v_c)(y) = A D G_m^h(v_c)(y) = m A L_m(y, v_c, ..., v_c).$$

Define a map $\xi : \mathbb{R}^n \to \mathbb{R}^n$ by

$$\xi(v) := v + G_m^h(\Pi_c v), \forall v \in \mathbb{R}^n.$$

Since the range of G_m^h is included in X_h, we obtain the following equivalence

$$u = \xi(v) \Leftrightarrow v = \xi^{-1}(u),$$

where

$$\xi^{-1}(u) := u - G_m^h(\Pi_c u), \forall u \in \mathbb{R}^n,$$

and

$$\Pi_c \xi^{-1}(u) = \Pi_c u, \forall u \in \mathbb{R}^n.$$

The following result justifies the change of variables (5.12).

Lemma 5.8 *Let Assumptions 5.1 be satisfied and let $G_m^h \in V^m(X_c, X_h)$. Assume that $u \in C([0, \tau], \mathbb{R}^n)$ is a solution of the Cauchy problem*

$$\frac{du(t)}{dt} = Au(t) + F(u(t)), \; t \in [0, \tau], \; u(0) = x \in \mathbb{R}^n. \tag{5.13}$$

Then $v(t) = \xi^{-1}(u(t))$ is the solution of the Cauchy problem

$$\frac{dv(t)}{dt} = Av(t) + H(v(t)),\ t \in [0, \tau],\ v(0) = \xi^{-1}(x) \in \mathbb{R}^n, \tag{5.14}$$

where $H : \mathbb{R}^n \to \mathbb{R}^n$ is the map defined by

$$H(v) = F(\xi(v)) - [A, G_m^h](\Pi_c v) - DG_m^h(\Pi_c v)[\Pi_c F(\xi(v))].$$

Conversely, if $v \in C([0, \tau], \mathbb{R}^n)$ is a solution of (5.14), then $u(t) = \xi(v(t))$ is the solution of (5.13).

Proof Assume that $u \in C([0, \tau], \mathbb{R}^n)$ is the solution of the system (5.13), that is,

$$u(t) = x + A\int_0^t u(l)dl + \int_0^t F(u(l))dl, \forall t \in [0, \tau].$$

Set

$$v(t) = \xi^{-1}(u(t)) = u - G_m^h(\Pi_c u), \forall t \in [0, \tau].$$

We have

$$\begin{aligned}
A\int_0^t v(l)dl &= A\int_0^t u(l)dl - \int_0^t AG_m^h(\Pi_c u(l))\,dl \\
&= u(t) - x - \int_0^t F(u(l))dl - \int_0^t AG_m^h(\Pi_c u(l))\,dl \\
&= u(t) - G_m^h(\Pi_c u(t)) - \left(x - G_m^h(\Pi_c x)\right) \\
&\quad + \left(G_m^h(\Pi_c u(t)) - G_m^h(\Pi_c x)\right) \\
&\quad - \int_0^t F(u(l))dl - \int_0^t AG_m^h(\Pi_c u(l))\,dl \\
&= v(t) - \xi^{-1}(x) + \left(G_m^h(\Pi_c u(t)) - G_m^h(\Pi_c x)\right) \\
&\quad - \int_0^t F(u(l))dl - \int_0^t AG_m^h(\Pi_c u(l))\,dl.
\end{aligned}$$

Since $\dim(X_c) < +\infty$, $t \to \Pi_c u(t)$ satisfies the following ordinary differential equations

$$\frac{d\Pi_c u(t)}{dt} = A_c \Pi_c u(t) + \Pi_c F(u(t)).$$

By integrating both sides of the above ordinary differential equations, we obtain

$$\begin{aligned}
G_m^h(\Pi_c u(t)) - G_m^h(\Pi_c x) &= \int_0^t DG_m^h(\Pi_c u(l))\left(\frac{d\Pi_c u(l)}{dl}\right)dl \\
&= \int_0^t DG_m^h(\Pi_c u(l))\,[A_c \Pi_c u(l) + \Pi_c F(u(l))]\,dl.
\end{aligned}$$

It follows that

$$
\begin{aligned}
A \int_0^t v(l) dl = v(t) - \xi^{-1}(x) \\
&+ \int_0^t DG_m^h(\Pi_c u(l)) [A_c \Pi_c u(l) + \Pi_c F(u(l))] \, dl \\
&- \int_0^t F(u(l)) dl - \int_0^t AG_m^h(\Pi_c u(l)) \, dl.
\end{aligned}
$$

Thus

$$
v(t) = \xi^{-1}(x) + A \int_0^t v(l) dl + \int_0^t H(v(l)) dl,
$$

in which

$$
\begin{aligned}
H(v) = F(\xi(v)) + AG_m^h(\Pi_c \xi(v)) \\
&- DG_m^h(\Pi_c \xi(v)) [A_c \Pi_c \xi(v) + \Pi_c F(\xi(v))].
\end{aligned}
$$

Since $\Pi_c \xi = \Pi_c$, the first implication follows. The converse follows from the first implication by replacing F by H and ξ^{-1} by ξ. □

Define $\Theta_m^h : V^m(X_c, X_h) \to V^m(X_c, X_h)$ by

$$
\Theta_m^h \left(G_m^h \right) := [A, G_m^h], \forall G_m^h \in V^m(X_c, X_h).
$$

We decompose $V^m(X_c, X_h)$ into the direct sum

$$
V^m(X_c, X_h) = \mathcal{R}_m \oplus \mathcal{C}_m,
$$

where

$$
\mathcal{R}_m := R(\Theta_m^h)
$$

is the range of Θ_m^h, and $\mathcal{C}_m$ is some complementary space of $\mathcal{R}_m$ into $V^m(X_c, X_h)$.

Set for each $\eta > 0$,

$$
BC^\eta(\mathbb{R}, \mathbb{R}^n) := \left\{ f \in C(\mathbb{R}, \mathbb{R}^n) : \sup_{t \in \mathbb{R}} e^{-\eta|t|} \|f(t)\| < +\infty \right\}.
$$

By using Lemma 4.16 and Lemma 4.18 of chapter 4, we now prove the following lemma.

Lemma 5.9 *If*

$$
f(t) = t^k e^{\lambda t} x
$$

for some $k \in \mathbb{N}$, $\lambda \in i\mathbb{R}$, *and* $x \in \mathbb{R}^n$, *then*

$$
(K_u + K_s)(\Pi_h f)(0) = (-1)^k k! (\lambda I - A_h)^{-(k+1)} \Pi_h x \in D(A_h) \subset \mathbb{R}^n,
$$

where K_u *and* K_s *are defined in Lemma 4.16 and Lemma 4.18 of chapter 4.*

Proof We have

$$
\begin{aligned}
K_u(f)(0) &= -\int_0^{+\infty} e^{\lambda l} l^k e^{-A_u l} \Pi_u x dl \\
&= -\frac{d^k}{d\lambda^k} \int_0^{+\infty} e^{\lambda l} e^{-A_u l} \Pi_u x dl \\
&= -\frac{d^k}{d\lambda^k} (-\lambda I + A_u)^{-1} \Pi_u x \\
&= \frac{d^k}{d\lambda^k} (\lambda I - A_u)^{-1} \Pi_u x \\
&= (-1)^k k! (\lambda I - A_u)^{-(k+1)} \Pi_u x.
\end{aligned}
$$

Similarly, for $\mu > \omega(A)(\omega(A) := \sup \{\text{Re}(\lambda) : \lambda \in \sigma(A)\})$, we have

$$
\begin{aligned}
(\mu I - A)^{-1} K_s(f)(0) &= \lim_{\tau \to -\infty} (\mu I - A)^{-1} \Pi_s \left(e^{A \cdot} * f(\tau + .)\right)(-\tau) \\
&= \int_0^{+\infty} e^{A_s l} (\mu I - A)^{-1} \Pi_s f(-l) dl.
\end{aligned}
$$

So we obtain that

$$
\begin{aligned}
(\mu I - A)^{-1} K_s(f)(0) &= \int_0^{+\infty} (-l)^k e^{-\lambda l} e^{A l} (\mu I - A)^{-1} \Pi_s x dl \\
&= \frac{d^k}{d\lambda^k} (\lambda I - A)^{-1} (\mu I - A)^{-1} \Pi_s x \\
&= (-1)^k k! (\lambda I - A)^{-(k+1)} (\mu I - A)^{-1} \Pi_s x \\
&= (\mu I - A)^{-1} (-1)^k k! (\lambda I - A_s)^{-(k+1)} \Pi_s x.
\end{aligned}
$$

Since $(\mu I - A)^{-1}$ is one to one, we deduce that

$$K_s(f)(0) = (-1)^k k! (\lambda I - A_s)^{-(k+1)} \Pi_s x$$

and the result follows. □

According to Lemma 5.2, when $F(0) = 0$, $DF(0) = 0$ and

$$\Pi_h D^m F(0) \mid_{X_c \times X_c \times \ldots \times X_c} = 0 \text{ for each } m = 2, \ldots, k,$$

we get

$$D^m \Psi(0) = 0 \text{ for each } m = 1, \ldots, k$$

and the Taylor's expansion of the reduced system (5.2) is simply given by

$$\frac{du_c(t)}{dt} = A_c u_c(t) + \sum_{l=1}^{k} \frac{1}{l!} \Pi_c D^l F(0) (u_c(t), \ldots, u_c(t)) + h.o.t.$$

In order to be able to make the terms $\Pi_h D^m F(0) \mid_{X_c \times X_c \times \ldots \times X_c}$ to vanish, the range of Θ_m^h must be the whole space $V^m(X_c, X_h)$ which is related to the nonresonant conditions, $m = 2, \ldots, k,$. From Lemma 5.5, we can obtain that the range of Θ_m^h must be the whole space $V^m(X_c, X_h)$. Here we use the following proposition to answer nonresonant conditions instead of Lemma 5.5 which provides a general setting to be naturally extended to infinite-dimensional space, such as the natural phase space for an abstract class of Cauchy problems on Banach spaces.

Proposition 5.10 *Let Assumptions 5.1 be satisfied. For each $R \in V^m(X_c, X_h)$, there exists a unique map $G^h_m \in V^m(X_c, X_h)$ such that*

$$\left[A, G^h_m\right](x_c) = R(x_c), \forall x_c \in X_c. \tag{5.15}$$

Moreover, (5.15) is equivalent to

$$G^h_m(x_c) = (K_u + K_s)\left(R(e^{A_c \cdot}x_c)\right)(0),$$

or

$$L_m(x_1, \ldots, x_m) = (K_u + K_s)\left(H\left(e^{A_c \cdot}x_1, \ldots, e^{A_c \cdot}x_m\right)\right)(0),$$

with $L_m := \frac{1}{m!}D^m G^h_m(0)$ and $H := \frac{1}{m!}D^m R(0)$.

Proof Let $G^h_m \in V^m(X_c, X_h)$. Then $L_m = \frac{1}{m!}D^m G^h_m(0) \in \mathcal{L}_s\left(X_c^m, X_h\right)$ and

$$[A, G^h_m](x_c) = \frac{d}{dt}\left[L_m(e^{A_c t}x_c, \ldots, e^{A_c t}x_c)\right](0) - A_h L_m(x_c, \ldots, x_c).$$

Assume that $G^h_m \in V^m(X_c, X_h)$ satisfies (5.15). Then $L_m = \frac{1}{m!}D^m G^h_m(0)$ satisfies

$$\frac{d}{dt}\left[L_m(e^{A_c t}x_1, ..., e^{A_c t}x_m)\right](0) = A_h L_m(x_1, ..., x_m) + H(x_1, ..., x_m),$$

where $H = \frac{1}{m!}D^m R(0) \in \mathcal{L}_s\left(X_c^m, X_h\right)$. Then (5.15) is satisfied if and only if for each $(x_1, ..., x_m) \in X_c^m$ and each $t \in \mathbb{R}$,

$$\begin{aligned}\frac{d}{dt}\left[L_m(e^{A_c t}x_1, ..., e^{A_c t}x_m)\right](t) = A_h L_m(e^{A_c t}x_1, ..., e^{A_c t}x_m)\\ +H(e^{A_c t}x_1, ..., e^{A_c t}x_m).\end{aligned} \tag{5.16}$$

Set

$$v(t) := L_m(e^{A_c t}x_1, ..., e^{A_c t}x_m), \ \forall t \in \mathbb{R}$$

and

$$w(t) := H(e^{A_c t}x_1, ..., e^{A_c t}x_m), \ \forall t \in \mathbb{R}.$$

The system (5.16) can be rewritten as

$$\frac{dv(t)}{dt} = A_h v(t) + w(t), \quad \forall t \in \mathbb{R}. \tag{5.17}$$

Since L_m and H are bounded multilinear maps and $\sigma(A_c) \subset i\mathbb{R}$, it follows that for each $\eta > 0$,

$$v \in BC^\eta(\mathbb{R}, \mathbb{R}^n) \text{ and } w \in BC^\eta(\mathbb{R}, \mathbb{R}^n).$$

Let $\eta \in \left(0, \min\left(\min_{\lambda\in\sigma(A_s)} -\mathrm{Re}\,(\lambda), \min_{\lambda\in\sigma(A_u)} \mathrm{Re}\,(\lambda)\right)\right)$. By projecting (5.17) on X_u, we have

$$\frac{d\Pi_u v(t)}{dt} = A_u \Pi_u v(t) + \Pi_u w(t),$$

or equivalently, $\forall t, s \in \mathbb{R}$ with $t \geq s$,

$$\Pi_u v(t) = e^{A_u(t-s)}\Pi_u v(s) + \int_s^t e^{A_u(t-l)}\Pi_u w(l)dl,$$

$$\Pi_u v(s) = e^{-A_u(t-s)}\Pi_u v(t) - \int_s^t e^{-A_u(l-s)}\Pi_u w(l)dl.$$

By using the fact that $v \in BC^\eta(\mathbb{R}, \mathbb{R}^n)$, we obtain when t goes to $+\infty$ that

$$\Pi_u v(s) = K_u(\Pi_u w)(s), \ \forall s \in \mathbb{R}.$$

Thus, for $s = 0$ we have

$$\Pi_u L_m(x_1, ..., x_m) = K_u(\Pi_u H(e^{A_c \cdot}x_1, ..., e^{A_c \cdot}x_m))(0). \tag{5.18}$$

By projecting (5.17) on X_s, we obtain

$$\frac{d\Pi_s v(t)}{dt} = A_s \Pi_s v(t) + \Pi_s w(t),$$

or equivalently, $\forall t, s \in \mathbb{R}$ with $t \geq s$,

$$\Pi_s v(t) = e^{A_s(t-s)}\Pi_s v(s) + \left(e^{A_s \cdot} * \Pi_s w(. + s)\right)(t - s).$$

By using the fact that $v \in BC^\eta(\mathbb{R}, \mathbb{R}^n)$, we have when s goes to $-\infty$ that

$$\Pi_s v(t) = K_s(\Pi_s w)(t), \ \forall t \in \mathbb{R}.$$

Thus, for $t = 0$ it follows that

$$\Pi_s L_m(x_1, ..., x_m) = K_s(\Pi_s H(e^{A_c \cdot}x_1, ..., e^{A_c \cdot}x_m))(0). \tag{5.19}$$

Summing up (5.18) and (5.19), we deduce that

$$L_m(x_1, ..., x_m) = (K_u + K_s)\,(H\left(e^{A_c \cdot}x_1, ..., e^{A_c \cdot}x_m\right))(0). \tag{5.20}$$

Conversely, assume that $L_m(x_1, ..., x_m)$ is defined by (5.20) and set

$$v(t) := (K_u + K_s)\left(H\left(e^{A_c \cdot}x_1, ..., e^{A_c \cdot}x_m\right)\right)(t), \ \forall t \in \mathbb{R}.$$

Then we have

$$v(t) = L_m(e^{A_c t}x_1, ..., e^{A_c t}x_m), \ \forall t \in \mathbb{R}.$$

Moreover, using Lemma 4.16-(iii) and Lemma 4.18-(iii) of chapter 4, we deduce that for each $t, s \in \mathbb{R}$ with $t \geq s$,

$$v(t) = e^{A(t-s)}v(s) + \left(e^{A\cdot} * w(. + s)\right)(t - s),$$

or equivalently,

$$v(t) = v(s) + A\int_s^t v(l)dl + \int_s^t w(l)dl$$

and then

$$\frac{dv(t)}{dt} = Av(t) + w(t),\ \forall t \in \mathbb{R}.$$

The result follows. □

Remark 5.11 (An explicit formula for L_m in proposition 5.10) Since $l := \dim(X_c) < +\infty$, we can find a basis $\{e_1, ..., e_l\}$ of X_c such that the matrix of A_c (with respect to this basis) is reduced to the Jordan's form. Then for each $x_c \in X_c$, $e^{A_c t}x_c$ is a linear combination of elements of the form

$$t^k e^{\lambda t} x_j$$

for some $k \in \{1, ..., l\}$, some $\lambda \in \sigma(A_c) \subset i\mathbb{R}$, and some $x_j \in \{e_1, ..., e_l\}$. Let $\lambda_1, ..., \lambda_m \in \sigma(A_c) \subset i\mathbb{R}$, $x_1, ..., x_m \in \{e_1, ..., e_l\}$, $k_1, ..., k_m \in \{1, ..., l\}$. Define

$$f(t) := H\left(t^{k_1}e^{\lambda_1 t}x_1, ..., t^{k_m}e^{\lambda_m t}x_m\right), \forall t \in \mathbb{R},$$

Since $H \in \mathcal{L}_s\left(X_c^m, X_h\right)$ is m-linear, we obtain

$$f(t) = t^k e^{\lambda t} y$$

with

$$k = k_1 + k_2 + ... + k_m,\ \ \lambda = \lambda_1 + ... + \lambda_m,$$

and

$$y = H(x_1, ..., x_m).$$

Now by using Lemma 5.9, we obtain the explicit formula

$$(K_u + K_s)\left(H\left((.)^{k_1}e^{\lambda_1 \cdot}x_1, ..., (.)^{k_m}e^{\lambda_m \cdot}x_m\right)\right)(0) = (-1)^k k!\,(\lambda I - A_h)^{-(k+1)}\Pi_h y \in \mathbb{R}^n.$$

5.2.0.2 $G_m \in V^m(X_c, \mathbb{R}^n)$

We consider the following change of variables

$$u(t) = w(t) + G_m(\Pi_c w(t)),\ G_m \in V^m(X_c, \mathbb{R}^n). \tag{5.21}$$

Then the map $I + G_m \circ \Pi_c : \mathbb{R}^n \to \mathbb{R}^n$ is locally invertible around 0. From (5.3), for each $H \in V^m(X_c, \mathbb{R}^n)$, to find $G_m \in V^m(X_c, \mathbb{R}^n)$ satisfying

$$[A, G_m](x_c) = H(x_c), \tag{5.22}$$

is equivalent to find $L_m \in \mathcal{L}_s\left(X_c^m, \mathbb{R}^n\right)$ satisfying

$$\frac{d}{dt}\left[L_m(e^{A_c t}x_1, \ldots, e^{A_c t}x_m)\right]_{t=0} = AL_m(x_1, \ldots, x_m) + \widehat{H}(x_1, \ldots, x_m) \tag{5.23}$$

for each $(x_1, \ldots, x_m) \in X_c^m$ with

$$\mathcal{G}(\widehat{H}) = H.$$

By projecting on X_c and X_h, it follows that solving system (5.22) is equivalent to find $G_m^c \in V^m(X_c, X_c)$ and $G_m^h \in V^m(X_c, X_h)$ satisfying

$$[A_c, G_m^c] = \Pi_c H \tag{5.24}$$

and

$$[A, G_m^h] = \Pi_h H. \tag{5.25}$$

Define $\Theta_m^c : V^m(X_c, X_c) \to V^m(X_c, X_c)$ by

$$\Theta_m^c\left(G_m^c\right) := [A_c, G_m^c], \forall G_m^c \in V^m(X_c, X_c) \tag{5.26}$$

and $\Theta_m^h : V^m(X_c, X_h) \to V^m(X_c, X_h)$ by

$$\Theta_m^h\left(G_m^h\right) := [A, G_m^h], \forall G_m^h \in V^m(X_c, X_h).$$

We decompose $V^m(X_c, X_c)$ into the direct sum

$$V^m(X_c, X_c) = \mathcal{R}_m^c \oplus C_m^c\,,$$

where

$$\mathcal{R}_m^c := R(\Theta_m^c)$$

is the range of Θ_m^c, and C_m^c is some complementary space of $\mathcal{R}_m^c$ into $V^m(X_c, X_c)$.

Proposition 5.12 *Let Assumptions 5.1 be satisfied. Let $H \in \mathcal{R}_m^c \oplus V^m(X_c, X_h)$. Then there exists $G_m \in V^m(X_c, \mathbb{R}^n)$ (non-unique in general) satisfying*

$$[A, G_m](x_c) = H(x_c). \tag{5.27}$$

Furthermore, if $N(\Theta_m^c) = \{0\}$ (the null space of Θ_m^c), then G_m is uniquely determined.

Proof It is clear that we can solve (5.24). Moreover, we can also apply Proposition 5.10 and deduce that (5.25) can be solved. □

Remark 5.13 In practice, we often have

$$N(\Theta_m^c) \cap R(\Theta_m^c) = \{0\}\,,$$

In this case, a natural splitting of $V^m(X_c, X_c)$ will be

$$V^m(X_c, X_c) = R(\Theta_m^c) \oplus N(\Theta_m^c).$$

Define $P_m : V^m(X_c, \mathbb{R}^n) \to V^m(X_c, \mathbb{R}^n)$ the bounded linear projector satisfying

$$\mathcal{P}_m\left(V^m(X_c, \mathbb{R}^n)\right) = \mathcal{R}_m^c \oplus V^m(X_c, X_h), \text{ and } (I - \mathcal{P}_m)\left(V^m(X_c, \mathbb{R}^n)\right) = \mathcal{C}_m^c.$$

Again consider the system (5.13). Assume that $DF(0) = 0$. Without loss of generality we also assume that for some $m \in \{2, ..., k\}$,

$$\Pi_h D^j F(0) \mid_{X_c \times X_c \times \ldots \times X_c} = 0, \ \mathcal{G}\left(\Pi_c D^j F(0) \mid_{X_c \times X_c \times \ldots \times X_c}\right) \in \mathcal{C}_j^c, \qquad ((C_{m-1}))$$

for each $j = 1, ..., m - 1$.

We will show that we can find $G_m \in V^m(X_c, \mathbb{R}^n)$ such that after the change of variables (5.21) we can rewrite the system (5.13) as

$$\frac{dw(t)}{dt} = Aw(t) + H(w(t)), \text{ for } t \geq 0, \text{ and } w(0) = (I + G_m \circ \Pi_c)^{-1} x \in \mathbb{R}^n, \quad (5.28)$$

where H satisfies the condition (C_m). This will provide a normal form method.

Lemma 5.14 *Let Assumptions 5.1 be satisfied. Let $G_m \in V^m(X_c, \mathbb{R}^n)$. Assume that $u \in C([0, \tau], \mathbb{R}^n)$ is a solution of the Cauchy problem (5.13). Then $w(t) = (I + G_m \circ \Pi_c)^{-1}(u(t))$ is a solution of the system (5.28), where $H : \overline{D(A)} \to \mathbb{R}^n$ is the map defined by*

$$H(w(t)) = F(w(t)) - [A, G_m](\Pi_c w(t)) + O(\|w(t)\|^{m+1}).$$

Conversely, if $w \in C([0, \tau], \mathbb{R}^n)$ is a solution of (5.28), then

$$u(t) = (I + G_m \circ \Pi_c)\, w(t)$$

is a solution of (5.13).

Lemma 5.14 can be proved similarly as Lemma 5.8, here we omit it.

Proposition 5.15 *Let Assumptions 5.1 be satisfied. Let $r > 0$ and let $F : B_{\mathbb{R}^n}(0, r) \to \mathbb{R}^n$ be a map. Assume that there exists an integer $k \geq 1$ such that F is k-time continuously differentiable in $B_{\mathbb{R}^n}(0, r)$ with $F(0) = 0$ and $DF(0) = 0$. Let $m \in \{2, ..., k\}$ be such that F satisfies the condition (C_{m-1}). Then there exists a map $G_m \in V^m(X_c, \mathbb{R}^n)$ such that after the change of variables*

$$u(t) = w(t) + G_m(\Pi_c w(t)),$$

we can rewrite system (5.13) as (5.28) and H satisfies the condition (C_m), where

$$H(w(t)) = F(w(t)) - [A, G_m](\Pi_c w(t)) + O(\|w(t)\|^{m+1}).$$

Proof Let $x_c \in X_c$. We have

$$H(x_c) = F(x_c) - [A, G_m](\Pi_c x_c) + O(\|x_c\|^{m+1}).$$

It follows that

$$\begin{aligned}H(x_c) &= \frac{1}{2!}D^2F(0)(x_c,x_c) + ... + \frac{1}{(m-1)!}D^{m-1}F(0)(x_c,\dots,x_c)\\ &+\mathcal{P}_m\left[\frac{1}{m!}D^mF(0)(x_c,\dots,x_c)\right] + (I-\mathcal{P}_m)\left[\frac{1}{m!}D^mF(0)(x_c,\dots,x_c)\right]\\ &-[A,G_m](x_c) + O(\|x_c\|^{m+1})\end{aligned}$$

since $DF(0) = 0$. Moreover, by using Proposition 5.12 we obtain that there exists a map $G_m \in V^m(X_c, \mathbb{R}^n)$ such that

$$[A, G_m](x_c) = \mathcal{P}_m\left[\frac{1}{m!}D^mF(0)(x_c,\dots,x_c)\right].$$

Hence,

$$\begin{aligned}H(x_c) &= \frac{1}{2!}D^2F(0)(x_c,x_c) + ... + \frac{1}{(m-1)!}D^{m-1}F(0)(x_c,\dots,x_c)\\ &+(I-\mathcal{P}_m)\left[\tfrac{1}{m!}D^mF(0)(x_c,\dots,x_c)\right] + O(\|x_c\|^{m+1}).\end{aligned} \tag{5.29}$$

By the assumption, we have for all $j = 1, ..., m-1$ that

$$\Pi_h D^j H(0)\mid_{X_c\times X_c\times\ldots\times X_c} = \Pi_h D^j F(0)\mid_{X_c\times X_c\times\ldots\times X_c} = 0$$

and

$$\mathcal{G}\left(\Pi_c D^j H(0)\mid_{X_c\times X_c\times\ldots\times X_c}\right) = \mathcal{G}\left(\Pi_c D^j F(0)\mid_{X_c\times X_c\times\ldots\times X_c}\right) \in C_j^c.$$

Now by using (5.29), we have

$$\frac{1}{m!}\Pi_h D^m H(0)\mid_{X_c\times X_c\times\ldots\times X_c} = \Pi_h\mathcal{G}^{-1}\left[(I-\mathcal{P}_m)\left(\frac{1}{m!}D^mF(0)(x_c,\dots,x_c)\right)\right] = 0$$

and

$$\begin{aligned}&\mathcal{G}\left(\Pi_c D^m H(0)\mid_{X_c\times X_c\times\ldots\times X_c}\right)\\ &\qquad= \mathcal{G}\left\{\Pi_c\mathcal{G}^{-1}\left[(I-\mathcal{P}_m)\left(D^mF(0)(x_c,\dots,x_c)\right)\right]\right\} \in C_m^c.\end{aligned}$$

The result follows. □

5.3 Normal Form Computations

In this section we provide the two methods which use the different types of change of variables $G_m^h \in V^m(X_c, X_h)$ and $G_m \in V^m(X_c, \mathbb{R}^n)$, respectively, to compute the normal form of the reduced system at any order of a system topologically equivalent to the original system:

$$\begin{cases} \dfrac{du(t)}{dt} = Au(t) + F(u(t)), \quad t \geq 0, \\ u(0) = x \in \mathbb{R}^n. \end{cases} \tag{5.30}$$

Assumption 5.16 Assume that $F \in C^k(\mathbb{R}^n, \mathbb{R}^n)$ for some integer $k \geq 2$ with

$$F(0) = 0 \text{ and } DF(0) = 0.$$

Set

$$F_1 := F.$$

5.3.1 $G_m^h \in V^m(X_c, X_h)$

For $j = 2, ..., k$, we apply Proposition 5.10. Then there exists a unique function $G_j^h \in V^j(X_c, X_h)$ satisfying

$$\left[A, G_j^h\right](x_c) = \frac{1}{j!}\Pi_h D^j F_{j-1}(0)(x_c, ..., x_c), \forall x_c \in X_c. \tag{5.31}$$

Define $\xi_j : \mathbb{R}^n \to \mathbb{R}^n$ and $\xi_j^{-1} : \mathbb{R}^n \to \mathbb{R}^n$ by

$$\xi_j(x) := x + G_j^h(\Pi_c x) \text{ and } \xi_j^{-1}(x) := x - G_j^h(\Pi_c x), \forall x \in \mathbb{R}^n.$$

Then

$$F_j(x) := F_{j-1}(\xi_j(x)) - \left[A, G_j^h\right](\Pi_c x) - DG_j^h(\Pi_c x)\left[\Pi_c F_{j-1}(\xi_j(x))\right].$$

Moreover, we have for $x \in \mathbb{R}^n$ that

$$\Pi_c F_j(x) = \Pi_c F_{j-1}(\xi_j(x)) = \Pi_c F_{j-1}\left(x + G_j^h(\Pi_c x)\right).$$

Since the range of G_j^h is included in X_h, by induction we have

$$\Pi_c F_j(x) = \Pi_c F\left(x + G_2^h(\Pi_c x) + G_3^h(\Pi_c x) + ... + G_j^h(\Pi_c x)\right).$$

Now, we obtain

$$\Pi_h D^j F_k(0)\mid_{X_c \times X_c \times ... \times X_c} = 0 \text{ for all } j = 1, ..., k.$$

Setting

$$\begin{aligned} u_k(t) &= \xi_k^{-1} \circ \xi_{k-1}^{-1} \circ ... \xi_2^{-1}(u(t)) \\ &= u(t) - G_2^h(\Pi_c u(t)) - G_3^h(\Pi_c u(t)) - ... - G_k^h(\Pi_c u(t)), \end{aligned}$$

we deduce that $u_k(t)$ is a solution of the system

$$\begin{cases} \dfrac{du_k(t)}{dt} = Au_k(t) + F_k(u_k(t)), \quad t \geq 0, \\ u_k(0) = x_k \in \overline{D(A)}. \end{cases} \tag{5.32}$$

Applying the center manifold theorem in Chapter 14 and Lemma 5.2 to system (5.32), we get the reduced system of Cauchy problem (5.32).

Theorem 5.17 *Let Assumptions 5.1, 5.16 be satisfied. Then by using the change of variables*

$$\begin{cases} u_k(t) = u(t) - G_2^h(\Pi_c u(t)) - G_3^h(\Pi_c u(t)) - \ldots - G_k^(h)(\Pi_c u(t)) \\ \Leftrightarrow \\ u(t) = u_k(t) + G_2^h(\Pi_c u_k(t)) + G_3^h(\Pi_c u_k(t)) + \ldots + G_k^h(\Pi_c u_k(t)), \end{cases}$$

the map $t \to u(t)$ is a solution of the Cauchy problem (5.30) if and only if $t \to u_k(t)$ is a solution of the Cauchy problem (5.32). Moreover, the reduced system of Cauchy problem (5.32) is given by the ordinary differential equations on X_c :

$$\frac{dx_c(t)}{dt} = A_c x_c(t) + \Pi_c F \begin{bmatrix} x_c(t) + G_2^h(x_c(t)) + \\ G_3^h(x_c(t)) + \ldots + G_k^h(x_c(t)) \end{bmatrix} + R_c(x_c(t)), \tag{5.33}$$

where the remainder term $R_c \in C^k(X_c, X_c)$ satisfies

$$D^j R_c(0) = 0 \textit{ for each } j = 1, \ldots, k,$$

or in other words $R_c(x_c(t))$ is a remainder term of order k. If we assume in addition that $F \in C^{k+2}(\mathbb{R}^n, \mathbb{R}^n)$, then the map $R_c \in C^{k+2}(X_c, X_c)$ and $R_c(x_c(t))$ is a remainder term of order $k+2$, that is

$$R_c(x_c) = \|x_c\|^{k+2} O(x_c), \tag{5.34}$$

where $O(x_c)$ is a function of x_c which remains bounded when x_c goes to 0, or equivalently,

$$D^j R_c(0) = 0 \textit{ for each } j = 1, \ldots, k+1.$$

Proof By applying the center manifold theorem in 4 and Lemma 5.2 to system (5.32), there exists $\Psi_k \in C^k(X_c, X_h)$ such that the reduced system of (5.32) is given by

$$\begin{aligned} \frac{dx_c(t)}{dt} &= A_c x_c(t) \\ &\quad + \Pi_c F\left[x_c(t) + G_2^h(x_c(t)) + G_3^h(x_c(t)) + \ldots + G_k^h(x_c(t)) + \Psi_k(x_c(t))\right] \end{aligned}$$

and

$$D^j \Psi_k(0) = 0 \text{ for } j = 1, \ldots, k.$$

By setting

$$\begin{aligned} R_c(x_c) = {} & \Pi_c F\left[x_c + G_2^h(x_c) + G_3^h(x_c) + \ldots + G_k^h(x_c) + \Psi_k(x_c)\right] \\ & - \Pi_c F\left[x_c + G_2^h(x_c) + G_3^h(x_c) + \ldots + G_k^h(x_c)\right], \end{aligned}$$

we obtain the first part of the theorem. If we assume in addition that $F \in C^{k+2}(\mathbb{R}^n, \mathbb{R}^n)$, then $\Psi_k \in C^{k+2}(X_c, X_h)$. Thus,

$$R_c \in C^{k+2}(X_c, X_c).$$

Set

$$h(x_c) := x_c + G_2^h(x_c) + G_3^h(x_c) + \dots + G_k^h(x_c).$$

We have

$$\begin{aligned} R_c(x_c) &= \Pi_c \{F[h(x_c) + \Psi_k(x_c)] - F[h(x_c)]\} \\ &= \Pi_c \int_0^1 DF(h(x_c) + s\Psi_k(x_c))(\Psi_k(x_c))\, ds. \end{aligned}$$

Define

$$\widehat{h}(x_c) := h(x_c) + s\Psi_k(x_c).$$

Since $DF(0) = 0$, we have

$$\begin{aligned} DF\left(\widehat{h}(x_c)\right)(\Psi_k(x_c)) &= DF(0)(\Psi_k(x_c)) \\ &\quad + \int_0^1 D^2F\left(l\widehat{h}(x_c)\right)\left(\widehat{h}(x_c), \Psi_k(x_c)\right) dl \\ &= \int_0^1 D^2F\left(l\widehat{h}(x_c)\right)\left(\widehat{h}(x_c), \Psi_k(x_c)\right) dl. \end{aligned}$$

Hence,

$$R_c(x_c) = \Pi_c \int_0^1 \int_0^1 D^2F(l(h(x_c) + s\Psi_k(x_c)))(h(x_c) + s\Psi_k(x_c), \Psi_k(x_c))\, dlds \tag{5.35}$$

and $h(x_c)$ is a term of order 1, $\Psi_k(x_c)$ is a term of order $k+1$, it follows that (5.35) holds. This completes the proof. □

Remark 5.18 In order to apply the above approach, we first need to compute Π_c and A_c, then $\Pi_h := I - \Pi_c$ can be derived. The point to apply the above procedure is to solve system (5.31). To do this, one may compute

$$(\lambda I - A_h)^{-k} \frac{1}{j!} \Pi_h D^j F(0) \tag{5.36}$$

for each $\lambda \in i\mathbb{R}$ and each $k \geq 1$ by using Remark 5.11, or one may directly solve system (5.31) by computing $\Pi_h \frac{1}{j!} D^j F_{j-1}$. This last approach will involve the computation of (5.36) for some specific values of $\lambda \in i\mathbb{R}$ and some specific values of $k \geq 1$. This turns out to be the main difficulty in applying the above method.

In the applications, we will use the last part of Theorem 5.17 to avoid some unnecessary computations. We will apply this theorem for $k = 2$, F in C^4, and the remainder term $R_c(x_c)$ of order 4. This means that if we want to compute the Taylor's expansion of the reduced system to the order 3 (which is very common in such a context), we only need to compute G_2. So in application the last part of Theorem 5.17 will help us to avoid a lot of computations.

Assume that the reduced system (5.33), expressed in terms of the basis $\Lambda = \{b_j\}_{j=1,\dots,l}$ of X_c, has the following form

$$x'(t) = \mathcal{A}x(t) + \mathcal{F}_2(x) + \mathcal{F}_3(x) + \cdots + \mathcal{F}_k(x) + O(|x|^{k+1}), \tag{5.37}$$

where $x \in \mathbb{R}^l$ and $\mathcal{F}_m(x) \in H^l_m(\mathbb{R}^l)$, $2 \le m \le k$. The desired simplification of (5.37) up to terms k will be obtained by performing inductively a sequence of near identity change of coordinates of the form

$$x = \xi^c_m(\overline{x}) = \overline{x} + G^c_m(\overline{x}), \overline{x} \in B^m(0, r), \tag{5.38}$$

where $G^c_m \in H^l_m(\mathbb{R}^l)$ and $B^m(0, r)$ is a small neighborhood of the origin, $2 \le m \le k$. Notice that the map $\xi^c_m(\overline{x})$ is a diffeomorphism in some neighborhood of the origin and we take $B^m(0, r)$ small enough such that $D\xi^c_m(\overline{x}) = I + DG^c_m(\overline{x})$ is invertible on it and

$$D\xi^c_m(\overline{x})^{-1} = I - DG^c_m(\overline{x}) + O(|\overline{x}|^{2m-2}), \overline{x} \in B^m(0, r),$$

here I is the identity matrix. Substituting (5.38) into (5.37) and dropping the hats for simplicity of notation, we get

$$\begin{aligned} x' = {} & \mathcal{A}x(t) + \mathcal{F}_2(x) + \mathcal{F}_3(x) + \cdots + \mathcal{F}_{m-1}(x) + \left\{\mathcal{F}_m(x) - [\mathcal{A}, G^c_m]\right\} \\ & + \overline{\mathcal{F}}_{m+1}(x) + \cdots + \overline{\mathcal{F}}_k(x) + O(|x|^{k+1}), x \in B^m(0, r). \end{aligned} \tag{5.39}$$

where $[\mathcal{A}, G^c_m]$ is the Lie bracket. Notice that after the change of coordinates (5.38), only terms of order higher than $m - 1$ are modified.

Define a linear operator $\Theta^c_m : H^l_m(\mathbb{R}^l) \to H^l_m(\mathbb{R}^l)$ by

$$\left(\Theta^c_m G^c_m\right)(x) := [\mathcal{A}, G^c_m](x) \equiv DG^c_m(x)\mathcal{A}x - \mathcal{A}G^c_m(x), G^c_m \in H^l_m(\mathbb{R}^l). \tag{5.40}$$

From elementary linear algebra, we know that $H^l_m(\mathbb{R}^l)$ can be (non uniquely) represented as the direct sum

$$H^l_m(\mathbb{R}^l) = \mathcal{R}_m \oplus C_m, \tag{5.41}$$

where

$$\mathcal{R}_m := R(\Theta^c_m)$$

is the range of Θ^c_m, and C_m is some complementary space of $\mathcal{R}_m$ into $H^l_m(\mathbb{R}^l)$. The range of the operators Θ^c_m, defined in the spaces $H^l_m(\mathbb{R}^l)$, contains exactly the terms that can be taken away from the equation in the computation of the normal form.

Let $\sigma(\mathcal{A}) = \{\lambda_1, \cdots, \lambda_l\}$. The spectrum of the linear operator Θ^c_m defined in (5.40) is the following set:

$$\sigma\left(\Theta^c_m\right) = \left\{\lambda \cdot d - \lambda_j : j = 1, \cdots, l, \ \sum_{j=1}^{l} d_j = m\right\}. \tag{5.42}$$

Let $\mathcal{A}$ be a Jordan canonical matrix and the decomposition (5.41) of $H^l_m(\mathbb{R}^l)$ be given for $m = 2, \ldots, k$. Then the appropriate transformations $x = \xi^c_m(\overline{x})$, where $\overline{x} \in B^m(0, r)$, $B^m(0, r)$ is a neighborhood of the origin and $B^{m+1}(0, r) \sqsubseteq B^m(0, r), m = 2, \cdots, k$, can be chosen so that the system (5.37) is transformed into

$$x' = \mathcal{A}x + g_2(x) + g_3(x) + \cdots + g_k(x) + O(|x|^{k+1}), x \in B^k(0, r), \quad (5.43)$$

where $g_m(x) \in C_m$, $m = 2, \cdots, k$. System (5.43) is said to be in normal form through order k for reduced differential equations on the center manifold of system (5.1). The terms $g_m(x), m = 2, \cdots, k$ in (5.43) are called resonance terms.

Remark 5.19 Since the change of coordinates (5.38) are local diffeomorphisms near the origin, the normal form theory presented in the above gives only local results of (5.37) near the origin. Furthermore the normal form theory above is about the normal form up to certain finite order k. If $F(u)$ is analytic in u, then it is clear that we can formally transform (5.37) to a normal form with $k = \infty$, that is,

$$x' = \mathcal{A}x + \sum_{m\geq 2} g_m(x).$$

However, the issue of convergence of the power series for the normal forms and the associated transformations of variables should be considered. The answer for the issue of convergence is positive under some conditions for ordinary differential equations, and we refer to Chow et al. [54]. In fact in the applications of normal forms, usually the flow is completely determined by the terms of a normal form up to a certain finite order, for example, we only need compute the normal form up to third order for Hopf bifurcations.

For this compution of normal form for (5.37), we refer to Vanderbauwhede [338], Arnold [11], Chow and Hale [53], Guckenheimer and Holmes [128], Meyer and Hall [253], Siegel and Moser [309], Chow et al. [54], Kuznetsov [201], Bibikov [30], Wiggins [352], and others for more details.

It is obvious that the simplified system (5.43) is strongly depending on the specific choice of the complementary spaces C_m and then the normal form of (5.37) is not unique. The key to compute the normal form of the vector field is to find C_m. One method to find the complementary subspace C_m or normal form is to use the matrix representation $\mathbf{M}_m$ of the linear operator Θ^c_m with respect to the given basis of $H^l_m(\mathbb{R}^l)$

$$\Phi_m = \left\{ x_1^{n_1} x_2^{n_2} \ldots x_l^{n_l} e_j, j = 1, \cdots, l, \sum_{j=1}^{l} n_j = m, n_j \geq 0 \text{ are integers} \right\},$$

where $e_1, \cdots, e_l$ is the standard basis of $\mathbb{R}^l$. The dimension of $H^l_m(\mathbb{R}^l)$ is

$$d_m := \dim H^l_m(\mathbb{R}^l) = \frac{l\,(l + m - 1)!}{m!\,(l - 1)!}.$$

We write the basis element of $H^l_m(\mathbb{R}^l)$ with respect to the reverse lexicographic ordering

$$\Phi_m = \left\{\varphi_1, \dots, \varphi_{d_m}\right\}.$$

Then $\mathbf{M}_m$ is a $d_m \times d_m$ matrix and $\mathbb{R}^{d_m} = \widetilde{C}_m \oplus \widetilde{\mathcal{R}}_m$, where $\widetilde{\mathcal{R}}_m$ is the range of $\mathbf{M}_m$ in $\mathbb{R}^{d_m}$ and $\widetilde{C}_m$ any complementary subspace. Our choice of $\widetilde{C}_m$ is certainly not unique. According to Fredholm Theorem, $\widetilde{C}_m = N((\mathbf{M}_m)^T)$ is a complementary subspace, where $N\left((\mathbf{M}_m)^T\right)$ is the null-space of the transpose of $\mathbf{M}_m$. It is obvious that

$$H^l_m(\mathbb{R}^l) = C_m \oplus \mathcal{R}_m$$

where

$$C_m := \left\{\kappa = \sum_{i=1}^{d_m} \xi_i\varphi_i \in H^l_m(\mathbb{R}^l) \middle| \varphi_i \in \Phi_m,\ (\xi_1, \dots, \xi_{d_m}) \in \widetilde{C}_m\right\},$$

$$\mathcal{R}_m := \left\{\kappa = \sum_{i=1}^{d_m} \xi_i\varphi_i \in H^l_m(\mathbb{R}^l) \middle| \varphi_i \in \Phi_m,\ (\xi_1, \dots, \xi_{d_m}) \in \widetilde{\mathcal{R}}_m\right\},$$

$\mathcal{R}_m$ is the range of Θ^c_m, and C_m is a complementary space of $\mathcal{R}_m$ into $H^l_m(\mathbb{R}^l)$. We refer to [54], [338] and [347] for more details and also for other methods to get the complementary subspace C_m or normal form. Here we will show an example to expain this method.

Example 5.20 (The Takens-Bogdanov Normal Form) We will compute the normal form up to terms of order 2 for the following vector field on $\mathbb{R}^2$ in the neighborhood of the origin

$$x'(t) = Ax(t) + \begin{pmatrix} c_1x_1^2 + c_2x_1x_2 + c_3x_2^2 \\ d_1x_1^2 + d_2x_1x_2 + d_3x_2^2 \end{pmatrix} + O(|x|^3),$$

where $A = \begin{pmatrix} 0 & 1 \\ 0 & 0 \end{pmatrix}$. Let $\{e_1, e_2\}$ be the standard basis of $\mathbb{R}^2$. The basis Φ_2 of $H^2_2(\mathbb{R}^2)$ with the elements in the reverse lexicographic ordering is

$$\Phi_2 = \left(\begin{pmatrix} 0 \\ x_1^2 \end{pmatrix}, \begin{pmatrix} 0 \\ x_1x_2 \end{pmatrix}, \begin{pmatrix} 0 \\ x_2^2 \end{pmatrix}, \begin{pmatrix} x_1^2 \\ 0 \end{pmatrix}, \begin{pmatrix} x_1x_2 \\ 0 \end{pmatrix}, \begin{pmatrix} x_2^2 \\ 0 \end{pmatrix}\right).$$

We first obtain the matrix representation $\mathbf{M}_2$ of the linear operator Θ_2 with respect to the given basis Φ_2:

$$\mathbf{M}_2 = \begin{pmatrix} 0 & 0 & 0 & 0 & 0 & 0 \\ 2 & 0 & 0 & 0 & 0 & 0 \\ 0 & 1 & 0 & 0 & 0 & 0 \\ -1 & 0 & 0 & 0 & 0 & 0 \\ 0 & -1 & 0 & 2 & 0 & 0 \\ 0 & 0 & -1 & 0 & 1 & 0 \end{pmatrix}.$$

Then $\mathbb{R}^6 = \widetilde{C}_2 \oplus \widetilde{\mathcal{R}}_2$, where $\widetilde{\mathcal{R}}_2$ is the range of $\mathbf{M}_2$ in $\mathbb{R}^6$ and $\widetilde{C}_2$ any complementary subspace. Let $(e_1, e_2, \dots, e_6)$ be the standard basis of $\mathbb{R}^6$ and then $N((\mathbf{M}_2)^T) =$

$\{e_1, e_2 + 2e_4\}$ is a complementary subspace of $\widetilde{\mathcal{R}}_2$. Therefore $H_2^2(\mathbb{R}^2) = C_2 \oplus \mathcal{R}_2$, where

$$C_2 = \text{span}\left\{\begin{pmatrix} 0 \\ x_1^2 \end{pmatrix}, \begin{pmatrix} 2x_1^2 \\ x_1 x_2 \end{pmatrix}\right\}.$$

Thus we obtain the normal form up to terms of order 2

$$x'(t) = Ax(t) + d_1 \begin{pmatrix} 0 \\ x_1^2 \end{pmatrix} + \frac{2c_1 + d_2}{5} \begin{pmatrix} 2x_1^2 \\ x_1 x_2 \end{pmatrix} - O(|x|^3).$$

5.3.1.1 $G_m \in V^m(X_c, \mathbb{R}^n)$

By using this change of variables, we can obtain the normal form directly through order k for reduced differential equations on the center manifold of system (5.1), without computing beforehand the reduced system on the center manifold of system (5.1). Now we apply Proposition 5.15 recursively to (5.30). Set

$$u_1 := u.$$

For $m = 2, \ldots, k$, let $G_m \in V^m(X_c, \mathbb{R}^n)$ be defined such that

$$[A, G_m](x_c) = \mathcal{P}_m \left[\frac{1}{m!} D^m F_{m-1}(0)(x_c, \ldots, x_c)\right] \text{ for each } x_c \in X_c.$$

We use the change of variables

$$u_{m-1} = u_m + G_m(\Pi_c u_m).$$

Then we consider F_m given by Proposition 5.15 and satisfying

$$F_m(u_m) = F_{m-1}(u_m) - [A, G_m](\Pi_c u_m) + O(\|u_m\|^{m+1}).$$

By applying Proposition 5.15, we have

$$\Pi_h D^j F_m(0) \mid_{X_c \times X_c \times \ldots \times X_c} = 0, \text{ for all } j = 1, \ldots, m,$$

and

$$\mathcal{G}\left(\Pi_c D^j F_m(0) \mid_{X_c \times X_c \times \ldots \times X_c}\right) \in C_j^c, \text{ for all } j = 1, \ldots, m.$$

Thus by using the change of variables locally around 0

$$u_k(t) = (I + G_k \Pi_c)^{-1} \ldots (I + G_3 \Pi_c)^{-1} (I + G_2 \Pi_c)^{-1} u(t),$$

we deduce that $u_k(t)$ is a solution of (5.32). By using the center manifold theorem in 4 and Lemma 5.2 to (5.32), we obtain the following result which is the main result of this Chapter and indicates that systems (5.32) and (5.30) are locally topologically equivalent around 0.

Theorem 5.21 *Let Assumptions 5.1 and 5.16 be satisfied. Then by using the change of variables locally around* 0

$$\begin{cases} u_k(t) = (I + G_k\Pi_c)^{-1} \dots (I + G_3\Pi_c)^{-1} (I + G_2\Pi_c)^{-1} u(t) \\ \Leftrightarrow \\ u(t) = (I + G_2\Pi_c)(I + G_3\Pi_c) \dots (I + G_k\Pi_c) u_k(t), \end{cases}$$

the map $t \to u(t)$ is a solution of the Cauchy problem (5.30) if and only if $t \to u_k(t)$ is a solution of the Cauchy problem (5.32). Moreover, the reduced equation in normal form of Cauchy problem (5.32) is given by the ordinary differential equations on X_c :

$$\frac{dx_c(t)}{dt} = A_c x_c(t) + \sum_{m=2}^{k} \frac{1}{m!} \Pi_c D^m F_k(0)(x_c(t), \dots, x_c(t)) + R_c(x_c(t)),$$

where

$$\mathcal{G}\left(\frac{1}{m!}\Pi_c D^m F_k(0) \mid_{X_c \times X_c \times \dots \times X_c}\right) \in C_m^c, \text{ for all } m = 1, \dots, k,$$

and the remainder term $R_c \in C^k(X_c, X_c)$ satisfies

$$D^j R_c(0) = 0 \text{ for each } j = 1, \dots, k,$$

or in other words $R_c(x_c(t))$ is a remainder term of order k. If we assume in addition that $F \in C^{k+2}(\mathbb{R}^n, \mathbb{R}^n)$. Then the reduced equation of Cauchy problem (5.32) is given by the ordinary differential equations on X_c :

$$\frac{dx_c(t)}{dt} = A_c x_c(t) + \sum_{m=2}^{k+1} \frac{1}{m!} \Pi_c D^m F_k(0)(x_c(t), \dots, x_c(t)) + R_c(x_c(t)),$$

the map $R_c \in C^{k+2}(X_c, X_c)$, and $R_c(x_c(t))$ is a remainder term of order $k+2$, that is

$$R_c(x_c) = \|x_c\|^{k+2} O(x_c),$$

where $O(x_c)$ is a function of x_c which remains bounded when x_c goes to 0, or equivalently,

$$D^j R_c(0) = 0 \text{ for each } j = 1, \dots, k+1.$$

Proof By the center manifold theorem in 4 and Lemma 5.2 to (5.32), there exists $\Psi_k \in C^k(X_c, X_h)$ such that the reduced system of (5.32) is given by

$$\frac{dx_c(t)}{dt} = A_c x_c(t) + \Pi_c F_k[x_c(t) + \Psi_k(x_c(t))]$$

and

$$D^j \Psi_k(0) = 0 \text{ for } j = 1, \dots, k.$$

By setting

$$R_c(x_c) = \Pi_c F_k[x_c + \Psi_k(x_c)] - \Pi_c F_k(x_c),$$

we obtain the first part of the Theorem. If we assume in addition that $F \in C^{k+2}(\mathbb{R}^n, \mathbb{R}^n)$, then $\Psi_k \in C^{k+2}(X_c, X_h)$. Thus, $R_c \in C^{k+2}(X_c, X_c)$ and

$$\begin{aligned}R_c(x_c) &= \Pi_c\{F_k[x_c+\Psi_k(x_c)]-F_k(x_c)\}\\ &= \Pi_c\int_0^1 DF_k(x_c+s\Psi_k(x_c))(\Psi_k(x_c))\,ds.\end{aligned}$$

Set

$$h(x_c) := x_c + s\Psi_k(x_c).$$

Since $DF(0) = 0$, we have

$$\begin{aligned}DF_k(h(x_c))(\Psi_k(x_c)) &= DF_k(0)(\Psi_k(x_c))\\ &\quad+\int_0^1 D^2F_k(lh(x_c))(h(x_c),\Psi_k(x_c))\,dl\\ &= \int_0^1 D^2F_k(lh(x_c))(h(x_c),\Psi_k(x_c))\,dl.\end{aligned}$$

Hence,

$$R_c(x_c) = \Pi_c\int_0^1\int_0^1 D^2F_k(l(x_c+s\Psi_k(x_c)))(x_c+s\Psi_k(x_c),\Psi_k(x_c))\,dlds$$

and $\Psi_k(x_c)$ is a term of order $k+1$, it follows that

$$R_c(x_c) = \|x_c\|^{k+2}O(x_c).$$

The result follows. □

5.4 Remarks and notes

A normal form theorem was obtained first by Poincaré [274] and later by Siegel [308] for analytic differential equations. Simpler proofs of Poincaré's theorem and Siegel's theorem were given in Arnold [11], Meyer [254], Moser [259], and Zehnder [363]. Normal form theory has been extended to various classes of differential equations. In the context of functional differential equations we refer to Faria [108, 109]. In the context of autonomous partial differential equations we refer to Ashwin and Mei [13] (PDEs on the square), Eckmann et al. [99] (abstract parabolic equations), Faou et al. [103, 104] (Hamiltonian PDEs), Hassard, Kazarinoff and Wan [151] (Functional Differential Equations), Faria [105, 106] (PDEs with delay), Foias et al. [114] (Navier-Stokes equation), Kokubu [192] (reaction-diffusion equations), McKean and Shatah [250] (Schrödinger equation and heat equations), Nikolenko [267] (abstract semi-linear equations), Shatah [302] (Klein-Gordon equation), Zehnder [364] (abstract parabolic equations), Chow et al. [59] (and references therein) for a normal form theory in quasi periodic partial differential equations.

5.4.1 Normal form theory for an abstract non-densely defined Cauchy problem on Banach space

In Liu et al. [223], a normal form theory for the following abstract Cauchy problem with non-dense domain is developed

$$u'(t) = Au(t) + F(u(t)), \text{ for } t \geq 0, \text{ and } u(0) = x \in \overline{D(A)},$$

where $A : D(A) \subset X \to X$ is a non-densely defined linear operator on a Banach space X and $F : \overline{D(A)} \to X$ is a k-time continuously differentiable function for some $k \geq 2$.

The main motivation comes from investigating the nonlinear dynamics of structured (age, size, space, etc.) population models described by various types of equations including partial differential equations with nonlinear and nonlocal boundary conditions (Cantrell and Cosner [42], Diekmann and Heesterbeek [84], Iannelli [182], Murray [265], Perthame [272], Thieme [331], Webb [349], etc.).

We would like to mention that since the normal form theory and the computation procedure developed in Liu et al. [223] is for general semilinear Cauchy problems, the theory and technique of computing the reduced system and normal form could be applied to other types of equations such as transport equations (Perthame [272]), reaction–diffusion equations (Kokubu [192], Eckmann et al. [99]), and partial differential equations with delay (Faria [105, 106]).

5.4.2 Bifurcations for age structured models

To determine the qualitative behavior of a nonlinear system in the neighborhood of a non-hyperbolic equilibrium point, center manifold reduction and the normal form theory, one reduces the dimension of the original system and the other eliminates the non-linearity of the reduced system, are conjunctly used to study bifurcations in nonlinear dynamical systems. In chapter 7, we apply the normal form theory developed in this chapter to one example of predator prey model, namely, we compute the Taylor expansion and normal form of the reduced system of the predator prey model on the center manifold, from which we will be able to determine the direction of the Hopf bifurcation and stability of the bifurcating periodic solutions in this model.

Liu et al. [220] study an example of Bogdanov-Takens normal form computation for a predator prey system with an age structure for the predator. The system considered is the following

$$\begin{cases} \dfrac{dU(t)}{dt} = rU(t)\left(1 - \kappa^{-1}U(t)\right) - \dfrac{A(t)U(t)}{\alpha + U(t)^2}, \\ \partial_t v(t,a) + \partial_a v(t,a) = -Dv(t,a), \text{ for } a \geq 0, \\ v(t,0) = \mu \dfrac{A(t)U(t)}{\alpha + U(t)^2}, \end{cases} \tag{5.44}$$

for $t \geq 0$ and with initial distribution

$$U(0) = U_0 \geq 0 \text{ and } v(0,a) = v_0(a) \in L^1_+(0,\infty).$$

In the model (5.44), $U(t)$ is the number of prey at time t, $a \to v(t,a)$ is the population density of predator structured with respect to the chronological age a of the predator. That is to say that

$$\int_{a_1}^{a_2} v(t,a)da$$

is the number of predator with age between a_1 and a_2. We assume that the predators become adults (i.e. can reproduce) when they reach the age $\tau > 0$ and we set

$$A(t) = \int_\tau^\infty v(t,a)da,$$

which is the number of adult predators.

Consider the new time variable $\widehat{t} = \dfrac{t}{\tau}$ and the new age variable $\widehat{a} = \dfrac{a}{\tau}$ and set

$$\widehat{U}\left(\widehat{t}\right) = U\left(\tau\widehat{t}\right) \text{ and } \widehat{v}\left(\widehat{t},\widehat{a}\right) = \tau v\left(\tau\widehat{t},\tau\widehat{a}\right).$$

Then

$$\widehat{A}(\widehat{t}) := \int_1^\infty \widehat{v}\left(\widehat{t},\widehat{a}\right) d\widehat{a} = \int_1^\infty \tau v\left(\tau\widehat{t},\tau\widehat{a}\right) d\widehat{a} = \int_\tau^\infty v\left(\tau\widehat{t},a\right) da = A(\tau\widehat{t}).$$

By using this rescaling (and dropping the hat notation simplicity) we obtain the following system for $t \geq 0$

$$\begin{cases} \dfrac{dU(t)}{dt} = \tau\left[rU(t)\left(1-\kappa^{-1}U(t)\right) - \dfrac{A(t)U(t)}{\alpha+U(t)^2}\right], \\ \partial_t v(t,a) + \partial_a v(t,a) = -\tau\, D\, v(t,a), \text{ for } a \geq 0, \\ v(t,0) = \tau\mu\dfrac{A(t)U(t)}{\alpha+U(t)^2}, \end{cases} \tag{5.45}$$

with

$$A(t) = \int_1^\infty v(t,a)da.$$

Now we will consider the rescaled system (5.45) instead of (5.44).

Consider $X = \mathbb{R}^2 \times L^1(0,\infty)$ endowed with the usual product norm

$$\left\|\begin{pmatrix} U \\ \gamma \\ v \end{pmatrix}\right\| = |U| + |\gamma| + \|v\|_{L^1}.$$

Consider the closed subspace of X

$$X_0 = \mathbb{R} \times \{0_{\mathbb{R}}\} \times \mathrm{L}^1(0,\infty) = \left\{ \begin{pmatrix} U \\ 0_{\mathbb{R}} \\ v \end{pmatrix} \in \mathrm{X} \right\}.$$

Let $L : D(L) \subset X \to X$ be the linear operator defined by

$$L \begin{pmatrix} U \\ 0_{\mathbb{R}} \\ v \end{pmatrix} = \begin{pmatrix} -\chi U \\ -v(0) \\ -v' - \tau D v \end{pmatrix}$$

and

$$D(L) := \mathbb{R} \times \{0_{\mathbb{R}}\} \times \mathrm{W}^{1,1}(0,\infty).$$

Then we can observe that $X_0 = \overline{D(L)}$ and L is non densely defined. Let $F : X_0 \to X$ be the nonlinear operator defined by

$$F \begin{pmatrix} U \\ 0_{\mathbb{R}} \\ v \end{pmatrix} = \begin{pmatrix} B_1(U,A) \\ B_2(U,A) \\ 0_{\mathrm{L}^1} \end{pmatrix}$$

where

$$A := \int_1^{+\infty} v(a)da, \; B_1(U,A) := \chi U + \tau \left[r\left(1 - \kappa^{-1} U\right) - \frac{A}{\alpha + U^2} \right] U,$$
$$B_2(U,A) := \tau\mu \frac{AU}{\alpha + U^2}.$$

Problem (5.45) can be rewritten as the following non-densely defined abstract Cauchy problem

$$\frac{dw(t)}{dt} = Lw(t) + F(w(t)) \text{ for } t \geq 0, \; w(0) = w_0 \in X_0.$$

Based on the normal form theory by Liu et al. [223] and the center manifold theory by Magal and Ruan [239] for semilinear equations with non-dense domain combined with integrated semigroup theory, the existence result of the Bogdanov-Takens singularity is studied. Furthermore, it is more interesting to determine a versal unfolding for the original system (5.45) with a Bogdanov–Takens singularity, i.e., to determine which of the parameters can be chosen as bifurcation parameters such that (5.45) exhibits Bogdanov–Takens bifurcation. For this, choose κ and D in (5.45) as the bifurcation parameters, i.e., consider $1/\kappa + \beta_1$ and $D + \beta_2$, where $\beta := (\beta_1, \beta_2)$ vary in a sufficiently small neighborhood of $(0,0)$. Adding these perturbations to system (5.45), we obtain

$$\frac{dw(t)}{dt} = Lw(t) + F(\beta, w(t)), t \geq 0, w(0) \in X_0, \tag{5.46}$$

which is corresponding to

$$
\begin{cases}
\dfrac{dU(t)}{dt} = \tau\left[rU(t)\left(1-\left(\kappa^{-1}+\beta_1\right)U(t)\right) - \dfrac{A(t)U(t)}{\alpha+U(t)^2}\right], \\
\partial_t v(t,a) + \partial_a v(t,a) = -\tau\left(D+\beta_2\right)v(t,a), \text{ for } a \geq 0, \\
v(t,0) = \tau\mu\dfrac{A(t)U(t)}{\alpha+U(t)^2},
\end{cases}
\tag{5.47}
$$

where L is defined by

$$
L\begin{pmatrix} U \\ 0 \\ v \end{pmatrix} = \begin{pmatrix} -\chi U \\ -v(0) \\ -v' - \tau D v \end{pmatrix}
$$

and $F : \overline{D(L)} \to X$ is defined by

$$
F(\beta, w) = \begin{pmatrix} B_1^*(\beta, w) \\ B_2^*(\beta, w) \\ -\tau\beta_2 v \end{pmatrix}
$$

for $w = \begin{pmatrix} U \\ 0 \\ v \end{pmatrix} \in \overline{D(L)}$, here

$$
\begin{aligned}
B_1^*(\beta, w) &= \chi U + \tau r U\left[1 - \left(1/\kappa + \beta_1\right)U\right] - \frac{\tau A U}{\alpha + U^2}, \\
B_2^*(\beta, w) &= \frac{\tau\mu A U}{\alpha + U^2}
\end{aligned}
$$

with $A := \int_1^{+\infty} v(a)da$. The bifurcation behavior of the parameterized predator prey model (5.47) with Bogdanov-Takens singularity could be studied by using the normal form theory [223] and the center manifold theory [239]. We refer to [220] for more details.

Based on the center manifold theorem proved in Magal and Ruan [239], a Hopf bifurcation theorem has been presented for abstract non- densely defined Cauchy problem in [221]. These theorems have been successfully applied to study the existence of Hopf bifurcation for some age/size-structured (see [219, 224, 225, 321]) and furthermore, by applying the normal form theory [223], the direction and stability of these Hopf bifurcations are also studied (see [60, 223, 240]).

Part II
Applications to Predator Prey Systems

Chapter 6
A Holling's predator-prey model with handling and searching predators

This chapter is based on the paper of Hsu, Liu, and Magal [171] and provides an example of a monotone ordinary differential equation in the context of a predator-prey system. The system is three-dimensional, and since the monotone property only reduces by one dimension of systems, Hopf bifurcation may occur.

6.1 Introduction

This chapter is devoted to the following predator-prey system with handling and searching predators

$$\begin{cases} N' = \underbrace{\beta_N N - \mu_N N - \delta N^2}_{\textbf{Logistic growth}} - \underbrace{N\kappa P_S}_{\substack{\textbf{Consumption}\\ \textbf{of prey}}} \\ P'_S = -\underbrace{(\mu_P + \eta) P_S}_{\textbf{Mortality}} - \underbrace{\rho N \kappa P_S}_{\substack{\textbf{Searching}\\ \textbf{becoming}\\ \textbf{handling}}} + \gamma P_H \\ P'_H = \underbrace{-\mu_P P_H}_{\textbf{Mortality}} + \rho N \kappa P_S - \underbrace{\gamma P_H}_{\substack{\textbf{Handling}\\ \textbf{becoming}\\ \textbf{seaching}}} + \underbrace{\beta_P (P_S + P_H)}_{\textbf{New born predator}}, \end{cases} \tag{6.1}$$

where $N(t)$ denotes the number of prey at time t, $P_S(t)$ is the number of predators searching for prey at time t, and $P_H(t)$ is the number of predators handling a prey at time t.

Here the terminology "handling and searching predators" refers to the work of Holling [161]. We should also mention that Metz and Diekmann developed similar

A. Ducrot et al., *Differential Equations and Population Dynamics II*,
Lecture Notes on Mathematical Modelling in the Life Sciences,
https://doi.org/10.1007/978-3-031-91409-6_6

ideas of searching and handling predator in their edited bock [252, pages 6-7]. In (6.1), the term $\beta_P\,(P_S(t) + P_H(t))$ corresponds to the flow of new born predators. Here we assume that all the new born predators are handlers. The parameter ρ should be interpreted as a conversion rate. The term $P_S(t)\,\rho\,\kappa\,N(t)$ (in the P_S-equation and in the P_H-equation) is a flow of searching predators becoming handling predators. The term $\gamma P_H(t)$ (in the P_S-equation and in the P_H-equation) is the flow of handling predators becoming searching predators. The term μ_P is the natural mortality of the predators and η is an extra mortality term for the searching predators only. The term $N(t)\,\kappa\,P_S(t)$ in the N-equation corresponds to the consumption of the prey by the predators. The part $\beta_N N(t) - \mu_N N(t) - \delta N(t)^2$ in the N-equation is the standard logistic equation.

The main idea about this model is to distinguish the vital dynamics (birth and death process) of the predators and their survival due to the consumption of prey. In the model the survival of predators will depend on the status searching or handling. The handling predators are satisfied with their consumption of prey and they don't need to find more prey to survive. At the opposite the searching predators are unsatisfied with their consumption of prey and they need to find some prey to survive. Once a searching predator finds a prey (or enough prey) it becomes a handling and after some time the handling predator becomes a searching predator again.

This process only influences the survival of predators which depends on their ability to find a prey. In our model, a predator can reproduce at time t because he found enough prey to survive from its birth until the time t. In section 6.2 we will first make some basic assumptions in order for the predators to extinct in absence of prey. Then based on these setting we will analyze the dynamical properties of the system (6.1). The main advantage with the model (6.1) is that we can separate the vital dynamic and the consumption of prey to describe the behavior of the predators. This will be especially convenient if we want to add an age or size structure to the predator population. This kind of question is left for future work.

In section 6.6 we will see that our model is also comparable to the standard predator-prey model whenever $\rho = \dfrac{\chi}{\varepsilon}$ and $\gamma = \dfrac{1}{\varepsilon}$ for $\varepsilon > 0$ small which means that predators are going back and forth from handling to searching very rapidly. In that case (as a singular limit) we obtain a convergence result to the standard Rosenweig-MacArthur model [285]

$$
\begin{cases}
N'(t) = \underbrace{r\,N(t)\left(1 - \dfrac{N(t)}{K}\right)}_{\textbf{The vital dynamic of the prey}} - \underbrace{P(t)\left(\dfrac{mN(t)}{a + N(t)}\right)}_{\textbf{Consumption of prey}}, \\[2ex]
P'(t) = \underbrace{-d\,P(t)}_{\substack{\textbf{The vital dynamic of the predator}\\ \textbf{in absence of prey}}} + \underbrace{c\,P(t)\left(\dfrac{mN(t)}{a + N(t)}\right)}_{\textbf{Converted new born predators}},
\end{cases}
\tag{6.2}
$$

which is one of the most popular predator-prey system discussed in the literature.

This model was proposed first in the context of biochemistry by Michaelis and Menten [255] in 1913 to describe enzyme kinetics. The model was also introduced by the French Nobel prize Jacques Monod [261] in 1950 to describe the consumption of nutrients by bacteria in a chemostat. This model also corresponds to predator-prey system introduced Holling [161] in 1959. This is the so-called Holling type II functional response. Rosenweig-MacArthur came last [285] in 1963.

In the model, the parameter $c > 0$ is the conversion (rate) of prey consumed in newborn predators, and the parameter a measures the "satiation" effect of the predator population.

We refer the reader to Holling [161] and Rosenzweig and MacArthur [285]. Let us recall that the derivation of Holling type II functional response

$$\frac{mN}{a+N},$$

can be found in Holling [160, 161] and Hsu, Hubbell and Waltman [170].

The origin of Liapunov's functional goes back to Volterra itself. In his article [344] in 1928, he already considered the functional

$$V(t) = \delta N(t) - \gamma \ln(N(t)) + \beta P(t) - \alpha \ln(P(t)),$$

which can be constant in time for a solution of the Lotka-Volterra [229, 343] system

$$\begin{cases} N'(t) = \alpha N(t) - \beta N(t) P(t), \\ P'(t) = -\gamma P(t) + \delta N(t) P(t), \end{cases} \tag{6.3}$$

where $\alpha > 0, \beta > 0, \gamma > 0$, and $\delta > 0$.

Indeed, we have (assuming formally that we are not dividing by 0)

$$\frac{N'}{P'} = \frac{N(\alpha - \beta P)}{P(-\gamma + \delta N)},$$

and we obtain

$$\delta N' - \gamma \frac{N'}{N} = \alpha \frac{P'}{P} - \beta P'$$

and the result follows.

In the 70th, Hsu (see [167] in 1978) was one of the pioneers providing a Lyapunov function for the Rosenweig-MacArthur model. Other classes of systems was also explored by Goh [123, 124, 125] in 1976-1978 and Harrison [150] in 1979 also studied Liapunov function for predator-prey systems. One can note that epidemic models can also be viewed as a special case of predator-prey systems with a conversion rate equal to 1, and such Liapunov functions have also been employed to understand the global dynamics of epidemic models.

There are several mathematical problems related to (6.2), namely, the global asymptotic stability of the locally asymptotically stable interior equilibrium (when it exists) and the uniqueness of the limit cycles when the interior equilibrium is unstable. For the global asymptotic stability of this equilibrium we may apply the Dulac's criterion (Hsu, Hubbell and Waltman [168]), weak negative Bendixson Lemma (Cheng, Hsu and Lin [48]) or the construction of suitable Lyapunov functions (see for instance Ardito and Ricciardi [7]).

As far as the uniqueness of limit cycle of Rosenzweig-MacArthur model (6.2) is concerned, Cheng [47] employed the symmetry of the prey isocline to prove the exponential asymptotic stability of each limit cycle. Kuang and Freedman [199] observed that (6.2) is nothing but a generalized Lienard equation which has a unique of limit cycle. Such a uniqueness result was proved for generalized Lienard equation by Zhang in [365]. The stability of periodic orbits was proved by Hsu, Hubbell and Waltman [168]. We refer to Murray [263], Hastings [152], Turchin [333] for more results about predator-prey models.

The plan of the chapter is the following. In section 6.2 we set some basic assumptions in order for the predators to extinct in absence of prey. In section 6.3 we prove that the system is dissipative. In section 6.4 we study the uniform persistence and extinction properties of the predators. We study the system in the interior region which corresponds to the region of co-existence of prey and predators in section 6.5. We should mention that we can obtain a rather complete description of the asymptotic behavior thanks to the fact the system is competitive (for a new partial order). In section 6.6 we prove the convergence of our model to the Rosenweig-MacArthur model. In section 6.7 we apply the model to the Canadian snowshoe Hares and the Lynx.

6.2 Basic assumptions

In this section, we set some basic assumptions in order for the predators to go to extinction in the absence of prey. Hence consider the total number of predators

$$P = P_H + P_S.$$

Then

$$P' = (\beta_P - \mu_P - \eta)\, P_S + (\beta_P - \mu_P)\, P_H.$$

The following assumptions mean that when $\frac{P_S}{P_H} > -\frac{\beta_P - \mu_P}{\beta_P - \mu_P - \eta}$, the total population of predators decreases. The total population of predators increases otherwise.

Assumption 6.1 We assume that all the parameters of the model (6.1) are strictly positive and

$$\beta_N - \mu_N > 0, \beta_P - \mu_P > 0 \text{ and } \beta_P - \mu_P - \eta < 0.$$

Note also that the first condition in the above assumption means that the growth rate for the prey is positive, while in absence of the prey, the dynamics of predator population is described by

$$\begin{cases} P'_S = -(\mu_P + \eta)P_S + \gamma P_H \\ P'_H = \beta_P P_S + (\beta_P - \mu_P - \gamma)\, P_H. \end{cases}$$

Define the matrix

$$M = \begin{bmatrix} -\mu_P - \eta & \gamma \\ \beta_P & \beta_P - \mu_P - \gamma \end{bmatrix}. \tag{6.4}$$

By using Assumption 6.1 we have

$$\mathrm{tr}\,(M) = (\beta_P - \mu_P - \gamma) - (\mu_P + \eta) < 0.$$

Therefore in absence of prey the population of predators goes to extinction if and only if

$$\det (M) = -\,(\mu_P + \eta)\,(\beta_P - \mu_P - \gamma) - \beta_P \gamma > 0.$$

This last inequality can be equivalently reformulated in the following assumption.

Assumption 6.2 (Extinction of the predators) We assume that

$$(\beta_P - \mu_P - \gamma) < -\frac{\beta_P \gamma}{\mu_P + \eta} \Leftrightarrow (\beta_P - \mu_P) < -\frac{\gamma}{\mu_P + \eta}\,(\beta_P - \mu_P - \eta)\,. \tag{6.5}$$

Remark 6.3 The first inequality in (6.5) implies that $(\beta_P - \mu_P - \gamma) < 0$. Moreover the second inequality in (6.5) and $(\beta_P - \mu_P) > 0$ imply that $(\beta_P - \mu_P - \eta) < 0$.

Lemma 6.4 *Let Assumptions 6.1 and 6.2 be satisfied. Then in absence of the prey, the population of predators goes to extinction.*

6.3 Dissipativity

In this section, we will prove that system (6.1) is dissipative. We look for a positive left eigenvector $(\widetilde{P}_S, \widetilde{P}_H) \in (0, +\infty)^2$ and an eigenvalue $\lambda > 0$ of matrix M, that is

$$(\widetilde{P}_S, \widetilde{P}_H) \begin{bmatrix} -\mu_P - \eta & \gamma \\ \beta_P & \beta_P - \mu_P - \gamma \end{bmatrix} = -\lambda (\widetilde{P}_S, \widetilde{P}_H),$$

which is equivalent to the following system

$$\begin{cases} -\,(\mu_P + \eta)\,\widetilde{P}_S + \beta_P \widetilde{P}_H = -\lambda \widetilde{P}_S \\ \gamma \widetilde{P}_S + (\beta_P - \mu_P - \gamma)\,\widetilde{P}_H = -\lambda \widetilde{P}_H \end{cases} \Leftrightarrow \begin{cases} \beta_P \widetilde{P}_H = [(\mu_P + \eta) - \lambda]\,\widetilde{P}_S \\ \gamma \widetilde{P}_S = [-\,(\beta_P - \mu_P - \gamma) - \lambda]\,\widetilde{P}_H. \end{cases}$$

Thus the sign $\widetilde{P}_S$ and $\widetilde{P}_H$ are the same if we impose

$$\lambda \in (0, \min\left((\mu_P + \eta)\,, -\,(\beta_P - \mu_P - \gamma)\right))$$

and λ must satisfy the following equation

$$1 = \frac{[(\mu_P + \eta) - \lambda]}{\beta_P}\,\frac{[-\,(\beta_P - \mu_P - \gamma) - \lambda]}{\gamma} =: \Psi\,(\lambda)\,.$$

The function $\lambda \to \Psi(\lambda)$ decreases between 0 and $\min\left(\left(\mu_P+\eta\right), -\left(\beta_P-\mu_P-\gamma\right)\right)$ and by using (6.5) we have $\Psi(0) > 1$. It follows that there exists a unique $\lambda^* \in (0, \min\left(\left(\mu_P+\eta\right), -\left(\beta_P-\mu_P-\gamma\right)\right))$ such that

$$1 = \frac{\left[\left(\mu_P+\eta\right)-\lambda^*\right]}{\beta_P} \frac{\left[-\left(\beta_P-\mu_P-\gamma\right)-\lambda^*\right]}{\gamma}. \tag{6.6}$$

Note that

$$\frac{\left[-\left(\beta_P-\mu_P-\gamma\right)-\lambda^*\right]}{\gamma} < 1 \Leftrightarrow -\left(\beta_P-\mu_P\right) < \lambda^*.$$

By assumption $\left(\beta_P-\mu_P\right) > 0$ it follows from (6.6) that

$$\frac{\left[\left(\mu_P+\eta\right)-\lambda^*\right]}{\beta_P} > 1.$$

Since

$$\gamma \widetilde{P}_S = \left[-\left(\beta_P-\mu_P-\gamma\right)-\lambda^*\right] \widetilde{P}_H,$$

it follows that

$$\widetilde{P}_H > \widetilde{P}_S > 0. \tag{6.7}$$

By using the P_S-equation and the P_H-equation of system (6.1) we obtain

$$\widetilde{P}_S P_S' + \widetilde{P}_H P_H' = -\lambda^* \left[\widetilde{P}_S P_S + \widetilde{P}_H P_H\right] - \left(\widetilde{P}_S - \widetilde{P}_H\right) P_S \rho \kappa N. \tag{6.8}$$

By using the N-equation and the comparison principle, one can find some constant $N^* = \max\left(N(0), \left(\beta_N-\mu_N\right)/\delta\right)$ such that

$$N(t) \le N^*, \forall t \ge 0,$$

where $N(0)$ denotes the initial value. Then it follows that

$$\begin{aligned}\rho\left(\widetilde{P}_H-\widetilde{P}_S\right) N' + \widetilde{P}_S P_S' + \widetilde{P}_H P_H' \le -\rho\left(\widetilde{P}_H-\widetilde{P}_S\right) \mu_N N - \lambda^* \left[\widetilde{P}_S P_S + \widetilde{P}_H P_H\right] \\ + \rho\left(\widetilde{P}_H-\widetilde{P}_S\right) \beta_N N^*\end{aligned}$$

and the dissipativity follows.

Set

$$M = \frac{\rho\left(\widetilde{P}_H-\widetilde{P}_S\right) \beta_N N^*}{\min\left(\mu_N, \lambda^*\right)} > 0.$$

As a consequence of the last inequality, we obtain the following results.

Proposition 6.5 *Let Assumptions 6.1 and 6.2 be satisfied. Then system* (6.1) *generates a unique continuous semiflow* $\{U(t)\}_{t\ge 0}$ *on* $[0,\infty)^3$. *Moreover the domain*

$$D = \left\{(N, P_S, P_H) \in [0,\infty)^3 : \rho\left(\widetilde{P}_H-\widetilde{P}_S\right) N + \widetilde{P}_S P_S + \widetilde{P}_H P_H \le M\right\}$$

is positively invariant with respect to the semiflow U, that is to say that

$$U(t)D \subset D, \forall t \geq 0.$$

Furthermore D attracts every point of $[0,\infty)^3$ *for U, that is*

$$\lim_{t\to\infty} \delta(U(t)x, D) = 0, \forall x \in [0,\infty)^3,$$

where $\delta(x, D) := \inf_{y\in D} \|x-y\|$ *is the Hausdorff's semi-distance. As a consequence the semiflow of U has a compact global attractor* $\mathcal{A} \subset [0,\infty)^3$.

6.4 Uniform persistence and extinction of the predator

In this section, we study the uniform persistence and extinction of the predators. Firstly we consider the existence of the equilibrium. An equilibrium point $\left(\overline{N}, \overline{P}_S, \overline{P}_H\right) \in [0,\infty)^3$ satisfies the following system of equation

$$\begin{cases} 0 = \overline{N}\left[\beta_N - \mu_N - \delta\overline{N} - \kappa\overline{P}_S\right], \\ 0 = -(\mu_P + \eta)\overline{P}_S - \overline{P}_S\,\rho\,\kappa\,\overline{N} + \gamma\overline{P}_H, \\ 0 = -\mu_P\overline{P}_H + \overline{P}_S\,\rho\,\kappa\,\overline{N} - \gamma\overline{P}_H + \beta_P\left(\overline{P}_S + \overline{P}_H\right). \end{cases}$$

Due to Assumptions 6.1 and 6.2, we deduce that the only equilibrium satisfying $\overline{N} = 0$ is $E_1 = (0,0,0)$. If we assume next that $\overline{N} > 0$, we obtain the system

$$\begin{cases} 0 = \beta_N - \mu_N - \delta\overline{N} - \kappa\overline{P}_S, \\ 0 = -(\mu_P + \eta)\overline{P}_S - \overline{P}_S\,\rho\,\kappa\,\overline{N} + \gamma\overline{P}_H, \\ 0 = -\mu_P\overline{P}_H + \overline{P}_S\,\rho\,\kappa\,\overline{N} - \gamma\overline{P}_H + \beta_P\left(\overline{P}_S + \overline{P}_H\right). \end{cases}$$

From the first equation we have

$$\overline{N} = \widehat{N} - \frac{\kappa}{\delta}\overline{P}_S$$

with $\widehat{N} > 0$ given by

$$\widehat{N} = \frac{\beta_N - \mu_N}{\delta}.$$

By adding-up the last two equations, we end-up with

$$\overline{P}_H = \frac{(\mu_P + \eta - \beta_P)}{(\beta_P - \mu_P)}\overline{P}_S.$$

Combining the two above equations with

$$-(\mu_P + \eta)\overline{P}_S - \overline{P}_S\,\rho\,\kappa\,\overline{N} + \gamma\overline{P}_H = 0,$$

yields

$$\left(-(\mu_P+\eta)-\frac{(\beta_N-\mu_N)\rho\kappa}{\delta}+\gamma\frac{(\mu_P+\eta-\beta_P)}{(\beta_P-\mu_P)}\right)\overline{P}_S+\frac{\kappa^2\rho}{\delta}\overline{P}_S^2=0$$

and then

$$\overline{P}_S=0 \text{ or } \overline{P}_S=\left((\mu_P+\eta)-\frac{\gamma(\mu_P+\eta-\beta_P)}{(\beta_P-\mu_P)}\right)\frac{\delta}{\kappa^2\rho}+\frac{(\beta_N-\mu_N)}{\kappa}.$$

Thus we get the following lemma.

Lemma 6.6 *Let Assumptions 6.1 and 6.2 be satisfied. System* (6.1) *always has the following two boundary equilibria*

$$E_1=(0,0,0),\quad E_2=\left(\widehat{N},0,0\right).$$

Moreover there exists a unique interior equilibrium $E^*=\left(N^*,P_S^*,P_H^*\right)$ *if and only if*

$$(\beta_N-\mu_N)(\beta_P-\mu_P)\kappa\rho+\delta\,(\beta_P-\mu_P)(\mu_P+\eta)>-\delta\gamma(\beta_P-\mu_P-\eta).\quad (6.9)$$

Furthermore, $\left(N^*,P_S^*,P_H^*\right)$ *are given by*

$$\begin{aligned}
N^* &= \frac{-(\beta_P-\mu_P)(\mu_P+\eta)-\gamma(\beta_P-\mu_P-\eta)}{(\beta_P-\mu_P)\kappa\rho}>0,\\
P_S^* &= \frac{\delta\,(\beta_P-\mu_P)(\mu_P+\eta)+\delta\gamma(\beta_P-\mu_P-\eta)+(\beta_N-\mu_N)(\beta_P-\mu_P)\kappa\rho}{(\beta_P-\mu_P)\kappa^2\rho}>0,\\
P_H^* &= -\frac{(\beta_P-\mu_P-\eta)}{\beta_P-\mu_P}P_S^*>0.
\end{aligned}$$

6.4.1 Stability of the equilibrium E_1

The Jacobian matrix at the equilibrium E_1 reads as

$$\begin{bmatrix}\beta_N-\mu_N & 0 & 0\\ 0 & -\mu_P-\eta & \gamma\\ 0 & \beta_P & \beta_P-\mu_P-\gamma\end{bmatrix}$$

and the characteristic equation is

$$[(\lambda+\mu_P+\eta)(\lambda-(\beta_P-\mu_P-\gamma))-\beta_P\gamma]\,[\lambda-(\beta_N-\mu_N)]=0.$$

So one of the eigenvalues is $\lambda_{1,E_1}=\beta_N-\mu_N>0$. Thus we can get that the equilibrium E_1 is unstable. The rest of the spectrum coincides with the spectrum of the matrix M defined in (6.4). Thus we obtain the following lemma.

Lemma 6.7 *Let Assumptions 6.1 and 6.2 be satisfied. The equilibrium* E_1 *is hyperbolic and the unstable manifold is one dimensional.*

6.4.2 Stability of the equilibrium E_2

The Jacobian matrix at the equilibrium E_2 is given by

$$\begin{bmatrix} -\left(\beta_N-\mu_N\right) & -\kappa\widehat{N} & 0 \\ 0 & -\left((\mu_P+\eta)+\rho\kappa\widehat{N}\right) & \gamma \\ 0 & \rho\,\kappa\,\widehat{N}+\beta_P & \beta_P-\mu_P-\gamma \end{bmatrix}$$

and the characteristic equation reads as

$$\left[\left(\lambda+\mu_P+\eta+\rho\kappa\widehat{N}\right)(\lambda-(\beta_P-\mu_P-\gamma))-\gamma\left(\rho\kappa\widehat{N}+\beta_P\right)\right][\lambda+(\beta_N-\mu_N)]=0.$$

So one of the eigenvalues is $\lambda_{1,E_2}=-\left(\beta_N-\mu_N\right)<0$ and the remaining part of the characteristic equation is

$$\lambda^2+a\lambda+b=0$$

with

$$a=\left(\mu_P+\eta+\rho\,\kappa\,\widehat{N}\right)-(\beta_P-\mu_P-\gamma)$$

and

$$b=(\mu_P+\gamma-\beta_P)\left(\mu_P+\eta+\rho\,\kappa\,\widehat{N}\right)-\gamma\left(\rho\,\kappa\,\widehat{N}+\beta_P\right).$$

By using Assumptions 6.1 and 6.2 we have $a>0$. Moreover by using the Routh-Hurwitz criterion E_2 is stable if and only if $b>0$ which corresponds to the following condition

$$\begin{aligned} &(\mu_P+\gamma-\beta_P)\left(\mu_P+\eta+\rho\kappa\widehat{N}\right)-\gamma\left(\rho\,\kappa\,\widehat{N}+\beta_P\right)>0 \\ &\Leftrightarrow (\beta_N-\mu_N)(\beta_P-\mu_P)\,\kappa\,\rho+\delta\,(\beta_P-\mu_P)(\mu_P+\eta)<-\delta\gamma(\beta_P-\mu_P-\eta) \end{aligned}$$

Now we obtain the following result.

Lemma 6.8 *Let Assumptions 6.1 and 6.2 be satisfied. E_2 is unstable if the interior equilibrium exits (i.e. the condition* (6.9) *is satisfied) and the unstable manifold is one dimensional and the stable manifold is two dimensional.*

6.4.3 Extinction of the predator and the global stability of E_2

We decompose the positive cone $M=\mathbb{R}_+^3$ into the interior region

$$\overset{\circ}{M}=\{(N,P_S,P_H)\in M:N>0 \text{ and } P_S+P_H>0\},$$

the boundary region with predators only

$$\partial M_P:=\{(N,P_S,P_H)\in M:N=0\}, \tag{6.10}$$

and the boundary region with prey only

$$\partial M_N := \{(N, P_S, P_H) \in M : P_S + P_H = 0\}. \tag{6.11}$$

Each sub domain $\overset{\circ}{M}$, ∂M_P and ∂M_N is positively invariant by the semiflow generated by (6.1).

Theorem 6.9 *Let Assumptions 6.1 and 6.2 be satisfied. Assume that E_2 is locally asymptotically stable (i.e. $\left(\mu_P + \eta + \rho\kappa\widehat{N}\right)(\mu_P + \gamma - \beta_P) > \gamma\left(\beta_P + \rho\kappa\widehat{N}\right)$). Then the predator goes to extinction. More precisely for each initial value in $M = (N(0), P_S(0), P_H(0)) \in [0, \infty)^3$, we have*

$$\lim_{t\to\infty} P_S(t) + P_H(t) = 0$$

and

$$\lim_{t\to\infty} N(t) = \begin{cases} \widehat{N}, & \text{if } N(0) > 0, \\ 0, & \text{if } N(0) = 0. \end{cases}$$

Proof The boundary region with the predator only, ∂M_P, is positively invariant by the semiflow generated by (6.1) and by Assumption 6.2 any solution starting from ∂M_P exponentially converges to E_1.

So it remains to investigate the limit of a solution starting from $\overset{\circ}{M} \cup \partial M_N \backslash \{E_1\}$. We consider the Lyapunov function

$$V(N, P_S, P_H) = \int_{\widehat{N}}^{N} \frac{\xi - \widehat{N}}{\xi} d\xi + c_1 P_S + c_2 P_H \tag{6.12}$$

where $c_1 > 0$ and $c_2 > 0$ have to be determined. We have

$$\begin{aligned}
\dot{V} = &\left(N - \widehat{N}\right)(\beta_N - \mu_N - \delta N - \kappa P_S) \\
&+ c_1\left(-(\mu_P + \eta)P_S - \rho\kappa P_S N + \gamma P_H\right) \\
&+ c_2\left(-\mu_P P_H + \rho\kappa P_S N - \gamma P_H + \beta_P\left(P_S + P_H\right)\right) \\
= &\left(N - \widehat{N}\right)\left(-\delta\left(N - \widehat{N}\right) - \kappa P_S\right) \\
&+ c_1\left(-(\mu_P + \eta)P_S - \rho\kappa P_S\left(N - \widehat{N}\right) - \rho\kappa P_S\widehat{N} + \gamma P_H\right) \\
&+ c_2\left(-\mu_P P_H + \rho\kappa P_S\left(N - \widehat{N}\right) + \rho\kappa P_S\widehat{N} - \gamma P_H + \beta_P\left(P_S + P_H\right)\right).
\end{aligned}$$

Thus we obtain

$$\begin{aligned}
\dot{V} = &-\delta\left(N - \widehat{N}\right)^2 + \kappa P_S\left(N - \widehat{N}\right)(-1 - c_1\rho + c_2\rho) \\
&+ P_S\left(-c_1(\mu_P + \eta) - c_1\rho\kappa\widehat{N} + c_2\rho\kappa\widehat{N} + c_2\beta_P\right) \\
&+ P_H\left(-c_2\mu_P + c_1\gamma - c_2\gamma + c_2\beta_P\right).
\end{aligned}$$

We claim that we can choose $c_1 > 0$ and $c_2 > 0$ such that $c_2 = c_1 + \dfrac{1}{\rho}$ and the following inequalities are satisfied

$$-c_1(\mu_P+\eta)-c_1\rho\kappa\widehat{N}+c_2\rho\kappa\widehat{N}+c_2\beta_P < 0 \text{ and } -c_2\mu_P+c_1\gamma-c_2\gamma+c_2\beta_P < 0. \quad (6.13)$$

In fact the inequalities in (6.13) lead us to consider the lines

$$c_2 = c_1 \frac{\gamma}{\mu_P + \gamma - \beta_P} \tag{L_1}$$

and

$$c_2 = c_1 \frac{(\mu_P + \eta) + \rho\kappa\widehat{N}}{\beta_P + \rho\kappa\widehat{N}}. \tag{L_2}$$

By Assumption 6.2 (see Remark 6.3) we have $\mu_P + \gamma - \beta_P > 0$ and by Assumption 6.1 we have $\mu_P + \eta > \beta_P$, so that we get

$$\frac{(\mu_P + \eta) + \rho\kappa\widehat{N}}{\beta_P + \rho\kappa\widehat{N}} > 1.$$

Note that

$$\begin{aligned}
&\frac{(\mu_P + \eta) + \rho\kappa\widehat{N}}{\beta_P + \rho\kappa\widehat{N}} > \frac{\gamma}{\mu_P + \gamma - \beta_P} \\
&\Leftrightarrow \left(\mu_P + \eta + \rho\kappa\widehat{N}\right)(\mu_P + \gamma - \beta_P) > \gamma\left(\beta_P + \rho\kappa\widehat{N}\right)
\end{aligned}$$

and thus we obtain that the slope of (L_2) is greater than the slope of (L_1). Finally we have

$$\lim_{N\to 0^+} V(N, P_S, P_N) = \lim_{N\to 0^+} (N - \widehat{N}) - \widehat{N}\ln\left(\frac{N}{\widehat{N}}\right) + c_1 P_S + c_2 P_N = +\infty.$$

By LaSalle's invariance principle we obtain that E_2 is globally asymptotically stable for the system restricted to $\overset{\circ}{M} \cup \partial M_N \backslash \{E_1\}$. □

6.4.4 Uniform persistence of the predator

We decompose the positive cone into

$$\mathbb{R}^3_+ = \partial M \cup \overset{\circ}{M}$$

where the boundary region is defined as

$$\partial M := \partial M_P \cup \partial M_N.$$

It is clear that both regions $\overset{\circ}{M}$ and ∂M are positively invariant by the semiflow generated by the system. Moreover we have the following result.

Theorem 6.10 *Let Assumptions 6.1 and 6.2 be satisfied. If the interior equilibrium exits then the predator uniformly persists with respect to the domain decomposition*

$\left(\partial M, \overset{\circ}{M}\right)$. *That is to say that there exists* $\nu > 0$ *such that for each initial value* $N(0) > 0$ *and* $P_S(0) + P_H(0) > 0$ *the corresponding solution satisfies*

$$\liminf_{t\to\infty} N(t) > \nu \text{ and } \liminf_{t\to\infty} (P_S(t) + P_H(t)) > \nu.$$

Proof The equilibrium $E_1 = (0,0,0)$ is clearly chained to $E_2 = (\widehat{N},0,0)$. By using Theorem 4.1 in [145], we only need to prove that

$$W^s(E_i) \cap \overset{\circ}{M} = \varnothing,$$

where $i = 1, 2$ and

$$W^s(E_i) = \{(N, P_S, P_H) \in M : \omega((N, P_S, P_H)) \neq \varnothing \text{ and } \omega((N, P_S, P_H)) \subset E_i\},$$

$$W^s(E_i) = \{(N, P_S, P_H) \in M : \omega((N, P_S, P_H)) = \{E_i\}\}, \tag{6.14}$$

where ω means the ω-limit set.
To prove these properties, assume by contradiction that there exists $E^0 = \left(N^0, P_S^0, P_H^0\right) \in \overset{\circ}{M}$ (which means $N^0 > 0$ and $P_S^0 + P_H^0 > 0$) such that $\omega(E^0) = \{E_1\}$. Then for any $\varepsilon > 0$, there exists $t_0 \geq 0$, such that

$$N(t) + P_S(t) + P_H(t) \leq \varepsilon, \forall t \geq t_0$$

where $(N(t), P_S(t), P_H(t)) = U(t)E^0$. By using the first equation of model (6.1)

$$N' = \beta_N N - \mu_N N - \delta N^2 - N\kappa P_S,$$

we have

$$N' \geq N\left(\beta_N - \mu_N - \delta\varepsilon - \kappa\varepsilon\right).$$

Therefore for $\varepsilon > 0$ small enough, we have $\beta_N - \mu_N - \delta\varepsilon - \kappa\varepsilon > 0$ and then

$$\lim_{t\to\infty} N(t) = \infty$$

which is in contradiction to the dissipativity of the model.
Assume now by contradiction that there exists $E^0 = \left(N^0, P_S^0, P_H^0\right) \in \overset{\circ}{M}$ such that $\omega(E^0) \subset E_2$. Then for any $\varepsilon > 0$, there exists $t_0 \geq 0$, such that

$$\left|N(t) - \widehat{N}\right| + P_S(t) + P_H(t) \leq \varepsilon, \forall t \geq t_0$$

where $(N(t), P_S(t), P_H(t)) = U(t)E^0$. By using the last two equation of system (6.1), we obtain

$$\begin{aligned} P_S' &\geq -(\mu_P + \eta)P_S - P_S\rho\kappa\left(\widehat{N} + \varepsilon\right) + \gamma P_H \\ P_H' &\geq -\mu_P P_H + P_S\rho\kappa\left(\widehat{N} - \varepsilon\right) - \gamma P_H + \beta_P\left(P_S + P_H\right). \end{aligned} \tag{6.15}$$

By using the fact that for $\varepsilon > 0$ small enough the right hand side of (6.15) is a cooperative system together with Lemma 6.8 we deduce that

$$\lim_{t\to\infty} (P_S(t) + P_H(t)) = \infty.$$

This contradicts the dissipativity of the system and the uniform persistence follows.□

As a consequence of the dissipativity as well as the uniform persistence (see Magal and Zhao [244]) we deduce the following result.

Theorem 6.11 *Let Assumptions 6.1 and 6.2 be satisfied. Assume in addition that the interior equilibrium exits. Then the system* (6.1) *has a global attractor* A_0 *in the interior region* $\overset{\circ}{M}$. *Namely* A_0 *is a compact invariant set by the semiflow generated by* (6.1) *on* $\overset{\circ}{M}$ *and* A_0 *is locally stable and attracts the compact subsets of* $\overset{\circ}{M}$.

6.5 Interior region

In this section, we will study the system in the interior region which corresponds to the region of co-existence of the prey and the predator.

6.5.1 Local stability of E^*

The Jacobian matrix at the equilibrium E^* is

$$\begin{bmatrix} (\beta_N - \mu_N - 2\delta N^* - \kappa P_S^*) & -N^*\kappa & 0 \\ -P_S^*\rho\kappa & -(\mu_P + \eta + \rho\kappa N^*) & \gamma \\ P_S^*\rho\kappa & \rho\kappa N^* + \beta_P & \beta_P - \mu_P - \gamma \end{bmatrix}$$

and the characteristic equation reads as

$$\lambda^3 + p_1\lambda^2 + p_2\lambda + p_3 = 0$$

with

$$\begin{aligned} p_1 &= -(\beta_N - \mu_N - 2\delta N^* - \kappa P_S^*) + (\mu_P + \eta + \rho\kappa N^*) - (\beta_P - \mu_P - \gamma), \\ p_2 &= -(\mu_P + \eta + \rho\kappa N^*)(\beta_N - \mu_N - 2\delta N^* - \kappa P_S^*) - N^*\kappa P_S^*\rho\kappa \\ &\quad + (\beta_N - \mu_N - 2\delta N^* - \kappa P_S^*)(\beta_P - \mu_P - \gamma) \\ &\quad - (\mu_P + \eta + \rho\kappa N^*)(\beta_P - \mu_P - \gamma) - (\rho\kappa N^* + \beta_P)\gamma, \\ p_3 &= (\beta_N - \mu_N - 2\delta N^* - \kappa P_S^*)(\mu_P + \eta + \rho\kappa N^*)(\beta_P - \mu_P - \gamma) \\ &\quad + N^*\kappa\gamma P_S^*\rho\kappa + \gamma(\beta_N - \mu_N - 2\delta N^* - \kappa P_S^*)(\rho\kappa N^* + \beta_P) + \\ &\quad P_S^*\rho\kappa N^*\kappa(\beta_P - \mu_P - \gamma). \end{aligned}$$

By using the Routh-Hurwitz criterion, we get that the equilibrium E^* is stable if and only if

$$p_1 > 0, p_1 p_2 - p_3 > 0 \text{ and } p_3 > 0.$$

After some algebra we have

$$\begin{aligned}
p_1 &= \frac{1}{\kappa\rho(\beta_P - \mu_P)}\Big[-\kappa\rho(\beta_P - \mu_P - \gamma)(\beta_P - \mu_P) \\
&\quad - \gamma(\delta + \kappa\rho)(\beta_P - \mu_P - \eta) - \delta(\beta_P - \mu_P)(\mu_P + \eta)\Big], \\
p_2 &= \frac{(\beta_P - \mu_P - \gamma)(\mu_P + \eta) + \gamma\beta_P}{\kappa\rho(\beta_P - \mu_P)^2}\Big[(\beta_P - \mu_P)[\delta(\beta_P + \eta + \gamma) + \kappa\rho(\beta_N - \mu_N)] \\
&\quad - 2\delta\gamma\eta\Big], \\
p_3 &= \frac{(\beta_P - \mu_P - \gamma)(\mu_P + \eta) + \gamma\beta_P}{\kappa\rho(\beta_P - \mu_P)}\Big\{ -\delta(\beta_P - \mu_P - \gamma)(\mu_P + \eta) - \gamma\delta\beta_P \\
&\quad - \kappa\rho(\beta_P - \mu_P)(\beta_N - \mu_N)\Big\}.
\end{aligned}$$

Assumption 6.12 Assume that $\gamma\eta - (\beta_P - \mu_P)^2 > 0$.

From Assumptions 6.1 and 6.2, we have

$$\frac{(\beta_P - \mu_P - \gamma)(\mu_P + \eta) + \gamma\beta_P}{\kappa\rho(\beta_P - \mu_P)} < 0.$$

Then if the inequality (6.9) is satisfied, we have $p_3 > 0$. Furthermore, let $\alpha := \beta_N - \mu_N$ and we have

$$\begin{aligned}
& p_1 p_2 - p_3 > 0 \\
&\Leftrightarrow [-\kappa\rho(\beta_P - \mu_P - \gamma)(\beta_P - \mu_P) - \gamma(\delta + \kappa\rho)(\beta_P - \mu_P - \eta) - \delta(\beta_P - \mu_P)(\mu_P + \eta)] \\
&\cdot[(\beta_P - \mu_P - \gamma)(\mu_P + \eta) + \gamma\beta_P]\{(\beta_P - \mu_P)[\delta(\beta_P + \eta + \gamma) + \kappa\rho\alpha] - 2\delta\gamma\eta\} \\
&-\kappa\rho(\beta_P - \mu_P)^2[(\beta_P - \mu_P - \gamma)(\mu_P + \eta) + \gamma\beta_P] \\
&\cdot\{-\delta(\beta_P - \mu_P - \gamma)(\mu_P + \eta) - \gamma\delta\beta_P - \alpha\kappa\rho(\beta_P - \mu_P)\} > 0 \\
&\Leftrightarrow \alpha < \tfrac{A}{B},
\end{aligned}$$

where

$$\begin{aligned}
A &:= \kappa\rho(\beta_P - \mu_P)^2[-\delta(\beta_P - \mu_P - \gamma)(\mu_P + \eta) - \gamma\delta\beta_P][(\beta_P - \mu_P - \gamma)(\mu_P + \eta) + \gamma\beta_P] \\
&\quad - [\delta(\beta_P - \mu_P)(\beta_P + \eta + \gamma) - 2\delta\gamma\eta][(\beta_P - \mu_P - \gamma)(\mu_P + \eta) + \gamma\beta_P] \\
&\quad \cdot[-\kappa\rho(\beta_P - \mu_P - \gamma)(\beta_P - \mu_P) - \gamma(\delta + \kappa\rho)(\beta_P - \mu_P - \eta) - \delta(\beta_P - \mu_P)(\mu_P + \eta)], \\
B &:= \kappa\rho(\beta_P - \mu_P)[(\beta_P - \mu_P - \gamma)(\mu_P + \eta) + \gamma\beta_P] \\
&\quad \cdot[-\kappa\rho(\beta_P - \mu_P - \gamma)(\beta_P - \mu_P) - \gamma(\delta + \kappa\rho)(\beta_P - \mu_P - \eta) - \delta(\beta_P - \mu_P)(\mu_P + \eta)] \\
&\quad + \kappa\rho(\beta_P - \mu_P)^2\kappa\rho(\beta_P - \mu_P)[(\beta_P - \mu_P - \gamma)(\mu_P + \eta) + \gamma\beta_P].
\end{aligned}$$

Note that

$$
\begin{aligned}
&-\kappa\rho(\beta_P-\mu_P-\gamma)(\beta_P-\mu_P)-\gamma(\delta+\kappa\rho)(\beta_P-\mu_P-\eta)-\delta(\beta_P-\mu_P)(\mu_P+\eta)\\
&=-\kappa\rho(\beta_P-\mu_P)^2+\kappa\rho\gamma(\beta_P-\mu_P)-\gamma\delta(\beta_P-\mu_P-\eta)-\gamma\kappa\rho(\beta_P-\mu_P)+\gamma\kappa\rho\eta\\
&\quad-\delta(\beta_P-\mu_P)(\mu_P+\eta)\\
&=-\kappa\rho(\beta_P-\mu_P)^2-\delta\left[\gamma(\beta_P-\mu_P-\eta)+(\beta_P-\mu_P)(\mu_P+\eta)\right]+\gamma\kappa\rho\eta\\
&=-\kappa\rho(\beta_P-\mu_P)^2-\delta[\gamma\beta_P-\gamma\mu_P-\gamma\eta+(\beta_P-\mu_P)(\mu_P+\eta)\\
&\quad-\gamma(\mu_P+\eta)+\gamma\mu_P+\gamma\eta]+\gamma\kappa\rho\eta\\
&=\kappa\rho[\gamma\eta-(\beta_P-\mu_P)^2]-\delta\left[(\beta_P-\mu_P-\gamma)(\mu_P+\eta)+\gamma\beta_P\right]
\end{aligned}
$$

and

$$
\begin{aligned}
&(\beta_P-\mu_P)(\beta_P+\eta+\gamma)-2\gamma\eta\\
&=(\beta_P-\mu_P)\beta_P+(\beta_P-\mu_P)\eta+(\beta_P-\mu_P)\gamma-\gamma\eta-\gamma\eta\\
&=(\beta_P-\mu_P)\beta_P+(\beta_P-\mu_P)\eta+(\beta_P-\mu_P)\mu_P-(\beta_P-\mu_P)\mu_P+\beta_P\gamma-\mu_P\gamma\\
&\quad-\gamma\eta-\gamma\eta\\
&=(\beta_P-\mu_P)\beta_P+(\beta_P-\mu_P)(\eta+\mu_P)-(\beta_P-\mu_P)\mu_P+\beta_P\gamma-\gamma(\eta+\mu_P)-\gamma\eta\\
&=(\beta_P-\mu_P)^2+(\beta_P-\mu_P-\gamma)(\eta+\mu_P)+\beta_P\gamma-\gamma\eta\\
&=(\beta_P-\mu_P-\gamma)(\eta+\mu_P)+\beta_P\gamma-[\gamma\eta-(\beta_P-\mu_P)^2].
\end{aligned}
$$

Then

$$
\begin{aligned}
A=&-\delta\kappa\rho(\beta_P-\mu_P)^2\left[(\beta_P-\mu_P-\gamma)(\mu_P+\eta)+\gamma\beta_P\right]^2\\
&-\delta\left[(\beta_P-\mu_P)(\beta_P+\eta+\gamma)-2\gamma\eta\right]\left[(\beta_P-\mu_P-\gamma)(\mu_P+\eta)+\gamma\beta_P\right]\\
&\cdot\left\{\kappa\rho[\gamma\eta-(\beta_P-\mu_P)^2]-\delta\left[(\beta_P-\mu_P-\gamma)(\mu_P+\eta)+\gamma\beta_P\right]\right\},
\end{aligned}
$$

$$
\begin{aligned}
B=&\,\kappa\rho(\beta_P-\mu_P)[(\beta_P-\mu_P-\gamma)(\mu_P+\eta)+\gamma\beta_P]\\
&\cdot\left[-\kappa\rho(\beta_P-\mu_P)^2-\delta\left[(\beta_P-\mu_P-\gamma)(\mu_P+\eta)+\gamma\beta_P\right]+\gamma\kappa\rho\eta\right]\\
&+\kappa\rho(\beta_P-\mu_P)^2\kappa\rho(\beta_P-\mu_P)[(\beta_P-\mu_P-\gamma)(\mu_P+\eta)+\gamma\beta_P]\\
=&\,\kappa\rho(\beta_P-\mu_P)[(\beta_P-\mu_P-\gamma)(\mu_P+\eta)+\gamma\beta_P]\\
&\cdot\left\{-\kappa\rho(\beta_P-\mu_P)^2-\delta\left[(\beta_P-\mu_P-\gamma)(\mu_P+\eta)+\gamma\beta_P\right]+\gamma\kappa\rho\eta+\kappa\rho(\beta_P-\mu_P)^2\right\}\\
=&\,\kappa\rho(\beta_P-\mu_P)[(\beta_P-\mu_P-\gamma)(\mu_P+\eta)+\gamma\beta_P]\\
&\cdot\left\{-\delta\left[(\beta_P-\mu_P-\gamma)(\mu_P+\eta)+\gamma\beta_P\right]+\gamma\kappa\rho\eta\right\},
\end{aligned}
$$

and

$$
\begin{aligned}
p_1&=\frac{-\kappa\rho(\beta_P-\mu_P-\gamma)(\beta_P-\mu_P)-\gamma(\delta+\kappa\rho)(\beta_P-\mu_P-\eta)-\delta(\beta_P-\mu_P)(\mu_P+\eta)}{\kappa\rho(\beta_P-\mu_P)}\\
&=\frac{\kappa\rho[\gamma\eta-(\beta_P-\mu_P)^2]-\delta[(\beta_P-\mu_P-\gamma)(\mu_P+\eta)+\gamma\beta_P]}{\kappa\rho(\beta_P-\mu_P)}.
\end{aligned}
$$

From Assumptions 6.1, 6.2 and 6.12, we have $A<0$, $B<0$ and $p_1>0$. Set $\alpha^\star=\frac{A}{B}$. Thus

$$
\begin{aligned}
\alpha^\star=&\frac{-\delta(\beta_P-\mu_P)[(\beta_P-\mu_P-\gamma)(\mu_P+\eta)+\gamma\beta_P-(\beta_P-\mu_P)(\beta_P+\eta+\gamma)+2\gamma\eta]}{-\delta[(\beta_P-\mu_P-\gamma)(\mu_P+\eta)+\gamma\beta_P]+\gamma\kappa\rho\eta}\\
&+\frac{-\delta[(\beta_P-\mu_P)(\beta_P+\eta+\gamma)-2\gamma\eta]}{\kappa\rho(\beta_P-\mu_P)}>0.
\end{aligned}
$$

Since

$$
\begin{aligned}
&(\beta_P-\mu_P-\gamma)(\mu_P+\eta)+\gamma\beta_P-(\beta_P-\mu_P)(\beta_P+\eta+\gamma)+2\gamma\eta\\
&=(\beta_P-\mu_P)\mu_P+(\beta_P-\mu_P)\eta-\gamma\mu_P-\gamma\eta+\gamma\beta_P-(\beta_P-\mu_P)\beta_P\\
&\quad-(\beta_P-\mu_P)(\eta+\gamma)+2\gamma\eta\\
&=-(\beta_P-\mu_P)^2+(\beta_P-\mu_P)\eta+(\beta_P-\mu_P)\gamma+\gamma\eta-(\beta_P-\mu_P)(\eta+\gamma)\\
&=-(\beta_P-\mu_P)^2+\gamma\eta.
\end{aligned}
$$

Then

$$\alpha^\star = \frac{\delta(\beta_P-\mu_P)\left[(\beta_P-\mu_P)^2-\gamma\eta\right]}{\gamma\kappa\rho\eta-\delta[(\beta_P-\mu_P-\gamma)(\mu_P+\eta)+\gamma\beta_P]} + \frac{\delta[2\gamma\eta-(\beta_P-\mu_P)(\beta_P+\eta+\gamma)]}{\kappa\rho(\beta_P-\mu_P)} > 0.$$

Thus we have the following result.

Lemma 6.13 *Let Assumptions 6.1, 6.2, 6.12 and inequality* (6.9) *be satisfied. The equilibrium E^* is stable if and only if $\alpha < \alpha^\star$.*

6.5.2 Three dimensional $\mathbb{K}$-competitive system

In this section we use a Poincaré-Bendixson theorem for three dimensional $\mathbb{K}$-competitive system. First recall that, for an equilibrium point p, the set $W^s(p)$ is the set of points whose ω-limit set is equal to $\{p\}$, see (6.14).

Theorem 6.14 *([313, Theorem 4.2 p. 43]) Let the autonomous system of ordinary differential equations $x' = f(x)$ be a competitive system, where f is continuously differentiable on an open subset $D \subset \mathbb{R}^3$ and suppose that D contains a unique equilibrium point p which is hyperbolic. Suppose further that $W^s(p)$ is one-dimensional and tangent at p to a vector $v \gg 0$. If $q \in D \setminus W^s(p)$ and $\gamma^+(q)$ has compact closure in D then $\omega(q)$ is a nontrival periodic orbit.*

By applying this theorem to the system (6.1) restricted to the interior global attractor A_0 we obtain the following result.

Theorem 6.15 *Suppose that $E^* = \left(N^*, P_S^*, P_H^*\right)$ exists and is hyperbolic and unstable for* (6.1). *Then the stable manifold $W^s(E^*)$ of E^* is one dimensional and the omega limit set $\omega(N(0), P_S(0), P_H(0))$ is a nontrivial periodic orbit in $\mathbb{R}^3_+$ for every $(N(0), P_S(0), P_H(0)) \in \mathbb{R}^3_+ \setminus W^s(E^*)$.*

Proof The Jacobian matrix of the vector field (6.1) at the point $(N, P_S, P_H) \in (0, \infty)^3$ is given by

$$J = \begin{pmatrix} (\beta_N - \mu_N) - 2\delta N - \kappa P_S & -\kappa N & 0 \\ -\rho\kappa P_S & -(\mu_P + \eta) - \rho\kappa N & \gamma \\ \rho\kappa P_S & \rho\kappa N + \beta_P & -\mu_P + \beta_P - \gamma \end{pmatrix}. \tag{6.16}$$

The off-diagonal entries of J are sign-stable and sign symmetric in $\mathbb{R}^3_+$.

Let

$$\mathbb{K} = \left\{(N, P_S, P_H) \in \mathbb{R}^3 : N \geq 0, P_S \geq 0, P_H \leq 0\right\}.$$

The system is $\mathbb{K}$-competitive, since the matrix of the time-reversed linearized system $-J$ is cooperative with respect to the cone $\mathbb{K}$. □

6.6 Convergence to the Rosenzweig-MacArthur model

The time scale for the life expectancy (as well as the time scale needed for the reproduction) is the year, while the time needed for the lynx to handle the rabbit is measured by days (no more than one week). Therefore there is a huge difference between the time scales for the vital dynamic and the consumption dynamic.

The consumption of prey by the predator is a fast process compared to the vital dynamic which is slow. In the model γ^{-1} is the average time spent by the predators to handle prey. γ^{-1} should be very small in comparison with the other parameters. Then it makes sense to make the following assumption.

Assumption 6.16 Assume that

$$\rho = \frac{\chi}{\varepsilon} \text{ and } \gamma = \frac{1}{\varepsilon}$$

with $\varepsilon \ll 1$ is a small parameter.

Under the above assumption, system (6.1) becomes

$$\begin{cases} \dot{N^\varepsilon} = (\beta_N - \mu_N) N^\varepsilon - \delta (N^\varepsilon)^2 - \kappa N^\varepsilon P_S^\varepsilon \\ \dot{P_S^\varepsilon} = -(\mu_P + \eta) P_S^\varepsilon - \frac{\chi}{\varepsilon} \kappa N^\varepsilon P_S^\varepsilon + \frac{1}{\varepsilon} P_H^\varepsilon \\ \dot{P_H^\varepsilon} = -\mu_P P_H^\varepsilon \qquad + \frac{\chi}{\varepsilon} \kappa N^\varepsilon P_S^\varepsilon - \frac{1}{\varepsilon} P_H^\varepsilon + \beta_P \left(P_S^\varepsilon + P_H^\varepsilon \right) \end{cases} \tag{6.17}$$

and we fix the initial value

$$N^\varepsilon(0) = N_0 \geq 0, \; P_S^\varepsilon(0) = P_{S0} \geq 0 \text{ and } P_H^\varepsilon(0) = P_{H0} \geq 0.$$

The first equation of (6.17) reads as

$$\dot{N^\varepsilon} = (\beta_N - \mu_N) N^\varepsilon - \delta (N^\varepsilon)^2 - \kappa N^\varepsilon P_S^\varepsilon. \tag{6.18}$$

Hence

$$\dot{N^\varepsilon} \leq (\beta_N - \mu_N) N^\varepsilon. \tag{6.19}$$

By adding-up the two last equations of (6.17) we obtain

$$\dot{P^\varepsilon} = (\beta_P - \mu_P) P^\varepsilon - \eta P_S^\varepsilon \tag{6.20}$$

and $P_S^\varepsilon \geq 0$ implies that

$$\dot{P^\varepsilon} \leq (\beta_P - \mu_P) P^\varepsilon. \tag{6.21}$$

Therefore by using (6.19) and (6.21) we obtain the following finite time estimation uniform in ε.

Lemma 6.17 *For each $\tau > 0$ we can find a constant $M = M(\tau, N_0, P_0) > 0$ (independent of $\varepsilon > 0$) such that*

$$0 \leq N^{\varepsilon}(t) \leq M \text{ and } 0 \leq P^{\varepsilon}(t) \leq M, \forall t \in [0, \tau]. \tag{6.22}$$

and

$$\sup_{t \in [0,\tau]} |\dot{N^{\varepsilon}}(t)| \leq M \text{ and } \sup_{t \in [0,\tau]} |\dot{P^{\varepsilon}}(t)| \leq M. \tag{6.23}$$

Proof We first deduce (6.22) by using the inequalities (6.19) and (6.21). By using the fact $P_S \geq 0$ and $P_H \geq 0$ we have

$$0 \leq P_S^{\varepsilon}(t) \leq M, \text{ and } 0 \leq P_H^{\varepsilon}(t) \leq M, \forall t \in [0, \tau]. \tag{6.24}$$

Therefore by injecting these estimations into (6.18) and (6.20) we deduce (6.23). □

By using Lemma 6.6, and the Arzela-Ascoli theorem we deduce that we can find a sequence $\varepsilon_n \to 0$ such that

$$\lim_{n \to \infty} N^{\varepsilon_n} = N \text{ and } \lim_{n \to \infty} P^{\varepsilon_n} = P$$

where the convergence is taking place in $C([0, \tau], \mathbb{R})$, that is with respect to the uniform convergence topology.

Moreover by using the fact that $P_H^{\varepsilon} = P^{\varepsilon} - P_S^{\varepsilon}$, the P_S^{ε}-equation can be rewritten as

$$\dot{P_S^{\varepsilon}} = -\left((\mu_P + \eta) + \frac{\chi}{\varepsilon}\kappa N^{\varepsilon}\right) P_S^{\varepsilon} + \frac{1}{\varepsilon}\left(P^{\varepsilon} - P_S^{\varepsilon}\right). \tag{6.25}$$

By using (6.24), the map $t \mapsto P_S^{\varepsilon}(t)$ is bounded uniformly in ε. So the family $\left(P_S^{\varepsilon_n}\right)_{n \geq 0}$ is bounded in $L^{\infty}((0, \tau), \mathbb{R})$ which is the dual space of $L^1((0, \tau), \mathbb{R})$. Therefore by using the Banach-Alaoglu-Bourbaki's theorem, we can find a subsequence (denoted with the same index) such that the sequence $\left(P_S^{\varepsilon_n}\right)_{n \geq 0}$ converges to $P_S \in L^{\infty}((0, \tau), \mathbb{R})$ for the weak star topology of $\sigma\left(L^{\infty}((0, \tau), \mathbb{R}), L^1((0, \tau), \mathbb{R})\right)$ as $n \to \infty$. That is to say that for each $\chi \in L^1((0, \tau), \mathbb{R})$

$$\lim_{n \to \infty} \int_0^{\tau} \chi(t)\left(P_S^{\varepsilon_n}(t) - P_S(t)\right) dt = 0.$$

By multiplying (6.25) by $\chi \in C_c^1((0, \tau), \mathbb{R})$ (the space C^1 functions with compact support in $(0, \tau)$) and by integrating over $[0, \tau]$ we obtain

$$-\int_0^{\tau} \dot{\chi}(t) P_S^{\varepsilon_n}(t) dt = \int_0^{\tau} \chi(t)\Bigg[-\left((\mu_P + \eta) + \frac{\chi}{\varepsilon_n}\kappa N^{\varepsilon_n}(t)\right) P_S^{\varepsilon_n}(t) + \frac{1}{\varepsilon_n}\left(P^{\varepsilon_n}(t) - P_S^{\varepsilon_n}(t)\right)\Bigg] dt.$$

Hence by multiplying both sides by ε_n and by taking the limit when n goes to infinity we obtain

$$0 = \int_0^{\tau} \chi(t)\left[-\left(\chi\kappa N(t)\right) P_S(t) + \left(P(t) - P_S(t)\right)\right] dt$$

and since $C_c^1((0,\tau),\mathbb{R})$ is dense in $L^1((0,\tau),\mathbb{R})$ we deduce that

$$P_S^{\varepsilon_n}(t) \overset{*}{\rightharpoonup} \frac{1}{1+\chi\kappa N(t)} P(t) \text{ as } n \to \infty.$$

By using the first equation of (6.17) and (6.20), we have

$$\begin{aligned} N^{\varepsilon_n}(t) &= \frac{e^{\int_0^t \beta_N - \mu_N - \kappa P_S^{\varepsilon_n}(\sigma)d\sigma} N_0}{1+\delta\int_0^t e^{\int_0^l \beta_N - \mu_N - \kappa P_S^{\varepsilon_n}(\sigma)d\sigma} N_0 dl}, \\ P^{\varepsilon_n}(t) &= e^{(\beta_P-\mu_P)t}P_0 - \int_0^t e^{(\beta_P-\mu_P)(t-s)}\eta P_S^{\varepsilon_n}(\sigma)d\sigma. \end{aligned}$$

By taking the limit $n \to \infty$ on both sides we deduce that

$$\begin{cases} \dot{N} = (\beta_N - \mu_N) N(t) - \delta N(t)^2 - \dfrac{\kappa N(t)}{1+\chi\kappa N(t)} P(t), \\ \dot{P} = (\beta_P - \mu_P) P - \eta \dfrac{1}{1+\chi\kappa N(t)} P. \end{cases}$$

Therefore we obtain the following theorem.

Theorem 6.18 *Let $\tau > 0$ be fixed. For each fixed initial values $N_0 \geq 0$, $P_{S0} \geq 0$ and $P_{H0} \geq 0$, we have the following results:*

$$\lim_{\varepsilon \to 0} N^\varepsilon(t) = N(t) \text{ and } \lim_{\varepsilon \to 0} P_S^\varepsilon(t) + P_H^\varepsilon(t) = P(t)$$

where the limit is uniform on $[0,\tau]$, $(N^\varepsilon(t), P_S^\varepsilon(t), P_H^\varepsilon(t))$ is the solution of (6.17) with the initial conditions

$$N^\varepsilon(0) = N_0 \geq 0,\ P_S^\varepsilon(0) = P_{S0} \geq 0 \text{ and } P_H^\varepsilon(0) = P_{H0} \geq 0$$

while $(N(t), P(t))$ is the solution of the Rosenzweig-MacArthur model

$$\begin{cases} \dot{N} = (\beta_N - \mu_N) N(t) - \delta N(t)^2 - \dfrac{\kappa N(t)}{1+\chi\kappa N(t)} P(t), \\ \dot{P} = (\beta_P - \mu_P - \eta) P + \eta \dfrac{\chi\kappa N(t)}{1+\chi\kappa N(t)} P \end{cases} \tag{6.26}$$

with the initial conditions

$$N(0) = N_0 \text{ and } P(0) = P_{S0} + P_{H0}.$$

Remark 6.19 If instead of model (6.1) we consider the following model

$$\begin{cases} \dot{N} = (\beta_N - \mu_N) N - \delta N^2 - \kappa N^l P_S \\ \dot{P_S} = -(\mu_P + \eta) P_S - \rho\kappa N^m P_S + \gamma P_H, \\ \dot{P_H} = \beta_P (P_S + P_H) - \mu_P P_H + \rho\kappa N^m P_S - \gamma P_H. \end{cases} \tag{6.27}$$

Then by using the same procedure as above we obtain a convergence result to the following predator-prey model

$$\begin{cases} \dot{N} = (\beta_N - \mu_N)\, N(t) - \delta N(t)^2 - \dfrac{\kappa N(t)^l}{1 + \chi\kappa N(t)^m} P(t), \\ \dot{P} = (\beta_P - \mu_P - \eta)\, P + \eta \dfrac{\chi\kappa N(t)^m}{1 + \chi\kappa N(t)^m} P. \end{cases} \tag{6.28}$$

And choosing $l = m$, we obtain the classical Holling's type functional response.

6.7 Application to the Snowshoe Hares and the Lynx

In this section we reconsider predator-prey system formed by the hares (prey) and lynxes (predator) in the years 1900-1920 recorded by the Hudson Bay Company. The data are available for example in [80].

Year	**Hares** (in thousands)	**Lynx** (in thousands)
1900	30	4
1901	47.2	6.1
1902	70.2	9.8
1903	77.4	35.2
1904	36.3	59.4
1905	20.6	41.7
1906	18.1	19
1907	21.4	13
1908	22	8.3
1909	25.4	9.1
1910	27.1	7.4
1911	40.3	8
1912	57	12.3
1913	76.6	19.5
1914	52.3	45.7
1915	19.5	51.1
1916	11.2	29.7
1917	7.6	15.8
1918	14.6	9.7
1919	16.2	10.1
1920	24.7	8.6

Table 6.1: *Numbers of hares (prey) and lynxes (predator) in the years 1900-1920 recorded by the Hudson Bay Company*

The limit model obtained for ε small enough is given by

$$\begin{cases} \dot{N} = (\beta_N - \mu_N)\, N \left(1 - \dfrac{\delta N}{\beta_N - \mu_N}\right) - \dfrac{\kappa P N}{1 + \chi\kappa N}, \\ \dot{P} = (\beta_P - \mu_P - \eta)\, P + \eta \dfrac{\chi\kappa P N}{1 + \chi\kappa N}, \end{cases} \tag{6.29}$$

with initial value

$$N(0) = N_0 = 30 \times 10^3 \text{ and } P(0) = P_0 = 4 \times 10^3.$$

Symbol	Interpretation	Value	Unit	Method
$1/\mu_N$	Life expectancy of hares	1	year	fixed
β_N	Birth rate of hares	1.6567	number of new born/year	fitted
δ	Inverse carrying capacity of hares	2.1×10^{-6}	year	fitted
κ		3.2×10^{-5}		fitted
χ		0.11		fitted
$1/\mu_P$	Life expectancy of Lynx	7	year	fixed
β_P	Birth rate of Lynx	8.5127	number of new born/year	fitted
η	Extra mortality of searching Lynx	9.24	year^{-1}	fitted
$\beta_P - \mu_P - \eta$	Growth of searching lynx	−0.8702		fitted
$\eta\chi$	Convertion rate	1.0164		fitted

Table 6.2: *List of parameters for the model* (6.29)*, their interpretations, values and symbols. In this table we have fixed* μ_N *and* μ_P *and we have obtained all the remaining parameters by using a least square method between the data in Table 6.1 and the solution of the model* (6.29)*. The life expectancy of Snowshoe Hares is not known [51, 120]. Here we fix the life expectancy of hares to be* 1 *year (similarly to [361]). In the wild a Canadian Lynx can live up to* 14 *years. Here we fix the life expectancy to be* 7 *years (see [101] for more result). A Canadian lynx can have between 1 and 8 new babies [188]. So the estimation obtained for the birth rate of lynx is still reasonable.*

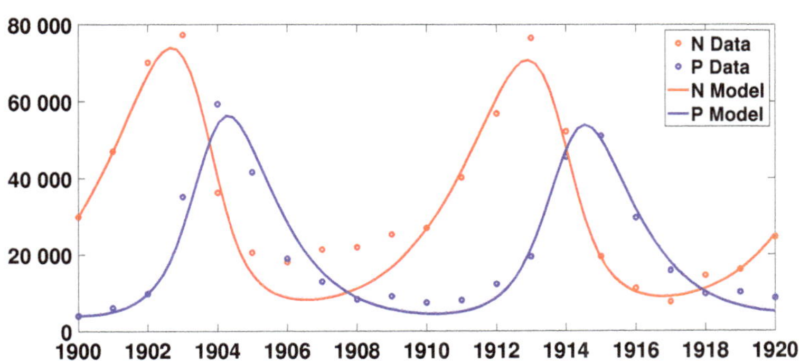

Fig. 6.1: *In this figure we run a simulation of the model* (6.29) *(solide lines) compared with the data (circles).*

In section 6.6, we proved that the model (6.29) can be obtained as the singular limit (when $\varepsilon \to 0$) of the following system

$$\begin{cases} \dot{N^\varepsilon} = (\beta_N - \mu_N)\, N^\varepsilon \left(1 - \dfrac{\delta N^\varepsilon}{\beta_N - \mu_N}\right) - \kappa N^\varepsilon\, P_S^\varepsilon \\ \dot{P_S^\varepsilon} = -(\mu_P + \eta) P_S^\varepsilon - \dfrac{\chi}{\varepsilon}\kappa N^\varepsilon P_S^\varepsilon + \dfrac{1}{\varepsilon} P_H^\varepsilon \\ \dot{P_H^\varepsilon} = -\mu_P P_H^\varepsilon \qquad + \dfrac{\chi}{\varepsilon}\kappa N^\varepsilon P_S^\varepsilon - \dfrac{1}{\varepsilon} P_H^\varepsilon + \beta_P \left(P_S^\varepsilon + P_H^\varepsilon\right) \end{cases} \tag{6.30}$$

and we fix the initial value

$$N^\varepsilon(0) = N_0 = 30 \times 10^3 \geq 0,\ P_S^\varepsilon(0) = P_{S0} \geq 0 \text{ and } P_H^\varepsilon(0) = P_{H0} \geq 0.$$

In the proof Theorem 6.18, we observed that for ε small enough

$$P_S^\varepsilon(t) \simeq \frac{1}{1 + \chi\kappa N(t)} P(t) \text{ and } P_R^\varepsilon(t) \simeq \left(1 - \frac{1}{1 + \chi\kappa N(t)}\right) P(t) = \frac{\chi\kappa N(t)}{1 + \chi\kappa N(t)} P(t). \tag{6.31}$$

By using the value for $\chi\kappa$ estimated in Table 6.2, we obtain the following initial values for the model (6.30)

$$P_{S0}^\varepsilon = \frac{P_0}{1 + \chi\kappa N_0} = \frac{4 \times 10^3}{1 + 1.0164 \times 30 \times 10^3}, \tag{6.32}$$

and

$$P_{R0}^\varepsilon = \frac{\chi\kappa N_0}{1 + \chi\kappa N_0} P_0 = \frac{1.0164 \times 30 \times 10^3}{1 + 1.0164 \times 30 \times 10^3} 4 \times 10^3. \tag{6.33}$$

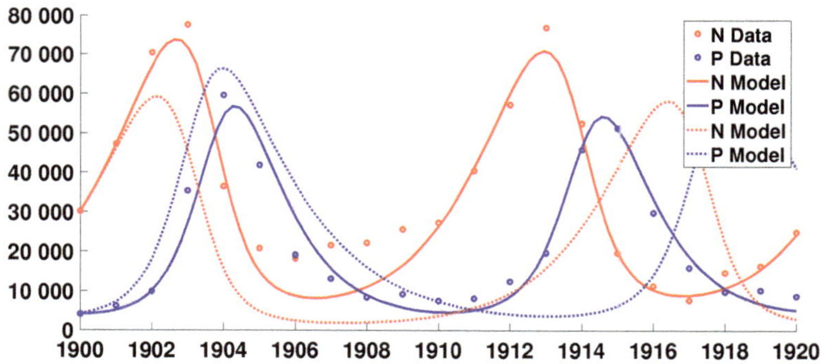

Fig. 6.2: *In this figure we run a simulation of the model* (6.30) *(solid and dotted lines) compared with the data (circles). The solid lines correspond to* $\varepsilon = 10^{-4}$ *and the dotted lines correspond to* $\varepsilon = 5.10^{-3}$.

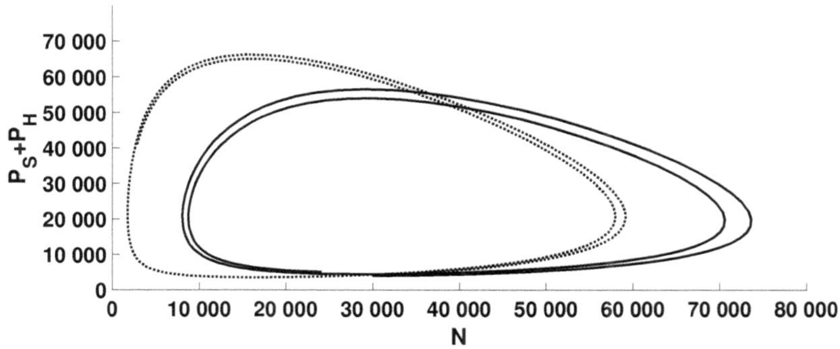

Fig. 6.3: *In this figure we run a simulation of the model* (6.30) *with* $\varepsilon = 10^{-4}$ *for solid line and with* $\varepsilon = 5.10^{-3}$ *for dotted line.*

From Figures 6.1 and 6.2 , we can see that ε does not need to be very small ($\varepsilon = 10^{-4}$) to get an almost perfect match of our model (6.30) with the Rosenzweig-MacArthur model (6.29). Our simulations for hares and lynxes fit the data reported by the Hudson Bay Company. As we mentioned the main advantage with the model (6.30) is that we can separate the vital dynamic and consumption of prey (hares) to describe the behavior of the predators (lynxes). From our model (6.30), people can study the interaction between predator and prey in detail and get more information.

6.8 Remarks and notes

We could address the same question for predator-prey models with Holling's type I, II, and III [160, 161] functional responses, and particularly for the following model with Beddington-DeAngelis functional response

$$\begin{cases} N'(t) = rN(t)\left(1 - \dfrac{N(t)}{K}\right) - P(t)\left(\dfrac{\alpha N(t)}{1 + \beta N(t) + \gamma P(t)}\right), \\ P'(t) = -d\,P(t) + c\;P(t)\left(\dfrac{\alpha N(t)}{1 + \beta N(t) + \gamma P(t)}\right), \end{cases} \tag{6.34}$$

In the 80th and 90th several groups were arguing about the most appropriate functional responses for predators model. Don DeAngelis invited them to publish a paper in the same volume of Ecology [6, 27, 121, 131]. Following these articles, Cosner, DeAngelis, Ault, and Olson [66] tried to provide a unified mechanistic approach for the derivation of various forms of functional responses proposed by those groups. They used the relative spatial distribution of the predator and prey to derive several functional responses, for example, a ratio-dependent version of the Beddington-DeAngelis model

$$\begin{cases} N'(t) = rN(t)\left(1 - \frac{N(t)}{K}\right) - P(t)\left(\frac{\alpha N(t)}{\beta N(t) + \gamma P(t)}\right), \\ P'(t) = -d\,P(t) + c\;P(t)\left(\frac{\alpha N(t)}{\beta N(t) + \gamma P(t)}\right), \end{cases} \tag{6.35}$$

An interesting question would be to reconsider the predator-prey model with handling and searching predator, and try to derive the different functional responses previously derived.

6.9 MATLAB codes

6.9.1 Figure 6.1

```
function   example
tspan = [1900 1920];
x=[1900 1901 1902 1903 1904 1905 1906 1907 1908 1909
    1910 1911 1912 1913 1914 1915 1916 1917 1918
    1919 1920]; % years
y=[3*1e4 4.72*1e4 70.2*1e3 77.4*1e3 36.3*1e3 20.6*1
    e3 18.1*1e3 21.4*1e3 22*1e3 25.4*1e3 27.1*1e3
    40.3*1e3 57*1e3 76.6*1e3 52.3*1e3 19.5*1e3 11.2*1
    e3 7.6*1e3 14.6*1e3 16.2*1e3 24.7*1e3];
z=[4*1e3 6.1*1e3 9.8*1e3 35.2*1e3 59.4*1e3 41.7*1e3
    19*1e3 13*1e3 8.3*1e3 9.1*1e3 7.4*1e3 8*1e3
    12.3*1e3 19.5*1e3 45.7*1e3 51.1*1e3 29.7*1e3
    15.8*1e3 9.7*1e3 10.1*1e3 8.6*1e3];
z0 = [30*1e3;(4*1e3)/(1+1.0164*30*1e3)+((1.0164*30*1
    e3)*(4*1e3))/(1+1.0164*30*1e3)];


sol = ode45(@myfun, tspan , z0);

t = linspace(tspan(1),tspan(2),100);

z_new = deval(sol , t);

plot(x,y,'ro','LineWidth',3);
hold on
plot(x,z,'bo','LineWidth',3);
hold on
plot(t,z_new(1,:),'r','linewidth',3);
hold on
plot(t,z_new(2,:),'b','linewidth',3);
hold on
```

```
legend('N Data', 'P Data','N Model','P Model')

set(gca, 'FontSize', 30);
set(gca, 'FontWeight', 'bold');

function dz = myfun(t,z)
muN = 1;
betaN = 1.6567;
delta = 0.6567/303000;
k = 3.2*1e-5;
X = 0.11;
muP = 1/7;
betaP = 8.5127;
eta = 9.24;

dz = zeros(2,1);
dz(1) = (betaN-muN)*z(1)*(1-(delta*z(1))/(betaN-muN)
    )-(k*z(1)*z(2))/(1+X*k*z(1));
dz(2) = (betaP-muP-eta)*z(2)+(eta*k*X*z(1)*z(2))/(1+
    X*k*z(1));
dz=dz(:);
```

6.9.2 Figure 6.2

```
function example
tspan = [1900 1920];
x=[1900 1901 1902 1903 1904 1905 1906 1907 1908 1909
     1910 1911 1912 1913 1914 1915 1916 1917 1918
    1919 1920];
y=[3*1e4 4.72*1e4 70.2*1e3 77.4*1e3 36.3*1e3 20.6*1
    e3 18.1*1e3 21.4*1e3 22*1e3 25.4*1e3 27.1*1e3
    40.3*1e3 57*1e3 76.6*1e3 52.3*1e3 19.5*1e3 11.2*1
    e3 7.6*1e3 14.6*1e3 16.2*1e3 24.7*1e3];
z=[4*1e3 6.1*1e3 9.8*1e3 35.2*1e3 59.4*1e3 41.7*1e3
    19*1e3 13*1e3 8.3*1e3 9.1*1e3 7.4*1e3 8*1e3
    12.3*1e3 19.5*1e3 45.7*1e3 51.1*1e3 29.7*1e3
    15.8*1e3 9.7*1e3 10.1*1e3 8.6*1e3];
y0 = [30*1e3;(4*1e3)/(1+1.0164*30*1e3);((1.0164*30*1
    e3)*(4*1e3))/(1+1.0164*30*1e3)];
z0 = [30*1e3;(4*1e3)/(1+1.0164*30*1e3);((1.0164*30*1
    e3)*(4*1e3))/(1+1.0164*30*1e3)];
sol1 = ode45(@myfun,tspan,y0);
sol2 = ode45(@myfun2,tspan,z0);
t = linspace(tspan(1),tspan(2),100);
y_new = deval(sol1,t);
z_new = deval(sol2,t);
```

```
plot(x,y,'ro','LineWidth',3);
hold on
plot(x,z,'bo','LineWidth',3);
hold on
plot(t,z_new(1,:),'r','linewidth',3);
hold on
plot(t,z_new(2,:)+z_new(3,:),'b','linewidth',3);
hold on
plot(t,y_new(1,:),'r:','linewidth',3);
hold on
plot(t,y_new(2,:)+y_new(3,:),'b:','linewidth',3);
hold on

legend('N Data', 'P Data','N Model','P Model','N Model', 'P Model')

set(gca, 'FontSize', 30);
set(gca, 'FontWeight', 'bold');

function dy = myfun(t,y)
muN = 1;
betaN = 1.6567;
delta = 0.6567/303000;
k = 3.2e-5;
X = 0.11;
muP = 1/7;
betaP = 8.5127;
eta = 9.24;
epison =5*1e-3;
rho=X/epison;
r=1/epison;
dy = zeros(3,1);
dy(1) = (betaN-muN)*y(1)*(1-(delta*y(1))/(betaN-muN))-k*y(1)*y(2);
dy(2) = -(muP+eta)*y(2)-rho*k*y(1)*y(2)+y(3)/epison;
dy(3) = -muP*y(3)+X/epison*k*y(1)*y(2)-y(3)*r+betaP*(y(2)+y(3));
hold on

function dz = myfun2(t,z)
muN = 1;
betaN = 1.6567;
delta = 0.6567/303000;
k = 3.2e-5;
X = 0.11;
```

```
muP = 1/7;
betaP = 8.5127;
eta = 9.24;
epison =1e−4;
rho=X/epison;
r=1/epison;
dz = zeros(3,1);
dz(1) = (betaN−muN)∗z(1)∗(1−(delta∗z(1))/(betaN−muN)
    )−k∗z(1)∗z(2);
dz(2) = −(muP+eta)∗z(2)−rho∗k∗z(1)∗z(2)+z(3)/epison;
dz(3) = −muP∗z(3)+X/epison∗k∗z(1)∗z(2)−z(3)∗r+betaP
    ∗(z(2)+z(3));
```

6.9.3 Figure 6.3

```
function example
tspan = [1900 1920];
x=[1900 1901 1902 1903 1904 1905 1906 1907 1908 1909
     1910 1911 1912 1913 1914 1915 1916 1917 1918
    1919 1920];
y=[3∗1e4 4.72∗1e4 70.2∗1e3 77.4∗1e3 36.3∗1e3 20.6∗1
    e3 18.1∗1e3 21.4∗1e3 22∗1e3 25.4∗1e3 27.1∗1e3
    40.3∗1e3 57∗1e3 76.6∗1e3 52.3∗1e3 19.5∗1e3 11.2∗1
    e3 7.6∗1e3 14.6∗1e3 16.2∗1e3 24.7∗1e3];
z=[4∗1e3 6.1∗1e3 9.8∗1e3 35.2∗1e3 59.4∗1e3 41.7∗1e3
    19∗1e3 13∗1e3 8.3∗1e3 9.1∗1e3 7.4∗1e3 8∗1e3
    12.3∗1e3 19.5∗1e3 45.7∗1e3 51.1∗1e3 29.7∗1e3
    15.8∗1e3 9.7∗1e3 10.1∗1e3 8.6∗1e3];
y0 = [30∗1e3;(4∗1e3)/(1+1.0164∗30∗1e3);((1.0164∗30∗1
    e3)∗(4∗1e3))/(1+1.0164∗30∗1e3)];
z0 = [30∗1e3;(4∗1e3)/(1+1.0164∗30∗1e3);((1.0164∗30∗1
    e3)∗(4∗1e3))/(1+1.0164∗30∗1e3)];
sol1 = ode45(@myfun,tspan,y0);
sol2 = ode45(@myfun2,tspan,z0);
t = linspace(tspan(1),tspan(2),200);
y_new = deval(sol1,t);
z_new = deval(sol2,t);


plot(y_new(1,:),y_new(2,:)+y_new(3,:),'k:','
    linewidth',3);
hold on
plot(z_new(1,:),z_new(2,:)+z_new(3,:),'k','linewidth
    ',3);
hold on
xlabel('N');
```

```
ylabel('P_S+P_H')


set(gca, 'FontSize', 30);
set(gca, 'FontWeight', 'bold');



function dy = myfun(t,y)
muN = 1;
betaN = 1.6567;
delta = 0.6567/303000;;
k = 3.2e-5;
X = 0.11;
muP = 1/7;
betaP = 8.5127;
eta = 9.24;
epison =5*1e-3;
rho=X/epison;
r=1/epison;
dy = zeros(3,1);
dy(1) = (betaN-muN)*y(1)*(1-(delta*y(1))/(betaN-muN))-k*y(1)*y(2);
dy(2) = -(muP+eta)*y(2)-rho*k*y(1)*y(2)+y(3)/epison;
dy(3) = -muP*y(3)+X/epison*k*y(1)*y(2)-y(3)*r+betaP*(y(2)+y(3));
hold on

function dz = myfun2(t,z)
muN = 1;
betaN = 1.6567;
delta = 0.6567/303000;
k = 3.2e-5;
X = 0.11;
muP = 1/7;
betaP = 8.5127;
eta = 9.24;
epison =1e-4;
rho=X/epison;
r=1/epison;
dz = zeros(3,1);
dz(1) = (betaN-muN)*z(1)*(1-(delta*z(1))/(betaN-muN))-k*z(1)*z(2);
dz(2) = -(muP+eta)*z(2)-rho*k*z(1)*z(2)+z(3)/epison;
dz(3) = -muP*z(3)+X/epison*k*z(1)*z(2)-z(3)*r+betaP*(z(2)+z(3));
```

Chapter 7
Hopf bifurcation for a Holling's predator-prey model with handling and searching predators

This chapter provides an application of Hopf bifurcation, center manifold and normal form theories in the context of predator prey system.

7.1 Introduction

We continue to consider the system (6.1) of Chapter 6:

$$\begin{cases} N' = \beta_N N - \mu_N N - \delta N^2 - N \kappa P_S \\ P_S' = -(\mu_P + \eta) P_S - \rho N \kappa P_S + \gamma P_H \\ P_H' = -\mu_P P_H + \rho N \kappa P_S - \gamma P_H + \beta_P (P_S + P_H), \end{cases} \tag{7.1}$$

and assume that system (7.1) satisfies Assumptions 6.1 and 6.2. For model (7.1), we get that there exists a unique interior equilibrium $E^* = (N^*, P_S^*, P_H^*)$ if and only if (6.9) holds in Chapter 6. Note that (6.9) is equivalent to the following inequality

$$\beta_N - \mu_N > \frac{\delta [\gamma\eta - (\beta_P - \mu_P)(\mu_P + \eta + \gamma)]}{\kappa\rho (\beta_P - \mu_P)}. \tag{7.2}$$

The goal of this chapter is to apply the bifurcation theory, the center manifold theory and the normal form method to investigate the existence and properties of the Hopf bifurcation around the unique interior equilibrium E^* for the model (7.1). For convenience, we summarize our assumptions into the following one, which contains Assumption 6.1, Assumption 6.2 and (7.2), and which we assume to hold throughout Chapter 7.

Assumption 7.1 We assume that all the parameters of (7.1) are strictly positive and

$$\beta_N - \mu_N > 0, \quad \beta_P - \mu_P > 0, \text{ and } \beta_P - \mu_P - \eta < 0. \tag{7.3}$$

We also impose the following inequality, which implies that the population of predators goes extinct in the absence of prey:

A. Ducrot et al., *Differential Equations and Population Dynamics II*,
Lecture Notes on Mathematical Modelling in the Life Sciences,
https://doi.org/10.1007/978-3-031-91409-6_7

$$\beta_P - \mu_P < -\frac{\gamma}{\mu_P + \eta}(\beta_P - \mu_P - \eta). \tag{7.4}$$

Finally we assume that the inequality (7.2) holds.

We choose the parameter $\alpha := \beta_N - \mu_N$ as the Hopf bifurcation parameter and start with the linearization of (7.1) around the positive equilibrium E^* in the sequel.

7.2 Linearized System

First of all, we translate the unique interior equilibrium E^* of system (7.1) (see Section 6.5 of Chapter 6) to the origin, by setting $\overline{N} = N - N^*$, $\overline{P}_S = P_S - P_S^*$, $\overline{P}_H = P_H - P_H^*$. Then after dropping the bar for notational simplicity, system (7.1) rewrites as

$$\frac{du(t)}{dt} = f(u(t)), \quad t \geq 0 \tag{7.5}$$

with $u = \begin{pmatrix} N \\ P_S \\ P_H \end{pmatrix} \in \mathbb{R}^3$, and

$$f(u) = \begin{pmatrix} (\beta_N - \mu_N)(N + N^*) - \delta(N + N^*)^2 - \kappa(N + N^*)(P_S + P_S^*) \\ -(\mu_P + \eta)(P_S + P_S^*) - \rho\kappa(P_S + P_S^*)(N + N^*) + \gamma(P_H + P_H^*) \\ -(\mu_P + \gamma)(P_H + P_H^*) + \rho\kappa(P_S + P_S^*)(N + N^*) + \beta_P(P_S + P_S^* + P_H + P_H^*) \end{pmatrix}.$$

By using the fact that (N^*, P_S^*, P_H^*) is an equilibrium, we can transform $f(u)$ into the following form:

$$f(u) = \begin{pmatrix} (\beta_N - \mu_N - 2\delta N^* - \kappa P_S^*)N - \kappa N^* P_S - \delta N^2 - \kappa N P_S \\ -\rho\kappa P_S^* N - (\mu_P + \eta + \rho\kappa N^*)P_S + \gamma P_H - \rho\kappa N P_S \\ \rho\kappa P_S^* N + (\rho\kappa N^* + \beta_P)P_S + (\beta_P - \mu_P - \gamma)P_H + \rho\kappa P_S N \end{pmatrix}.$$

Next linearization at the zero equilibrium of (7.5) yields

$$\frac{du(t)}{dt} = Au(t), \quad t \geq 0 \tag{7.6}$$

where A denotes the Jacobian matrix of f around the zero equilibrium, which reads as

$$A = \begin{pmatrix} \beta_N - \mu_N - 2\delta N^* - \kappa P_S^* & -\kappa N^* & 0 \\ -\rho\kappa P_S^* & -(\mu_P + \eta + \rho\kappa N^*) & \gamma \\ \rho\kappa P_S^* & \rho\kappa N^* + \beta_P & \beta_P - \mu_P - \gamma \end{pmatrix}, \tag{7.7}$$

and $u = \begin{pmatrix} N \\ P_S \\ P_H \end{pmatrix} \in \mathbb{R}^3$. Next (7.5) can then be written as

$$\frac{du(t)}{dt} = Au(t) + F(u(t)), \quad t \geq 0 \tag{7.8}$$

with

$$F(u) = f(u) - Au = \begin{pmatrix} -\delta N^2 - \kappa N P_S \\ -\rho\kappa N P_S \\ \rho\kappa N P_S \end{pmatrix} \tag{7.9}$$

which satisfies $F(0) = 0$ and $DF(0) = 0$. The characteristic equation reads

$$\Delta(\lambda)_\alpha := \lambda^3 + p_1\lambda^2 + p_2\lambda + p_3 = 0 \tag{7.10}$$

which is defined in section 6.5 of Chapter 6 and the stability of the equilibrium E^* is also obtained in lemma 6.13 of Chapter 6. Here we explicitly write down the subscript α to indicate the dependency with respect to α, which is our bifurcation parameter.

7.3 Existence of Hopf bifurcation

Recall that $\alpha = \beta_N - \mu_N$ is our bifurcation parameter. Under Assumption 7.1, we get $p_3 > 0$ and then $\lambda = 0$ is not a eigenvalue of (7.10). Let $\lambda = i\omega, \omega > 0$ be a pure imaginary root of (7.10). Then we have

$$-i\omega^3 - p_1\omega^2 + ip_2\omega + p_3 = 0.$$

Separating the real part and the imaginary part of the above equation, we get

$$p_2 = \omega^2, \ \frac{p_3}{p_1} = \omega^2.$$

Thus (7.10) has a pair of pure imaginary roots $\pm i\omega$ with $\omega = \sqrt{p_2}$ if and only if

$$p_2 > 0 \text{ and } p_3 = p_1 p_2.$$

After some algebra, we obtain that under the following Assumption 7.2, (7.10) has a pair of pure imaginary roots $\pm i\omega$ with $\omega = \sqrt{p_2}$ if and only if $\alpha = \alpha^\star$ with

$$\alpha^\star = \frac{\delta\left[2\gamma\eta - (\beta_P - \mu_P)(\beta_P + \eta + \gamma)\right]}{\kappa\rho(\beta_P - \mu_P)} + \frac{\delta(\beta_P - \mu_P)\left[(\beta_P - \mu_P)^2 - \gamma\eta\right]}{\gamma\kappa\rho\eta - \delta\left\{(\beta_P - \mu_P - \gamma)(\mu_P + \eta) + \gamma\beta_P\right\}}. \tag{7.11}$$

Assumption 7.2 Assume that $\gamma\eta - (\beta_P - \mu_P)^2 > 0$.

Note that we impose the condition in the above Assumption 7.2 to make sure $\alpha^\star$ is positive.

Let $\lambda_{\pm}(\alpha) := \sigma(\alpha) \pm i\nu(\alpha)$ with $\sigma(\alpha^\star) = 0, \nu(\alpha^\star) = \omega > 0$ are the eigenvalues of (7.10) in a neighborhood of $\alpha = \alpha^\star$. Next we get

$$\left.\frac{d\mathrm{Re}\,(\lambda_{\pm}(\alpha))}{d\alpha}\right|_{\alpha=\alpha^\star} = -\left.\frac{[\,(\beta_P - \mu_P - \gamma)\,(\mu_P + \eta) + \gamma\beta_P]\left(\frac{2p_1p_2}{(\beta_P - \mu_P)} + 2p_2\right)}{\left[p_2 - 3\omega^2\right]^2 + [2p_1\omega]^2}\right|_{\alpha=\alpha^\star}$$
$$\neq 0$$

and then we derive the following Theorem.

Theorem 7.3 *Let Assumptions 7.1 and 7.2 be satisfied. Then system (7.1) undergoes a Hopf bifurcation at the positive equilibrium E^* when $\alpha = \alpha^\star$, where $\alpha := \beta_N - \mu_N$ and $\alpha^\star$ is defined in (7.11).*

7.4 Computation of the Normal Form

In this section, we study the direction and the stability of the Hopf bifurcation by applying the normal form theory developed in Chapter 5 to system (7.8). In order to apply the center manifold theory and normal form theory, we include the parameter α into the state variable of system (7.8) and consider the system

$$\begin{cases} \dfrac{d\alpha}{dt} = 0, \\ \dfrac{du}{dt} = A(\alpha)u + F(\alpha, u), \end{cases}$$

where

$$A(\alpha) = \begin{bmatrix} \alpha - 2\delta N^* - \kappa P_S^* & -N^*\kappa & 0 \\ -P_S^*\rho\kappa & -(\mu_P + \eta + \rho\kappa N^*) & \gamma \\ P_S^*\rho\kappa & \rho\kappa N^* + \beta_P & \beta_P - \mu_P - \gamma \end{bmatrix}, \tag{7.12}$$

and $F(\alpha, u) = f(\alpha, u) - A(\alpha)u$ with

$$f(\alpha, u) = \begin{pmatrix} \alpha\,(N + N^*) - \delta\,(N + N^*)^2 - \kappa\,(N + N^*)\,(P_S + P_S^*) \\ -(\mu_P + \eta)\,(P_S + P_S^*) - \rho\kappa\,(P_S + P_S^*)\,(N + N^*) + \gamma\,(P_H + P_H^*) \\ -(\mu_P + \gamma)\,(P_H + P_H^*) + \rho\kappa\,(P_S + P_S^*)\,(N + N^*) + \beta_P\,(P_S + P_S^* + P_H + P_H^*) \end{pmatrix}.$$

Making the change of variables

$$\alpha = \overline{\alpha} + \alpha^\star,$$

we obtain the system after dropping the hat

$$\begin{cases} \dfrac{d\alpha}{dt} = 0, \\ \dfrac{du}{dt} = A(\alpha + \alpha^\star)u + F(\alpha + \alpha^\star, u). \end{cases}$$

Separating the linear and nonlinear part, we have

$$\begin{cases} \dfrac{d\alpha}{dt} = 0, \\ \dfrac{du}{dt} = A(\alpha^\star)u + W(\alpha, u), \end{cases} \tag{7.13}$$

with $W(\alpha, u) = F(\alpha + \alpha^\star, u) + A(\alpha + \alpha^\star)u - A(\alpha^\star)u$.

Consider now the linear operator $\mathcal{A} : \mathbb{R}^4 \to \mathbb{R}^4$ defined by

$$\mathcal{A}\begin{pmatrix} \alpha \\ u \end{pmatrix} = \begin{pmatrix} 0 \\ A(\alpha^\star)u \end{pmatrix}$$

and the map $\mathcal{H} : \mathbb{R}^4 \to \mathbb{R}^4$ defined by

$$\mathcal{H}\begin{pmatrix} \alpha \\ u \end{pmatrix} = \begin{pmatrix} 0 \\ W(\alpha, u) \end{pmatrix}.$$

Then we have

$$\mathcal{H}\begin{pmatrix} 0 \\ 0 \end{pmatrix} = 0 \text{ and } D\mathcal{H}\begin{pmatrix} 0 \\ 0 \end{pmatrix} = 0.$$

Now setting $w = \begin{pmatrix} \alpha \\ u \end{pmatrix}$ with $\alpha \in \mathbb{R}$ and $u \in \mathbb{R}^3$, (7.13) can reformulated as the following system of equations

$$\frac{dw(t)}{dt} = \mathcal{A}w(t) + \mathcal{H}\left(w(t)\right). \tag{7.14}$$

In the sequel we shall study this system around the zero equilibrium in $\mathbb{R}^4$.

7.4.1 Projectors on the Eigenspaces

In order to use the center manifold reduction and changes of variables to compute the Taylor's expansion of the reduced system of a system topologically equivalent to the original problem (7.14), we need to compute the spectral projectors on the generalized eigenspaces associated with the eigenvalues of $\mathcal{A}$. Note that for $\lambda \in \rho(\mathcal{A})$ we have

$$(\lambda I - A(\alpha^\star))^{-1} = \frac{\begin{pmatrix} A_{11} & A_{21} & A_{31} \\ A_{12} & A_{22} & A_{32} \\ A_{13} & A_{23} & A_{33} \end{pmatrix}_{\alpha^\star, \lambda}}{\det\left(\lambda I - A(\alpha^\star)\right)} = \frac{\begin{pmatrix} A_{11} & A_{21} & A_{31} \\ A_{12} & A_{22} & A_{32} \\ A_{13} & A_{23} & A_{33} \end{pmatrix}_{\alpha^\star, \lambda}}{\Delta(\lambda)_{\alpha^\star}}, \tag{7.15}$$

where $\Delta(\lambda)_{\alpha^\star}$ is defined in (7.10), and $\begin{pmatrix} A_{11} & A_{21} & A_{31} \\ A_{12} & A_{22} & A_{32} \\ A_{13} & A_{23} & A_{33} \end{pmatrix}_{\alpha^\star, \lambda}$ is the adjoint matrix of $(\lambda I - A(\alpha^\star))$ and A_{ij} is the cofactor of the element in the ith row and jth column of $(\lambda I - A(\alpha^\star))$, $i, j = 1, 2, 3$. Furthermore the resolvent of $A(\alpha^\star)$ has the following Laurent's expansion around $\widehat{\lambda} \in \sigma(A(\alpha^\star))$ with $\widehat{\lambda} = i\omega$ or $-i\omega$:

$$(\lambda I - A(\alpha^\star))^{-1} = \sum_{n=-1}^{+\infty} (\lambda - \widehat{\lambda})^n B_n^{\widehat{\lambda}}.$$

The projector $\Pi_{\widehat{\lambda}}^{A(\alpha^\star)}$ on the generalized eigenspace associated with $\widehat{\lambda} = i\omega$ or $-i\omega$ is $B_{-1}^{\widehat{\lambda}}$. Since

$$(\lambda - \widehat{\lambda})(\lambda I - A(\alpha^\star))^{-1} = \sum_{m=0}^{+\infty} (\lambda - \widehat{\lambda})^m B_{m-1}^{\widehat{\lambda}}$$

and

$$\lim_{\lambda \to \widehat{\lambda}} \frac{(\lambda - \widehat{\lambda})}{\Delta(\lambda)_{\alpha^\star}} = \lim_{\lambda \to \widehat{\lambda}} \frac{1}{\Delta'(\lambda)_{\alpha^\star}} = \frac{1}{\Delta'(\widehat{\lambda})_{\alpha^\star}},$$

we have

$$\begin{aligned} \Pi_{\widehat{\lambda}}^{A(\alpha^\star)} &= B_{-1}^{\widehat{\lambda}} = \lim_{\lambda \to \widehat{\lambda}} (\lambda - \widehat{\lambda})(\lambda I - A(\alpha^\star))^{-1} \\ &= \lim_{\lambda \to \widehat{\lambda}} \left(\frac{(\lambda - \widehat{\lambda})}{\Delta(\lambda)_{\alpha^\star}} \begin{pmatrix} A_{11} & A_{21} & A_{31} \\ A_{12} & A_{22} & A_{32} \\ A_{13} & A_{23} & A_{33} \end{pmatrix}_{\alpha^\star, \lambda} \right) = \frac{\begin{pmatrix} A_{11} & A_{21} & A_{31} \\ A_{12} & A_{22} & A_{32} \\ A_{13} & A_{23} & A_{33} \end{pmatrix}_{\alpha^\star, \widehat{\lambda}}}{\Delta'(\widehat{\lambda})_{\alpha^\star}}. \end{aligned}$$

Then we end-up with the following expression

$$\Pi_{i\omega}^{A(\alpha^\star)} \begin{pmatrix} 0 \\ 0 \\ 1 \end{pmatrix} = \frac{\begin{pmatrix} A_{31} \\ A_{32} \\ A_{33} \end{pmatrix}_{\alpha^\star, i\omega}}{\Delta'(i\omega)_{\alpha^\star}}.$$

Define $\Pi_c^{A(\alpha^\star)} : \mathbb{R}^3 \to \mathbb{R}^3$ by

$$\Pi_c^{A(\alpha^\star)} \begin{pmatrix} N \\ P_S \\ P_H \end{pmatrix} = \Pi_{i\omega}^{A(\alpha^\star)} \begin{pmatrix} N \\ P_S \\ P_H \end{pmatrix} + \Pi_{-i\omega}^{A(\alpha^\star)} \begin{pmatrix} N \\ P_S \\ P_H \end{pmatrix}, \ \forall \begin{pmatrix} N \\ P_S \\ P_H \end{pmatrix} \in \mathbb{R}^3.$$

Set

$$\Pi_h^{A(\alpha^\star)} := I - \Pi_c^{A(\alpha^\star)}.$$

Denote

$$X_c := \Pi_c^{A(\alpha^\star)} \left(\mathbb{R}^3 \right), \ X_h := \Pi_h^{A(\alpha^\star)} \left(\mathbb{R}^3 \right),$$

and

$$A_c := A(\alpha^\star)\mid_{X_c},\ A_h := A(\alpha^\star)\mid_{X_h}.$$

Let

$$b_1 = \begin{pmatrix} A_{31} \\ A_{32} \\ A_{33} \end{pmatrix}_{\alpha^\star,\, i\omega},\quad b_2 = \overline{b}_1 = \begin{pmatrix} A_{31} \\ A_{32} \\ A_{33} \end{pmatrix}_{\alpha^\star,\, -i\omega}. \tag{7.16}$$

Then $\{b_1, b_2\}$ forms a basis of X_c. In the following, we set

$$e_1 = \begin{pmatrix} 1 \\ 0 \\ 0 \end{pmatrix}, e_2 = \begin{pmatrix} 0 \\ 1 \\ 0 \end{pmatrix}, e_3 = \begin{pmatrix} 0 \\ 0 \\ 1 \end{pmatrix}.$$

Then the following lemma holds true.

Lemma 7.4 *Let Assumptions 7.1 and 7.2 be satisfied. Then for each $\lambda \in i\mathbb{R} \setminus \{-i\omega, i\omega\}$, we have*

$$\left(\lambda I - A(\alpha^\star)\big|_{\Pi_h^{A(\alpha^\star)}(\mathbb{R}^3)}\right)^{-1} \Pi_h^{A(\alpha^\star)} e_k = \Phi_{k,\lambda}$$

with

$$\Phi_{k,\lambda} = \begin{pmatrix} \Phi^1_{k,\lambda} \\ \Phi^2_{k,\lambda} \\ \Phi^3_{k,\lambda} \end{pmatrix} = \frac{\begin{pmatrix} A_{k1} \\ A_{k2} \\ A_{k3} \end{pmatrix}_{\alpha^\star,\, \lambda}}{\Delta(\lambda)_{\alpha^\star}} - \frac{\begin{pmatrix} A_{k1} \\ A_{k2} \\ A_{k3} \end{pmatrix}_{\alpha^\star,\, i\omega}}{(\lambda - i\omega)\,\Delta'(i\omega)_{\alpha^\star}} - \frac{\begin{pmatrix} A_{k1} \\ A_{k2} \\ A_{k3} \end{pmatrix}_{\alpha^\star,\, -i\omega}}{(\lambda + i\omega)\,\Delta'(-i\omega)_{\alpha^\star}},$$
$$k = 1, 2, 3.$$

For $\lambda = i\omega$,

$$\left(i\omega I - A(\alpha^\star)\big|_{\Pi_h^{A(\alpha^\star)}(\mathbb{R}^3)}\right)^{-1} \Pi_h^{A(\alpha^\star)} e_k = \Phi_{k,i\omega}$$

with

$$\Phi_{k,i\omega} = \begin{pmatrix} \Phi^1_{k,i\omega} \\ \Phi^2_{k,i\omega} \\ \Phi^3_{k,i\omega} \end{pmatrix} = \frac{1}{2i\omega\,(\Delta'(i\omega)_{\alpha^\star})^2\,\Delta'(-i\omega)_{\alpha^\star}} \times$$
$$\left(2i\omega\Delta'(i\omega)_{\alpha^\star}\Delta'(-i\omega)_{\alpha^\star} \begin{pmatrix} \frac{dA_{k1}}{d\lambda} \\ \frac{dA_{k2}}{d\lambda} \\ \frac{dA_{k3}}{d\lambda} \end{pmatrix}_{\alpha^\star,\, i\omega} \right.$$
$$\left. -i\omega\Delta''(i\omega)_{\alpha^\star}\Delta'(-i\omega)_{\alpha^\star} \begin{pmatrix} A_{k1} \\ A_{k2} \\ A_{k3} \end{pmatrix}_{\alpha^\star,\, i\omega} - (\Delta'(i\omega)_{\alpha^\star})^2 \begin{pmatrix} A_{k1} \\ A_{k2} \\ A_{k3} \end{pmatrix}_{\alpha^\star,\, -i\omega} \right),$$
$$k = 1, 2, 3.$$

For $\lambda = -i\omega$,

$$\left(-i\omega I - A(\alpha^\star)|_{\Pi_h^{A(\alpha^\star)}(\mathbb{R}^3)}\right)^{-1} \Pi_h^{A(\alpha^\star)} e_k = \Phi_{k,-i\omega}$$

with

$$\Phi_{k,-i\omega} = \begin{pmatrix} \Phi^1_{k,-i\omega} \\ \Phi^2_{k,-i\omega} \\ \Phi^3_{k,-i\omega} \end{pmatrix} = \frac{1}{-2i\omega\left(\Delta'(-i\omega)\right)^2_{\alpha^\star} \Delta'(i\omega)_{\alpha^\star}} \times$$

$$\left(-2i\omega\Delta'(i\omega)_{\alpha^\star}\Delta'(-i\omega)_{\alpha^\star} \begin{pmatrix} \frac{dA_{k1}}{d\lambda} \\ \frac{dA_{k2}}{d\lambda} \\ \frac{dA_{k3}}{d\lambda} \end{pmatrix}_{\alpha^\star,\,-i\omega} \right.$$

$$+ i\omega\Delta''(-i\omega)_{\alpha^\star}\Delta'(i\omega)_{\alpha^\star} \begin{pmatrix} A_{k1} \\ A_{k2} \\ A_{k3} \end{pmatrix}_{\alpha^\star,\,-i\omega}$$

$$\left. - \left(\Delta'(-i\omega)_{\alpha^\star}\right)^2 \begin{pmatrix} A_{k1} \\ A_{k2} \\ A_{k3} \end{pmatrix}_{\alpha^\star,\,i\omega} \right), k = 1, 2, 3.$$

Proof Since

$$(\lambda I - A(\alpha^\star))^{-1} \left(\frac{\begin{pmatrix} A_{11} \\ A_{12} \\ A_{13} \end{pmatrix}_{\alpha^\star,\,i\omega}}{\Delta'(i\omega)_{\alpha^\star}} \right) = \frac{1}{\lambda - i\omega} \left(\frac{\begin{pmatrix} A_{11} \\ A_{12} \\ A_{13} \end{pmatrix}_{\alpha^\star,\,i\omega}}{\Delta'(i\omega)_{\alpha^\star}} \right)$$

and

$$(\lambda I - A(\alpha^\star))^{-1} \left(\frac{\begin{pmatrix} A_{11} \\ A_{12} \\ A_{13} \end{pmatrix}_{\alpha^\star,\,-i\omega}}{\Delta'(-i\omega)_{\alpha^\star}} \right) = \frac{1}{\lambda + i\omega} \left(\frac{\begin{pmatrix} A_{11} \\ A_{12} \\ A_{13} \end{pmatrix}_{\alpha^\star,\,-i\omega}}{\Delta'(-i\omega)_{\alpha^\star}} \right),$$

for each $\lambda \in i\mathbb{R} \setminus \{-i\omega, i\omega\}$, we have

$$
\begin{aligned}
&\left(\lambda I - A(\alpha^\star)\big|_{\Pi_h^{A(\alpha^\star)}(\mathbb{R}^3)}\right)^{-1} \Pi_h^{A(\alpha^\star)} e_1 \\
&= \left(\lambda I - A(\alpha^\star)\right)^{-1}\left(e_1 - \Pi_c^{A(\alpha^\star)} e_1\right) \\
&= \frac{\begin{pmatrix} A_{11} \\ A_{12} \\ A_{13} \end{pmatrix}_{\alpha^\star,\,\lambda}}{\Delta(\lambda)_{\alpha^\star}} - \frac{\begin{pmatrix} A_{11} \\ A_{12} \\ A_{13} \end{pmatrix}_{\alpha^\star,\,i\omega}}{(\lambda - i\omega)\,\Delta'(i\omega)_{\alpha^\star}} - \frac{\begin{pmatrix} A_{11} \\ A_{12} \\ A_{13} \end{pmatrix}_{\alpha^\star,\,-i\omega}}{(\lambda + i\omega)\,\Delta'(-i\omega)_{\alpha^\star}}.
\end{aligned}
$$

For $\lambda = i\omega$,

$$
\begin{aligned}
&\left(i\omega I - A(\alpha^\star)\big|_{\Pi_h^{A(\alpha^\star)}(\mathbb{R}^3)}\right)^{-1} \Pi_h^{A(\alpha^\star)} e_1 \\
&= \lim_{\substack{\lambda \to i\omega \\ \lambda \in \rho(A(\alpha^\star))}} \left(\lambda I - A(\alpha^\star)\big|_{\Pi_h^{A(\alpha^\star)}(\mathbb{R}^3)}\right)^{-1} \Pi_h^{A(\alpha^\star)} e_1 \\
&= \lim_{\substack{\lambda \to i\omega \\ \lambda \in i\mathbb{R}\setminus\{-i\omega, i\omega\}}} \left(\frac{\begin{pmatrix} A_{11} \\ A_{12} \\ A_{13} \end{pmatrix}_{\alpha^\star,\,\lambda}}{\Delta(\lambda)} - \frac{\begin{pmatrix} A_{11} \\ A_{12} \\ A_{13} \end{pmatrix}_{\alpha^\star,\,i\omega}}{(\lambda - i\omega)\,\Delta'(i\omega)_{\alpha^\star}} - \frac{\begin{pmatrix} A_{11} \\ A_{12} \\ A_{13} \end{pmatrix}_{\alpha^\star,\,-i\omega}}{(\lambda + i\omega)\,\Delta'(-i\omega)_{\alpha^\star}} \right) \\
&= \lim_{\substack{\lambda \to i\omega \\ \lambda \in i\mathbb{R}\setminus\{-i\omega, i\omega\}}} \left\{ \frac{1}{\Delta(\lambda)\,(\lambda - i\omega)\,(\lambda + i\omega)\,\Delta'(-i\omega)_{\alpha^\star}\Delta'(i\omega)_{\alpha^\star}} \times \right. \\
&\quad \left((\lambda - i\omega)\,(\lambda + i\omega)\,\Delta'(i\omega)_{\alpha^\star}\Delta'(-i\omega)_{\alpha^\star} \begin{pmatrix} A_{11} \\ A_{12} \\ A_{13} \end{pmatrix}_{\alpha^\star,\,\lambda} \right. \\
&\quad -\Delta(\lambda)\,(\lambda + i\omega)\,\Delta'(-i\omega)_{\alpha^\star} \begin{pmatrix} A_{11} \\ A_{12} \\ A_{13} \end{pmatrix}_{\alpha^\star,\,i\omega} \\
&\quad \left.\left. -\Delta(\lambda)\,(\lambda - i\omega)\,\Delta'(i\omega)_{\alpha^\star} \begin{pmatrix} A_{11} \\ A_{12} \\ A_{13} \end{pmatrix}_{\alpha^\star,\,-i\omega} \right) \right\}
\end{aligned}
$$

so that

$$
\begin{aligned}
&\left(i\omega I - A(\alpha^\star)|_{\Pi_h^{A(\alpha^\star)}(\mathbb{R}^3)}\right)^{-1} \Pi_h^{A(\alpha^\star)} e_1 \\
&= \lim_{\substack{\lambda\to i\omega \\ \lambda\in i\mathbb{R}\setminus\{-i\omega, i\omega\}}} \frac{(\lambda - i\omega)^2}{\Delta(\lambda)(\lambda - i\omega)(\lambda + i\omega)\Delta'(-i\omega)_{\alpha^\star}\Delta'(i\omega)_{\alpha^\star}} \times \\
&\lim_{\substack{\lambda\to i\omega \\ \lambda\in i\mathbb{R}\setminus\{-i\omega, i\omega\}}} \left\{ \frac{1}{(\lambda - i\omega)^2} \left((\lambda - i\omega)(\lambda + i\omega)\Delta'(i\omega)_{\alpha^\star}\Delta'(-i\omega)_{\alpha^\star} \begin{pmatrix} A_{11} \\ A_{12} \\ A_{13} \end{pmatrix}_{\alpha^\star,\,\lambda} \right.\right. \\
&\left.\left. -\Delta(\lambda)(\lambda + i\omega)\Delta'(-i\omega)_{\alpha^\star} \begin{pmatrix} A_{11} \\ A_{12} \\ A_{13} \end{pmatrix}_{\alpha^\star,\, i\omega} \right)\right\} \\
&- \lim_{\substack{\lambda\to i\omega \\ \lambda\in i\mathbb{R}\setminus\{-i\omega, i\omega\}}} \left(\Delta(\lambda)(\lambda - i\omega)\Delta'(i\omega)_{\alpha^\star} \begin{pmatrix} A_{11} \\ A_{12} \\ A_{13} \end{pmatrix}_{\alpha^\star,\, -i\omega} \right) \\
&= \frac{1}{2i\omega\left(\Delta'(i\omega)_{\alpha^\star}\right)^2 \Delta'(-i\omega)_{\alpha^\star}} \times \\
&\left(2i\omega\Delta'(i\omega)_{\alpha^\star}\Delta'(-i\omega)_{\alpha^\star} \begin{pmatrix} \frac{dA_{11}}{d\lambda} \\ \frac{dA_{12}}{d\lambda} \\ \frac{dA_{13}}{d\lambda} \end{pmatrix}_{\alpha^\star,\, i\omega} - i\omega\Delta''(i\omega)_{\alpha^\star}\Delta'(-i\omega)_{\alpha^\star} \begin{pmatrix} A_{11} \\ A_{12} \\ A_{13} \end{pmatrix}_{\alpha^\star,\, i\omega} \right. \\
&\left. - \left(\Delta'(i\omega)_{\alpha^\star}\right)^2 \begin{pmatrix} A_{11} \\ A_{12} \\ A_{13} \end{pmatrix}_{\alpha^\star,\, -i\omega} \right).
\end{aligned}
$$

Similarly we can prove the other results. □

Next the following result hold.

Lemma 7.5 *Let Assumptions 7.1 and 7.2 be satisfied. We have*

$$\sigma(\mathcal{A}) = \sigma\left(A(\alpha^\star)\right) \cup \{0\}.$$

Moreover, for $\lambda \in \rho(\mathcal{A}) = \mathbb{C} \setminus \left(\sigma\left(A(\alpha^\star)\right) \cup \{0\}\right)$, *we have*

$$(\lambda I - \mathcal{A})^{-1} \begin{pmatrix} \alpha \\ u \end{pmatrix} = \begin{pmatrix} \frac{\alpha}{\lambda} \\ (\lambda I - A(\alpha^\star))^{-1} u \end{pmatrix}$$

and the eigenvalues 0 *and* $\pm i\omega$ *of* $\mathcal{A}$ *are simple. The corresponding projectors* $\Pi_0, \Pi_{\pm i\omega} : \mathbb{R}^4 \to \mathbb{R}^4$ *on the generalized eigenspace of* $\mathcal{A}$ *associated to* $0, \pm i\omega$, *respectively, are given for* $r \in \mathbb{R}$ *and* $u \in \mathbb{R}^3$ *by*

$$\Pi_0 \begin{pmatrix} r \\ u \end{pmatrix} = \begin{pmatrix} r \\ 0 \end{pmatrix},$$

$$\Pi_{\pm i\omega} \begin{pmatrix} r \\ u \end{pmatrix} = \begin{pmatrix} 0 \\ \Pi_{\pm i\omega}^{A(\alpha^\star)} u \end{pmatrix}.$$

In this context, the projectors $\Pi_c : \mathbb{R}^4 \to \mathbb{R}^4$ and $\Pi_h : \mathbb{R}^4 \to \mathbb{R}^4$ are defined by

$$\Pi_c(y) = (\Pi_0 + \Pi_{i\omega} + \Pi_{-i\omega})(y), \quad \forall y \in \mathbb{R}^4,$$
$$\Pi_h(y) = (I - \Pi_c)(y), \quad \forall y \in \mathbb{R}^4.$$

Denote

$$Y_c := \Pi_c\left(\mathbb{R}^4\right), \quad Y_h := \Pi_h\left(\mathbb{R}^4\right),$$

and

$$\mathcal{A}_c := \mathcal{A}\,|_{Y_c}, \ \mathcal{A}_h := \mathcal{A}\,|_{Y_h}.$$

Then we have

$$\Pi_c \begin{pmatrix} 0_{\mathbb{R}} \\ e_k \end{pmatrix} = \begin{pmatrix} 0_{\mathbb{R}} \\ \Pi_c^{A(\alpha^\star)} e_k \end{pmatrix}, k = 1, 2, 3.$$

Recalling that the basis $\{b_1, b_2\}$ is defined in (7.16), let us define the basis of $Y_c = \Pi_c\left(\mathbb{R}^4\right)$ by

$$\widehat{b}_1 = \begin{pmatrix} 1 \\ 0_{\mathbb{R}^3} \end{pmatrix}, \ \widehat{b}_2 = \begin{pmatrix} 0_{\mathbb{R}} \\ b_1 \end{pmatrix}, \ \widehat{b}_3 = \begin{pmatrix} 0_{\mathbb{R}} \\ b_2 \end{pmatrix}. \tag{7.17}$$

This basis will allow us to compute the Taylor expansion of the reduced system. Next we have the following lemma for the hyperbolic part of $\mathcal{A}$.

Lemma 7.6 *Let Assumptions 7.1 and 7.2 be satisfied. For $\lambda \in i\mathbb{R}$ we have*

$$(\lambda I - \mathcal{A}_h)^{-1} \Pi_h \begin{pmatrix} 0 \\ e_k \end{pmatrix} = \begin{pmatrix} 0 \\ \left(\lambda I - A(\alpha^\star)|_{\Pi_h^{A(\alpha^\star)}(\mathbb{R}^3)}\right)^{-1} \Pi_h^{A(\alpha^\star)} e_k \end{pmatrix}, \ k = 1, 2, 3.$$

7.4.2 Change of Variables

In the following, we will compute the Taylor expansion of the reduced system of a system topologically equivalent to the original system (7.14). We apply the procedure described in section 15.3.2 of Chapter 5 and apply the method with $k = 2$ in Theorem 5.17 of Chapter 5. Therefore we must find $L_2 \in \mathcal{L}_s\left(Y_c^2, Y_h\right)$ by solving the following equation for each $(w_1, w_2) \in Y_c^2$:

$$\frac{d}{dt}\left[L_2(e^{\mathcal{A}_c t} w_1, e^{\mathcal{A}_c t} w_2)\right](0) = \mathcal{A}_h L_2(w_1, w_2) + \frac{1}{2!}\Pi_h D^2 \mathcal{H}(0)(w_1, w_2). \tag{7.18}$$

Then define $G_2 \in V^2(Y_c, Y_h)$ by

$$G_2(\Pi_c w) := L_2(\Pi_c w, \Pi_c w), \quad \forall w \in \mathbb{R}^4,$$

and the change of variables $\varsigma_2 : \mathbb{R}^4 \to \mathbb{R}^4$ and $\varsigma_2^{-1} : \mathbb{R}^4 \to \mathbb{R}^4$ by

$$\varsigma_2(w) := w + G_2(\Pi_c w) \text{ and } \varsigma_2^{-1}(w) := w - G_2(\Pi_c w), \quad \forall w \in \mathbb{R}^4.$$

By applying Theorem 5.17 to (7.14) with $k = 2$, we obtain the following theorem.

Theorem 7.7 *Let Assumptions 7.1 and 7.2 be satisfied. By using the change of variables*

$$\begin{aligned} & w_2(t) = \varsigma_2^{-1}(w(t)) = w(t) - G_2(\Pi_c w(t)) \\ \Leftrightarrow \quad & w(t) = \varsigma_2(w_2(t)) = w_2(t) + G_2(\Pi_c w_2(t)), \end{aligned}$$

the map $t \mapsto w(t)$ is a solution of the system (7.14) if and only if $t \mapsto w_2(t)$ is a solution of the system

$$\frac{dw_2(t)}{dt} = \mathcal{A} w_2(t) + \mathcal{H}_2(w_2(t)), \quad t \geq 0, \tag{7.19}$$

where $\mathcal{H}_2 : \mathbb{R}^4 \to \mathbb{R}^4$ is defined by

$$\begin{aligned} \mathcal{H}_2(w) = & \mathcal{H}(\varsigma_2(w)) + \mathcal{A} G_2(\Pi_c w) - DG_2(\Pi_c w)\mathcal{A}_c \Pi_c w \\ & - DG_2(\Pi_c w)\Pi_c \mathcal{H}(\varsigma_2(w)). \end{aligned}$$

Moreover, the reduced system of (7.19) is given by the ordinary differential equations on $\mathbb{R} \times X_c$:

$$\begin{cases} \dfrac{d\alpha(t)}{dt} = 0, \\ \dfrac{dx_c(t)}{dt} = A_c x_c(t) + \Pi_c^{A(\alpha^\star)} W (I + G_2) \begin{pmatrix} \alpha(t) \\ x_c(t) \end{pmatrix} + R_c \begin{pmatrix} \alpha(t) \\ x_c(t) \end{pmatrix}, \end{cases} \tag{7.20}$$

where $R_c \in C^4(\mathbb{R} \times X_c, X_c)$, and $R_c \begin{pmatrix} \alpha(t) \\ x_c(t) \end{pmatrix}$ is a remainder of order 4, that is,

$$R_c \begin{pmatrix} \alpha \\ x_c \end{pmatrix} = \|(\alpha, x_c)\|^4 O(\alpha, x_c),$$

where $O(\alpha, x_c)$ is a function of (α, x_c) which remains bounded when (α, x_c) goes to 0, or equivalently,

$$D^j R_c(0) = 0 \text{ for each } j = 1, ..., 3.$$

Furthermore,

$$\frac{\partial^j R_c(0)}{\partial^j \alpha} = 0, \quad \forall j = 1, ..., 4,$$

which implies that

$$R_c\begin{pmatrix}\alpha\\ x_c\end{pmatrix} = O\left(\alpha^3\,\|x_c\| + \alpha^2\,\|x_c\|^2 + \alpha\,\|x_c\|^3 + \|x_c\|^4\right).$$

Proof We apply Theorem 5.17 to (7.14) and deduce that there exists $\Psi_2 \in C^2(Y_c, Y_h)$ such that the reduced system of (7.19) consists of ordinary differential equations on Y_c of the form

$$\frac{dw_c(t)}{dt} = \mathcal{A}_c w_c(t) + \Pi_c \mathcal{H}\left[w_c(t) + G_2\left(w_c(t)\right)\right] + \widehat{R}_c\left(w_c(t)\right),$$

where $\widehat{R}_c \in C^4\left(Y_c, Y_c\right)$ is the remainder term of order 4 and

$$\widehat{R}_c\left(w_c\right) := \Pi_c\left\{\mathcal{H}\left[w_c + G_2\left(w_c\right) + \Psi_2\left(w_c\right)\right] - \mathcal{H}\left[w_c + G_2\left(w_c\right)\right]\right\},$$

with $D^j\widehat{R}_c(0) = 0, \forall j = 1, 2, 3$. Since the first component of $\mathcal{H}$ is 0, by using the formula for Π_c we deduce that $\widehat{R}_c\left(w_c\right)$ has the following form

$$\widehat{R}_c\begin{pmatrix}\alpha\\ x_c\end{pmatrix} = \begin{pmatrix}0_{\mathbb{R}}\\ R_c\begin{pmatrix}\alpha\\ x_c\end{pmatrix}\end{pmatrix}.$$

Moreover, since for each $\alpha \in \mathbb{R}$ small enough, $\begin{pmatrix}\alpha\\ 0_{\mathbb{R}_3}\end{pmatrix} \in Y_c$ is an equilibrium solution of (7.14) and belongs to the center manifold. It follows that

$$\begin{pmatrix}\alpha\\ 0_{\mathbb{R}_3}\end{pmatrix} = \Pi_c\begin{pmatrix}\alpha\\ 0_{\mathbb{R}_3}\end{pmatrix} + \Psi_2\left(\Pi_c\begin{pmatrix}\alpha\\ 0_{\mathbb{R}_3}\end{pmatrix}\right).$$

Thus $\Pi_c\begin{pmatrix}\alpha\\ 0_{\mathbb{R}_3}\end{pmatrix} = \begin{pmatrix}\alpha\\ 0_{\mathbb{R}_3}\end{pmatrix}$ and

$$\Psi_2\begin{pmatrix}\alpha\\ 0_{\mathbb{R}_3}\end{pmatrix} = 0, \quad \text{for all } \alpha \in \mathbb{R} \text{ small enough.}$$

So we must have

$$\widehat{R}_c\begin{pmatrix}\alpha\\ 0_{\mathbb{R}_3}\end{pmatrix} = 0, \quad \text{for all } \alpha \in \mathbb{R} \text{ small enough}$$

and we deduce that

$$\frac{\partial^j \widehat{R}_c(0)}{\partial^j \alpha} = 0, \forall j = 1, \ldots, 4.$$

This completes the proof. □

In order to apply the above theorem to compute the Taylor expansion of the reduced system it only remains to compute L_2.

Set

$$w_1 := \begin{pmatrix} \alpha_1 \\ u_1 \end{pmatrix}, \ w_2 := \begin{pmatrix} \alpha_2 \\ u_2 \end{pmatrix}, \ w_3 := \begin{pmatrix} \alpha_3 \\ u_3 \end{pmatrix} \in \mathbb{R}^4$$

with $u_i = \begin{pmatrix} N_i \\ P_{S,i} \\ P_{H,i} \end{pmatrix}, i = 1, 2, 3$. We have

$$D^2\mathcal{H}(0)(w_1, w_2) = \begin{pmatrix} 0_{\mathbb{R}} \\ D^2W(0)(w_1, w_2) \end{pmatrix}$$

and

$$D^3\mathcal{H}(0)(w_1, w_2, w_3) = \begin{pmatrix} 0_{\mathbb{R}} \\ D^3W(0)(w_1, w_2, w_3) \end{pmatrix},$$

where

$$\begin{aligned} & D^2W(0)(w_1, w_2) \\ &= \begin{pmatrix} \alpha_1 N_2 + \alpha_2 N_1 - 2\delta N_2 N_1 - \kappa N_1 P_{S,2} - \kappa N_2 P_{S,1} \\ -\rho\,\kappa P_{S,1} N_2 - \rho\,\kappa P_{S,2} N_1 \\ \rho\,\kappa P_{S,1} N_2 + \rho\,\kappa P_{S,2} N_1 \end{pmatrix} \end{aligned}$$

and

$$D^3W(0)(w_1, w_2, w_3) = 0.$$

Next recall that

$$\frac{d}{dt}\left[L_2(e^{\mathcal{A}_c t}w_1, e^{\mathcal{A}_c t}w_2)\right](0) = L_2(\mathcal{A}_c w_1, w_2) + L_2(w_1, \mathcal{A}_c w_2).$$

Hence system (7.18) can be rewritten as

$$L_2(\mathcal{A}_c w_1, w_2) + L_2(w_1, \mathcal{A}_c w_2) = \mathcal{A}_h L_2(w_1, w_2) + \frac{1}{2!}\Pi_h D^2\mathcal{H}(0)(w_1, w_2).$$

Recalling the definition of the basis $\left\{\widehat{b}_1, \widehat{b}_2, \widehat{b}_3\right\}$ defined in (7.17), note that $L_2(\widehat{b}_1, \widehat{b}_2)=L_2(\widehat{b}_2, \widehat{b}_1)$, $L_2(\widehat{b}_1, \widehat{b}_3)=L_2(\widehat{b}_3, \widehat{b}_1)$, $L_2(\widehat{b}_2, \widehat{b}_3)=L_2(\widehat{b}_3, \widehat{b}_2)$.

(i) Computation of $L_2(\widehat{b}_1, \widehat{b}_1)$. We have

$$\Pi_h D^2\mathcal{H}(0)(\widehat{b}_1, \widehat{b}_1) = 0, \quad \mathcal{A}_c\widehat{b}_1 = 0.$$

So the equation

$$L_2\left(\mathcal{A}_c\widehat{b}_1, \widehat{b}_1\right) + L_2\left(\widehat{b}_1, \mathcal{A}_c\widehat{b}_1\right) = \mathcal{A}_h L_2(\widehat{b}_1, \widehat{b}_1) + \frac{1}{2!}\Pi_h D^2\mathcal{H}(0)(\widehat{b}_1, \widehat{b}_1)$$

is equivalent to

$$0 = \mathcal{A}_h L_2(\widehat{b}_1, \widehat{b}_1).$$

Since 0 belongs to the resolvent set of $\mathcal{A}_h$, we obtain that

$$L_2(\widehat{b}_1, \widehat{b}_1) = 0. \tag{7.21}$$

(ii) Computation of $L_2(\widehat{b}_1, \widehat{b}_2)$. We have

$$\frac{1}{2!} D^2\mathcal{H}(0)(\widehat{b}_1, \widehat{b}_2) = \frac{1}{2}\begin{pmatrix} 0_{\mathbb{R}} \\ \begin{pmatrix} (A_{31})_{\alpha^\star, i\omega} \\ 0 \\ 0 \end{pmatrix} \end{pmatrix}$$

and

$$\mathcal{A}_c \widehat{b}_1 = 0, \mathcal{A}_c \widehat{b}_2 = i\omega \widehat{b}_2.$$

As a consequence, system (7.18) becomes

$$i\omega L_2\left(\widehat{b}_1, \widehat{b}_2\right) = \mathcal{A}_h L_2(\widehat{b}_1, \widehat{b}_2) + \frac{1}{2!}\Pi_h D^2\mathcal{H}(0)(\widehat{b}_1, \widehat{b}_2).$$

Now by using Lemmas 7.4 and 7.6, we obtain

$$\begin{aligned} L_2(\widehat{b}_1, \widehat{b}_2) &= \frac{1}{2}(i\omega I - \mathcal{A}_h)^{-1}\Pi_h D^2\mathcal{H}(0)(\widehat{b}_1, \widehat{b}_2) \\ &= \frac{1}{2}(i\omega I - \mathcal{A}_h)^{-1}\Pi_h \left(\begin{pmatrix} 0_{\mathbb{R}} \\ \begin{pmatrix} (A_{31})_{\alpha^\star, i\omega} \\ 0 \\ 0 \end{pmatrix} \end{pmatrix}\right) \\ &= \begin{pmatrix} 0 \\ \frac{1}{2}(A_{31})_{\alpha^\star, i\omega}\Phi_{1,i\omega} \end{pmatrix}, \end{aligned}$$

where $\Phi_{1,i\omega}$ is defined in Lemma 7.4.

(iii) Computation of $L_2(\widehat{b}_1, \widehat{b}_3)$. We have

$$\frac{1}{2!} D^2\mathcal{H}(0)(\widehat{b}_1, \widehat{b}_3) = \frac{1}{2!}\begin{pmatrix} 0 \\ \begin{pmatrix} (A_{31})_{\alpha^\star, -i\omega} \\ 0 \\ 0 \end{pmatrix} \end{pmatrix}$$

and

$$\mathcal{A}_c \widehat{b}_1 = 0, \mathcal{A}_c \widehat{b}_3 = -i\omega \widehat{b}_3.$$

So in this case system (7.18) becomes

$$-i\omega L_2\left(\widehat{b}_1, \widehat{b}_3\right) = \mathcal{A}_h L_2(\widehat{b}_1, \widehat{b}_3) + \frac{1}{2!}\Pi_h D^2\mathcal{H}(0)(\widehat{b}_1, \widehat{b}_3).$$

Now by using Lemmas 7.4 and 7.6, we obtain

$$L_2(\widehat{b}_1, \widehat{b}_3) = \frac{1}{2!}(-i\omega I - \mathcal{A}_h)^{-1}\Pi_h D^2\mathcal{H}(0)(\widehat{b}_1, \widehat{b}_3) = \begin{pmatrix} 0 \\ \frac{1}{2}(A_{31})_{\alpha^\star, -i\omega}\Phi_{1,-i\omega} \end{pmatrix},$$

where $\Phi_{1,-i\omega}$ is defined in Lemma 7.4.

(iv) Computation of $L_2(\widehat{b}_2, \widehat{b}_2)$. We have

$$\frac{1}{2}D^2\mathcal{H}(0)(\widehat{b}_2, \widehat{b}_2) = \begin{pmatrix} 0 \\ \begin{pmatrix} -\delta (A_{31})^2_{\alpha^\star, i\omega} - \kappa (A_{31})_{\alpha^\star, i\omega} (A_{32})_{\alpha^\star, i\omega} \\ -\rho\, \kappa (A_{31})_{\alpha^\star, i\omega} (A_{32})_{\alpha^\star, i\omega} \\ \rho\, \kappa (A_{31})_{\alpha^\star, i\omega} (A_{32})_{\alpha^\star, i\omega} \end{pmatrix} \end{pmatrix}$$

and

$$\mathcal{A}_c \widehat{b}_2 = i\omega \widehat{b}_2,$$

so that system (7.18) becomes

$$2i\omega L_2(\widehat{b}_2, \widehat{b}_2) = \mathcal{A}_h L_2(\widehat{b}_2, \widehat{b}_2) + \frac{1}{2!}\Pi_h D^2\mathcal{H}(0)(\widehat{b}_2, \widehat{b}_2),$$

and Lemmas 7.4 and 7.6 imply

$$L_2(\widehat{b}_2, \widehat{b}_2) = \frac{1}{2}(2i\omega I - \mathcal{A}_h)^{-1}\Pi_h D^2\mathcal{H}(0)(\widehat{b}_2, \widehat{b}_2) = \begin{pmatrix} 0 \\ \frac{1}{2}F_1 \end{pmatrix}, \tag{7.22}$$

with

$$F_1 = \begin{pmatrix} F_{11} \\ F_{12} \\ F_{13} \end{pmatrix} = 2\left(-\delta (A_{31})^2_{\alpha^\star, i\omega} - \kappa (A_{31})_{\alpha^\star, i\omega} (A_{32})_{\alpha^\star, i\omega}\right)\Phi_{1,2i\omega}$$
$$+2\rho\, \kappa (A_{31})_{\alpha^\star, i\omega} (A_{32})_{\alpha^\star, i\omega}\left(\Phi_{3,2i\omega} - \Phi_{2,2i\omega}\right),$$

and $\Phi_{j,2i\omega}$, $j = 1, 2, 3$, defined in Lemma 7.4.

(v) Computation of $L_2(\widehat{b}_2, \widehat{b}_3)$. We have

$$D^2\mathcal{H}(0)(\widehat{b}_2, \widehat{b}_3) = \begin{pmatrix} 0 \\ \varsigma \end{pmatrix}$$

with

$$\varsigma = \begin{pmatrix} -2\delta (A_{31})_{\alpha^\star, i\omega} (A_{31})_{\alpha^\star, -i\omega} - \kappa (A_{31})_{\alpha^\star, i\omega} (A_{32})_{\alpha^\star, -i\omega} - \kappa (A_{32})_{\alpha^\star, i\omega} (A_{31})_{\alpha^\star, -i\omega} \\ -\rho\, \kappa \left((A_{32})_{\alpha^\star, i\omega} (A_{31})_{\alpha^\star, -i\omega} + (A_{31})_{\alpha^\star, i\omega} (A_{32})_{\alpha^\star, -i\omega}\right) \\ \rho\, \kappa \left((A_{32})_{\alpha^\star, i\omega} (A_{31})_{\alpha^\star, -i\omega} + (A_{31})_{\alpha^\star, i\omega} (A_{32})_{\alpha^\star, -i\omega}\right) \end{pmatrix}$$

and

$$\mathcal{A}_c \widehat{b}_2 = i\omega \widehat{b}_2, \ \mathcal{A}_c \widehat{b}_3 = -i\omega \widehat{b}_3.$$

In this case system (7.18) reduces to

$$L_2\left(\mathcal{A}_c \widehat{b}_2, \widehat{b}_3\right) + L_2\left(\widehat{b}_2, \mathcal{A}_c \widehat{b}_3\right) = \mathcal{A}_h L_2(\widehat{b}_2, \widehat{b}_3) + \frac{1}{2!}\Pi_h D^2\mathcal{H}(0)(\widehat{b}_2, \widehat{b}_3)$$

and then

$$0 = \mathcal{A}_h L_2(\widehat{b}_2, \widehat{b}_3) + \frac{1}{2!}\Pi_h D^2\mathcal{H}(0)(\widehat{b}_2, \widehat{b}_3).$$

Thus

$$-\mathcal{A}_h L_2(\widehat{b}_2, \widehat{b}_3) = \frac{1}{2!}\Pi_h D^2\mathcal{H}(0)(\widehat{b}_2, \widehat{b}_3),$$

and by Lemmas 7.4 and 7.6, we obtain

$$L_2(\widehat{b}_2, \widehat{b}_3) = \frac{1}{2}(-\mathcal{A}_h)^{-1}\Pi_h D^2\mathcal{H}(0)(\widehat{b}_2, \widehat{b}_3) = \begin{pmatrix} 0 \\ \frac{1}{2}F_2 \end{pmatrix}, \tag{7.23}$$

with

$$F_2 = \begin{pmatrix} F_{21} \\ F_{22} \\ F_{23} \end{pmatrix} = \varkappa\Phi_{1,0}$$
$$+\rho\,\kappa\left((A_{32})_{\alpha^\star, i\omega}(A_{31})_{\alpha^\star, -i\omega} + (A_{31})_{\alpha^\star, i\omega}(A_{32})_{\alpha^\star, -i\omega}\right)\left(\Phi_{3,0} - \Phi_{2,0}\right),$$

with

$$\varkappa = -2\delta\,(A_{31})_{\alpha^\star, i\omega}(A_{31})_{\alpha^\star, -i\omega} - \kappa\,(A_{31})_{\alpha^\star, i\omega}(A_{32})_{\alpha^\star, -i\omega}$$
$$- \kappa\,(A_{32})_{\alpha^\star, i\omega}(A_{31})_{\alpha^\star, -i\omega},$$

and $\Phi_{j,0}$, $j = 1, 2, 3$, given in Lemma 7.4.

(vi) Computation of $L_2(\widehat{b}_3, \widehat{b}_3)$. We have

$$D^2\mathcal{H}(0)(\widehat{b}_3, \widehat{b}_3) = \begin{pmatrix} 0 \\ \begin{pmatrix} -2\delta\,(A_{31})^2_{\alpha^\star, -i\omega} - 2\kappa\,(A_{31})_{\alpha^\star, -i\omega}(A_{32})_{\alpha^\star, -i\omega} \\ -2\rho\,\kappa\,(A_{31})_{\alpha^\star, -i\omega}(A_{32})_{\alpha^\star, -i\omega} \\ 2\rho\,\kappa\,(A_{31})_{\alpha^\star, -i\omega}(A_{32})_{\alpha^\star, -i\omega} \end{pmatrix} \end{pmatrix}$$

and

$$\mathcal{A}_c\widehat{b}_3 = -i\omega\widehat{b}_3.$$

Hence system (7.18) rewrites as

$$-2i\omega L_2(\widehat{b}_3, \widehat{b}_3) = \mathcal{A}_h L_2(\widehat{b}_3, \widehat{b}_3) + \frac{1}{2!}\Pi_h D^2\mathcal{H}(0)(\widehat{b}_3, \widehat{b}_3).$$

It follows from Lemmas 7.4 and 7.6 that

$$L_2(\widehat{b}_3, \widehat{b}_3) = \frac{1}{2}(-2i\omega I - \mathcal{A}_h)^{-1}\Pi_h D^2\mathcal{H}(0)(\widehat{b}_3, \widehat{b}_3) = \begin{pmatrix} 0 \\ \frac{1}{2}F_3 \end{pmatrix}$$

with

$$F_3 = \begin{pmatrix} F_{31} \\ F_{32} \\ F_{33} \end{pmatrix} = \left(-2\delta\,(A_{31})^2_{\alpha^\star, -i\omega} - 2\kappa\,(A_{31})_{\alpha^\star, -i\omega}(A_{32})_{\alpha^\star, -i\omega}\right)\Phi_{1,-2i\omega}$$
$$+ 2\rho\,\kappa\,(A_{31})_{\alpha^\star, -i\omega}(A_{32})_{\alpha^\star, -i\omega}\left(\Phi_{3,-2i\omega} - \Phi_{2,-2i\omega}\right), \tag{7.24}$$

and wherein $\Phi_{j,-2i\omega}$, $j = 1, 2, 3$, are defined in Lemma 7.4.

7.4.3 Computation of the Taylor's Expansion of the Reduced System

By using the formula obtained for L_2 and G_2, we get the Taylor's expansion of the reduced system (7.20) up to terms of order 3.

Theorem 7.8 *Let Assumptions 7.2 and 7.2 be satisfied. The reduced system (7.20) expressed in terms of the basis* $\left\{\widehat{b}_1, \widehat{b}_2, \widehat{b}_3\right\}$ *has the following form*

$$\begin{cases} \dfrac{d\alpha(t)}{dt} = 0, \\ \dfrac{d}{dt}\begin{pmatrix} x(t) \\ y(t) \end{pmatrix} = M_c \begin{pmatrix} x(t) \\ y(t) \end{pmatrix} + \left(H_2 + H_3 + \widetilde{R}_c\right)\left(\begin{pmatrix} \alpha(t) \\ x(t) \\ y(t) \end{pmatrix}\right), \end{cases} \tag{7.25}$$

where

$$M_c = \begin{bmatrix} i\omega & 0 \\ 0 & -i\omega \end{bmatrix},$$

the maps H_2 and $H_3 : \mathbb{C}^3 \to \mathbb{C}^2$ are defined by

$$H_2\left(\begin{pmatrix} \alpha \\ x \\ y \end{pmatrix}\right) = \left(\alpha\xi_{11} - \delta\xi_{11}^2\right)\Upsilon_1 - \kappa\xi_{21}\,\xi_{11}\Upsilon_2$$

and

$$H_3\left(\begin{pmatrix} \alpha \\ x \\ y \end{pmatrix}\right) = \left(\alpha\xi_{12} - 2\delta\xi_{11}\xi_{12}\right)\Upsilon_1 - \kappa\,\left(\xi_{11}\xi_{22} + \xi_{12}\xi_{21}\right)\Upsilon_2,$$

the remainder term $\widetilde{R}_c \in C^4\left(\mathbb{R}^3, \mathbb{R}^2\right)$ is given by

$$\widetilde{R}_c\left(\begin{pmatrix} \alpha(t) \\ x(t) \\ y(t) \end{pmatrix}\right) = R_c\left(\begin{pmatrix} \alpha(t) \\ x(t) \\ y(t) \end{pmatrix}\right) - \delta\xi_{12}^2\Upsilon_1 - \kappa\xi_{22}\,\xi_{12}\Upsilon_2$$

and thus

$$\widetilde{R}_c\left(\begin{pmatrix} \alpha \\ x \\ y \end{pmatrix}\right) = O\left(\alpha^3\left\|\begin{pmatrix} x \\ y \end{pmatrix}\right\| + \alpha^2\left\|\begin{pmatrix} x \\ y \end{pmatrix}\right\|^2 + \alpha\left\|\begin{pmatrix} x \\ y \end{pmatrix}\right\|^3 + \left\|\begin{pmatrix} x \\ y \end{pmatrix}\right\|^4\right).$$

Here the vectors Υ_1 and Υ_2 are given by

$$\Upsilon_1 = \begin{pmatrix} \Upsilon_{11} \\ \Upsilon_{12} \end{pmatrix} = \begin{pmatrix} \frac{\vartheta_1}{\Delta'(i\omega)_{\alpha^\star}} + \frac{\overline{\vartheta}_2}{\Delta'(-i\omega)_{\alpha^\star}} \\ \frac{\overline{\vartheta}_1}{\Delta'(-i\omega)_{\alpha^\star}} + \frac{\vartheta_2}{\Delta'(i\omega)_{\alpha^\star}} \end{pmatrix}, \tag{7.26}$$

$$\Upsilon_2 = \begin{pmatrix} \Upsilon_{21} \\ \Upsilon_{22} \end{pmatrix} = \rho \begin{pmatrix} \frac{\vartheta_3}{\Delta'(i\omega)_{\alpha^\star}} + \frac{\overline{\vartheta}_4}{\Delta'(-i\omega)_{\alpha^\star}} - \frac{1}{\Delta'(i\omega)_{\alpha^\star}} \\ \frac{\overline{\vartheta}_3}{\Delta'(-i\omega)_{\alpha^\star}} + \frac{\vartheta_4}{\Delta'(i\omega)_{\alpha^\star}} - \frac{1}{\Delta'(-i\omega)_{\alpha^\star}} \end{pmatrix}$$
$$+ \begin{pmatrix} \frac{\vartheta_1}{\Delta'(i\omega)_{\alpha^\star}} + \frac{\overline{\vartheta}_2}{\Delta'(-i\omega)_{\alpha^\star}} \\ \frac{\overline{\vartheta}_1}{\Delta'(-i\omega)_{\alpha^\star}} + \frac{\vartheta_2}{\Delta'(i\omega)_{\alpha^\star}} \end{pmatrix},$$

while $\xi_{ij}, i, j = 1, 2$ and $\vartheta_i, i = 1, 2, 3, 4$ are defined in (7.27) and (7.28), respectively.

Proof Since

$$D^j W(0)(w_1, w_2, \dots w_j) = 0, \ j \geq 3, \ w_i := \begin{pmatrix} \begin{pmatrix} \alpha_i \\ N_i \\ P_{S,i} \\ P_{H,i} \end{pmatrix} \end{pmatrix} \in \mathbb{R}^4,$$

the reduced system (7.20) can be rewritten as follows by using the Taylor's expansion of W around 0:

$$\begin{cases} \frac{d\alpha(t)}{dt} = 0, \\ \frac{dx_c(t)}{dt} = A_c x_c(t) + \frac{1}{2}\Pi_c^{A(\alpha^\star)} D^2 W(0) \left((I + G_2) \begin{pmatrix} \alpha(t) \\ x_c(t) \end{pmatrix} \right)^2 + R_c \begin{pmatrix} \alpha(t) \\ x_c(t) \end{pmatrix}, \end{cases}$$

where $R_c \in C^4(\mathbb{R} \times X_c, X_c)$ is defined in (7.20) and is a remainder of order 4. Set

$$x_c = x b_1 + y b_2, (x, y) \in \mathbb{C}^2.$$

Since $\left\{\widehat{b}_1, \widehat{b}_2, \widehat{b}_3\right\}$ is used as the basis for $Y_c = \Pi_c(Y)$, i.e., $\{b_1, b_2\}$ is a basis of $X_c := \Pi_c^{A(\alpha^\star)}(\mathbb{R}^3)$, we obtain that

$$M_c = \begin{bmatrix} i\omega & 0 \\ 0 & -i\omega \end{bmatrix}$$

and

$$\begin{aligned} (I + G_2) \begin{pmatrix} \alpha \\ x_c \end{pmatrix} &= \alpha\widehat{b}_1 + x\widehat{b}_2 + y\widehat{b}_3 + L_2\left(\alpha\widehat{b}_1 + x\widehat{b}_2 + y\widehat{b}_3, \alpha\widehat{b}_1 + x\widehat{b}_2 + y\widehat{b}_3\right) \\ &= \alpha\widehat{b}_1 + x\widehat{b}_2 + y\widehat{b}_3 + x^2 L_2\left(\widehat{b}_2, \widehat{b}_2\right) + y^2 L_2\left(\widehat{b}_3, \widehat{b}_3\right) \\ &\quad + 2\alpha x L_2\left(\widehat{b}_1, \widehat{b}_2\right) + 2\alpha y L_2\left(\widehat{b}_1, \widehat{b}_3\right) + 2xy L_2\left(\widehat{b}_2, \widehat{b}_3\right). \end{aligned}$$

Then it follows that

$$(I + G_2) \begin{pmatrix} \alpha \\ x_c \end{pmatrix} = \begin{pmatrix} \alpha \\ \xi \end{pmatrix}, \tag{7.27}$$

where $\xi = \begin{pmatrix} \xi_1 \\ \xi_2 \\ \xi_3 \end{pmatrix}$ with

$$\xi_1 = \xi_{11} + \xi_{12}, \xi_{11} = x\,(A_{31})_{\alpha^\star,\, i\omega} + y\,(A_{31})_{\alpha^\star,\, -i\omega},$$

$$\xi_{12} = \alpha x\,(A_{31})_{\alpha^\star,\, i\omega}\,\Phi^1_{1,i\omega} + \alpha y\,(A_{31})_{\alpha^\star,\, -i\omega}\,\Phi^1_{1,-i\omega} + xyF_{21} + \frac{1}{2}x^2F_{11} + \frac{1}{2}y^2F_{31},$$

$$\xi_2 = \xi_{21} + \xi_{22}, \xi_{21} = x\,(A_{32})_{\alpha^\star,\, i\omega} + y\,(A_{32})_{\alpha^\star,\, -i\omega},$$

$$\xi_{22} = \alpha x\,(A_{31})_{\alpha^\star,\, i\omega}\,\Phi^2_{1,i\omega} + \alpha y\,(A_{31})_{\alpha^\star,\, -i\omega}\,\Phi^2_{1,-i\omega} + xyF_{22} + \frac{1}{2}x^2F_{12} + \frac{1}{2}y^2F_{32}.$$

$$\xi_3 = \xi_{31} + \xi_{32}, \xi_{31} = x\,(A_{33})_{\alpha^\star,\, i\omega} + y\,(A_{33})_{\alpha^\star,\, -i\omega},$$

$$\xi_{32} = \alpha x\,(A_{31})_{\alpha^\star,\, i\omega}\,\Phi^3_{1,i\omega} + \alpha y\,(A_{31})_{\alpha^\star,\, -i\omega}\,\Phi^3_{1,-i\omega} + xyF_{23} + \frac{1}{2}x^2F_{13} + \frac{1}{2}y^2F_{33}.$$

Then we deduce that

$$\frac{1}{2!}D^2W\,(0)\left(\begin{pmatrix} \alpha \\ \xi \end{pmatrix}\right)^2 = \begin{pmatrix} \alpha\xi_1 - \delta\xi_1^2 - \kappa\,\xi_1\xi_2 \\ -\rho\,\kappa\xi_2\,\xi_1 \\ \rho\,\kappa\xi_2\xi_1 \end{pmatrix}.$$

Note that

$$\det\begin{pmatrix} (A_{31})_{\alpha^\star,\, i\omega} & (A_{31})_{\alpha^\star,\, -i\omega} \\ (A_{32})_{\alpha^\star,\, i\omega} & (A_{32})_{\alpha^\star,\, -i\omega} \end{pmatrix} \neq 0$$

and

$$\begin{aligned}\Pi_c^{A(\alpha^\star)}\begin{pmatrix} 1 \\ 0 \\ 0 \end{pmatrix} &= \frac{\begin{pmatrix} A_{11} \\ A_{12} \\ A_{13} \end{pmatrix}_{\alpha^\star,\, i\omega}}{\Delta'(i\omega)_{\alpha^\star}} + \frac{\begin{pmatrix} A_{11} \\ A_{12} \\ A_{13} \end{pmatrix}_{\alpha^\star,\, -i\omega}}{\Delta'(-i\omega)_{\alpha^\star}} \\ &= \left(\frac{\vartheta_1}{\Delta'(i\omega)_{\alpha^\star}} + \frac{\overline{\vartheta}_2}{\Delta'(-i\omega)_{\alpha^\star}}\right)b_1 + \left(\frac{\overline{\vartheta}_1}{\Delta'(-i\omega)_{\alpha^\star}} + \frac{\vartheta_2}{\Delta'(i\omega)_{\alpha^\star}}\right)b_2,\end{aligned}$$

$$\begin{aligned}\Pi_c^{A(\alpha^\star)}\begin{pmatrix} 0 \\ 1 \\ 0 \end{pmatrix} &= \frac{\begin{pmatrix} A_{21} \\ A_{22} \\ A_{23} \end{pmatrix}_{\alpha^\star,\, i\omega}}{\Delta'(i\omega)_{\alpha^\star}} + \frac{\begin{pmatrix} A_{21} \\ A_{22} \\ A_{23} \end{pmatrix}_{\alpha^\star,\, -i\omega}}{\Delta'(-i\omega)_{\alpha^\star}} \\ &= \left(\frac{\vartheta_3}{\Delta'(i\omega)_{\alpha=\alpha^\star}} + \frac{\overline{\vartheta}_4}{\Delta'(-i\omega)_{\alpha=\alpha^\star}}\right)b_1 + \left(\frac{\overline{\vartheta}_3}{\Delta'(-i\omega)_{\alpha=\alpha^\star}} + \frac{\vartheta_4}{\Delta'(i\omega)_{\alpha=\alpha^\star}}\right)b_2,\end{aligned}$$

$$\Pi_c^{A(\alpha^\star)}\begin{pmatrix}0\\0\\1\end{pmatrix} = \frac{\begin{pmatrix}A_{31}\\A_{32}\\A_{33}\end{pmatrix}_{\alpha^\star,\, i\omega}}{\Delta'(i\omega)_{\alpha^\star}} + \frac{\begin{pmatrix}A_{31}\\A_{32}\\A_{33}\end{pmatrix}_{\alpha^\star,\, -i\omega}}{\Delta'(-i\omega)_{\alpha^\star}}$$
$$= \frac{1}{\Delta'(i\omega)_{\alpha^\star}} b_1 + \frac{1}{\Delta'(-i\omega)_{\alpha^\star}} b_2$$

with

$$\vartheta_1 = \frac{\det\begin{pmatrix}(A_{11})_{\alpha^\star,\, i\omega} & (A_{31})_{\alpha^\star,\, -i\omega}\\ (A_{12})_{\alpha^\star,\, i\omega} & (A_{32})_{\alpha^\star,\, -i\omega}\end{pmatrix}}{\det\begin{pmatrix}(A_{31})_{\alpha^\star,\, i\omega} & (A_{31})_{\alpha^\star,\, -i\omega}\\ (A_{32})_{\alpha^\star,\, i\omega} & (A_{32})_{\alpha^\star,\, -i\omega}\end{pmatrix}},\ \vartheta_2 = \frac{\det\begin{pmatrix}(A_{31})_{\alpha^\star,\, i\omega} & (A_{11})_{\alpha^\star,\, i\omega}\\ (A_{32})_{\alpha^\star,\, i\omega} & (A_{12})_{\alpha^\star,\, i\omega}\end{pmatrix}}{\det\begin{pmatrix}(A_{31})_{\alpha^\star,\, i\omega} & (A_{31})_{\alpha^\star,\, -i\omega}\\ (A_{32})_{\alpha^\star,\, i\omega} & (A_{32})_{\alpha^\star,\, -i\omega}\end{pmatrix}}, \tag{7.28}$$

$$\vartheta_3 = \frac{\det\begin{pmatrix}(A_{21})_{\alpha^\star,\, i\omega} & (A_{31})_{\alpha^\star,\, -i\omega}\\ (A_{22})_{\alpha^\star,\, i\omega} & (A_{32})_{\alpha^\star,\, -i\omega}\end{pmatrix}}{\det\begin{pmatrix}(A_{31})_{\alpha^\star,\, i\omega} & (A_{31})_{\alpha^\star,\, -i\omega}\\ (A_{32})_{\alpha^\star,\, i\omega} & (A_{32})_{\alpha^\star,\, -i\omega}\end{pmatrix}},\ \vartheta_4 = \frac{\det\begin{pmatrix}(A_{31})_{\alpha^\star,\, i\omega} & (A_{21})_{\alpha^\star,\, i\omega}\\ (A_{32})_{\alpha^\star,\, i\omega} & (A_{22})_{\alpha^\star,\, i\omega}\end{pmatrix}}{\det\begin{pmatrix}(A_{31})_{\alpha^\star,\, i\omega} & (A_{31})_{\alpha^\star,\, -i\omega}\\ (A_{32})_{\alpha^\star,\, i\omega} & (A_{32})_{\alpha^\star,\, -i\omega}\end{pmatrix}}.$$

By projecting on X_c, we obtain that

$$\frac{1}{2!}\Pi_c^{A(\alpha^\star)} D^2W(0)\left((I+G_2)\left(\begin{pmatrix}\alpha\\x_c\end{pmatrix}\right)\right)^2$$
$$= \Pi_c^{A(\alpha^\star)}\begin{pmatrix}\alpha\xi_1 - \delta\xi_1^2 - \kappa\xi_1\xi_2\\ -\rho\,\kappa\xi_2\,\xi_1\\ \rho\,\kappa\xi_2\xi_1\end{pmatrix}$$
$$= \left(\alpha\xi_1 - \delta\xi_1^2 - \kappa\xi_1\xi_2\right)\Pi_c^{A(\alpha^\star)}\begin{pmatrix}1\\0\\0\end{pmatrix} - \rho\,\kappa\xi_2\,\xi_1\left[\Pi_c^{A(\alpha^\star)}\begin{pmatrix}0\\1\\0\end{pmatrix} - \Pi_c^{A(\alpha^\star)}\begin{pmatrix}0\\0\\1\end{pmatrix}\right].$$

Thus we obtain H_2, H_3 and $\widetilde{R}_c$ as stated in Theorem 7.8. □

7.4.4 Computation of the Normal Form of the Reduced System

We will compute the normal form of the reduced system (7.25) up to terms of order 3 by performing inductively a sequence of change of coordinates of the form

$$u = \varsigma_{m,c}(\overline{u}) = \overline{u} + G_{m,c}(\overline{u}), \overline{u} \in B^m(0,r), \tag{7.29}$$

where $G_{m,c}(\overline{u}) \in H_m^3(\mathbb{C}^3)$ and $B^m(0,r)$ is a small neighborhood of the origin in $\mathbb{C}^3$, $2 \le m \le 3$. Consider the linear operator $\Theta_{m,c} : H_m^3(\mathbb{C}^3) \to H_m^3(\mathbb{C}^3)$ by

$$\left(\Theta_{m,c}G_{m,c}\right)(u) = [B, G_{m,c}](u) = D_u G_{m,c}(u)Bu - BG_{m,c}(u), \tag{7.30}$$

$$u = \begin{pmatrix} \alpha \\ x \\ y \end{pmatrix}, \; G_{m,c} = \begin{pmatrix} G^1_{m,c} \\ G^2_{m,c} \\ G^3_{m,c} \end{pmatrix} \in H^3_m(\mathbb{C}^3), \; B = \begin{bmatrix} 0 & 0 & 0 \\ 0 & i\omega & 0 \\ 0 & 0 & -i\omega \end{bmatrix}.$$

A basis denoted by Φ_m for $H^3_m(\mathbb{C}^3), 2 \leq m \leq 3$, consists of all possible vector-valued homogeneous polynomials of degree m in 3 variables α, x, y with coefficients in $\{e_1, e_2, e_3\}$, that is

$$\Phi_m = \left\{ \begin{array}{c} u^d e_j \big|, \; u^d = \alpha^{d_1} x^{d_2} y^{d_3}, \; d = (d_1, d_2, d_3), \; \sum\limits_{i=1}^{3} d_i = m, \; d_i \geq 0 \text{ are integers}, \\ j = 1, \cdots, 3. \end{array} \right\}.$$

The dimension of $H^3_m(\mathbb{C}^3)$ is

$$d_m := \dim H^3_m(\mathbb{C}^3) = \frac{3\,(m+2)!}{2!m!}.$$

A direct calculation shows that for any $u^d e_j \in \Phi_m$,

$$\Theta_{m,c}\left(u^d e_j\right) = \left(i\omega d_2 - i\omega d_3 - \lambda_j\right) u^d e_j, \lambda_j \in \sigma(B),$$
$$\text{where } \lambda_1 = 0, \lambda_2 = i\omega, \lambda_3 = -i\omega.$$

Hence we know that $H^3_m(\mathbb{C}^3)$ can be represented as the direct sum

$$H^3_m(\mathbb{C}^3) = R(\Theta_{m,c}) \oplus \ker(\Theta_{m,c}),$$

where $R(\Theta_{m,c})$ is the range of $\Theta_{m,c}$, and $\ker(\Theta_{m,c})$ is the null-space of $\Theta_{m,c}$ and spanned by all resonant monomials of order m, $2 \leq m \leq 3$, that is,

$$\ker(\Theta_{m,c}) = \text{span}\{u^d e_k : \lambda \cdot d = \lambda_k, \; k = 1, 2, 3, \; u^d = \alpha^{d_1} x^{d_2} y^{d_3}, d = (d_1, d_2, d_3),$$
$$\sum_{j=1}^{3} d_j = m, \; d_j \geq 0 \text{ are integers}, \lambda \cdot d = \sum_{j=1}^{3} d_j \lambda_j\},$$

where $\{e_1, e_2, e_3\}$ is the standard basis of $\mathbb{C}^3$ and $\lambda = (\lambda_1, \lambda_2, \lambda_3)$ with $\lambda_1 = 0$, $\lambda_2 = i\omega$, $\lambda_3 = -i\omega$. Then it follows that

$$\ker(\Theta_{2,c}) = \text{span}\left\{ \begin{pmatrix} \alpha^2 \\ 0 \\ 0 \end{pmatrix}, \begin{pmatrix} xy \\ 0 \\ 0 \end{pmatrix}, \begin{pmatrix} 0 \\ \alpha x \\ 0 \end{pmatrix}, \begin{pmatrix} 0 \\ 0 \\ \alpha y \end{pmatrix} \right\},$$

$$\ker(\Theta_{3,c}) = \text{span}\left\{ \begin{pmatrix} \alpha^3 \\ 0 \\ 0 \end{pmatrix}, \begin{pmatrix} \alpha xy \\ 0 \\ 0 \end{pmatrix}, \begin{pmatrix} 0 \\ \alpha^2 x \\ 0 \end{pmatrix}, \begin{pmatrix} 0 \\ x^2 y \\ 0 \end{pmatrix}, \begin{pmatrix} 0 \\ 0 \\ \alpha^2 y \end{pmatrix}, \begin{pmatrix} 0 \\ 0 \\ xy^2 \end{pmatrix} \right\}.$$

By using the change of coordinates

$$u = \varsigma_{2,c}(\overline{u}) = \overline{u} + G_{2,c}(\overline{u}), \overline{u} \in B^2(0, r) \tag{7.31}$$

and dropping the hat as well as the auxiliary equation introduced to handle the parameter, we get

$$\frac{d}{dt}\begin{pmatrix} x(t) \\ y(t) \end{pmatrix} = M_c \begin{pmatrix} x(t) \\ y(t) \end{pmatrix} + \begin{pmatrix} C_1 x\alpha \\ \overline{C}_1 y\alpha \end{pmatrix} + \left(\widetilde{H}_3 + \widetilde{R}_{1c}\right)\left(\begin{pmatrix} \alpha(t) \\ \begin{pmatrix} x(t) \\ y(t) \end{pmatrix} \end{pmatrix}\right),$$

with

$$C_1 = (A_{31})_{\alpha^\star, i\omega} \left(\frac{\vartheta_1}{\Delta'(i\omega)_{\alpha^\star}} + \frac{\overline{\vartheta}_2}{\Delta'(-i\omega)_{\alpha^\star}} \right). \tag{7.32}$$

In order to get the third order terms of the normal form, we must compute $G_{2,c}$ and the third order terms

$$\widetilde{H}_3 = H_3(u) + D_u H_2(u)\, G_2(u) - D_u G_2(u) \begin{pmatrix} C_1 x\alpha \\ \overline{C}_1 y\alpha \end{pmatrix}$$

with $u = \left(\begin{pmatrix} \alpha \\ \begin{pmatrix} x \\ y \end{pmatrix} \end{pmatrix}\right)$. It is sufficient to get the coefficients of $\begin{pmatrix} x^2 y \\ 0 \end{pmatrix}$ and $\begin{pmatrix} 0 \\ xy^2 \end{pmatrix}$ for studying the Hopf bifurcation. Since

$$H_2\left(\begin{pmatrix} 0 \\ \begin{pmatrix} x \\ y \end{pmatrix} \end{pmatrix}\right) = \varpi_1 x^2 + \varpi_2 xy + \varpi_3 y^2,$$

with

$$\begin{aligned}
\varpi_1 &= \begin{pmatrix} \varpi_{11} \\ \varpi_{12} \end{pmatrix} \qquad (7.33)\\
&= -\delta\, (A_{31})^2_{\alpha^\star, i\omega} \begin{pmatrix} \Upsilon_{11} \\ \Upsilon_{12} \end{pmatrix} - \kappa\, (A_{31})_{\alpha^\star, i\omega} (A_{32})_{\alpha^\star, i\omega} \begin{pmatrix} \Upsilon_{21} \\ \Upsilon_{22} \end{pmatrix},\\
\varpi_2 &= \begin{pmatrix} \varpi_{21} \\ \varpi_{22} \end{pmatrix}\\
&= -2\delta\, (A_{31})_{\alpha^\star, i\omega} (A_{31})_{\alpha^\star, -i\omega} \begin{pmatrix} \Upsilon_{11} \\ \Upsilon_{12} \end{pmatrix}\\
&\quad -\kappa \left[(A_{32})_{\alpha^\star, i\omega} (A_{31})_{\alpha^\star, -i\omega} + (A_{32})_{\alpha^\star, -i\omega} (A_{31})_{\alpha^\star, i\omega} \right] \begin{pmatrix} \Upsilon_{21} \\ \Upsilon_{22} \end{pmatrix},\\
\varpi_3 &= \begin{pmatrix} \varpi_{31} \\ \varpi_{32} \end{pmatrix}\\
&= -\delta\, (A_{31})^2_{\alpha^\star, -i\omega} \begin{pmatrix} \Upsilon_{11} \\ \Upsilon_{12} \end{pmatrix} - \kappa\, (A_{32})_{\alpha^\star, -i\omega} (A_{31})_{\alpha^\star, -i\omega} \begin{pmatrix} \Upsilon_{21} \\ \Upsilon_{22} \end{pmatrix} \qquad (7.34)
\end{aligned}$$

and

$$\begin{pmatrix} G^2_{2,c}\left(\begin{pmatrix} 0 \\ x \\ y \end{pmatrix}\right) \\ G^3_{2,c}\left(\begin{pmatrix} 0 \\ x \\ y \end{pmatrix}\right) \end{pmatrix} = \frac{1}{i\omega}\begin{pmatrix} \varpi_{11}x^2 - \varpi_{21}xy - \frac{1}{3}\varpi_{31}y^2 \\ \frac{1}{3}\varpi_{12}x^2 + \varpi_{22}xy - \varpi_{32}y^2 \end{pmatrix},$$

we get that the normal form has the following form after dropping the auxiliary equation introduced for handling the parameter,

$$\begin{aligned} \frac{d}{dt}\begin{pmatrix} x(t) \\ y(t) \end{pmatrix} = M_c \begin{pmatrix} x(t) \\ y(t) \end{pmatrix} + \begin{pmatrix} C_1 x\alpha \\ \overline{C}_1 y\alpha \end{pmatrix} + \begin{pmatrix} C_2 x^2 y \\ \overline{C}_2 xy^2 \end{pmatrix} \\ + O(|(x,y)|\alpha^2 + |(\alpha,(x,y))|^4), \end{aligned} \tag{7.35}$$

with

$$\begin{aligned} C_2 = & -\left[2\delta F_{21}(A_{31})_{\alpha^\star, i\omega} + F_{11}\delta (A_{31})_{\alpha^\star, -i\omega}\right]\Upsilon_{11} \\ & -\kappa\left[F_{22}(A_{31})_{\alpha^\star, i\omega} + \frac{1}{2}F_{12}(A_{31})_{\alpha^\star, -i\omega} + F_{21}(A_{32})_{\alpha^\star, i\omega}\right. \\ & \left. + \frac{1}{2}F_{11}(A_{32})_{\alpha^\star, -i\omega}\right]\Upsilon_{21} + \frac{1}{i\omega}\left(\frac{2}{3}\varpi_{31}\varpi_{12} - \varpi_{11}\varpi_{21} + \varpi_{22}\varpi_{21}\right). \end{aligned}$$

The normal form (7.35) can be written in real coordinates (w_1, w_2) through the change of variables $x = v_1 - iv_2$, $y = v_1 + iv_2$. Setting $v_1 = \rho\cos\theta$, $v_2 = \rho\sin\theta$, it becomes

$$\begin{cases} \dot{\rho} = l_1\alpha\rho + l_2\rho^3 + O(\alpha^2\rho + |(\rho,\alpha)|^4), \\ \dot{\theta} = -\omega + O(|(\rho,\alpha)|), \end{cases} \tag{7.36}$$

where

$$l_1 = \operatorname{Re}(C_1) = 2p_2\left(\frac{\left(\gamma\eta - (\beta_P - \mu_P)^2\right)p_1}{\beta_P - \mu_P} + p_2\right) > 0, \; l_2 = \operatorname{Re}(C_2).$$

From Chapter 3 or [54], we know that the sign of l_2 determines the direction of the bifurcation and the stability of the nontrivial periodic orbits. In summary we have the following theorem.

Theorem 7.9 *Let Assumptions 7.1 and 7.2 be satisfied. The direction of the Hopf bifurcation described in Theorem 7.3 and the stability of the bifurcating periodic solutions are determined by the sign of* l_2 *: Hopf bifurcation is supercritical and the bifurcating periodic solutions are stable if* $l_2 < 0$*, and subcritical and unstable if* $l_2 > 0$.

7.5 Numerical Simulations and Conclusions

By carrying out bifurcation analysis and normal form computation of the model (7.1), we discussed the existence and properties of Hopf bifurcation. In this section, we provide some numerical simulations to illustrate the results. We firstly choose parameters $\delta = 8$, $\kappa = 1/2$, $\rho = 2$, $\mu_P = 1/7, \beta_P = 8.5127$, $\eta = 9.24$, $\gamma = 91$. In order to get the properties of Hopf bifurcation using Theorem 7.9, we need compute l_1 and l_2. Using Matlab we get $\alpha^\star = 676.347, l_1 = \mathrm{Re}\,(C_1) = 0.496645, l_2 = \mathrm{Re}\,(C_2) = -1.61528$. Then we choose $\alpha = \beta_N - \mu_N = 1.6567 - 1 = 0.6567 < \alpha^\star$, $N(0) = 0.7$, $P_S(0) = 1.40684$, $P_H(0) = 0.0071$ in Figure 7.1 and $\alpha = \beta_N - \mu_N = 700 - 1 = 699 > \alpha^\star$, $N(0) = 2.70 \times 10^{-81}$, $P_S(0) = 1467, P_H(0) = 145.7$ in Figure 7.2, respectively, and obtain graphs of $N(t)$, $P_S(t)$ and $P_H(t)$ by using Matlab.

Figure 7.1 and Figure 7.2, demonstrate that the positive equilibrium E^* of system (7.1) is asymptotically stable when the intrinsic growth rate α of the prey species is less than its critical value $\alpha^\star$ and system (7.1) undergoes a Hopf bifurcation and a stable non-trivial periodic solution bifurcates from the positive equilibrium when the intrinsic growth rate α of the prey species passes through the critical value $\alpha^\star$ which in accordance with Theorem 7.9.

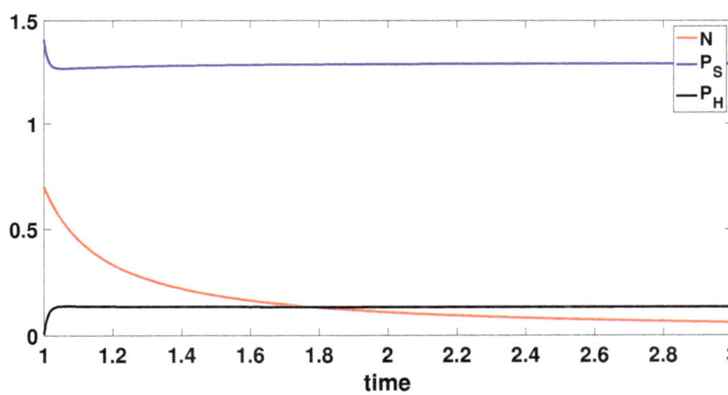

Fig. 7.1: *In this figure we run a simulation of the model* (7.2) *with stable positive equilibrium* E^*.

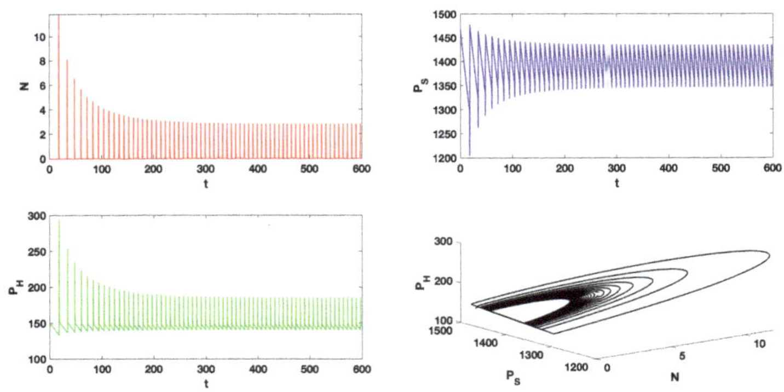

Fig. 7.2: *In this figure we run a simulation of the model* (7.2) *with stable periodic solution around the positive equilibrium* E^*.

7.6 Remarks and notes

By Chapter 5, we have two ways to compute the normal form of the reduced system of a system topologically equivalent to the original system (7.14). One is to compute the Taylor's expansion of the reduced system first and then we can compute the normal form of the reduced system. The other way is to compute the normal form of the reduced system directly. We use the first way to compute the normal form for Hopf bifurcation in this Chapter. We refer to the books [151] by Hassard, Kazarinoff and Wan, [53] by Chow and Hale, [54] by Chow et al., [201] by Kuznetsov, [352] by Wiggins and others for more results on the existence and properties of Hopf bifurcation.

7.7 MATLAB codes

7.7.1 Figure 7.1

```

tspan = [1 600];

y0 = [2.70*1e-81;1467;145.7];
[T,Y] = ode45(@myfun, tspan ,y0);
subplot(2,2,1)
plot(T,Y(:,1),'r','linewidth',1.5);
xlabel('t')
ylabel('N')
```

```
set(gca, 'FontSize', 15);
set(gca, 'FontWeight', 'bold');
subplot(2,2,2)
plot(T,Y(:,2),'b','linewidth',1.5);
xlabel('t')
ylabel('P_S')
set(gca, 'FontSize', 15);
set(gca, 'FontWeight', 'bold');
subplot(2,2,3)
plot(T,Y(:,3),'g','linewidth',1.5);
xlabel('t')
ylabel('P_H')
set(gca, 'FontSize', 15);
set(gca, 'FontWeight', 'bold');
subplot(2,2,4)
plot3(Y(:,1),Y(:,2),Y(:,3),'k','linewidth',1.5);

xlabel('N')
ylabel('P_S')
zlabel('P_H')
set(gca, 'FontSize', 15);
set(gca, 'FontWeight', 'bold');

function dy = myfun(t,y)
muN = 1;
betaN = 700;
delta = 8;
k=1/2;
rho=2;
muP = 1/7;
betaP = 8.5127;
eta = 9.24;
r=91;
dy = zeros(3,1);
dy(1) = (betaN-muN)*y(1)-delta*y(1)*y(1)-k*y(1)*y(2)
    ;
dy(2) = -(muP+eta)*y(2)-rho*k*y(1)*y(2)+y(3)*r;
dy(3) = -muP*y(3)+k*y(1)*y(2)*rho-y(3)*r+betaP*(y(2)
    +y(3));
dy=dy(:);
end
```

7.7.2 Figure 7.2

```

tspan = [1 600];
```

```

y0 = [2.70*1e-81;1467;145.7];
[T,Y] = ode45(@myfun,tspan,y0);
subplot(2,2,1)
plot(T,Y(:,1),'r','linewidth',1.5);
xlabel('t')
ylabel('N')
set(gca, 'FontSize', 15);
set(gca, 'FontWeight', 'bold');
subplot(2,2,2)
plot(T,Y(:,2),'b','linewidth',1.5);
xlabel('t')
ylabel('P_S')
set(gca, 'FontSize', 15);
set(gca, 'FontWeight', 'bold');
subplot(2,2,3)
plot(T,Y(:,3),'g','linewidth',1.5);
xlabel('t')
ylabel('P_H')
set(gca, 'FontSize', 15);
set(gca, 'FontWeight', 'bold');
subplot(2,2,4)
plot3(Y(:,1),Y(:,2),Y(:,3),'k','linewidth',1.5);

xlabel('N')
ylabel('P_S')
zlabel('P_H')
set(gca, 'FontSize', 15);
set(gca, 'FontWeight', 'bold');

function dy = myfun(t,y)
muN = 1;
betaN = 700;
delta = 8;
k=1/2;
rho=2;
muP = 1/7;
betaP = 8.5127;
eta = 9.24;
r=91;
dy = zeros(3,1);
dy(1) = (betaN-muN)*y(1)-delta*y(1)*y(1)-k*y(1)*y(2)
    ;
dy(2) = -(muP+eta)*y(2)-rho*k*y(1)*y(2)+y(3)*r;
dy(3) = -muP*y(3)+k*y(1)*y(2)*rho-y(3)*r+betaP*(y(2)
    +y(3));
```

```
dy=dy(:);
end
```

Chapter 8
Large Speed Traveling Waves for the Rosenzweig-MacArthur predator-prey Model with Spatial Diffusion

This chapter focuses on traveling wave solutions for the so-called Rosenzweig-MacArthur predator-prey model with spatial diffusion. The main results of this note are concerned with the existence and uniqueness of traveling wave solution as well as periodic wave train solution in the large wave speed asymptotic. Depending on the model parameters we more particularly study the existence and uniqueness of a traveling wave connecting two equilibria or connecting an equilibrium point and a periodic wave train. We also discuss the existence and uniqueness of such a periodic wave train. Our analysis is based on ordinary differential equation techniques by coupling the theories of invariant manifolds together with those of global attractors.

8.1 Introduction

In this work we study the traveling wave solutions for the so-called diffusive Rosenzweig-MacArthur predator-prey system that reads as follows

$$\begin{cases} u_t = \delta_1 u_{xx} + Au\left(1 - \dfrac{u}{K}\right) - B\dfrac{uv}{1+Eu}, \\ v_t = \delta_2 v_{xx} - Cv + D\dfrac{uv}{1+Eu}. \end{cases} \tag{8.1}$$

This system is posed for the one-dimensional spatial variable $x \in \mathbb{R}$ while t denotes time.

In the above system of equations $u = u(t,x)$ denotes the density of the prey population while $v = v(t,x)$ corresponds to those of the predator, at time $t > 0$ and spatial location $x \in \mathbb{R}$. The positive parameters δ_1 and δ_2 represent the diffusion coefficients for the prey and the predator, respectively. The underlying kinetic system describes the dynamics of the populations as well as their interactions and reads as the following ordinary differential equations (ODE for short)

$$\begin{cases} u'(t) = Au\left(1 - \dfrac{u}{K}\right) - B\dfrac{uv}{1+Eu}, \\ v'(t) = -Cv + D\dfrac{uv}{1+Eu}, \end{cases} \tag{8.2}$$

A. Ducrot et al., *Differential Equations and Population Dynamics II*,
Lecture Notes on Mathematical Modelling in the Life Sciences,
https://doi.org/10.1007/978-3-031-91409-6_8

wherein A, B, C, D, E and K are given positive constants. More precisely A stands for the growth factor for the prey species, K denotes its carrying capacity, B and D are the interaction rates for the two species while C corresponds to the natural death rate for the predator. Finally the parameter E measures the "satiation" effect of the predator population. We refer the reader to Holling [161] and Rosenzweig and and MacArthur [285], Rosenzweig [284] for more details on this model.

The aim of this chapter is to discuss the existence and qualitative properties of the traveling wave and the periodic wave train solutions for (8.1) . To discuss this issue, we first rescale the system by introducing

$$U = Eu,\ V = Bv/C,\ t' = Ct,\ x' = (C/\delta_2)^{1/2}x,$$
$$d = \frac{\delta_1}{\delta_2},\ \alpha = A/(ECK),\ \gamma = EK,\ \beta = D/(EC).$$

With these new variables and normalized parameters, (8.1) rewrites, omitting the prime for notational simplicity, as the following reaction-diffusion system

$$\begin{cases} U_t = dU_{xx} + \alpha U(\gamma - U) - \dfrac{UV}{1+U}, \\ V_t = V_{xx} - V + \beta\dfrac{UV}{1+U}, \end{cases} \tag{8.3}$$

while the underlying kinetic system, namely (8.2), becomes

$$\begin{cases} U' = \alpha U(\gamma - U) - \dfrac{UV}{1+U}, \\ V' = -V + \beta\dfrac{UV}{1+U}. \end{cases} \tag{8.4}$$

As mentioned above, the goal of this chapter is to discuss some properties of the traveling wave and periodic wave train solutions for the reaction-diffusion system (8.3). Here recall that a traveling wave solution corresponds to a complete orbit of (8.3) (that is a solution defined for all time $t \in \mathbb{R}$) of the form

$$U(t,x) = u(s),\quad V(t,x) = v(s) \text{ with } s = x + ct,$$

where $c \in \mathbb{R}$ is some constant that stands for the wave speed. When the profile $s \mapsto (u(s), v(s))$ is periodic we speak about periodic wave train with speed c. Plugging this specific form into (8.3) yields the following ODE system for the wave profiles $(u, v) = (u(s), v(s))$ for $s \in \mathbb{R}$

$$\begin{cases} cu' = du'' + \alpha u(\gamma - u) - \dfrac{uv}{1+u}, \\ cv' = v'' - v + \beta\dfrac{uv}{1+u}. \end{cases} \tag{8.5}$$

Traveling wave solutions for the above system or more generally for predator-prey systems have been widely investigated in the last decades. One may refer the reader to the works of Dunbar [95, 96, 97] who proposed ODE methods coupled with topological arguments to prove the existence of such special solutions. One may also refer to Gardner [118] who developed topological arguments based on the Conley index to obtain the existence of solutions with suitable behaviour at $s = \pm\infty$. We also refer to Huang, Lu and Ruan [173] for more general results also based on a coupling between ODE methods and topological arguments. We refer to Ruan [289] for some results on the existence of periodic wave train by using using Hopf bifurcation method. We refer the reader to the work of Hosono [164] and the references cited therein for results about the Lotka-Volterra predator-prey system as well as to the work of Li and Xiao [214] (see also the references therein) for results about the existence of traveling waves for more general functional responses and also for a nice review on this topic. The connexion between wave solutions and the asymptotic behaviour of the Cauchy problem (8.3) (when equipped with suitable initial data) has been scarcely studied. One may refer the reader to Gardner [117] who studied the local stability of wave solutions and to Ducrot, Giletti and Matano [88] (and the references therein) for results related to the so-called asymptotic speed of spread. We also refer to [4, 85, 165, 175] for other results.

One important difficulty when studying traveling wave solutions for predator-prey interactions relies on the ability of the underlying kinetic to develop sustained oscillations, typically through Hopf bifurcation. Hence the behaviour of the solutions of the corresponding reaction-diffusion system are expected to exhibit somehow complex spatio-temporal oscillations. Therefore the traveling wave solutions describing for instance the spatial invasion of a predator is also expected to exhibit oscillating patterns, connecting a predator-free equilibrium and some oscillating state, such as a periodic wave train (see [173, 289] for results about the existence of such periodic solutions using bifurcation methods). According to our knowledge, this question related to the shape and the behaviour of traveling waves remains largely open. It has been addressed by Dunbar in [97] and further developed by Huang [174]. In this aforementioned work, the author developed refined singular perturbation analysis based on the hyperbolicity of the periodic solutions of the kinetic system to construct oscillating traveling wave in the large speed asymptotic. In this work we revisit this issue by developing a dynamical system approach to obtain a complete picture of the traveling wave solutions for system (8.5), in the large wave speed asymptotic, $c \gg 1$. Our methodology also allows us to provide uniqueness results, on the one hand for traveling waves and, on the other hand, also for periodic wave trains with large wave speed.

In this chapter, we describe in particular sharp conditions on the parameters of the system that ensure the existence of a unique traveling wave solution for (8.5) connecting the predator free equilibrium to the interior equilibrium or to a unique periodic wave train. To reach such a refined description, we develop a methodology based on dynamical system arguments. Here we will more precisely couple center manifold and more generally invariant manifold reduction together with the global attractor theory and qualitative analysis for ODE. In fact, the method presented here is rather general and can be used to work on the traveling waves with other and more general nonlinearities. It can also be extended to handle infinite dimensional problems such as reaction-diffusion systems with time delay.

To perform our analysis, we make use of successive rescaling arguments to restrict our analysis to a system of two ordinary differential equations. Firstly let us set

$$\widehat{u}(s) = u(-cs), \widehat{v}(s) = v(-cs).$$

Then, dropping the hats on u and v for notational convenience, (8.5) becomes

$$\begin{cases} -u' = \dfrac{d}{c^2}u'' + \alpha u(\gamma - u) - \dfrac{uv}{1+u}, \\ -v' = \dfrac{1}{c^2}v'' - v + \beta\dfrac{uv}{1+u}. \end{cases} \tag{8.6}$$

Next let us set $\varepsilon = \frac{1}{c^2}$, $u_1 = u, u_2 = u', v_1 = v$ and $v_2 = v'$, so that the above problem (8.6) rewrites as

$$\begin{cases} u_1' = u_2, \\ d\varepsilon u_2' = -u_2 - \alpha u_1(\gamma - u_1) + \dfrac{u_1 v_1}{1+u_1}, \\ v_1' = v_2, \\ \varepsilon v_2' = -v_2 + v_1 - \beta\dfrac{u_1 v_1}{1+u_1}. \end{cases} \tag{8.7}$$

Set

$$\widehat{u}_1(s) = u_1(\varepsilon s), \widehat{u}_2(s) = u_2(\varepsilon s), \widehat{v}_1(s) = v_1(\varepsilon s), \widehat{v}_2(s) = v_2(\varepsilon s),$$

and (8.7) becomes (dropping the hats for notational convenience)

$$\begin{cases} u_1' = \varepsilon u_2, \\ du_2' = -u_2 - \alpha u_1(\gamma - u_1) + \dfrac{u_1 v_1}{1+u_1}, \\ v_1' = \varepsilon v_2, \\ v_2' = -v_2 + v_1 - \beta\dfrac{u_1 v_1}{1+u_1}, \end{cases} \tag{8.8}$$

where all the parameters d, α, γ, β and ε are strictly positive.

As mentioned above, in this chapter we will investigate traveling waves and periodic wave trains for (8.5), that correspond to heteroclinic connexions and periodic orbits, respectively, for system (8.7) or equivalently (8.8). Here we focus our study on the large speed asymptotic, namely $c \gg 1$, that is $0 < \varepsilon = \frac{1}{c^2} \ll 1$. To study this problem we will use center manifold reduction arguments to rewrite (8.8) on a suitable invariant set as a small perturbation of the kinetic system (8.4). The description of the heteroclinic and periodic orbits of the perturbed problem are then investigated using global attractor theory.

The organization of this chapter is as follows. Section 8.2 is devoted to the description of the global attractor for the Rosenzweig-MacArthur model (8.4) with a particular attention paid on the heteroclinic orbits and their uniqueness. Section 8.3 is concerned with the study of some complete orbit of (8.8), in the regime $0 < \varepsilon \ll 1$. We first reformulate this problem as a small perturbation of (8.4). We then study its global attractor and derive existence and uniqueness results for the traveling waves and periodic wave trains for (8.5) whenever c is large enough. In the last section we present some numerical simulations for the system in order illustrate our results.

8.2 Global attractors for the Rosenzweig-MacArthur model

In this section we propose a refine description of the global attractor of the Rosenzweig-MacArthur model

$$\begin{cases} U'(t) = \alpha U(\gamma - U) - \dfrac{UV}{1+U}, \\ V'(t) = -V + \beta \dfrac{UV}{1+U}. \end{cases} \tag{8.9}$$

The results presented in this section are mainly due to Hsu [167][Theorem 3.3], Hsu, Hubbell and Waltman [169, Lemma 4] where the global stability of the interior equilibrium is obtained by using the Dulac criterium, and to Cheng [47] who proved the uniqueness of the periodic orbit. In this section, we reformulate these results using the theory of the global attractor and as mentioned above we propose a refine description of this object by studying the existence and uniqueness of heteroclinic orbit starting from the no predator region ($V = 0$) to the interior global attractor (where $U > 0$ and $V > 0$). The results presented in the next main section, about (8.8), will make use of the refined description presented in this section.

To study (8.9) let us first observe that this system admits the following equilibrium points. The boundary equilibria are given by

$$(\overline{U}_0, \overline{V}_0) = (0,0) \text{ and } (\overline{U}_1, \overline{V}_1) = (\gamma, 0) \tag{8.10}$$

and the unique *interior equilibrium* whenever $\gamma(\beta - 1) > 1$, is given by

$$(\overline{U}_2, \overline{V}_2) = \left(\frac{1}{\beta - 1}, \frac{\alpha\beta\left[\gamma(\beta-1) - 1\right]}{(\beta-1)^2} \right). \tag{8.11}$$

Next define the functions

$$F(U,V) = \alpha U(\gamma - U) - \frac{UV}{1+U} = \frac{U}{1+U}[f(U) - V]$$

and

$$G(U,V) = V\left(\frac{\beta U}{1+U} - 1\right) = (\beta - 1)\frac{V\left(U - \overline{U}_2\right)}{1+U}$$

with the nullclines

$$f(U) = \alpha(\gamma - U)(1+U) \text{ and } U = \overline{U}_2$$

for U-equation and V-equation, respectively. Note that the map $f(U)$ is symmetric with respect to the vertical line $U = \frac{\gamma-1}{2}$.

We make the following parameter hypothesis for convenience and will refer these conditions throughout the chapter:
(H1): $\gamma(\beta - 1) > 1$;
(H2): $\gamma(\beta - 1) > \beta + 1$;

(H3): $1 < \gamma(\beta - 1) < \beta + 1$.

We now discuss the existence of global attractor for (8.9) and those existence is ensured by the next lemma.

Lemma 8.1 (Global attractor) *Let (H1) be satisfied. There exists $R > 0$ such that the triangle*

$$\mathbb{T} = \left\{(U, V) \in [0, \infty)^2 : \beta U + V \leq R\right\}$$

is positively invariant by the semiflow generated by (8.9). *The (positive) semiflow generated by* (8.9) *in $\mathbb{R}^2_+$ admits a global attractor, denoted by $\mathcal{A}_{\mathbb{R}^2_+}$, that is contained in $\mathbb{T}$. Furthermore the triangle $\mathbb{T}$ contains all the non negative equilibria of* (8.9).

Note that the invariance property follows from the fact that for all $R > 0$ large enough $\beta F(U, V) + G(U, V) < 0$ whenever $\beta U + V = R$ and $U \geq 0$ and $V \geq 0$.

We now focus on the precise description $\mathcal{A}_{\mathbb{R}^2_+}$. To that aim we first discuss the existence of an interior attractor by considering the regions

$$\partial_U \mathbb{R}^2_+ = \left\{(U, V) \in \mathbb{R}^2_+ : V = 0\right\} \text{ and } \partial_V \mathbb{R}^2_+ = \left\{(U, V) \in \mathbb{R}^2_+ : U = 0\right\},$$

and leading to the state space (disjoint) decomposition $\mathbb{R}^2_+ = \text{Int}\left(\mathbb{R}^2_+\right) \cup \left(\partial_U \mathbb{R}^2_+ \cup \partial_V \mathbb{R}^2_+\right)$. In the following lemma we are using the notion of global attractor considered first by Hale [138, 139]. We refer to Magal and Zhao [244] and Magal [236] for more results and examples about global attractors only attracting compact subsets (see also Chapter 2).

Lemma 8.2 *Let (H1) be satisfied. The semiflow generated by* (8.9) *restricted to $\mathbb{R}^2_+$ (respectively $\partial_U \mathbb{R}^2_+$, $\partial_V \mathbb{R}^2_+$ and $\text{Int}\left(\mathbb{R}^2_+\right)$) has a global attractor $\mathcal{A}_{\mathbb{R}^2_+}$ (respectively $\mathcal{A}_{\partial_U \mathbb{R}^2_+}$, $\mathcal{A}_{\partial_V \mathbb{R}^2_+}$ and $\mathcal{A}_{\text{Int}\left(\mathbb{R}^2_+\right)}$) which is a compact and connected subset which attracts all the compact subsets of $\mathbb{R}^2_+$ (respectively $\partial_U \mathbb{R}^2_+$, $\partial_V \mathbb{R}^2_+$ and $\text{Int}\left(\mathbb{R}^2_+\right)$).*

Remark 8.3 The global attractor $\mathcal{A}_{\text{Int}\left(\mathbb{R}^2_+\right)}$ only attracts the compact subsets of $\text{Int}\left(\mathbb{R}^2_+\right)$. That is to say that $\mathcal{A}_{\text{Int}\left(\mathbb{R}^2_+\right)}$ does not attract the bounded subsets of the interior region $\text{Int}\left(\mathbb{R}^2_+\right)$ (see Magal and Zhao [244] for more examples).

It is readily checked that the global attractor in $\partial_V \mathbb{R}^2_+$ is $\mathcal{A}_{\partial_V \mathbb{R}^2_+} = \{(0, 0)\}$ while the global attractor in $\partial_U \mathbb{R}^2_+$ is

$$\mathcal{A}_{\partial_U \mathbb{R}^2_+} = \left\{(U, V) \in \mathbb{R}^2_+ : U \in [0, \gamma] \text{ and } V = 0\right\}.$$

Indeed $\mathcal{A}_{\partial_U \mathbb{R}^2_+}$ contains the two equilibria in $\partial_U \mathbb{R}^2_+$ as well as the heteroclinic orbit joining these two equilibria.

Proof (Proof of Lemma 8.2) Note that the existence of the boundary attractors follows from the invariance of these sets together with the dissipativity stated in Lemma 8.1. Next to prove the existence of the interior attractor $\mathcal{A}_{\text{Int}\left(\mathbb{R}^2_+\right)}$, it is sufficient (see for instance Hale and Waltman [145]) to show that the state space decomposition $\left(\partial_U \mathbb{R}^2_+ \cup \partial_V \mathbb{R}^2_+; \text{Int}\left(\mathbb{R}^2_+\right)\right)$ is uniformly persistent, namely there exists some constant $\Theta > 0$, such that for each $(U_0, V_0) \in [0, \infty)^2$ with $U_0 > 0$ and $V_0 > 0$

$$\exists\Theta > 0, \ \forall (U_0, V_0) \in \mathrm{Int}\left(\mathbb{R}_+^2\right), \ \liminf_{t\to\infty} \ \min\left(U(t), V(t)\right) \geq \Theta. \tag{8.12}$$

To prove this property, since the two equilibria, $M_1 = (0, 0)$ and $M_2 = (\gamma, 0)$, on the boundary $\partial_U \mathbb{R}_+^2 \cup \partial_V \mathbb{R}_+^2$ are chained in the sense of Hale and Waltman's [145], it is sufficient to prove the local repulsivity of each of these equilibria with respect to the interior region $\mathrm{Int}\left(\mathbb{R}_+^2\right)$. Assume by contradiction that for some $\varepsilon > 0$ small enough one has

$$U_0 > 0 \text{ and } V_0 > 0 \text{ and } U(t) + V(t) \leq \varepsilon, \forall t \geq 0.$$

Then by using the U-equation of (8.9) we obtain

$$U' \geq [\alpha(\gamma - \varepsilon) - \varepsilon]\, U.$$

Therefore by choosing $\varepsilon > 0$ small enough (so that $[\alpha(\gamma - \varepsilon) - \varepsilon] > 0$) we deduce that $U(t) \to \infty$ as $t \to \infty$. which is impossible since the system is dissipative. Similarly assume that

$$V_0 > 0 \text{ and } |U(t) - \gamma| + V(t) \leq \varepsilon, \forall t \geq 0.$$

Then by using the V-equation of (8.9) we obtain

$$V' \geq -V + \beta \frac{(\gamma - \varepsilon)V}{1 + (\gamma - \varepsilon)}.$$

Choosing $\varepsilon > 0$ small enough (so that $\beta \dfrac{(\gamma - \varepsilon)}{1 + (\gamma - \varepsilon)} > 1 \Leftrightarrow (\beta - 1)(\gamma - \varepsilon) > 1$) one deduces that $V(t) \to \infty$ as $t \to \infty$, which is a contradiction, that completes the proof of (8.12). □

Before going to the description of the global attractor, let us first describe the interior attractor. To do so we summarize in the next theorem some important known results about (8.9).

Theorem 8.4 *System* (8.9) *enjoys the following properties.*

(i) If **(H3)** *is satisfied, then the interior equilibrium is globally asymptotically stable for system* (8.9) *restricted to* $\mathrm{Int}(\mathbb{R}_+^2)$.
(ii) If **(H2)** *is satisfied, then* (8.9) *admits a unique stable periodic orbit surrounding the interior equilibrium and the system has no other periodic orbit.*

Note that (i) has been proved by Hsu, Hubbell and Waltman [169, Lemma 4] using Dulac's criterion which proved that the system has no periodic orbit whenever **(H3)** holds. More precisely, setting $\varphi(U, V) = \frac{1+U}{U} V^{\xi+1}$ for some constant $\xi > 0$ such that $\frac{(\gamma-1)\alpha}{\beta-1} < \xi < \left(\frac{1}{\beta-1} - \frac{\gamma-1}{4}\right)\frac{4\alpha}{\beta-1}$, then, the aforementioned works proved that for each $0 < \eta < 1$ there exists $m_\eta > 0$ such that

$$\partial_U(\varphi F) + \partial_U(\varphi G) \leq -m_\eta, \ \forall (U, V) \in \left[\eta, \eta^{-1}\right]^2. \tag{8.13}$$

As far as the second point (ii) is concerned, the uniqueness of the periodic orbit was proved by Cheng in [47] while its stability was proved in [169]. Note also that

(8.9) undergoes a Hopf bifurcation around the interior equilibrium whenever we choose the bifurcation parameter γ and this Hopf bifurcation occurs at $\gamma = \gamma^\star$ with $\gamma^\star(\beta - 1) = \beta + 1$.

These results are related to the interior attractor and can be reformulated as follows.

Corollary 8.5 (Interior attractor) *The following holds.*

(i) *Assume that **(H3)** holds. Then the interior attractor $\mathcal{A}_{\text{Int}(\mathbb{R}^2_+)}$ reduces to the interior equilibrium.*
(ii) *Assume that **(H2)** holds. Then the interior attractor $\mathcal{A}_{\text{Int}(\mathbb{R}^2_+)}$ consists of the unique interior equilibrium, the unique interior periodic orbit and an infinite number of heteroclinic orbits joining the unique interior equilibrium and the unique periodic orbit.*

To complete this section, we are able to describe the global attractor $\mathcal{A}_{\mathbb{R}^2_+}$. Our result reads as follows.

Theorem 8.6 (Global attractor) *Let **(H1)** be satisfied. Then system* (8.9) *admits a unique heteroclinic orbit (U, V) joining $(\gamma, 0)$ to the boundary of the interior attractor $\mathcal{A}_{\text{Int}(\mathbb{R}^2_+)}$. The global attractor $\mathcal{A}_{\mathbb{R}^2_+}$ is composed of three disjoint parts*

$$\mathcal{A}_{\mathbb{R}^2_+} = [0, \gamma] \times \{0\} \bigcup \{(U(t), V(t)),\ t \in \mathbb{R}\} \bigcup \mathcal{A}_{\text{Int}(\mathbb{R}^2_+)}.$$

Proof The proof of this result requires three steps. We firstly derive the existence of heteroclinic orbits using a connectedness argument for the global attractor. Then we show that heteroclinic orbits starts from the stationary point $(\gamma, 0)$ and finally we conclude to the uniqueness of such heroclinic orbit by using a center unstable manifold argument (see [92] where a rather similar argument was used to derive a uniqueness property for traveling wave solutions arising in some epidemic problem).

Connectedness arguments: The largest global attractor $\mathcal{A}_{\mathbb{R}^2_+}$ is connected since it attracts the convex subset $\mathbb{T}$. It follows that the projection of $\mathcal{A}_{\mathbb{R}^2_+}$ on the horizontal and vertical axis is a compact interval.

The global attractor $\mathcal{A}_{\mathbb{R}^2_+}$ contains the interior global attractor $\mathcal{A}_{\text{Int}(\mathbb{R}^2_+)}$ which is compact, connected and locally stable. The global attractor $\mathcal{A}_{\mathbb{R}^2_+}$ also contains the boundary attractor $\mathcal{A}_{\partial_U \mathbb{R}^2_+}$. The connectedness of $\mathcal{A}_{\mathbb{R}^2_+}$ and compactness of $\mathcal{A}_{\text{Int}(\mathbb{R}^2_+)}$ and $\mathcal{A}_{\partial_U \mathbb{R}^2_+}$ imply

$$\mathcal{A}_{\mathbb{R}^2_+} - \left(\mathcal{A}_{\text{Int}(\mathbb{R}^2_+)} \bigcup \mathcal{A}_{\partial_U \mathbb{R}^2_+}\right) \neq \emptyset.$$

Moreover by using Theorem 3.2 due to Hale and Waltman [145], one deduces that for each point $(U, V) \in \mathcal{A}_{\mathbb{R}^2_+} - \left(\mathcal{A}_{\text{Int}(\mathbb{R}^2_+)} \cup \mathcal{A}_{\partial_U \mathbb{R}^2_+}\right)$ the alpha and the omega limit sets satisfy the following property:

$$\alpha(U, V) \in \mathcal{A}_{\partial_U \mathbb{R}^2_+} \text{ and } \omega(U, V) \in \mathcal{A}_{\text{Int}(\mathbb{R}^2_+)}.$$

Finally since the boundary attractor has a Morse decomposition $M_1 = \{(0, 0)\}$ and $M_2 = \{(\gamma, 0)\}$ we have either

$$\alpha(U,V) = M_1 \text{ or } \alpha(U,V) = M_2, \ \ \forall (U,V) \in \mathcal{A}_{\mathbb{R}^2_+} - \left(\mathcal{A}_{\mathrm{Int}(\mathbb{R}^2_+)} \bigcup \mathcal{A}_{\partial_U \mathbb{R}^2_+}\right).$$

No existence of heteroclinic orbit starting from $(0,0)$**:** Assume by contradiction that there exists such an orbit. Next by looking at the V-equation we deduce that

$$V(t) = \exp\left(\int_{t_0}^{t} \left(-1 + \frac{U(s)}{1+U(s)}\right) ds\right) V(t_0), \ \forall t, t_0 \in \mathbb{R}.$$

Since $V(t_0) > 0$ and there exists $T < 0$ such that $U(t)$ remains sufficiently small for all negative times $t < T$, so that $V(t) \to \infty$ as $t \to -\infty$, which contradicts the boundedness of the global attractor.

Existence and uniqueness of an heteroclinic orbit starting from $(\gamma,0)$**:** We only need to prove the uniqueness. The linearized equation around $(\gamma,0)$ has two eigenvalues: $\lambda_1 = -\alpha\gamma < 0$ and $\lambda_2 = \frac{\beta\gamma}{1+\gamma} - 1 > 0$ with eigenspaces

$$E_{\lambda_1} = \left\{(U,V) \in \mathbb{R}^2 : V = 0\right\}$$
and
$$E_{\lambda_2} = \left\{(U,V) \in \mathbb{R}^2 : U - \gamma = -\frac{\gamma}{\gamma(\beta-1)-1+\alpha\gamma(1+\gamma)} V\right\}$$

with

$$\frac{\gamma}{\gamma(\beta-1)-1+\alpha\gamma(1+\gamma)} > 0.$$

Note that

$$\mathbb{R}^2 = E_{\lambda_1} \bigoplus E_{\lambda_2}.$$

The center-unstable manifold at $(\gamma,0)$ is one dimensional. Let $\psi_{cu} : E_{\lambda_2} \to E_{\lambda_1}$ be a C^1 center-unstable manifold and consider the one dimensional manifold defined by

$$M_{cu} := \{x_{cu} + \psi_{cu}(x_{cu}) : x_{cu} \in E_{\lambda_2}\}.$$

It is locally invariant under the semiflow generated by (8.9) around $(\gamma,0)$. Since $D_{x_{cu}}\psi_{cu}(0) = 0$, the manifold M_{cu} is tangent to E_{λ_2} at $(\gamma,0)$. Moreover we know that there exists $\varepsilon > 0$, such that M_{cu} contains all negative orbits of the semiflow generated by (8.9) staying in the ball $B_{\mathbb{R}^2}((\gamma,0),\varepsilon)$ for all negative times.

In order to prove the uniqueness, we assume that there exists two heteroclinic orbits

$$O_1 = (U_1(t), V_1(t))_{t\in\mathbb{R}} \subset \mathrm{Int}\left(\mathbb{R}^2_+\right) \text{ and } O_2 = (U_2(t), V_2(t))_{t\in\mathbb{R}} \subset \mathrm{Int}\left(\mathbb{R}^2_+\right)$$

going from $(\gamma,0)$ to the interior attractor $\mathcal{A}_{\mathrm{Int}(\mathbb{R}^2_+)}$. Since

$$\lim_{t\to-\infty} (U_1(t), V_1(t)) = (\gamma,0) \text{ and } \lim_{t\to-\infty} (U_2(t), V_2(t)) = (\gamma,0),$$

without loss of generality, one may assume that

$$(U_1(t), V_1(t))_{t\le 0} \subset B_{\mathbb{R}^2}((\gamma,0),\varepsilon) \text{ and } (U_2(t), V_2(t))_{t\le 0} \subset B_{\mathbb{R}^2}((\gamma,0),\varepsilon)$$

which imply that $(U_1(t), V_1(t))_{t\leq 0} \subset M_{cu}$ and $(U_2(t), V_2(t))_{t\leq 0} \subset M_{cu}$.
Let Π_{λ_1} and Π_{λ_2} be the linear projectors from $\mathbb{R}^2$ to E_{λ_1} and E_{λ_2}, respectively. We can find $t_1 < 0$ and $t_2 < 0$ such that $\Pi_{\lambda_2}(U_1(t_1), V_1(t_1)) = \Pi_{\lambda_2}(U_2(t_2), V_2(t_2))$ and then $\psi_{cu}(\Pi_{\lambda_2}(U_1(t_1), V_1(t_1))) = \psi_{cu}(\Pi_{\lambda_2}(U_2(t_2), V_2(t_2)))$. Thus $(U_1(t_1), V_1(t_1)) = (U_2(t_2), V_2(t_2))$. By the uniqueness of the solutions for system (8.9), we get $(U_1(t_1 + \cdot), V_1(t_1 + \cdot)) = (U_2(t_2 + \cdot), V_2(t_2 + \cdot))$ and thus $O_1 = O_2$. The uniqueness of the heteroclinic orbit starting from $(\gamma, 0)$ follows and this completes the proof of the theorem.

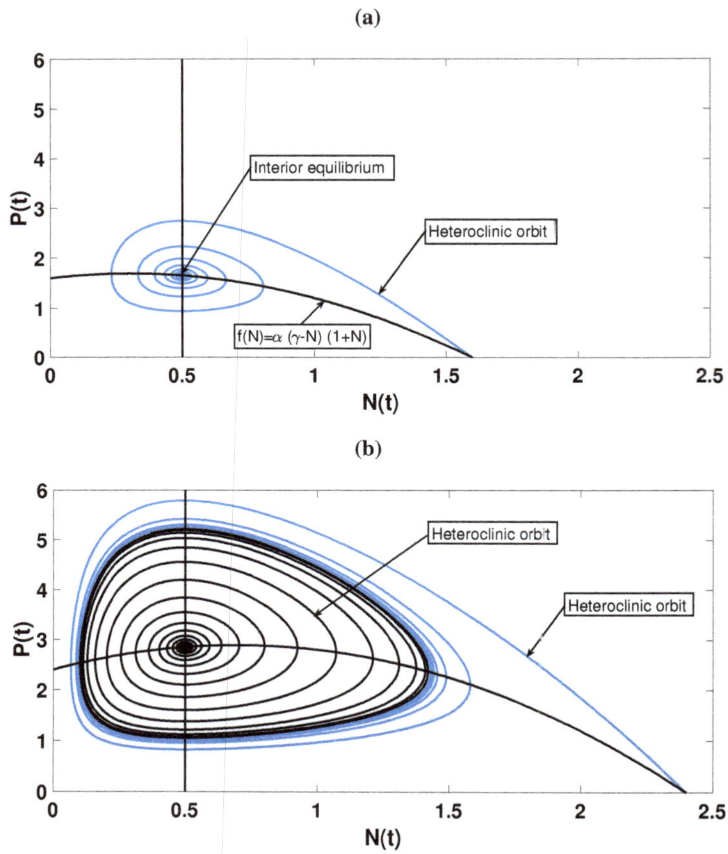

Fig. 8.1: *In this figure we run a simulation of the Rosenzweig-MacArthur model with* $\alpha = 1$, $\beta = 3$ *and* $\gamma = 18.7$ *(in Figure (a)) and* $\gamma = 2.4$ *(in Figure (b)). In both figures we plot the heteroclinic orbit joining the boundary equilibrium and the interior equilibrium (in Figure (a)) and the interior limit cycle which is a stable periodic orbit (in Figure (b)). In this figure we also plot the nullclines* $f(U) = \alpha(\gamma - U)(1 + U)$ *and* $U = \overline{U}_2$.

8.3 Application of a center manifold theorem to the traveling wave problem

This section is devoted to the study of traveling wave profile system of equations (8.8) for $\varepsilon \ll 1$. We will firstly apply a center manifold reduction on a suitable invariant region. The reduced system will be analysed. In the same spirit as in the previous section we will describe its global and interior attractor to obtain various results about the existence and uniqueness of traveling wave solutions as well as refined information about periodic wave trains.

8.3.1 Reduction of the traveling wave problem

Transformed system: In order to work with a subspace of equilibria for $\varepsilon = 0$ we use the following change of variable

$$\begin{cases} U_1 = u_1, \\ U_2 = u_2 + F(u_1, v_1), \\ V_1 = v_1, \\ V_2 = v_2 + G(u_1, v_1) \end{cases} \Leftrightarrow \begin{cases} u_1 = U_1, \\ u_2 = U_2 - F(U_1, V_1), \\ v_1 = V_1, \\ v_2 = V_2 - G(U_1, V_1). \end{cases} \tag{8.14}$$

By using this change of variables system (8.8) becomes

$$\begin{cases} U_1' = \varepsilon u_2 = \varepsilon\left[U_2 - F(U_1, V_1)\right], \\ dU_2' = du_2' + dF(U_1, V_1)' = -U_2 + d\partial_u F(U_1, V_1)U_1' + d\partial_v F(U_1, V_1)V_1', \\ V_1' = \varepsilon v_2 = \varepsilon\left[V_2 - G(U_1, V_1)\right], \\ V_2' = v_2' + G(U_1, V_1)' = -V_2 + \partial_u G(U_1, V_1)U_1' + \partial_v G(U_1, V_1)V_1' \end{cases}$$

and therefore we obtain

$$\begin{cases} U_1' = \varepsilon\left[U_2 - F(U_1, V_1)\right], \\ dU_2' = -U_2 + \varepsilon d P(U_1, U_2, V_1, V_2), \\ V_1' = \varepsilon\left[V_2 - G(U_1, V_1)\right], \\ V_2' = -V_2 + \varepsilon Q(U_1, U_2, V_1, V_2), \end{cases} \tag{8.15}$$

wherein P and Q are given by

$$P(U_1, U_2, V_1, V_2) = \partial_u F(U_1, V_1)\left[U_2 - F(U_1, V_1)\right] + \partial_v F(U_1, V_1)\left[V_2 - G(U_1, V_1)\right]$$

and

$$Q(U_1, U_2, V_1, V_2) = \partial_u G(U_1, V_1)\left[U_2 - F(U_1, V_1)\right] + \partial_v G(U_1, V_1)\left[V_2 - G(U_1, V_1)\right].$$

Truncated system: Let $\rho : \mathbb{R} \to \mathbb{R}$ be a C^∞ function such that

$$\rho(x) = \begin{cases} 1, \text{ if } x \geq 0, \\ \in [0,1], \text{ if } x \in [-1/2, 0], \\ 0, \text{ if } x \leq -1/2. \end{cases}$$

Define the set

$$\mathbb{E} = \Big\{ (U_1, U_2, V_1, V_2) \in \mathbb{R}^4 : (U_1, V_1) \in \mathbb{T}, \, |U_2 - F(U_1, V_1)| \leq 1, \\ \text{and } |V_2 - G(U_1, V_1)| \leq 1 \Big\}.$$

Let $L > 0$ be given large enough such that

$$L \geq 2 + \max_{(U_1,V_1)\in\mathbb{T}} |F(U_1, V_1)| + \max_{(U_1,V_1)\in\mathbb{T}} |G(U_1, V_1)| \\ + \max_{(U_1,U_2,V_1,V_2)\in\mathbb{E}} |P(U_1, U_2, V_1, V_2)| + \max_{(U_1,U_2,V_1,V_2)\in\mathbb{E}} |Q(U_1, U_2, V_1, V_2)|.$$

Let $\chi : \mathbb{R} \to \mathbb{R}$ be a C^∞ function such that

$$\chi(x) = \begin{cases} 1, \text{ if } x \in [-L, L], \\ \in [0,1], \text{ if } x \in [-(L+1), -L] \cup [L, (L+1)], \\ 0, \text{ if } x \notin [-(L+1), (L+1)]. \end{cases}$$

Then we have

$$\begin{cases} U_1' &= \varepsilon \left[U_2 - F(U_1, V_1)\right] \chi(U_2 - F(U_1, V_1))\rho(U_1), \\ dU_2' &= -U_2 + \varepsilon d P(U_1, U_2, V_1, V_2)\chi(P(U_1, U_2, V_1, V_2))\rho(U_1), \\ V_1' &= \varepsilon \left[V_2 - G(U_1, V_1)\right] \chi(V_2 - G(U_1, V_1))\rho(U_1), \\ V_2' &= -V_2 + \varepsilon Q(U_1, U_2, V_1, V_2)\chi(Q(U_1, U_2, V_1, V_2))\rho(U_1). \end{cases} \tag{8.16}$$

Define

$$h(x) = x\chi(x), \; x \in \mathbb{R}.$$

Then system (8.16) can be rewritten as

$$\begin{cases} U_1' &= \varepsilon \, h\left(U_2 - F(U_1, V_1)\right) \rho(U_1), \\ dU_2' &= -U_2 + \varepsilon \, d \, h\left(P(U_1, U_2, V_1, V_2)\right) \rho(U_1), \\ V_1' &= \varepsilon \, h\left(V_2 - G(U_1, V_1)\right) \rho(U_1), \\ V_2' &= -V_2 + \varepsilon \, h\left(Q(U_1, U_2, V_1, V_2)\right) \rho(U_1). \end{cases} \tag{8.17}$$

Remark 8.7 In this truncation procedure the function $\rho(U_1)$ serves to avoid the singularity at $U_1 = -1$ in F and G. The function $h(.)$ is used to obtain a bounded Lipschitz perturbation of the system with $\varepsilon = 0$.

By setting $X(t) = (U_1(t), V_1(t))$ and $Y(t) = (U_2(t), V_2(t))$, system (8.17) takes the following form

$$\begin{cases} X'(t) = \varepsilon \widetilde{F}(X(t), Y(t)), \\ Y'(t) = -DY(t) + \varepsilon \widetilde{G}(X(t), Y(t)), \end{cases} \tag{8.18}$$

where $\widetilde{F}, \widetilde{G} \in C^\infty\left(\mathbb{R}^2 \times \mathbb{R}^2, \mathbb{R}^2\right)$ are bounded and Lipschitz continuous functions and we have set $D = \operatorname{diag}(d^{-1}, 1)$. Therefore the central space is given by

$$X_c = \left\{(X,Y) \in \mathbb{R}^2 \times \mathbb{R}^2 : Y = 0\right\},$$

while the stable space reads as

$$X_s = \left\{(X,Y) \in \mathbb{R}^2 \times \mathbb{R}^2 : X = 0\right\}.$$

Remark 8.8 Due to the choice of the constant $L > 0$ the truncated system (8.17) coincides with the original system (8.15) whenever $(U_1, U_2, V_1, V_2) \in \mathbb{E}$. Moreover the equilibria of system (8.15) belong to $\mathbb{E}$ (since $U_1 = u_1$ and $V_1 = v_1$ and the equilibria of (8.8) satisfy $u_2 = v_2 = 0$ and (u_1, v_1) must be an equilibrium of (8.9)). Conversely the equilibria of (8.17) in $\mathbb{E}$ must satisfy

$$\widetilde{U}_2 = F(\widetilde{U}_1, \widetilde{V}_1) = 0 \text{ and } \widetilde{V}_2 = G(\widetilde{U}_1, \widetilde{V}_1) = 0.$$

Now by using Lemma 8.1 we have $(\widetilde{U}_1, \widetilde{V}_1) \in \mathbb{T}$.

For $\eta > 0$ and $p \in \mathbb{N} \setminus \{0\}$ we define the weighted spaces

$$BC^\eta(\mathbb{R}; \mathbb{R}^p) = \left\{u \in C(\mathbb{R}, \mathbb{R}^p) : \sup_{t \in \mathbb{R}} e^{-\eta|t|} \|u(t)\| < \infty\right\}.$$

Moreover for $\varepsilon > 0$ small enough we can apply the smooth center manifold theorem proved by Vanderbauwhede [338, Theorem 3.1] and Vanderbauwhede and Iooss [340, Theorem 1], see also Chapter 4. This yields the following reduction result.

Theorem 8.9 *Let $\eta \in (0, \min(1, 1/d))$ be given and fixed. Then there exists $\widetilde{\varepsilon}_0 > 0$ such that for each $\varepsilon \in [0, \widetilde{\varepsilon}_0]$ we can find a map $\Phi_\varepsilon = \left(\Phi_\varepsilon^1, \Phi_\varepsilon^2\right) \in C^k(\mathbb{R}^2, \mathbb{R}^2)$, for each integer $k > 0$, satisfying the following properties*

$$\Phi_\varepsilon(0_\mathbb{R}) = 0_{\mathbb{R}^2} \text{ and } D\Phi_\varepsilon(0_\mathbb{R}) = 0_{\mathcal{L}(\mathbb{R}^2)},$$

and Φ_ε is bounded as well as its derivatives up to the order k and

$$\lim_{\varepsilon \to 0} \|\Phi_\varepsilon\|_\infty = 0 \text{ and } \lim_{\varepsilon \to 0} \|\Phi_\varepsilon\|_{\text{Lip}} = 0.$$

Moreover we have the following properties:

(i) *The **global center manifold** $M_\varepsilon = \{(X,Y) : Y = \Phi_\varepsilon(X)\}$ is invariant by the semiflow generated by* (8.17) *(forward and backward in time). Namely if $t \mapsto X(t)$ is a solution of the **reduced system** on some interval $I \subset \mathbb{R}$*

$$X'(t) = \varepsilon \widetilde{F}(X(t), \Phi_\varepsilon(X(t))), \forall t \in I, \tag{8.19}$$

then $t \mapsto (X(t), \Phi_\varepsilon(X(t)))$ is a solution of (8.17) *on I.*

(ii) *If $t \mapsto (X(t), Y(t))$ is a solution of* (8.17) *on $\mathbb{R}$ which belongs to $BC^\eta(\mathbb{R}; \mathbb{R}^4)$, then*

$$(X(t), Y(t)) \in M_\varepsilon, \forall t \in \mathbb{R} \Leftrightarrow Y(t) = \Phi_\varepsilon(X(t)), \forall t \in \mathbb{R}.$$

Now let us prove the following invariance result.

Proposition 8.10 *There exists* $\varepsilon_0 \in (0, \widetilde{\varepsilon}_0]$ *such that triangle* $\mathbb{T}$ *is negatively invariant by the flow generated by the reduced system* (8.19). *That is to say that*

$$X'(t) = \varepsilon \widetilde{F}(X(t), \Phi_\varepsilon(X(t))), \forall t \in \mathbb{R} \text{ and } X(0) = X_0 \in \mathbb{T} \Rightarrow X(t) \in \mathbb{T} \text{ for all } t \leq 0.$$

Furthermore the following subsets are both negatively invariant by the flow generated by the reduced system (8.19).

$$\partial_u \mathbb{T} = \{(U, V) \in \mathbb{T} : V = 0\} \text{ and } \partial_v \mathbb{T} = \{(U, V) \in \mathbb{T} : U = 0\}.$$

Proof In the first step, we investigate the invariance for the boundary regions $\partial_u \mathbb{T}$ and $\partial_v \mathbb{T}$. To that aim we claim that

$$\begin{pmatrix} U_2 \\ V_2 \end{pmatrix} = \Phi_\varepsilon \begin{pmatrix} 0 \\ V_1 \end{pmatrix} \Rightarrow U_2 = 0. \tag{8.20}$$

Indeed, assume that $U_1 = U_2 = 0$ in system, then

$$U_1 = U_2 = 0 \Rightarrow U_2 - F(U_1, V_1) = 0 \text{ and } P(U_1, U_2, V_1, V_2) = 0.$$

Therefore the two last components of the truncated system (8.16) become

$$\begin{cases} V_1' = \varepsilon\, h\, (V_2 - G(0, V_1)), \\ V_2' = -V_2 + \varepsilon\, h\, (Q(0, 0, V_1, V_2)). \end{cases} \tag{8.21}$$

Now by applying the center manifold theorem to (8.21) (which applies for the value of $\varepsilon \in (0, \widetilde{\varepsilon}_0)$ since the estimations for systems (8.17) and (8.21) remain unchanged in the proof of the center manifold theorem), we deduce that we can find a map $\Psi_\varepsilon \in C^k(\mathbb{R}, \mathbb{R})$ such that the center manifold of the two dimensional system (8.21)

$$V_2 = \Psi_\varepsilon(V_1)$$

and the solution $t \mapsto (V_1^\star(t), V_2^\star(t))$ of (8.21) starting from an initial value $(V_1, \Psi_\varepsilon(V_1))$ satisfies

$$(V_1^\star, V_2^\star) \in BC^\eta(\mathbb{R}; \mathbb{R}^2).$$

We conclude that

$$(U_1, U_2, V_1, V_2) = (0, 0, V_1^\star, V_2^\star) \in BC^\eta(\mathbb{R}; \mathbb{R}^4)$$

is a solution of the truncated system (8.17). This completes the proof of the claim.

By using similar argument one deduces that

$$\begin{pmatrix} U_2 \\ V_2 \end{pmatrix} = \Phi_\varepsilon \begin{pmatrix} U_1 \\ 0 \end{pmatrix} \text{ and } U_1 \geq 0 \Rightarrow V_2 = 0. \tag{8.22}$$

We now turn to the invariance of the triangle $\mathbb{T}$. Note that one has

$$\begin{cases} U_1' = \varepsilon\, h\left(\Phi_\varepsilon^1(U_1, V_1) - F(U_1, V_1)\right)\rho(U_1), \\ V_1' = \varepsilon\, h\left(\Phi_\varepsilon^2(U_1, V_1) - G(U_1, V_1)\right)\rho(U_1). \end{cases} \tag{8.23}$$

Next whenever $\beta U_1 + V_1 = R$ and $U_1 \geq 0$ and $V_1 \geq 0$ in system (8.22), then $\rho(U_1) = 1$ and for $\varepsilon > 0$ small enough (h coincides with identity)

$$\beta U_1' + V_1' = \varepsilon\left(\Phi_\varepsilon^1(U_1, V_1) + \Phi_\varepsilon^2(U_1, V_1) - F(U_1, V_1) - G(U_1, V_1)\right) > 0.$$

Therefore by combining this fact together with (8.19) and (8.21), we deduce that the triangle $\mathbb{T}$ is negatively invariant by the reduced system. □

8.3.2 Global attractors

We investigate preliminary properties of the perturbed two-dimensional (reduced) system (8.19). Recall that **(H1)** is satisfied along this chapter and $\mathbb{T}$, $\partial_u\mathbb{T}$, $\partial_v\mathbb{T}$ are negatively invariant with respect to this system for all $\varepsilon \in (0, \varepsilon_0]$. Before going further, by setting $t = -\varepsilon s$ and $\left(\tilde{U}, \tilde{V}\right)(s) = (U_1, U_2)(t)$ the above system (8.19) becomes, dropping the tilde for notational simplicity

$$\begin{cases} U' = \left[-\Phi_\varepsilon^1(U, V) + F(U, V)\right] := F_\varepsilon(U, V), \\ V' = \left[-\Phi_\varepsilon^2(U, V) + G(U, V)\right] := G_\varepsilon(U, V). \end{cases} \tag{8.24}$$

Notice that $\mathbb{T}$, $\partial_u\mathbb{T}$ and $\partial_v\mathbb{T}$ become positively invariant with respect to the above system. Then, for each such $\varepsilon \in [0, \varepsilon_0]$, we denote by $\{T_\varepsilon(t)\}_{t\geq 0}$ the strongly continuous semiflow on the triangle $\mathbb{T}$ generated by (8.24). One may also observe that it continuously depends on ε, namely the map $(\varepsilon, t, X) \mapsto T_\varepsilon(t)X$ is continuous from $[0, \varepsilon_0] \times [0, \infty) \times \mathbb{T}$ into $\mathbb{T}$. Our first result reads as follows:

Lemma 8.11 *Let $\varepsilon \in [0, \varepsilon_0]$ be given. Then the semiflow $\{T_\varepsilon(t)\}_{t\geq 0}$ possesses a compact and connected global attractor $\mathcal{A}_\varepsilon \subset \mathbb{T}$ attracting $\mathbb{T}$ in the sense that*

$$\text{dist}\,(T_\varepsilon(t)X, \mathcal{A}_\varepsilon) \to 0 \text{ as } t \to \infty \text{ uniformly for } X \in \mathbb{T},$$

wherein $\text{dist}\,(X, \mathcal{A}_\varepsilon) = \inf_{Y\in\mathcal{A}_\varepsilon} \|X - Y\|$ *denotes the Euclidean distance from* $X \in \mathbb{R}^2$ *to* $\mathcal{A}_\varepsilon$.

Proof Fix $\varepsilon \in (0, \varepsilon_0]$. Since, for each $t \geq 0$, $T_\varepsilon(t) : \mathbb{T} \to \mathbb{T}$ is completely continuous and bounded dissipative ($\mathbb{T}$ is compact), Theorem 3.4.8 in [139] ensures the existence of a global attractor for the semiflow T_ε. In addition, since $\mathbb{T}$ is connected, the result of Gobbino and Sardella [122] applies and ensures that $\mathcal{A}_\varepsilon$ is connected. □

Lemma 8.12 *The family $(\mathcal{A}_\varepsilon)_{\varepsilon\in[0,\varepsilon_0]}$ is upper semi-continuous, in the sense that for each $\widehat{\varepsilon} \in [0, \varepsilon_0]$ one has*

$$\lim_{\varepsilon\to\widehat{\varepsilon}} \delta\,(\mathcal{A}_\varepsilon, \mathcal{A}_{\widehat{\varepsilon}}) = 0,$$

wherein $\delta\,(\mathcal{A}_\varepsilon, \mathcal{A}_{\widehat{\varepsilon}})$ is given by

$$\delta\left(\mathcal{A}_\varepsilon, \mathcal{A}_{\widehat{\varepsilon}}\right) = \sup_{y \in \mathcal{A}_\varepsilon} \operatorname{dist}\left(y, \mathcal{A}_{\widehat{\varepsilon}}\right).$$

Proof Since the map $(\varepsilon, t, X) \mapsto T_\varepsilon(t)X$ is continuous from $[0, \varepsilon_0] \times [0, \infty) \times \mathbb{T}$ into the compact set $\mathbb{T}$, Theorem 3.5.2 in [139] ensures that the family $\{A_\varepsilon\}_{\varepsilon \in [0, \varepsilon_0]}$ is upper semi-continuous. □

We continue this section by further studying some properties of the global attractor $\mathcal{A}_\varepsilon$. To that aim, we define

$$\partial\mathbb{T}^0 = \partial_u \mathbb{T} \cup \partial_v \mathbb{T} \text{ and } \mathbb{T}^0 = \mathbb{T} \setminus \partial\mathbb{T}^0 = \{(U, V) \in \mathbb{T} :\ U > 0 \text{ and } V > 0\}.$$

Here let us recall that, for all $\varepsilon \in [0, \varepsilon_0]$ and $t \geq 0$, one has

$$T_\varepsilon(t)\mathbb{T}^0 \subset \mathbb{T}^0 \text{ and } T_\varepsilon(t)\partial\mathbb{T}^0 \subset \partial\mathbb{T}^0. \tag{8.25}$$

We prove the following uniform persistence result for T_ε.

Lemma 8.13 *There exists $\varepsilon_1 \in (0, \varepsilon_0]$ and $\Theta > 0$ such that for all $\varepsilon \in (0, \varepsilon_1]$ and $X \in \mathbb{T}^0$ one has*

$$\liminf_{t \to \infty} \operatorname{dist}\left(T_\varepsilon(t)X, \partial\mathbb{T}^0\right) \geq \Theta.$$

The proof of this lemma relies on the application of the results of Hale and Waltman in [145].

Proof Firstly recall that

$$(F_\varepsilon, G_\varepsilon) \to (F, G) \text{ as } \varepsilon \to 0 \text{ in } C^1(\mathbb{T}).$$

Next fix $\varepsilon_1 \in (0, \varepsilon_0]$ such that

$$\begin{aligned}
&\partial_U F_\varepsilon(0,0) > \frac{1}{2}\partial_U F(0,0) > 0, \quad \partial_U F_\varepsilon(\gamma,0) < \frac{1}{2}\partial_U F(\gamma,0) < 0,\\
&\partial_V G_\varepsilon(0,0) < \frac{1}{2}\partial_V G(0,0) < 0, \quad \partial_V G_\varepsilon(\gamma,0) > \frac{1}{2}\partial_V G(\gamma,0) > 0.
\end{aligned} \tag{8.26}$$

Now, in order to apply the result of Hale and Waltman, consider the extended semiflow $U(t) : \mathbb{T} \times [0, \varepsilon_1] \to \mathbb{T} \times [0, \varepsilon_1]$ given by

$$U(t)\begin{pmatrix} X \\ \varepsilon \end{pmatrix} := \begin{pmatrix} T_\varepsilon(t)X \\ \varepsilon \end{pmatrix}, \ \forall \begin{pmatrix} X \\ \varepsilon \end{pmatrix} \in \mathbb{T} \times [0, \varepsilon_1].$$

Then U becomes a strongly continuous semiflow on the compact set $X := \mathbb{T} \times [0, \varepsilon_1]$. Next consider the two positively invariant sets (see (8.24))

$$X^0 := \mathbb{T}^0 \times [0, \varepsilon_1] \text{ and } \partial X^0 = \partial\mathbb{T}^0 \times [0, \varepsilon_1].$$

Now in order to prove the lemma, we will show that the pair $\left(\partial X^0, X^0\right)$ is uniformly persistent with respect to the extended semiflow U. To that aim, observe that U possesses a compact global attractor, denoted by A. Then $U|_{\partial X^0}$ also admits a global attractor $A_\partial = ([0,\gamma]\times\{0\})\times[0,\varepsilon_1]$ while $\tilde{A}_\partial := \bigcup_{Z\in A_\partial}\omega(X)$ can be decomposed as follows

$$\tilde{A}_\partial = M_1 \bigcup M_2 \text{ with } M_1 := \left\{\begin{pmatrix}0\\0\end{pmatrix}\right\}\times[0,\varepsilon_1] \text{ and } M_2 := \left\{\begin{pmatrix}\gamma\\0\end{pmatrix}\right\}\times[0,\varepsilon_1],$$

that corresponds to a covering of $\tilde{A}_\partial$ by disjoint compact isolated invariant sets M_1 and M_2 for $U|_{\partial X^0}$. Furthermore M_1 is chained to M_2 and this covering is acyclic (see [145]), since $\partial_U F_\varepsilon(0,0) > 0$ and $\partial_U F_\varepsilon(\gamma,0) < 0$.

Next since $\{U(t)\}_{t\geq 0}$ is bounded dissipative and completely continuous on X for each $t \geq 0$, in view of Theorem 4.1 in [145] to prove that the pair $\left(\partial X^0, X^0\right)$ is uniformly persistent, it is sufficient to check that

$$W^s(M_i)\cap X^0 = \emptyset,\ \forall i = 1,2.$$

This latter property follows from the same repulsiveness arguments as the ones developed in Lemma 8.2 using the inequalities in (8.25). □

Using the above lemma one obtains the following decomposition result.

Proposition 8.14 *For each $\varepsilon\in[0,\varepsilon_1]$, there exist a global attractor $\mathcal{A}_{0,\varepsilon}\subset\mathbb{T}^0$ and a global attractor $\mathcal{A}_{\partial,\varepsilon}$ in $\partial\mathbb{T}^0$ for T_ε and the following decomposition for the global attractor $\mathcal{A}_\varepsilon$ (provided by Lemma 8.12) holds true*

$$\mathcal{A}_\varepsilon = \mathcal{A}_{0,\varepsilon}\bigcup W^u\left(\mathcal{A}_{\partial,\varepsilon}\right), \tag{8.27}$$

where $W^u\left(\mathcal{A}_{\partial,\varepsilon}\right) = \left\{X\in\mathcal{A}_\varepsilon : \alpha(X)\subset\mathcal{A}_{\partial,\varepsilon}\right\}$. Furthermore the family $\left(\mathcal{A}_{0,\varepsilon}\right)_{\varepsilon\in[0,\varepsilon_0]}$ is upper semi-continuous.

Proof The proof of the above result relies on the application of Theorem 3.2 in [145] and Theorem 18.1 in [236]. To see this, let us first observe that the result in Lemma 8.13 can be reformulated as follows:

$$\liminf_{t\to\infty} \operatorname{dist}\left(T_\varepsilon(t)X, \partial\mathbb{T}^0\right) \geq \Theta,$$

for all $X\in\mathbb{T}^0$ and all $\varepsilon\in[0,\varepsilon_1]$. Hence, since for each $\varepsilon\in[0,\varepsilon_1]$, T_ε is completely continuous and bounded dissipative and satisfies (8.24), the existence $\mathcal{A}_{0,\varepsilon}$ $\mathcal{A}_{\partial,\varepsilon}$ together with the decomposition (8.26) follows from the results in [145]. Next, using Lemma 8.12 and 8.13, the results of Magal in [236] applies and ensures the upper semi-continuity for the family of interior attractors $\left\{\mathcal{A}_{0,\varepsilon}\right\}_{\varepsilon\in[0,\varepsilon_1]}$. This completes the proof of the proposition. □

Remark 8.15 One may notice that, for all $\varepsilon\in[0,\varepsilon_1]$ one has $\mathcal{A}_{\partial,\varepsilon} = [0,\gamma]\times\{0\}$. This point has – implicitly – already been used in the proof of Lemma 8.13.

In the following, we discuss some properties of the interior attractor $\mathcal{A}_{0,\varepsilon}$ for $\varepsilon\in(0,\varepsilon_1]$. Our first result consists in the perturbation of Corollary 8.5 (i) and it reads as follows.

Theorem 8.16 *Assume that **(H3)** holds. Then there exists* $\varepsilon_2 \in (0, \varepsilon_1]$ *such that*

$$\mathcal{A}_{0,\varepsilon} = \left\{ \begin{pmatrix} \overline{U}_2 \\ \overline{V}_2 \end{pmatrix} \right\}, \ \forall \varepsilon \in [0, \varepsilon_2].$$

In other words, the interior attractor reduces to the interior equilibrium for all $\varepsilon > 0$ *small enough.*

Proof The proof of this result relies on the application of Dulac's criterion. Note that due to Lemma 8.13, one has

$$\inf_{X \in \partial \mathbb{T}^0} \text{dist}\left(X, \mathcal{A}_{0,\varepsilon}\right) \geq \Theta, \ \forall \varepsilon \in [0, \varepsilon_0].$$

Let $K \subset \mathbb{T}$ be compact such that

$$\inf_{X \in \partial \mathbb{T}^0} \text{dist}\,(X, K) \geq \frac{\Theta}{2} \text{ and } A_{0,\varepsilon} \subset K \ \forall \varepsilon \in [0, \varepsilon_0].$$

As for the proof of Theorem 8.4 (i) given in [169], we consider the function $\varphi(U, V) = \frac{1+U}{U} V^{\xi+1}$ for some suitable ξ. Then, since $(F_\varepsilon, G_\varepsilon) \to (F, G)$ as $\varepsilon \to 0$ for the topology of $C^1(\mathbb{T})$, one has

$$\left[\partial_U(\varphi F_\varepsilon) + \partial_V(\varphi G_\varepsilon)\right] \to \left[\partial_U(\varphi F) + \partial_V(\varphi G)\right],$$

uniformly for $(U, V) \in K$ as $\varepsilon \to 0$. According to the computations (8.13) recalled in Theorem 8.4 one has

$$\max_{(U,V) \in K} \left[\partial_U(\varphi F) + \partial_V(\varphi G)\right] < 0.$$

As a consequence, there exists $\varepsilon_2 \in (0, \varepsilon_1]$ small enough and $\delta > 0$ such that, for all $\varepsilon \in [0, \varepsilon_2]$ one has

$$\left[\partial_U(\varphi F_\varepsilon) + \partial_V(\varphi G_\varepsilon)\right] \leq -\delta, \ \forall (U, V) \in K.$$

Since $\mathcal{A}_{0,\varepsilon} \subset K$ for all ε small enough, the result follows using Dulac's criterion. □

Lemma 8.17 *Assume that **(H2)** holds. Then there exists* $\varepsilon_3 \in (0, \varepsilon_1]$ *such that the interior equilibrium* $\left(\overline{U}_2, \overline{V}_2\right)$ *is an unstable spiral points for the semiflow* T_ε, *for all* $\varepsilon \in [0, \varepsilon_3]$. *More precisely, the linearized equation of system* (8.24) *around the interior equilibrium has two complex conjugated eigenvalues with strictly positive real parts, that is a two dimensional unstable manifold.*

Proof Consider the Jacobian matrix, denoted by J_ε, associated to (8.24) at $\left(\overline{U}_2, \overline{V}_2\right)$. Since $(F_\varepsilon, G_\varepsilon)$ is $C^1(\mathbb{T})$–close to (F, G) as $\varepsilon \to 0$, one has

$$J_\varepsilon = J + o(1) \text{ as } \varepsilon \to 0.$$

Herein J is the Jacobian matrix at $\left(\overline{U}_2, \overline{V}_2\right)$ of (8.24) with $\varepsilon = 0$ (that corresponds to system (8.9)). It is readily checked that the eigenvalues $\lambda_\pm$ of J are simple so that the eigenvalues of J_ε, $\lambda_{\pm,\varepsilon}$ are simple and continuous with respect to ε. Hence $\lambda_{\pm,\varepsilon} = \lambda_\pm + o(1)$. This completes the proof of the result since $\lambda_\pm$ are conjugated complex numbers with positive real parts. □

Note that system (8.24) has the same equilibria as system (8.9) and system (8.24) has the boundary equilibria $(\overline{U}_0, \overline{V}_0)$ and $(\overline{U}_1, \overline{V}_1)$ given in (8.10) and the unique *interior equilibrium* $(\overline{U}_2, \overline{V}_2)$ defined in (8.11), whenever **(H1)** holds.

As a consequence of the Poincaré-Bendixon theorem, one obtains the following corollary.

Corollary 8.18 *Assume that* **(H2)** *holds. Then there exists* $\varepsilon_4 \in (0, \varepsilon_3]$ *such that for all* $\varepsilon \in [0, \varepsilon_4]$, *the interior attractor* $\mathcal{A}_{0,\varepsilon}$ *contains a (non-trivial) periodic orbit surrounding the interior equilibrium* $\left(\overline{U}_2, \overline{V}_2\right)$.

8.3.3 Uniqueness of the periodic orbit and interior attractor

In this section we discuss the uniqueness of the periodic orbit for system (8.24) and its relationship with the global interior attractor when the parameters satisfy the condition

$$\textbf{(H2)} \Leftrightarrow \overline{U}_2 < \frac{\gamma - 1}{2}. \tag{8.28}$$

The aim of this section is to prove the following uniqueness result.

Theorem 8.19 (Unique stable periodic orbit) *Under condition* (8.28), *for all* $\varepsilon > 0$ *small enough, there exists a unique stable periodic orbit surrounding the interior equilibrium and the system has no other periodic orbit.*

According to Corollary 8.18, for each $\varepsilon > 0$ small enough, let $(U_\varepsilon(t), V_\varepsilon(t))$ denotes any non constant periodic orbit of (8.24) and $T_\varepsilon > 0$ its period. The associated closed curve is denoted by Γ_ε, that is

$$\Gamma_\varepsilon = \{(U_\varepsilon(t), V_\varepsilon(t)),\ t \in [0, T_\varepsilon]\}.$$

Recall that Γ_ε encloses the interior equilibrium $(\overline{U}_2, \overline{V}_2)$. Note also that $\Gamma_\varepsilon \subset \mathcal{A}_{0,\varepsilon}$. Hence Proposition 8.14 ensures that there exists $\theta > 0$ such that for all $\varepsilon > 0$ small enough

$$U_\varepsilon(t) \geq \theta,\ V_\varepsilon(t) \geq \theta,\ \forall t \in \mathbb{R}.$$

Throughout this section we also denote by Γ_0 the unique non constant periodic orbit of (8.9), that corresponds to (8.24) with $\varepsilon = 0$ (see Theorem 8.4 (*ii*)). The corresponding periodic solution of (8.9) is denoted by $(U_0(t), V_0(t))$ while $T_0 > 0$ is its period.

The idea of this proof is to show that Γ_ε becomes close to Γ_0 as $\varepsilon \to 0$. Then, as in [47] for the unperturbed system, we will prove that for all $\varepsilon > 0$ small enough,

$$\int_0^{T_\varepsilon} [\partial_U F_\varepsilon(U_\varepsilon(t), V_\varepsilon(t)) + \partial_V G_\varepsilon(U_\varepsilon(t), V_\varepsilon(t))]\, dt < 0.$$

According to Hale [135], the latter condition means that Γ_ε is locally asymptotically stable and then it follows that Γ_ε is unique when $\varepsilon > 0$ is small enough.

To prove Theorem 8.19, let us firstly prove the following lemma.

Lemma 8.20 *Let condition* (8.28) *be satisfied. Then, for each* $\delta \in \left(\overline{U}_2, \frac{\gamma-1}{2}\right)$, *there exists* $\varepsilon(\delta) > 0$ *small enough such for all* $\varepsilon \in (0, \varepsilon(\delta)]$ *the curve* Γ_ε *intersects the line* $U = \delta$. *In other words, one has* $\max\{U_\varepsilon(t) : t \in [0, T_\varepsilon]\} \geq \delta$ *for all* $\varepsilon \in (0, \varepsilon(\delta)]$.

Proof Consider the function

$$\mathcal{F}(U, V) = \int_{\overline{U}_2}^{U} \frac{\left(\frac{\beta\xi}{1+\xi} - 1 + \Psi_\varepsilon(\xi, V)\right)}{\frac{\beta\xi}{1+\xi}} d\xi + \frac{1}{\beta}\int_{\overline{V}_2}^{V} \frac{\eta - \overline{V}_2}{\eta} d\eta,$$

wherein we have set $\Psi_\varepsilon(U, V) = -V^{-1}\Phi_\varepsilon^2(U, V)$. Note that since $\Phi_\varepsilon(U, V)$ is C^1−small uniformly on the compact set $(U, V) \in \mathbb{T}$ with $V \geq \theta > 0$ and $U \geq \theta > 0$ then Ψ_ε is also C^1−small on the same compact set.

Next let us compute the derivative of function $\mathcal{F}(U_\varepsilon, V_\varepsilon)$ with respect to t along the periodic orbit Γ_ε, that yields

$$\begin{aligned}
\frac{d\mathcal{F}(U_\varepsilon(t), V_\varepsilon(t))}{dt} &= \frac{\left(\frac{\beta U_\varepsilon}{1+U_\varepsilon} - 1 + \Psi_\varepsilon(U_\varepsilon, V_\varepsilon)\right)}{\frac{\beta U_\varepsilon}{1+U_\varepsilon}} \left[-\Phi_\varepsilon^1(U_\varepsilon, V_\varepsilon) + F(U_\varepsilon, V_\varepsilon)\right] \\
&\quad + \frac{1}{\beta}\frac{V_\varepsilon - \overline{V}_2}{V_\varepsilon} V'_\varepsilon \\
&\quad + V'_\varepsilon \int_{\overline{U}_2}^{U_\varepsilon} \frac{(1+\xi)\partial_V \Psi_\varepsilon(\xi, V_\varepsilon)}{\beta\xi} d\xi.
\end{aligned}$$

This rewrites as

$$\begin{aligned}
\frac{d\mathcal{F}(U_\varepsilon, V_\varepsilon)}{dt} &= \frac{V'_\varepsilon}{V_\varepsilon}\frac{1+U_\varepsilon}{\beta U_\varepsilon}\left[-\Phi_\varepsilon^1(U_\varepsilon, V_\varepsilon) + F(U_\varepsilon, V_\varepsilon) + \frac{U_\varepsilon}{1+U_\varepsilon}(V_\varepsilon - \overline{V}_2)\right] \\
&\quad + V' \int_{\overline{U}_2}^{U_\varepsilon} \frac{1+\xi}{\beta\xi} \partial_V \Psi_\varepsilon(\xi, V_\varepsilon) d\xi,
\end{aligned}$$

and denoting by $\tilde{\Psi}_\varepsilon(U, V) = -\frac{1+U}{\beta U}\Phi_\varepsilon^1(U, V)$, this yields

$$\frac{d\mathcal{F}(U_\varepsilon, V_\varepsilon)}{dt} = \frac{V'_\varepsilon}{\beta V_\varepsilon}\left[\tilde{\Psi}_\varepsilon((U_\varepsilon, V_\varepsilon) + f(U_\varepsilon) - \overline{V}_2\right] + V'_\varepsilon \int_{\overline{U}_2}^{U_\varepsilon} \frac{(\partial_V \Psi_\varepsilon(\xi, V_\varepsilon))}{\frac{\beta\xi}{1+\xi}} d\xi.$$

Integrating the above equality on $[0, T_\varepsilon]$ leads

$$
\begin{aligned}
0 &= \int_0^{T_\varepsilon} \frac{d\mathcal{F}(U_\varepsilon(s), V_\varepsilon(s))}{dt} ds \\
&= \oint_{\Gamma_\varepsilon} \left\{ \frac{1}{\beta V} \left(\tilde{\Psi}_\varepsilon(U,V) + f(U) - \overline{V}_2 \right) + \int_{\overline{U}_2}^{U} \frac{1+\xi}{\beta \xi} \partial_V \Psi_\varepsilon(\xi, V) d\xi \right\} dV.
\end{aligned}
$$

Now denoting by Ω_ε the interior of the periodic curve Γ_ε and using the Green-Riemann formula, we infer the following identity

$$
0 = \int_{\Omega_\varepsilon} \left\{ \frac{1}{\beta V} \left(\partial_U \tilde{\Psi}_\varepsilon(U,V) + f'(U) \right) + \frac{1+U}{\beta U} \partial_V \Psi_\varepsilon(U,V) \right\} dU dV. \tag{8.29}
$$

Now fix $\delta \in \left(\overline{U}_2, \frac{\gamma-1}{2}\right)$ and recall that $\inf_{U \le \delta} f'(U) = f'(\delta) > 0$. Set $K = \{(U,V) \in \mathbb{T} : U \in [\theta, \delta], V \ge \theta\}$ and observe that $\partial_V \Psi_\varepsilon$ and $\tilde{\Psi}_\varepsilon$ tend to 0 as $\varepsilon \to 0$, uniformly for $(U,V) \in K$. Hence since K is bounded by some constant $R > 0$, we obtain uniformly for $(U,V) \in K$ and for all $0 < \varepsilon \ll 1$

$$
\frac{1}{V} \left(\partial_U \tilde{\Psi}_\varepsilon(U,V) + f'(U) \right) + \frac{1+U}{\beta U} \partial_V \Psi_\varepsilon(U,V) \ge f'(\delta) + o(1).
$$

Hence there exists $\varepsilon(\delta) > 0$ such that for all $\varepsilon \in (0, \varepsilon(\delta)]$ one has

$$
\sup_{(U,V) \in K} \frac{1}{V} \left(\partial_U \tilde{\Psi}_\varepsilon(U,V) + f'(U) \right) + \frac{1+U}{\beta U} \partial_V \Psi_\varepsilon(U,V) > 0.
$$

As a consequence, since $\Gamma_\varepsilon \cup \Omega_\varepsilon \subset K$ and Γ_ε encloses the equilibrium, if the curve Γ_ε does not intersect the line $U = \delta$ for all $\varepsilon > 0$ small enough then the integral on the right hand side of (8.29) would be positive which is a contradiction and we complete the proof of the lemma. □

We continue the proof of Theorem 8.19 by showing the following lemma.

Lemma 8.21 *Let condition* (8.28) *be satisfied. Let* Γ_0 *denote the unique non-constant periodic orbit of system* (8.9). *Then the following convergence holds*

$$
\lim_{\varepsilon \to 0} \mathrm{d}(\Gamma_\varepsilon, \Gamma_0) = 0
$$

where $\mathrm{d}(\Gamma_\varepsilon, \Gamma_0)$ *denotes the Hausdorff's semi-distance given by*

$$
\mathrm{d}(\Gamma_\varepsilon, \Gamma_0) := \sup_{x \in \Gamma_\varepsilon} \delta(x, \Gamma_0) \textit{ with } \delta(x, \Gamma_0) = \inf_{y \in \Gamma_0} \|x - y\|.
$$

In other words, for each neighbourhood V *of* Γ_0 *there exists* $\varepsilon_V > 0$ *such that*

$$
\Gamma_\varepsilon \subset V, \forall \varepsilon \in (0, \varepsilon_V].
$$

Furthermore the period $T_\varepsilon > 0$ *of* Γ_ε *converges to* T_0*, the period of* Γ_0*, as* $\varepsilon \to 0$.

Proof Fix $U^\star = \frac{1}{2}\left[\overline{U}_2 + \frac{\gamma-1}{2}\right] \in \left(\overline{U}_2, \frac{\gamma-1}{2}\right)$.
Step 1: From Lemma 8.20 for all $\varepsilon > 0$ small enough, there exists $t_\varepsilon \in \mathbb{R}$ such that

$$U_\varepsilon(t_\varepsilon) > U^\star.$$

Step 2: Using Arzela-Ascoli's theorem we can find a sequence $\varepsilon_n \to 0$ and $t \mapsto (U(t), V(t))$ a complete orbit of the unperturbed system (8.9) such that

$$\left(U_{\varepsilon_n}(t+t_\varepsilon), V_{\varepsilon_n}(t+t_\varepsilon)\right) \to (U(t), V(t))$$

for the topology of the local uniform convergence for $t \in \mathbb{R}$. The definition of t_ε above ensures that

$$U(0) > U^\star. \tag{8.30}$$

Moreover, since $(U_\varepsilon(t), V_\varepsilon(t)) \in \mathbb{T}$ and $U_\varepsilon(t) \geq \theta$ and $V_\varepsilon(t) \geq \theta$ for all ε small enough and $\forall t \in \mathbb{R}$, one obtains that

$$(U, V)(t) \in \mathbb{T}, \ \ \forall t \in \mathbb{R} \text{ and } U(t) \geq \theta, \ V(t) \geq \theta, \ \forall t \in \mathbb{R}.$$

Hence the limit orbit (U, V) lies in the interior attractor $\mathcal{A}_{\text{Int}(\mathbb{R}^2_+)}$ of (8.9) while (8.30) implies that the complete orbit is not reduced to the interior equilibrium, therefore

$$\lim_{t\to\infty} \delta\left(\left(U(t), V(t)\right), \Gamma_0\right) = 0.$$

Step 3: Let us fix $M^0 = \left(U^0, V^0\right) \in \Gamma_0$ such that

$$F(M^0) > 0 \text{ and } G(M^0) > 0.$$

In order to simplify the rest of the proof, we fix the norm $\|\cdot\|_1$ in $\mathbb{R}^2$ given by

$$\|(U, V)\|_1 = |U| + |V|, \ \forall (U, V) \in \mathbb{R}^2.$$

Let $\eta > 0$ be small enough and let $\varepsilon_0 = \varepsilon_0(\eta) > 0$ be small enough (depending on η) such that

$$F_\varepsilon(M) > F(M^0)/2 \text{ and } G_\varepsilon(M) > G(M^0)/2,$$

whenever $\|M - M^0\|_1 \leq \eta$ and $\varepsilon \in (0, \varepsilon_0)$.

By using the sign of F and G around M^0, we can find $M^1 = (U^1, V^1)$ a point on Γ_0 such that

$$M^0 < M^1 (\text{that is, } U^0 < U^1 \text{ and } V^0 < V^1) \text{ and } \|M^1 - M^0\|_1 < \eta.$$

Let $\delta \in (0, \eta)$ be such that

$$\left\{M \in \mathbb{R}^2 : \|M - M^1\|_1 \leq \delta\right\} \subset \left\{M \in \mathbb{R}^2 : M \geq M^0 \text{ and } \|M - M^0\|_1 \leq \eta\right\}.$$

Step 4: By using the continuous dependency of the semiflow generated by (8.24) with respect to the initial condition and with respect to the parameter ε we deduce that we can find $\widehat{\delta} \in (0, \delta)$ and $\varepsilon_1 \in (0, \varepsilon_0)$ such that every solution of (8.24) starting in the ball

$$B(M^1, \widehat{\delta}) := \left\{M \in \mathbb{R}^2 : \|M - M^1\|_1 \leq \widehat{\delta}\right\}$$

will belong to the larger ball

$$B(M^1,\delta) := \left\{M \in \mathbb{R}^2 : \|M - M^1\|_1 \le \delta\right\}$$

at time $t = T_0$.

Step 5: By using the Step 2, for all n large enough, we find $M_{\varepsilon_n} \in \Gamma_{\varepsilon_n}$ belonging in the ball $B(M^1, \widehat{\delta})$ and the solution of the approximated system (8.24) starting from M_{ε_n} belongs to the ball $B(M^1, \delta)$ at $t = T_0$.

Assume by contradiction that this solution leaves the triangle

$$T = \left\{M \in \mathbb{R}^2 : M \ge M^0 \text{ and } \|M - M^0\|_1 \le \eta\right\}$$

without intersecting the point M_{ε_n}. By using Jordan's theorem, we obtain a contradiction since the closed curve Γ_{ε_n} cannot return back through the triangle T from the "exit segment"

$$S = \left\{M \in \mathbb{R}^2 : M \ge M^0 \text{ and } \|M - M^0\|_1 = \eta\right\}.$$

This completes the proof of the lemma. □

We now complete the proof of Theorem 8.19 by proving, as announced above that for all $\varepsilon > 0$ small enough

$$\int_0^{T_\varepsilon} \left[\partial_U F_\varepsilon(U_\varepsilon(t), V_\varepsilon(t)) + \partial_V G_\varepsilon(U_\varepsilon(t), V_\varepsilon(t))\right] dt < 0.$$

However this estimate follows from some properties of the unique periodic orbit (U_0, V_0) of (8.9) together with the convergence result stated in Lemma 8.21. Indeed, note that Cheng [47] proved that, the unique unperturbed periodic orbit Γ_0 satisfies

$$\int_0^{T_0} \left[\partial_U F(U_0(t), V_0(t)) + \partial_V G(U_0(t), V_0(t))\right] dt < 0,$$

while Lemma 8.21 ensures that

$$\begin{aligned}&\lim_{\varepsilon\to 0} \int_0^{T_\varepsilon} \left[\partial_U F_\varepsilon(U_\varepsilon(t), V_\varepsilon(t)) + \partial_V G_\varepsilon(U_\varepsilon(t), V_\varepsilon(t))\right] dt\\ &= \int_0^{T_0} \left[\partial_U F(U_0(t), V_0(t)) + \partial_V G(U_0(t), V_0(t))\right] dt < 0.\end{aligned}$$

This completes the proof of the estimate and thus the one of Theorem 8.19.

As a consequence of the above result, we now can state the following properties of the interior attractor $\mathcal{A}_{0,\varepsilon}$ for all $0 < \varepsilon \ll 1$.

Theorem 8.22 *Assume that* ***(H2)*** *holds. Then for all $\varepsilon > 0$ small enough, the interior global attractor $\mathcal{A}_{0,\varepsilon}$ consists of the unique interior equilibrium $\left(\overline{U}_2, \overline{V}_2\right)$ and the interior of the unique periodic orbit surrounding the interior equilibrium, and an infinite number of heteroclinic orbits joining the unique interior equilibrium and the unique periodic orbit.*

8.3.4 Existence and uniqueness of a traveling wave joining $(\gamma, 0)$ and the interior global attractor

In this section, we use the previous results to provide a description of the heteroclinic orbits for (8.24) as well as their uniqueness.

Lemma 8.23 *Assume that **(H1)** holds. Then, for all* $\varepsilon > 0$ *small enough the equilibria* $(0, 0)$ *and* $(\gamma, 0)$ *are saddle points for the semiflow* T_ε. *More precisely, the linearized equation of system* (8.24) *around the equilibrium* $(0, 0)$ *(or* $(\gamma, 0)$*) has one eigenvalue with positive real part and one with negative real part.*

Proof Let us denote by J_ε the Jacobian matrix associated to (8.24) at $(0, 0)$. Since $(F_\varepsilon, G_\varepsilon)$ is $C^1(\mathbb{T})$–close to (F, G) as $\varepsilon \to 0$, one has

$$J_\varepsilon = J + o(1) \text{ as } \varepsilon \to 0,$$

where J denotes the Jacobian matrix at $(0, 0)$ of (8.24) with $\varepsilon = 0$ (that corresponds to system (8.9)). It is easy to check that the eigenvalues of J are the following: $\lambda_{+,J} = \alpha\gamma > 0$ and $\lambda_{-,J} = -1 < 0$. The eigenvalues $\lambda_{\pm,J_\varepsilon}$ of J_ε are continuous with respect to ε. Hence $\lambda_{\pm,J_\varepsilon} = \lambda_{\pm,J} + o(1)$. This completes the proof of the result. □

Proposition 8.24 *Assume that **(H1)** holds. Then system* (8.24) *admits a unique heteroclinic orbit going from* $(0, 0)$ *to* $(\gamma, 0)$, *for all* $0 < \varepsilon \ll 1$ *small enough.*

Proof Since $\partial_u \mathbb{T}$ is positively invariant with respect to the system (8.24) and

$$F(U, V) = \alpha U(\gamma - U) - \frac{UV}{1+U} = \frac{U}{1+U}[f(U) - V],$$

by using the following fact

$$\lim_{\varepsilon \to 0} \|\Phi_\varepsilon\|_\infty = 0,$$

we can deduce that there exists a unique heteroclinic orbit of system (8.24) going from $(0, 0)$ to $(\gamma, 0)$. □

We now discuss the existence and uniqueness of heteroclinc orbits for (8.24) joining the boundary to the interior attractor. As in the previous section, we make use of the connectedness of the global attractor to derive the existence of such connections. We then discuss further properties.

Connectedness arguments: The largest global attractor $\mathcal{A}_\varepsilon$ is connected. Since any continuous map maps a connected set into a connected set, it follows that the projection of $\mathcal{A}_\varepsilon$ on the horizontal and vertical axis is a compact interval.

The global attractor $\mathcal{A}_\varepsilon$ contains the interior global attractor $\mathcal{A}_{0,\varepsilon}$ which is compacts connected and locally stable and also contains the boundary attractor $\mathcal{A}_{\partial,\varepsilon}$. The connectedness of $\mathcal{A}_\varepsilon$ and compactness of $\mathcal{A}_{0,\varepsilon}$ and $\mathcal{A}_{\partial,\varepsilon}$ imply

$$\mathcal{A}_\varepsilon - \mathcal{A}_{0,\varepsilon} \bigcup \mathcal{A}_{\partial,\varepsilon} \neq \emptyset.$$

We deduce that for each point $(U, V) \in \mathcal{A}_\varepsilon - \mathcal{A}_{0,\varepsilon} \bigcup \mathcal{A}_{\partial,\varepsilon}$ the α and ω limit sets satisfy the following

$$\alpha(U,V) \in \mathcal{A}_{\partial,\varepsilon} \text{ and } \omega(U,V) \in \mathcal{A}_{0,\varepsilon}.$$

Finally since the boundary attractor has a Morse decomposition $M_1 = \{(0,0)\}$ and $M_2 = \{(\gamma,0)\}$ we have either

$$\alpha(U,V) = M_1 \text{ or } \alpha(U,V) = M_2, \forall (U,V) \in \mathcal{A}_\varepsilon - \mathcal{A}_{0,\varepsilon} \bigcup \mathcal{A}_{\partial,\varepsilon}.$$

Proposition 8.25 *Assume that* ***(H1)*** *holds. There for all ε small enough, system* (8.24) *does not admit any heteroclinic orbit going from* $(0,0)$ *to the interior global attractor.*

Proof Assume by contradiction that there exists one. Note that

$$V' = -\Phi_\varepsilon^2(U,V) + \left(-1 + \frac{\beta U}{1+U}\right)V.$$

We deduce that

$$V(t) = \exp\left(\int_{t_0}^t -1 + \frac{\beta U(s)}{1+U(s)}ds\right) \times \left(V(t_0) + \int_{t_0}^t -\Phi_\varepsilon^2(U(t),V(t)) \exp\left(\int_{t_0}^s 1 - \frac{\beta U(t)}{1+U(t)}dt\right)ds\right).$$

Since there exists $T < 0$ such that $U(t)$ remains sufficiently small for all negative times $t < T$ and $V(t_0) > 0$, we deduce that

$$\lim_{t\to-\infty} V(t) = +\infty$$

which contradicts the fact that the solution belongs to the global attractor and is therefore bounded. □

We complete this section by proving the uniqueness of the traveling wave solution connecting $(\gamma,0)$ to the interior global attractor. The arguments of this proof extend those used in [92].

Proposition 8.26 *Assume that* ***(H1)*** *holds. Then for all $\varepsilon > 0$ small enough, system* (8.24) *admits a unique heteroclinic orbit going from* $(\gamma,0)$ *to the interior global attractor.*

Proof We only need to prove the uniqueness. The center-unstable manifold at $(\gamma,0)$ is a one dimensional locally invariant manifold. By using the same arguments as in section 8.2 for the uniqueness of the heteroclinic orbit starting from $(\gamma,0)$ for system (8.9), we can prove the uniqueness of the heteroclinic orbit going from $(\gamma,0)$ to the interior global attractor for system (8.24). □

8.4 Numerical simulations

In this section we intend to observe the previous results numerically. We run some numerical simulations for the system

$$\begin{cases} U_t = dU_{xx} + \alpha U(\gamma - U) - \dfrac{UV}{1+U}, & \text{for } x \in [0, 1000] \\ V_t = V_{xx} - V + \beta \dfrac{UV}{1+U}, & \text{for } x \in [0, 1000] \end{cases}$$

with Neumann boundary conditions

$$U_x(t,x) = V_x(t,x) = 0, \text{ for } x = 0 \text{ and } x = 1000,$$

and the initial values

$$U(0,x) = \gamma \text{ and } V(0,x) = 0.1 * \exp(-\delta x).$$

Throughout the simulations the parameters will be unchanged and fixed as follows

$$d = 1,\ \alpha = 1/4,\ \gamma = 4,\ \beta = 2.$$

In order to capture large speed traveling waves, we impose exponentially decaying initial distribution for the predator while the initial distribution for the prey is set to its carrying capacity γ. As for the usual Fisher-KPP equation, we expect that reducing the exponential decay rate δ increases the wave speed of the predator invasion. Using formal computations around the unstable predator free equilibrium $(\gamma, 0)$ and using the usual antsatz $V(t,x) = e^{-\delta(x-ct)}$, we obtain the following formula for the wave speed depending on δ

$$c = \begin{cases} 2\sqrt{K} \text{ if } \delta > \sqrt{K}, \\ \delta + \frac{K}{\delta} \text{ if } \delta \in (0, \sqrt{K}), \end{cases} \quad \text{with } K = \frac{\gamma(\beta - 1) - 1}{\gamma + 1} > 0.$$

With the above parameter set, we have $\sqrt{K} \approx 0.77$. Below we perform numerical simulations of the model with two different decay rates $\delta = 0.1$ and $\delta = 1$. According to our formal wave speed computations, we expect to obtain a predator invasion with a small speed (in fact the minimal wave speed) for $\delta = 1$ and with a larger wave speed with $\delta = 0.1$.

In Figure 8.2, we observe the traveling wave joining $(\gamma, 0)$ and periodic wave train when we start from a $V(0,x)$ with $\delta = 0.1$.

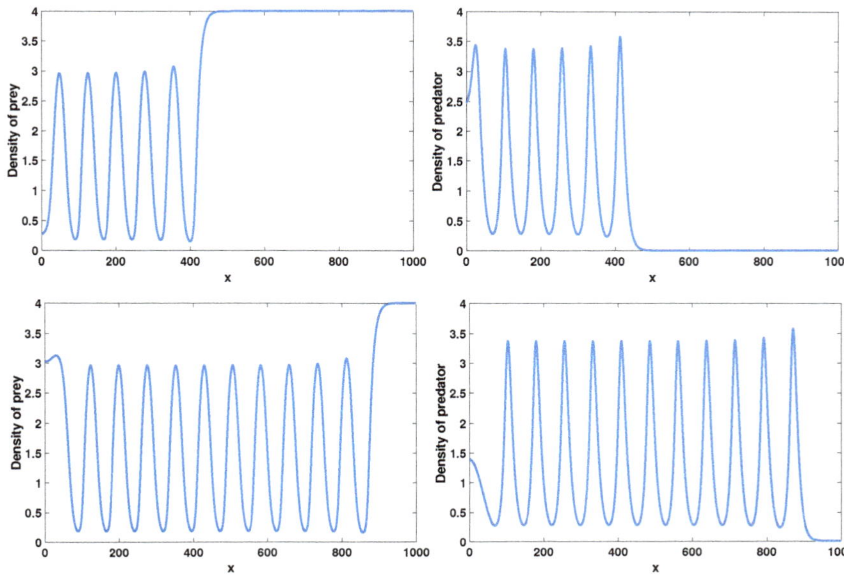

Fig. 8.2: *In this figure we plot* $U(t,x)$ *(left handside) and* $V(t,x)$ *(right handside) whenever the parameter* $\delta = 0.1$ *and* $t = 75$ *(top) and* $t = 150$ *(bottom). The initial distribution* $U(0,x) = \gamma$ *and* $V(0,x) = 0.1 * \exp(-0.1x)$. *We observe a traveling wave joining* $(\gamma, 0)$ *and a periodic waves train with both predator and prey oscillating periodically.*

In Figure 8.3, we observe slower predator invasion followed by some more complex behaviours whenever we start from a $V(0,x)$ with $\delta = 1$. The complexity in such a problem was already observed by Sherratt, Smith and Rademacher [303] in the multi-dimensional case.

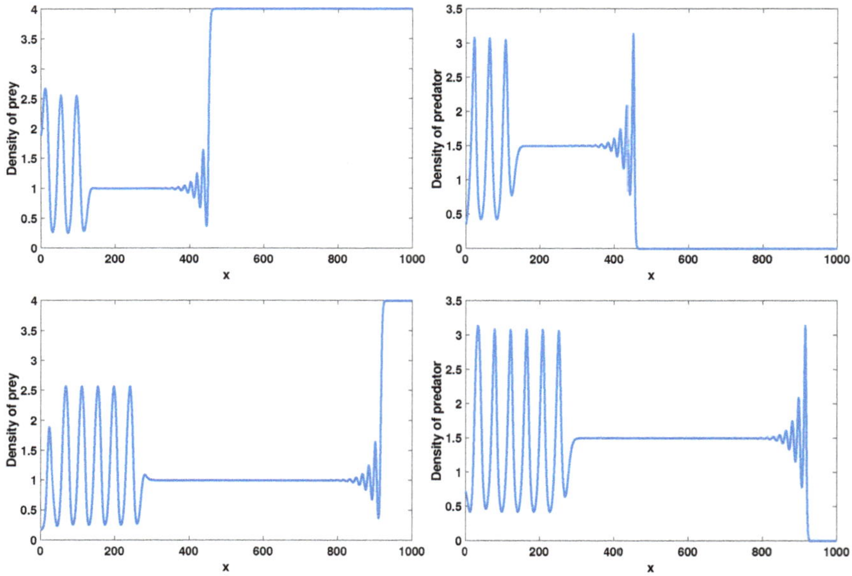

Fig. 8.3: *In this figure we plot* $U(t,x)$ *(left handside) and* $V(t,x)$ *(right handside) whenever the parameter* $\delta = 1$ *and* $t = 300$ *(top) and* $t = 600$ *(bottom). The initial distribution* $U(0,x) = \gamma$ *and* $V(0,x) = 0.1 * \exp(-x)$*. We observe a traveling wave joining* $(\gamma, 0)$ *and the positive equilibrium superposed with a periodic traveling pulse.*

The Figure 8.2 corresponds to our results when the speed of the traveling wave (with $\delta = 0.1$) is larger than in Figure 8.3 (with $\delta = 1$). The description of the small speed traveling waves is difficult question and left for further studies.

8.5 Remarks and Notes

In this chapter we have studied all the large speed traveling solutions for the one-dimensional diffusive Rosenzweig-MacArthur or-prey system. The methodology developed in this chapter is based on the center manifold reduction coupled with global attractor theory and its topological properties to understand some properties of the entire orbits (traveling waves, periodic wave train), for which the above mentioned analytical tools are particularly well adapted. Note that global attractor allows to obtain the existence of complete orbits, since every point of global attractor belongs to such a bounded complete orbit. Furthermore, the connectedness of global attractor allows in particular to prove the existence of complete orbits joining the boundary region and the interior equilibrium or periodic orbit. These complete orbits are nothing but traveling waves we are interested in.

As far as the reduction procedure is concerned, Fenichel's results about the persistence of normally hyperbolic invariant manifold may also be used. The two reduction techniques, based on Fenichel's theorem and center manifold theorem are comparable. Both methodologies provides the existence of manifold that is only locally invariant by the system (see [111, 338]).

In order to describe the system restricted to such a locally invariant manifold, one needs to carefully consider the truncation of the original system. It is important to observe that we consider the complete orbits that remain into the un-truncated region of the state space, where the two systems (with and without truncation) coincide. These tools permit to reduce the dimension as long as some bounded complete orbits remain in the un-truncated region of the state space.

Let us mention that such reduction arguments can be extended to infinite dimensional systems. We refer the reader to the books of Haragus and Iooss [148] and of Magal and Ruan [240] for infinite dimensional extension of the center manifold theorem. We also refer to Bates, Lu and Zeng [22, 23] and to Magal and Seydi [241] for results about the perturbation of normally hyperbolic manifold in infinite dimensional dynamical systems. More specifically the recent work of Magal and Seydi [241] deals with small perturbation of normally hyperbolic manifold, extended in particular the center manifold by allowing a nonlinear dynamic in the unperturbed central part of the system, that is roughly speaking for rather general – infinite dimensional – systems before perturbation with the form

$$\begin{cases} U'(t) = F(U(t)), \\ V'(t) = B(U(t))V(t), \end{cases}$$

wherein U corresponds to the central part and V corresponds to the hyperbolic part of the system.

This latest form is not only well adapted for the traveling wave problem considered in this chapter, but similar method can be applied to fast traveling wave for general reaction-diffusion systems (see [148, 240] for more results). As mentioned in the introduction this method can be developed in particular to handle more general nonlinearities. On the other hand, as mention above, since the center manifold is a flexible analytical tool with infinite dimensional extensions, the methodology developed in this work can also be used to study large speed traveling wave solutions for some infinite dimensional problems, including as a special case reaction-diffusion systems with time delay.

To conclude we provide in this chapter a new combination of reduction techniques and global analysis based on global attractors theory. Such a method can be employed in many other contexts to obtain the existence of large speed traveling waves.

8.6 MATLAB codes

8.6.1 Figure 8.1-(a)

```
close all;
clear all;
```

```
T= 10^3;
deltaT=10e-3;

alpha=1;
beta=3;
gamma=1.6;


x0=[gamma, 1e-3]

A=[alpha beta gamma]
t0=0:deltaT:T;
[t,x]= ode45(@model,t0,x0,[],A);
p=plot(x(:,1),x(:,2),'-','Color','[ 0 , 0.4470 ,
    0.7410]','LineWidth',3)

p.MarkerFaceColor = [ 0 , 0.4470 , 0.7410];
hold on


% we plot the isocline

x1=0:10e-3:gamma
y1=alpha*(gamma-x1).*(x1+1);
plot(x1,y1,'k','LineWidth',3)


% interior equilibrium
U2=1/(beta-1);
V2=(alpha*beta*[gamma*beta-1])/(beta-1)^2;



y2=0:6;

x2=U2*ones(size(y2));
plot(x2,y2,'k','LineWidth',3)

set(gca,'fontweight','bold','FontSize',30);

xlabel('N(t)');
ylabel('P(t)')

xlim([0 2.5])
ylim([0 6])

```

```
set(gca,'fontweight','bold','FontSize',30);

axis tight

set(gcf,'Position',[10 10 1000 700])

saveas(gcf,'Figure16_5','epsc')


function dz = model(t,z,A)
alpha=A(1);
beta=A(2);
gamma=A(3);
dz = zeros(2,1);
dz(1) = alpha*z(1)*(gamma-z(1))-z(1)*z(2)/(1+z(1));
dz(2) = -z(2)+beta*z(1)*z(2)/(1+z(1));
end
```

8.6.2 Figure 8.1-(b)

```
 close all;
 clear all;
T= 1.4*10^2;
deltaT=10e-3;

alpha=1;
beta=3;
gamma=2.4;


x0=[gamma, 1e-3]

A=[alpha beta gamma]
t0=0:deltaT:T;
[t,x]= ode45(@model,t0,x0,[],A);
plot(x(:,1),x(:,2),'-','Color','[ 0 , 0.4470 ,
    0.7410]','LineWidth',3)


hold on


% we plot the isocline

x1=0:10e-3:gamma
y1=alpha*(gamma-x1).*(x1+1);
```

```
plot(x1,y1,'k','LineWidth',3)


% interior equilibrium
U2=1/(beta-1);
V2=(alpha*beta*[gamma*(beta-1)-1])/(beta-1)^2;



y2=0:6;

x2=U2*ones(size(y2));
plot(x2,y2,'k','LineWidth',3)


x0=[U2, V2+1e-10]
[t,x]= ode45(@model,t0,x0,[],A);
p=plot(x(:,1),x(:,2),'k','LineWidth',3)


xlabel('N(t)');
ylabel('P(t)')

xlim([0 2.5])
ylim([0 6])

set(gca,'fontweight','bold','FontSize',30);

axis tight

set(gcf,'Position',[10 10 1000 700])

saveas(gcf,'Figure16_5','epsc')

function dz = model(t,z,A)
alpha=A(1);
beta=A(2);
gamma=A(3);
dz = zeros(2,1);
dz(1) = alpha*z(1)*(gamma-z(1))-z(1)*z(2)/(1+z(1));
dz(2) = -z(2)+beta*z(1)*z(2)/(1+z(1));
end
```

8.6.3 Figure 8.2

```
close all;
clear all;

xmin=0;
xmax=1e+3;
n=1e+4;
dx=(xmax-xmin)/(n-1);
x=xmin:dx:xmax;
x=x';
A=Laplacian(x);



% parameters of the model

d=1;
alpha=1/4;
gamma=4;
beta=2;
delta=0.1;



% initial distribution

u0=gamma*ones(size(x))


v0=0.1*exp(-delta*x);

% ---

I=eye(n,n);

dt=dx^2/4;

B1=I-(d*dt/dx^2)*A;
B1=sparse(B1);

B2=I-(dt/dx^2)*A;
B2=sparse(B2);


tmax=75;
t=0;
```

```
    while (t <=tmax+dt)
        u1=alpha*u0.*(gamma-u0)-(u0.*v0)./(1+u0);
        u1=u0+dt*u1;

        v1=-v0+beta*u0.*v0./(1+u0);
        v1=v0+dt*v1;

        u2=B1\u1;
        v2=B2\v1;

        t=t+dt;
        u0=u2;
        v0=v2;
    end
  figure(1)
    plot(x,u0,'LineWidth',5)

     set(gca,'fontweight','bold','FontSize',30)
     xlabel('x');
     ylabel('Density of prey');
     set(gcf,'Position',[10 10 1200 1000])
      saveas(gcf,'Figure1A','epsc')

    figure(2)
    plot(x,v0,'LineWidth',5)

    set(gca,'fontweight','bold','FontSize',30)
     xlabel('x');
     ylabel('Density of predator');
     set(gcf,'Position',[10 10 1200 1000])
      saveas(gcf,'Figure1B','epsc')


  function A=Laplacian(X1)


 N=length(X1);
 dx=X1(2)-X1(1);
 aux1=1/dx^2;

 %%%%%%%%%%%%%%%%%%%%%%%%%%%%%%%%%%%%%%%%%%%%%%%%%%%%%%%%%%
 % 1 dimensional matrix for Neumann BC
 %%%%%%%%%%%%%%%%%%%%%%%%%%%%%%%%%%%%%%%%%%%%%%%%%%%%%%%%%
```

```
B=diag(ones(N-1,1),-1)+diag(ones(N-1,1),1)-2*diag(
    ones(N,1));
B(1,1)=-1;
B(1,2)=1;
B(N,N-1)=1;
B(N,N)=-1;





A=B;
end
```

Chapter 9
A predator-prey model with infinitely many variants

9.1 Introduction

In this chapter we investigate the large time behavior of the predator-prey model

$$
\begin{cases}
\dfrac{\mathrm{d}}{\mathrm{d}t}N(t) = \beta - \mu^N N(t) - \displaystyle\sum_{k=0}^{+\infty} m_k N(t) P_k(t), & t > 0, \\
\dfrac{\mathrm{d}}{\mathrm{d}t}P_n(t) = m_n N(t) P_n(t) - \mu_n P_n(t), & t > 0, n \in \mathbb{N},
\end{cases}
\tag{9.1a}
$$

with the initial data

$$
N(0) = N_0 \in (0, +\infty), \qquad (P_n(0))_{n\in\mathbb{N}} = (P_n^0)_{n\in\mathbb{N}} \in \ell^1_+, \tag{9.1b}
$$

where ℓ^1_+ denotes the space of non-negative summable sequences.

In this model, $N(t)$ is the population of prey available at time $t > 0$ and, for each $n \in \mathbb{N}$, $P_n(t)$ is the quantity of predator of type n. The parameter $\beta > 0$ models a constant influx of prey, $\mu^N > 0$ is the death rate of the prey, m_n the consumption rate of the predator of type n, and μ_n the death rate of the predator of type n. Both m_n and μ_n are bounded sequences.

In this chapter we consider that many variants of the predator (possibly inifinitely many) compete for the limited resource that is the population of prey. We will talk about different *genotype* to denote predators with different n value. All of the possible genotypes are listed in an infinite sequence indexed by $n \in \mathbb{N}$, and for each genotype we denote (m_n, μ_n) the associated phenotype composed of the consumption rate m_n and the death rate μ_n of the predator. We do not specify the particular mechanisms that link the underlying variable $n \in \mathbb{N}$ and the phenotype (m_n, μ_n) but focus on the dynamics of the population under (9.1) conditionally to the knowledge of these mechanisms. For simplicity we consider that the predator is asexual; therefore, (9.1) can be considered a pure competition model where the predators compete for a single resource (the prey).

When P_n^0 is zero except for a finite number of indices, our problem is reduced to a system of ordinary differential equations:

A. Ducrot et al., *Differential Equations and Population Dynamics II*,
Lecture Notes on Mathematical Modelling in the Life Sciences,
https://doi.org/10.1007/978-3-031-91409-6_9

$$\begin{cases} \dfrac{\mathrm{d}}{\mathrm{d}t}N(t) = \beta - \mu^N N(t) - N(t)\big(m_1 P_1(t) + m_2 P_2(t) + \ldots + m_n P_n(t)\big) \\ \dfrac{\mathrm{d}}{\mathrm{d}t}P_1(t) = m_1 N(t) P_1(t) - \mu_1 P_1(t) \\ \quad\vdots \\ \dfrac{\mathrm{d}}{\mathrm{d}t}P_n(t) = m_n N(t) P_n(t) - \mu_n P_n(t), \end{cases} \tag{9.2a}$$

with the initial data

$$N(0) = N_0 > 0, \quad P_1(0) = P_1^0 > 0, \quad \ldots, \quad P_n(0) = P_n^0 > 0. \tag{9.2b}$$

In this context, Hsu, Hubbell and Waltman (1977) and later Hsu (1978) [167, 169] showed for a similar system of ordinary differential equations that the solution eventually converges to an equilibrium which may not be unique but is always concentrated on the equations that maximize the fitness m_n/μ_n. Here we extend these results to an infinite-dimensional dynamical system (we consider infinitely many variants) and prove that, for some well-chosen coefficients, the system stays eternally in a transient state and never converges to a single equilibrium. Thieme (2011) [332] considers a related model in the context of epidemiology in which a continuous distribution of host classes is infected by a pathogen that can be transmitted across classes; he proves, among other results and under very general assumptions, the global stability of the endemic equilibrium.

The problem of several species competing for a single resource has received a lot of attention in the literature. In this context, the "Competitive exclusion principle" states that "Complete competitors cannot coexist", which means that given a number of species competing for the same resource in the same place, only one can survive in the long run. This idea was already present to some extent in the book of Darwin, and is sometimes referred to as Gause's law [149]. This problem of survival of competitors has attracted the attention of mathematicians since the '70s and many studies have proved this property in many different contexts – let us mention the seminal works of Hsu, Hubbel and Waltman [167, 169] followed by Armstrong and McGehee [9, 249], Butler and Wolkowicz [40], Wolkowicz and Lu [354], Hsu, Smith and Waltman, [172] Wolkowicz and Xia [355] and Li [213], to cite a few – and also disproved in other contexts, for instance in fluctuating environments, see Cushing [70] and Smith [312].

In our model, the fitness of a variant with genotype n is given by the formula $\mathcal{R}_0^n = \frac{\beta m_n}{\mu^N \mu_n} = \frac{\beta}{\mu^N}\alpha_n$, where $\alpha_n := \frac{m_n}{\mu_n}$; the competitive exclusion principle implies that the only genotypes that eventually remain are the ones that maximize $\mathcal{R}_0^n$. We prove that this is correct - asymptotically, only the genotypes that maximize the fitness survive - but incomplete. Indeed it does not suffice to describe the asymptotic behavior of the population, especially when there are equality cases in the fitness of the variants (i.e. $\frac{m_{n_1}}{\mu_{n_1}} = \frac{m_{n_2}}{\mu_{n_2}}$ with $n_1 \neq n_2$, and possibly $\mu_1 \neq \mu_2$), of the maximal fitness is not reached by any genotype, or both. In the latter case we can even observe an alternation of the prevalent variant and eternal oscillations in the total number of infected, see the second example in section 9.3.2. Similar behaviors have been observed in the literature for related models (among others, [167] already gives a similar description). For example in the epidemiological context of [73], it has been observed that a strain 1 with a higher value of μ and a slightly lower $\mathcal{R}_0$ value than a strain 2 may nevertheless be dominant for some time. These borderline cases shed light on our understanding of transient dynamics, see also [39] where estimates for the transient dynamics for a related evolutionary model are provided based on the local flatness of the fitness function.

Quantitative traits such as the consumption rate or the mortality rate of a predator, and more generally any observable feature such as height, weight, muscular mass, speed, size of legs, etc. are naturally represented using continuous or discrete variables. Such a description of a population seems highly relevant and has been used mostly in modelling studies involving some kind of evolution [18, 19, 33, 79, 89, 127, 187, 228, 235, 243, 277]. In many of those models, mutation is considered as a process that is continuously occurring through time. That assumption may or may not be realistic depending on the context. It may also be realistic to model mutation as discrete events in time; in that case, the behavior of the population between such mutation events is correctly described by pure competition equations like (9.1). In the same spirit, we provide in this chapter a precise description of what happens in the vanishing mutation limit, and the trajectories of a model with a small but non-zero mutation operator are expected to be close, at least transiently, to those of the limit.

The structure of the chapter is as follows. In section 9.2 we present our main results. In section 9.3 we provide some examples and numerical simulations that present different asymptotic behaviors. In section 9.4 we propose a discussion of our results. Finally in section 9.5 we prove the results we claimed in section 9.2.

9.2 Main results

In this chapter we study the solutions of (9.1a) supplemented with the initial data (9.1b). Before starting, let us precise that we use the notation ℓ^1 to denote the Banach space of absolutely summable real sequences equipped with the norm

$$\|(a_n)_{n\in\mathbb{N}}\|_{\ell^1} := \sum_{n=0}^{+\infty} |a_n|,$$

and ℓ^1_+ is the positive cone of ℓ^1, that is to say the set of non-negative summable sequences. Similarly, ℓ^∞ denotes the Banach space of bounded sequences equipped with the norm

$$\|(a_n)_{n\in\mathbb{N}}\|_{\ell^\infty} := \sup_{n\in\mathbb{N}} |a_n|.$$

We will make the following assumption on the parameters arising in (9.1).

Assumption 9.1 The constants $\beta > 0$ and $\mu^N > 0$ are given. The sequences $(m_n)_{n\in\mathbb{N}} \in \ell^\infty$ and $(\mu_n)_{n\in\mathbb{N}} \in \ell^\infty$ are bounded and we assume that there exist constants $0 < m_0 < m^\infty$ and $0 < \mu_0 < \mu^\infty$ such that

$$0 < m_0 \le m_n \le m^\infty \qquad \text{and} \qquad 0 < \mu_0 \le \mu_n \le \mu^\infty, \qquad \text{for all } n \in \mathbb{N}.$$

As a consequence of this assumption, the sequence $(m_n/\mu_n)_{n\in\mathbb{N}}$ is bounded. We let α^* be the maximal fitness defined by

$$\alpha^* := \sup_{k\in\mathbb{N}} \frac{m_k}{\mu_k}.$$

Our next assumptions ensures that the maximal fitness α^* is effectively attained (possibly at infinity) by a non-neglibile population of predator. In other words, our model (9.1) is not equivalent to another model with a strictly lower maximal fitness.

Assumption 9.2 We let $N_0 > 0$, $(P_n^0)_{n\in\mathbb{N}} \in \ell_+^1$ be given and assume that there exists a sequence of indices $n_k \in \mathbb{N}$ with

$$P_{n_k}^0 > 0 \text{ and } \lim_{k\to+\infty} \frac{m_{n_k}}{\mu_{n_k}} = \alpha^*.$$

Let us precise that the sequence n_k in Assumption 9.2 can be eventually stationary; in particular we do not assume that the set of positive components is infinite.

We define the basic reproductive number at equilibrium $\mathcal{R}_0$ by

$$\mathcal{R}_0 := \frac{\beta}{\mu^N}\alpha^*. \tag{9.3}$$

In the rest of the chapter we will refer to this quantity as the basic reproductive number and we omit the mention "at equilibrium".

Finally, we make a technical assumption to avoid unnecessary theoretical discussions.

Assumption 9.3 (Finite ω-limit sets) We assume that the ω-limit sets of the sequences μ_n and $\alpha_n := \frac{m_n}{\mu_n}$, defined by

$$\omega\left((\mu_n)_{n\in\mathbb{N}}\right) := \bigcap_{n\in\mathbb{N}} \overline{\{\mu_k \,:\, k \ge n\}} \text{ and } \omega\left((\alpha_n)_{n\in\mathbb{N}}\right) := \bigcap_{n\in\mathbb{N}} \overline{\{\alpha_k \,:\, k \ge n\}},$$

are finite.

9.2.1 The Cauchy problem: existence and uniqueness

Our first result concerns the existence and uniqueness of the solution to the Cauchy problem (9.1). We show that the system (9.1) is well posed in the sense of Hadamard.

Proposition 9.4 (The Cauchy problem) *Let Assumption 9.1 hold, and $N_0 > 0$ and $(P_n^0)_{n\in\mathbb{N}} \in \ell_+^1$ be given. Then the system (9.1) has a unique global mild solution $\left(N(t), (P_n(t))_{n\in\mathbb{N}}\right) \in C^0\left([0,+\infty),\mathbb{R}\right) \times C^0\left([0,+\infty),\ell_+^1\right)$, which is also a classical solution:*

$$N(t) \in C^1\big([0,+\infty),\mathbb{R}\big) \text{ and } (P_n(t))_{n\in\mathbb{N}} \in C^1\big([0,+\infty),\ell^1_+\big).$$

Moreover for all $t > 0$ the map

$$\Big(N_0,\big(P_n^0\big)_{n\in\mathbb{N}}\Big) \in \mathbb{R}\times\ell^1_+ \longmapsto \big(N(t),\big(P_n(t)\big)_{n\in\mathbb{N}}\big) \in \mathbb{R}\times\ell^1_+$$

is continuous.

As an important consequence of Proposition 9.4, we note that each component of $(P_n(t))$ can be computed from $N(t)$. More precisely, if we set

$$\overline{N}(t) := \frac{1}{t}\int_0^t N(s)\mathrm{d}s,$$

then we have the following formula for $P_n(t)$:

$$P_n(t) = e^{t\left(m_n\overline{N}(t)-\mu_n\right)}, \qquad \text{for all } n\in\mathbb{N} \text{ and } t\geq 0. \tag{9.4}$$

We will use equation (9.4) repeatedly in the proofs of our results.

9.2.2 Persistence and asymptotic behavior

Next we investigate the asymptotic behavior of the solutions to (9.1) when $t \longrightarrow +\infty$. We first show that the population of pathogens gets extinct if $\mathcal{R}_0 \leq 1$.

Proposition 9.5 (Extinction) *Let Assumption 9.1 hold, and $N_0 > 0$ and $(P_n^0)_{n\in\mathbb{N}} \in \ell^1_+$ be given. Suppose that either $\mathcal{R}_0 < 1$, or $\mathcal{R}_0 = 1$ and Assumption 9.3 holds true. Then we have*

$$\lim_{t\to+\infty}\sum_{n=0}^{+\infty} P_n(t) = 0.$$

When $\mathcal{R}_0 > 1$, on the contrary, we can show that the pathogen survives in large time.

Theorem 9.6 (Persistence) *Suppose that the Assumptions 9.1, 9.2 and 9.3 hold true, and assume that $\mathcal{R}_0 > 1$. Then we have*

$$N(t) \xrightarrow[t\to+\infty]{} \frac{1}{\alpha^*} \text{ and } N'(t) \xrightarrow[t\to+\infty]{} 0.$$

Concerning the behavior of $P_n(t)$, we distinguish two cases.

i) *Suppose that there is some $i\in\mathbb{N}$ such that $\frac{m_i}{\mu_i} = \alpha^*$, possibly for multiple indices. Then $P_n(t)$ converges in ℓ^1 to the following asymptotic stationary state*

$$P_n^\infty = \begin{cases} 0 & \text{if } \frac{m_n}{\mu_n} < \alpha^*, \\ e^{\tau\mu_n}P_n^0 & \text{if } \frac{m_n}{\mu_n} = \alpha^*, \end{cases} \quad \text{for all } n\in\mathbb{N},$$

where the constant $\tau \in \mathbb{R}$ is the unique solution of the equation:

$$\sum_{\{n\in\mathbb{N}:\, \frac{m_n}{\mu_n}=\alpha^*\}} \mu_n P_n^0 e^{\tau\mu_n} = \frac{\mu^N}{\alpha^*}(\mathcal{R}_0 - 1).$$

ii) *Suppose that for all $n \in \mathbb{N}$, we have $\frac{m_n}{\mu_n} < \alpha^*$. Then one has $P_n(t) \to 0$ for all $n \in \mathbb{N}$ as $t \to \infty$, while*

$$\liminf_{t\to+\infty} \sum_{n\in\mathbb{N}} P_n(t) > 0. \tag{9.5}$$

Moreover if $(n_k)_{k\in\mathbb{N}}$ is a sequence of integers such that

$$\sup_{k\in\mathbb{N}} \frac{m_{n_k}}{\mu_{n_k}} < \alpha^*,$$

then we have

$$\limsup_{t\to+\infty} \sum_{k\in\mathbb{N}} P_{n_k}(t) = 0. \tag{9.6}$$

The proof of Theorem 9.6 will be given in section 9.5.3.

Let us explain in a few words the content of Theorem 9.6. Our basic assumption is that the basic reproduction number $\mathcal{R}_0$ is greater than one, because the predator population gets extinct otherwise. There are two typical situation. Case i) corresponds to the case when the maximal fitness is attained by at least one variant; in this case, the behavior of the infinite system is similar to the one of the finite system: we observe the unconditional convergence to an equilibrium state, that can be computed from the initial data. The case ii), when none of the variants attain the maximal fitness, is more intricate. In general, we can only draw two conclusions: the first is that the pathogen persists in large time as a whole (that's (9.5)), and the second is that any family of variants whose fitness is uniformly dominated, eventually gets extinct (that's (9.6)). In section 9.3 we will give a counterexample showing that, in some cases, the total pathogen population does not converge to a limit.

To go deeper in our analysis, we can be somewhat more precise on the behavior of the total pathogen population at the expense of a slightly stronger assumption on the coefficients. When the phenotypic values m_n and μ_n are in some sense uniformly represented in the initial state, the pathogen strains that win the competition are the ones that maximize $\frac{m_n}{\mu_n}$ first, and then μ_n, as we will show in Proposition 9.9. We precise now what we mean by "in some sense uniformly represented in the initial state". First we properly define the notion of "maximal reachable mortality rate".

Definition 9.7 (Maximal reachable mortality rate) Let Assumptions 9.1 and 9.3 hold. Then the *maximal reachable mortality rate* is

$$\mu^* := \sup\left\{\mu \in \omega\left((\mu_n)_{n\in\mathbb{N}}\right) \,:\, (\alpha^*, \mu) \in \omega\left(\left((\alpha_n, \mu_n)\right)_{n\in\mathbb{N}}\right)\right\},$$

wherein $\omega\left(\left((\alpha_n, \mu_n)\right)_{n\in\mathbb{N}}\right)$ is the ω-limit set of the joint sequence $\left((\alpha_n, \mu_n)\right)_{n\in\mathbb{N}}$. In other words, μ^* is the maximal limit value of a subsequence of μ_n such that $\frac{m_{n_k}}{\mu_{n_k}} \to \alpha^*$.

In the following assumption we impose that the initial mass of pathogens is never negligible around the maximal value μ^*.

Assumption 9.8 Suppose that the Assumptions 9.1, 9.2 and 9.3 hold true, and let μ^* be the maximal reachable mortality rate as in Definition 9.7. We assume that for each $\varepsilon > 0$ sufficiently small, there exist constants $\delta > 0$ and $m > 0$ such that for each value $\alpha \in [\alpha^* - \delta, \alpha^*]$ such that there exists $n \in \mathbb{N}$ with $\frac{m_n}{\mu_n} = \alpha$, we have

$$\frac{\sum\limits_{\left\{n:\, \frac{m_n}{\mu_n} = \alpha \text{ and } \mu_n \geq \mu^* - \varepsilon\right\}} P_n^0}{\sum\limits_{\left\{n:\, \frac{m_n}{\mu_n} = \alpha\right\}} P_n^0} \geq m.$$

In other words, the probability of "picking" a pathogen with μ_n close to μ^* conditionally to the fact that $\alpha_n = \alpha$ with α close to α^* has a uniform positive lower bound.

Let us give a few examples of initial data that do or do not satisfy the assumption 9.8. If μ_n converges to its limit, then the limit is necessarily μ^* and assumption 9.8 holds independently of the initial data $P_n^0 > 0$. If $\omega\big((\mu_n)_{n\in\mathbb{N}}\big) = \{\mu^1, \mu^2\}$ has exactly two elements $\mu^1 < \mu^2$, we have to look at the equality cases $\mathcal{A}_n := \{k : \alpha_k = \alpha_n\}$. For simplicity, suppose that $\mu_{2n} \to \mu^1$ and $\mu_{2n+1} \to \mu^2$ and $\mathcal{A}_n = \{2n, 2n+1\}$, then Assumption 9.8 is satisfied if, and only if, there is $m > 0$ such that

$$\frac{P_{2n+1}^0}{P_{2n}^0 + P_{2n+1}^0} \geq m.$$

When Assumption 9.8 is satisfied, we can prove the following result.

Proposition 9.9 *Suppose that Assumption 9.8 holds and that $\frac{m_n}{\mu_n} < \alpha^*$ for all $n \in \mathbb{N}$. Then the total pathogen population converges to a positive limit*

$$\lim_{t\to+\infty} \sum_{n=0}^{+\infty} P_n(t) = \frac{\mu^N}{\alpha^* \mu^*}(\mathcal{R}_0 - 1). \tag{9.7}$$

Moreover if $n_k \in \mathbb{N}$ is a sequence of integers such that

$$\sup_{k\in\mathbb{N}} \frac{m_{n_k}}{\mu_{n_k}} = \alpha^* \text{ and } \sup_{k\in\mathbb{N}} \mu_{n_k} < \mu^*, \tag{9.8}$$

then we have

$$\limsup_{t\to+\infty} \sum_{k\in\mathbb{N}} P_{n_k}(t) = 0. \tag{9.9}$$

The proof of the results of this Section will be given in section 9.5. First we present some parameter choices for which the replacement dynamics of the variants can be analytically understood.

9.3 Examples

In this section we provide examples of explicit choices of the coefficients for which the asymptotic dynamics can be understood analytically. We also provide numerical simulations of the corresponding set of ODEs, with a particular attention to the dynamics of variants replacement.

We subdivide further the section in two subsections: in section 9.3.1 we deal with examples for which $\mu_n \equiv \mu$ is a constant sequence, while in section 9.3.2 we deal with examples for which μ_n takes alternatively two values.

9.3.1 Replacement dynamics 1: Monovalent μ_n

In this section we investigate the asymptotic transition time between the prevalence of two given variants, depending on their characteristics. We place ourselves in the case when μ_n is *monovalent*, that is to say, $\mu_n \equiv \mu > 0$ is independent of $n \in \mathbb{N}$. We also assume that $m_n < m^*$ for all $n \geq 0$, so that we already know that the mass converges thanks to Proposition 9.9:

$$\mathcal{P}(t) := \sum_{n=0}^{+\infty} P_n(t) \xrightarrow[t\to+\infty]{} \mathcal{P}_\infty. \tag{9.10}$$

Now let us observe that the total mass rewrites as

$$\begin{aligned}\sum_{n=0}^{+\infty} P_n(t) = \sum_{n=0}^{+\infty} P_n^0 e^{m_n \overline{N}(t) - \mu_n t} &= \sum_{n=0}^{+\infty} P_n^0 e^{(m_n - m^*)\overline{N}(t)t + (m^*\overline{N}(t)-\mu)t} \\ &= e^{(m^*\overline{N}(t)-\mu)t} \sum_{n=0}^{+\infty} P_n^0 e^{-(m^*-m_n)\overline{N}(t)t},\end{aligned}$$

and in particular:

$$\sum_{n=0}^{+\infty} P_n(t) = e^{t\left(m^*\overline{N}(t)-\mu\right)} F\left(t\overline{N}(t)\right), \tag{9.11}$$

where F is the function defined by

$$F(\xi) := \sum_{n=0}^{+\infty} P_n^0 e^{-(m^*-m_n)\xi}. \tag{9.12}$$

Notice that, since $m_n < m^*$ for all $n \geq 0$, by (9.12) we have

$$\lim_{\xi\to+\infty} F(\xi) = 0.$$

so that, using the fact that $\sum P_n(t) = \mathcal{P}_\infty + o(1) > 0$, we get

$$\lim_{t\to+\infty} e^{(m^*\overline{N}(t)-\mu)t} = +\infty.$$

Moreover, we have

$$\ln\left(\sum_{n=0}^{+\infty} P_n(t)\right) = t\left(m^*\overline{N}(t) - \mu\right) + \ln\left(F(t\overline{N}(t))\right)$$

so that thanks to Theorem 9.6 we have

$$\frac{1}{t}\ln\left(F(\overline{N}(t))\right) = -\left(m^*\overline{N}(t) - \mu\right) + \frac{1}{t}\ln\left(\sum_{n=0}^{+\infty} P_n(t)\right) \xrightarrow[t\to+\infty]{} 0.$$

These algebraic remarks will serve to estimate the replacement speed of the variants. In particular, we will use the key relation

$$\overline{N}(t) = \frac{\mu}{m^*} - \frac{1}{tm^*}\ln\left(F(t\overline{N}(t))\right) + \frac{1}{tm^*}\ln\left(\sum_{n=0}^{+\infty} P_n(t)\right). \tag{9.13}$$

9.3.1.1 Monovalent example 1: the algebraic-algebraic case.

In this subsection we assume the following framework.

Assumption 9.10 We let $\mu_n \equiv \mu > 0$ be a constant sequence and assume that the initial data is algebraic and the convergence of the fitness to its maximum is algebraic:

$$P_n^0 = \frac{1}{(n+1)^A} \text{ with } A > 1 \text{ while } m_n = m^* - \frac{1}{n+1} \text{ for all } n \geq 0. \tag{9.14}$$

Claim Under assumption 9.10, there exist constants $C > 1$ and $\hat{t} > 0$ large enough such that for all $t \geq \hat{t}$ and for all $n \geq 0$ one has

$$\frac{1}{Ct} f\left(\frac{(n+1)}{\mu t + (A-1)\ln t}\right) \leq P_n(t) \leq \frac{C}{t} f\left(\frac{(n+1)}{\mu t + (A-1)\ln t}\right),$$

where the profile function f is given by

$$f(X) = \exp\left(-A\ln X - \frac{1}{Xm^*}\right). \tag{9.15}$$

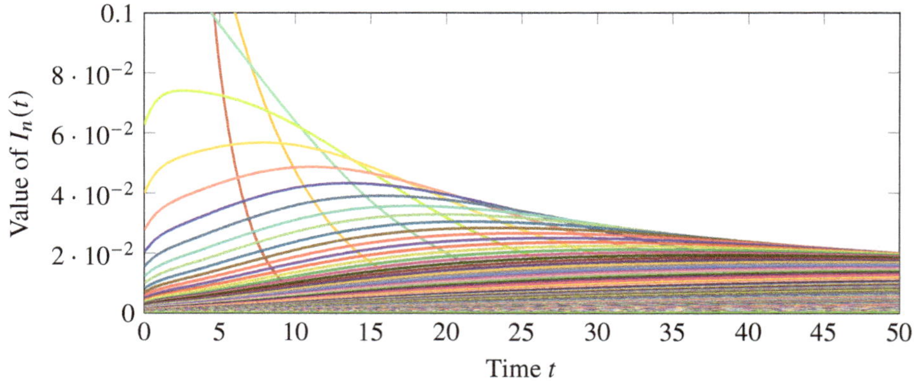

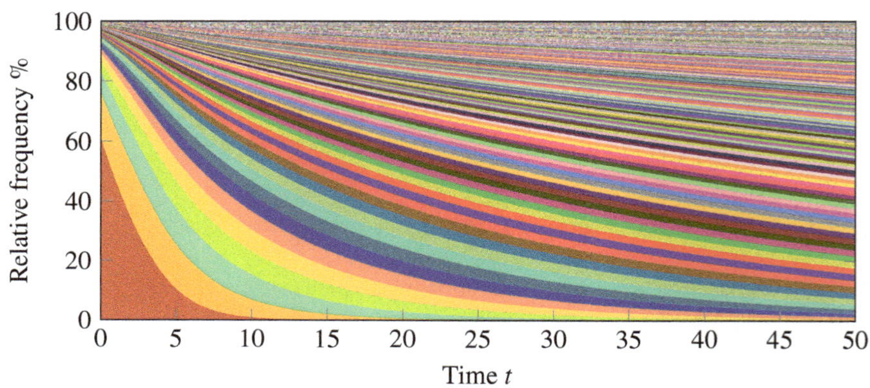

Fig. 9.1: Plots of the solution $P_n(t)$ in the monovalent case 1. Colors are chosen at random so that each variant has a different color. The bottom figure suggests that the diversity of variants increases with time, since there are more and more lines with a noticeable width. **Top figure:** Value of $P_n(t)$. **Bottom figure:** Relative frequencies of the variants as a function of time. This is the proportion of each variant in the global population. **Parameters:** $\beta = 1$, $\mu^N = 1$, $N_0 = 1$, $\mu = \frac{1}{2}$, $A = 2$, $m^* = 1$. We used a total of $N = 300$ variants for this simulation.

We start with a technical lemma.

Lemma 9.11 *The function* $F(\xi)$ *defined by* (9.12) *satisfies*

$$\ln F(\xi) = -(A-1)\ln \xi + O(1) \text{ as } \xi \to \infty.$$

Proof Let $u_n(\xi) := \exp\left(-A\ln(n+1) - \frac{\xi}{n+1}\right)$ so that

$$F(\xi) = \sum_{n=0}^{\infty} u_n(\xi).$$

For $\xi > 0$ we define $x(\xi) > 0$ and $n(\xi) \in \mathbb{N}$ by

$$x(\xi) + 1 = \frac{\xi}{A} \text{ and } n(\xi) = \lfloor x(\xi) \rfloor \text{ the integer part of } x(\xi).$$

Next we have for $p \in \mathbb{N}$ and $\xi > 0$:

$$\frac{u_{n(\xi)+p}(\xi)}{u_{n(\xi)}(\xi)} = \left(\frac{n(\xi)+1}{n(\xi)+1+p}\right)^A \exp\left(\xi \frac{p}{(n(\xi)+1)(n(\xi)+1+p)}\right)$$

Hence we get

$$\begin{aligned}\sum_{p=1}^{\infty} \frac{u_{n(\xi)+p}(\xi)}{u_{n(\xi)}(\xi)} &\le \sum_{p=1}^{\infty} \left(\frac{n(\xi)+1}{n(\xi)+1+p}\right)^A \exp\left(\frac{\xi}{n(\xi)+1}\right) \\ &\le (n(\xi)+1)^A \exp\left(\frac{\xi}{n(\xi)+1}\right) \sum_{k=n(\xi)+2}^{\infty} \frac{1}{k^A}.\end{aligned}$$

Since

$$\sum_{k=n(\xi)+2}^{+\infty} \frac{1}{k^A} \le \int_{n(\xi)+1}^{+\infty} \frac{1}{y^A} \mathrm{d}y = \frac{1}{A-1} (n(\xi)+1)^{-(A-1)},$$

we obtain that there exists some constant $C > 0$ and ξ_0 large enough such that

$$\sum_{p=1}^{\infty} \frac{u_{n(\xi)+p}(\xi)}{u_{n(\xi)}(\xi)} \le C n(\xi) \exp\left(\frac{\xi}{n(\xi)+1}\right) \quad \forall \xi > \xi_0.$$

This also rewrites as

$$\sum_{p=1}^{\infty} u_{n(\xi)+p}(\xi) \le C n(\xi) u_{n(\xi)}(\xi) \exp\left(\frac{\xi}{n(\xi)+1}\right) \quad \forall \xi > \xi_0.$$

On the other hand we have

$$\sum_{k=0}^{n(\xi)} \frac{u_k(\xi)}{u_{n(\xi)}(\xi)} = \sum_{k=0}^{n(\xi)} \left(\frac{n(\xi)+1}{k+1}\right)^A \exp\left(\frac{\xi}{n(\xi)+1} \frac{k-n(\xi)}{k+1}\right).$$

Next due to the definition of $n(\xi)$ we have

$$n(\xi) \le x(\xi) = \frac{\xi}{A} - 1 < n(\xi) + 1,$$

so that

$$A \frac{n(\xi)}{n(\xi)+1} \le \frac{\xi - A}{n(\xi)+1} \le A.$$

Hence there exists $\xi_1 > 0$ large enough so that

$$2A \ge \frac{\xi}{n(\xi)+1} \ge \frac{A}{2}, \ \forall \xi > \xi_1.$$

As a consequence for all $\xi > \xi_1$ we have

$$\sum_{k=0}^{n(\xi)} \frac{u_k(\xi)}{u_{n(\xi)}(\xi)} \geq \sum_{k=0}^{n(\xi)} \left(\frac{n(\xi)+1}{k+1}\right)^A \exp\left(-2A\frac{n(\xi)-k}{k+1}\right)$$
$$\geq e^{2A} \sum_{k=0}^{n(\xi)} \left(\frac{n(\xi)+1}{k+1}\right)^A \exp\left(-2A\frac{n(\xi)+1}{k+1}\right)$$

and similarly

$$\sum_{k=0}^{n(\xi)} \frac{u_k(\xi)}{u_{n(\xi)}(\xi)} \leq \sum_{k=0}^{n(\xi)} \left(\frac{n(\xi)+1}{k+1}\right)^A \exp\left(-\frac{A}{2}\frac{n(\xi)+1}{k+1}\right).$$

Now using Riemann sums with the continuous function

$$x \mapsto \begin{cases} 0 \text{ if } x \leq 0 \\ \frac{1}{x^A} \exp\left(-\frac{\alpha}{x}\right) \text{ if } x > 0, \end{cases}$$

let us observe that we have

$$\frac{1}{n}\sum_{k=0}^{n} \left(\frac{n+1}{k+1}\right)^A \exp\left(-\alpha\frac{n+1}{k+1}\right) \to \int_0^1 \frac{1}{x^A} \exp\left(-\frac{\alpha}{x}\right) dx \text{ as } n \to \infty.$$

As a consequence of this convergence applied for $\alpha = 2A$ and $\alpha = \frac{1}{2A}$, there exists some constant, still denoted by $C > 1$ large enough and $\xi_2 > 0$ large enough such that

$$C^{-1} n(\xi) u_{n(\xi)}(\xi) \leq \sum_{k=0}^{n(\xi)} u_k(\xi) \leq C n(\xi) u_{n(\xi)}(\xi), \ \forall \xi > \xi_2.$$

Coupling the two above estimates ensures that there exists $C > 1$ and $\hat{\xi} \gg 1$ large enough so that

$$F(\xi) \leq C n(\xi) u_{n(\xi)}(\xi) \leq C \xi^{1-A}, \ \forall \xi > \hat{\xi},$$

while

$$F(\xi) \geq \sum_{k=0}^{n(\xi)} u_k(\xi) \geq C^{-1} n(\xi) u_{n(\xi)}(\xi).$$

The completes the proof of the lemma.

Proof (Proof of Claim 9.3.1.1) As before the proof relies on Lemma 9.11 and (9.13). Indeed using these we have

$$\begin{aligned}\overline{N}(t) &= \frac{\mu}{m^*} - \frac{1}{tm^*}\ln\left(F(t\overline{N}(t))\right) + \frac{1}{tm^*}\ln\left(\sum_{n=0}^{+\infty} P_n(t)\right)\\ &= \frac{\mu}{m^*} + \frac{A-1}{m^*}\frac{\ln(t)}{t} + O\left(\frac{1}{t}\right).\end{aligned}$$

As a consequence we obtain

$$\begin{aligned}P_n(t) &= P_n^0 \exp\left(\left(m^* - \frac{1}{n+1}\right)\left(\frac{\mu}{m^*} + \frac{A-1}{m^*}\frac{\ln(t)}{t} + O\left(\frac{1}{t}\right)\right)t - \mu t\right)\\ &= P_n^0 \exp\left((A-1)\ln(t) - \frac{1}{n+1}\left(\frac{\mu}{m^*} + \frac{A-1}{m^*}\frac{\ln(t)}{t}\right)t + O(1)\right)\\ &= \exp\left(-A\ln(n+1) + (A-1)\ln(t) - \frac{1}{n+1}\left(\frac{\mu}{m^*} + \frac{A-1}{m^*}\frac{\ln(t)}{t}\right)t + O(1)\right).\end{aligned}$$

This rewrites uniformly for $n \geq 0$ as follows

$$P_n(t) = e^{O(1)}\frac{1}{t}\exp\left(-A\ln\frac{(n+1)}{\mu t + (A-1)\ln t} - \frac{1}{m^*}\frac{\mu t + (A-1)\ln t}{n+1}\right).$$

Recalling the definition of the function f in (9.15), the above equality becomes

$$P_n(t) \asymp \frac{1}{t} f\left(\frac{(n+1)}{\mu t + (A-1)\ln t}\right) \text{ with } f(X) = \exp\left(-A\ln X - \frac{1}{Xm^*}\right),$$

that completes the proof of the lemma. □

9.3.1.2 Monovalent example 2: the exponential-exponential case.

We work under the following assumption.

Assumption 9.12 We assume that $\mu_n \equiv \mu > 0$ is a positive constant, $P_n^0 = e^{-An}$ and $m_n = m^* - Be^{-Cn}$ for some constants $A > 0$, $B > 0$, and $C > 0$. In other words, the initial data is exponential and the convergence of the fitness to its maximum is also exponential.

In this case we can show that the function $F(\xi)$ behaves like $\xi^{-\frac{A}{C}}$ and that $P_n(t)$ eventually reaches a fixed shape shifting toward $+\infty$ like $\frac{1}{C}\ln(t)$. More precisely $P_n(t)$ behaves asymptotically like $\mathcal{E}\big(n - n_0(t)\big)$ as $t \to +\infty$, where $n_0(t) := \frac{1}{C}\ln(t)$ and

$$\mathcal{E}(\nu) := \exp\left(-\frac{\mu}{m^*}Be^{-C\nu} - A\nu\right). \tag{9.16}$$

Claim (Asymptotic behavior) Let Assumption 9.12 hold and $\mathcal{E}$ be defined by (9.16). Let $n_0(t) := \frac{1}{C}\ln(t)$ and $\mu(t) := \mu_0 \dfrac{\ln\left(\ln t\right)}{\ln t}$ for $\mu_0 > \max(C, 1)$ and $t > 1$. Then there exists a constant $K > 0$ such that for all $n \geq \mu(t)n_0(t)$ we have

$$\frac{1}{K}\mathcal{E}\big(n - n_0(t)\big) \leq P_n(t) \leq K\mathcal{E}\big(n - n_0(t)\big).$$

and $\sum_{k=0}^{\lfloor \mu(t) n_0(t) \rfloor} P_k(t) \xrightarrow[t \to +\infty]{} 0$ (where $\lfloor \mu(t) n_0(t) \rfloor$ is the biggest integer smaller than $\mu(t) n_0(t)$).

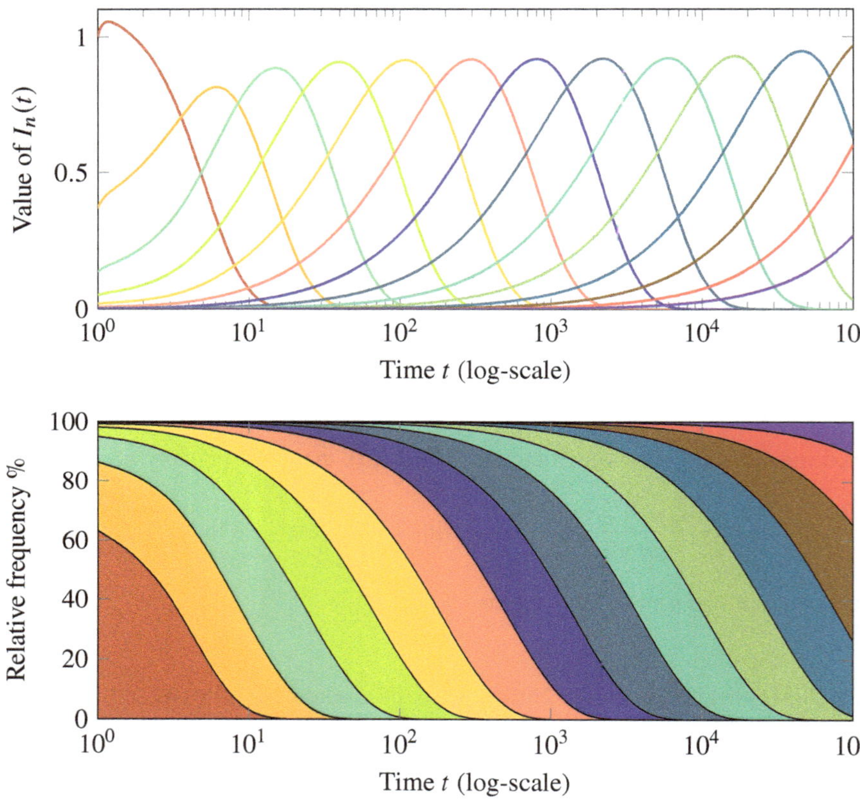

Fig. 9.2: Plots of the solution $P_n(t)$ in the monovalent case 2. Colors are chosen at random so that each variant has a unique color. The top figure suggests that each variant has a similar behavior in time. The bottom figure suggests that the diversity of variants remains approximately constant in time (more precisely, the number of non-negligible variants in the population is uniformly greater than a positive constant). **Top figure:** Value of $P_n(t)$. **Bottom figure:** Relative frequencies of the variants as a function of time. **Parameters:** $\beta = 10$, $\mu^N = 10$, $N_0 = 1$, $\mu = 2$, $m^* = 4$, $A = 1$, $B = 1$, $C = 1$. We used a total of $N = 14$ variants for this simulation. For interpretation of the colors in the figure(s), the reader is referred to the online version of this article.

We begin with the asymptotic expansion of F.

Lemma 9.13 *Let Assumption 9.12 hold and let $F(\xi)$ be defined by* (9.12). *Then*

$$F(\xi) = \xi^{-\frac{A}{C}} \times e^{O(1)} \ as\ \xi \to \infty.$$

Proof We claim that F satisfies the functional equation:

$$F(\xi) = e^{-B\xi} + e^{-A} F(e^{-C}\xi).$$

Indeed,

$$\begin{aligned} F(\xi) &= \sum_{n=0}^{+\infty} e^{-B\xi e^{-Cn} - An} = \sum_{n=0}^{+\infty} e^{-An} \sum_{k=0}^{+\infty} \frac{(-B\xi)^k}{k!} \left(e^{-Cn}\right)^k = \sum_{k=0}^{+\infty} \frac{(-B\xi)^k}{k!} \sum_{n=0}^{+\infty} e^{-(A+kC)n} \\ &= \sum_{k=0}^{+\infty} \frac{(-B\xi)^k}{k!} \cdot \frac{1}{1 - e^{-A-kC}} \end{aligned} \tag{9.17}$$

and then

$$\begin{aligned} F(\xi) &= \left(\sum_{k=0}^{+\infty} \frac{(-B\xi)^k}{k!} \right) + \sum_{k=0}^{+\infty} \frac{(-B\xi)^k}{k!} \left(\frac{1}{1 - e^{-A-kC}} - 1 \right) \\ &= e^{-B\xi} + \sum_{k=0}^{+\infty} \frac{(-B\xi)^k}{k!} \frac{e^{-A-kC}}{1 - e^{-A-kC}} \\ &= e^{-B\xi} + e^{-A} \sum_{k=0}^{+\infty} \frac{(-Be^{-C}\xi)^k}{k!} \frac{e^{-A-kC}}{1 - e^{-A-kC}} \\ &= e^{-B\xi} + e^{-A} F\left(e^{-C}\xi\right), \end{aligned}$$

where we have used the identity (9.17) for the last equality. We deduce that

$$\begin{aligned} F(e^C\xi) &= e^{-Be^C\xi} + e^{-A} F(\xi), \\ F(e^{2C}\xi) &= e^{-Be^{2C}\xi} + e^{-A} F(e^C\xi) = e^{-Be^{2C}\xi} + e^{-A-Be^C\xi} + e^{-2A} F(\xi), \\ F(e^{2C}\xi) &= e^{-Be^{3C}\xi} + e^{-A} F(e^{2C}\xi) = e^{-Be^{3C}\xi} + e^{-A-Be^{2C}\xi} + e^{-2A-Be^C\xi} + e^{-3A} F(\xi), \\ &\ \vdots \\ F(e^{nC}\xi) &= e^{-nA} F(\xi) + \sum_{k=1}^{n} e^{-(n-k)A - Be^{kC}\xi} = e^{-nA} \left(F(\xi) + \sum_{k=1}^{n} e^{kA - Be^{kC}\xi} \right). \end{aligned}$$

For $\xi \in \left[1, e^C\right)$, we let $X = \xi e^{nC}$ so that

$$n = \left\lfloor \frac{1}{C} \ln(X) \right\rfloor \text{ and } \xi = X e^{-C\lfloor \ln(X)/C \rfloor}.$$

We get:

$$F(X) = \frac{1}{\exp\left(A \left\lfloor \frac{1}{C} \ln(X) \right\rfloor\right)} \left(F\left(\xi\right) + \sum_{k=1}^{n} e^{kA - Be^{kC}\xi} \right),$$

and finally, as $X \longrightarrow +\infty$,

$$F(X) = X^{-\frac{A}{C}} \times e^{O(1)}.$$

Proof (Proof of Claim 9.3.1.2) Thanks to Claim 9.13 we have, recalling (9.13)

$$\begin{aligned}\overline{N}(t) &= \frac{\mu}{m^*} - \frac{1}{tm^*}\ln\left(F(t\overline{N}(t))\right) + \frac{1}{tm^*}\ln\left(\sum_{n=0}^{+\infty} P_n(t)\right) \\ &= \frac{\mu}{m^*} + \frac{A}{Cm^*}\frac{\ln(t)}{t} + O\left(\frac{1}{t}\right).\end{aligned}$$

Therefore,

$$\begin{aligned}P_n(t) &= P_n^0 \exp\left(\left(m^* - Be^{-Cn}\right)\left(\frac{\mu}{m^*} + \frac{A}{Cm^*}\frac{\ln(t)}{t} + O\left(\frac{1}{t}\right)\right)t - \mu t\right) \\ &= P_n^0 \exp\left(\frac{A}{C}\ln(t) - Be^{-Cn}\left(\frac{\mu}{m^*} + \frac{A}{Cm^*}\frac{\ln(t)}{t}\right)t + O(1)\right) \\ &= \exp\left(-Be^{-Cn+\ln(t)}\left(\frac{\mu}{m^*} + \frac{A}{Cm^*}\frac{\ln(t)}{t}\right) + \frac{A}{C}\ln(t) - An + O(1)\right) \\ &= \exp\left(-\left(1 + \frac{A}{C\mu}\frac{\ln(t)}{t}\right)\frac{\mu}{m^*}Be^{-C(n-\frac{1}{C}\ln(t))} - A\left(n - \frac{1}{C}\ln(t)\right) + O(1)\right) \\ &= \exp\left(-\left(1 + \frac{A}{C\mu}\frac{\ln(t)}{t}\right)\frac{\mu}{m^*}Be^{-C(n-n_0(t))} - A\big(n - n_0(t)\big) + O(1)\right)\end{aligned}$$

with $n_0(t) = \frac{1}{C}\ln(t)$. Thus for $(n,t) \in \{(\tilde{n},\tilde{t}) \ : \ \tilde{n} \geq \mu(\tilde{t})n_0(\tilde{t})\}$ we have

$$\frac{P_n(t)}{\mathcal{E}(n - n_0(t))} = \exp\left(-\frac{A}{C\mu}\frac{\ln(t)}{t}\frac{\mu}{m^*}Be^{-C(n-n_0(t))} + O(1)\right) = e^{O(1)},$$

since

$$0 \leq \frac{\ln(t)}{t}e^{-C(n-n_0(t))} \leq e^{\ln(\ln t)-\ln(t)-(\mu(t)-1)\ln(t)} = e^{(1-\mu_0)\ln(\ln t)} \xrightarrow[t\to+\infty]{} 0.$$

On the other hand if $n \leq \mu(t)n_0(t)$ then

$$\begin{aligned}P_n(t) &= \exp\left[-\frac{\mu}{m^*}Bte^{-Cn} - \frac{A}{Cm^*}\ln(t)e^{-Cn} - An + \frac{A}{C}\ln(t) + O(1)\right] \\ &\leq \exp\left[-\left(\frac{\mu}{m^*}Bt - \frac{A}{C}\right)e^{-Cn} - An + \frac{A}{C}\left(\ln(t) - e^{-n}\right) + O(1)\right] \leq e^{-An+O(1)},\end{aligned}$$

because $n \leq \mu(t)n_0(t) = \frac{\mu_0}{C}\ln(\ln t)$ with $\mu_0 > C$ implies $e^{-n} \geq \ln(t)^{\frac{2}{C}} \geq \ln(t)$ for t large, whence

$$\sum_{n=0}^{\lfloor \mu(t)n_0(t)\rfloor} P_n(t) \xrightarrow[t\to+\infty]{} 0$$

by Lebesgue's dominated convergence Theorem. This proves Claim 9.3.1.2. □

9.3.1.3 Monovalent example 3: the Gaussian-exponential case.

Assumption 9.14 Assume that the initial data is Gaussian and the convergence of the fitness to its maximum is exponential:

$$P_n^0 = e^{-An^2} \text{ while } m^* - m_n = Be^{-Cn} \text{ for all } n \geq 0, \tag{9.18}$$

with $A > 0$, $B > 0$ and $C > 0$ given constants.

Under this Assumption we can prove that the dynamics of the entire family of variants is actually guided by only three variants at a time at most (and most of the times by only one variant at a time) provided C is sufficiently small. We let W_0 be the principal branch of the Lambert-W function, that is to say the smooth real function satisfying $W_0(x)e^{W_0(x)} = x$ for all $x \geq e^{-1}$ and $\lim_{x\to+\infty} W_0(x) = +\infty$; see [64] for details. We define:

$$Y(t) = \frac{1}{C} W_0 \left(\frac{BC^2}{2A} t\overline{N}(t) \right). \tag{9.19}$$

Claim Let Assumption 9.14 hold. If C is sufficiently small, then there exists $\delta > 0$ small enough such that:

$$\sum_{n=0}^{\infty} P_n(t) \sim \begin{cases} P_{I(t)-1}(t) + P_{I(t)}(t) \text{ for } t \gg 1 \text{ and } -1/2 \leq R(t) \leq -1/2 + \delta, \\ P_{I(t)} \text{ for } t \gg 1 \text{ and } -1/2 + \delta \leq R(t) \leq 1/2 - \delta, \\ P_{I(t)}(t) + P_{I(t)+1}(t) \text{ for } t \gg 1 \text{ and } 1/2 - \delta < R(t) \leq 1/2, \end{cases}$$

where $I(t) = \lfloor Y(t) + \frac{1}{2} \rfloor$, $R(t) = Y(t) - I(t)$ and $Y(t)$ is defined in (9.19).

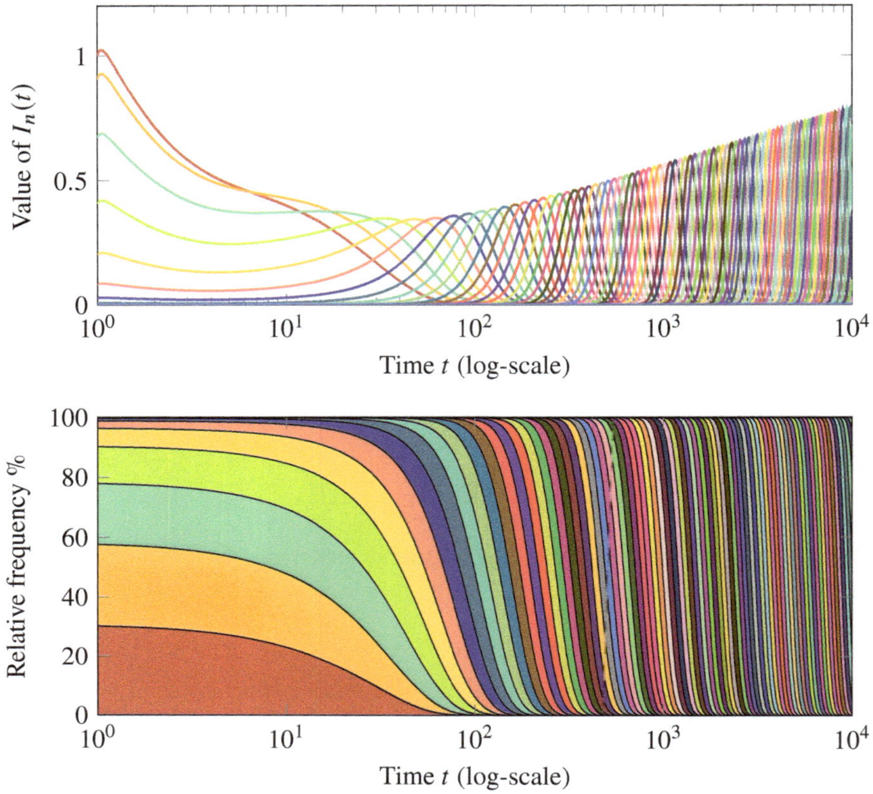

Fig. 9.3: Plots of the solution $P_n(t)$ in the monovalent case 3. Colors are chosen at random so that each variant has a unique color. The bottom figure suggests that the diversity of variants diminishes in time since the steepness of the separation between variants is slowly increasing (and our theoretical predictions show that only one variant remains at carefully chosen times). **Top figure:** Value of $P_n(t)$. **Bottom figure:** Relative frequencies of the variants as a function of time. **Parameters:** $\beta = 10$, $\mu^N = 10$, $N_0 = 1$, $\mu = 2$, $m^* = 1$, $A = 0.1$, $B = 1$, $C = 0.03$. We used a total of $N = 100$ variants for this simulation. For interpretation of the colors in the figure(s), the reader is referred to the online version of this article.

We define for $\xi > 0, x(\xi) > 0$ at the solution of the so-called transcendental Lambert equation

$$xe^{Cx} = \frac{BC}{2A}\xi.$$

Remark that $x(\xi)$ can be expressed thanks to the principal branch of the Lambert-W function, $W_0(z)$. More precisely, $x(\xi) = \frac{1}{C} W_0\left(\frac{BC^2}{2A}\xi\right)$.

We denote by $n(\xi)$ the integer which is the closest of $x(\xi)$, that is

$$x(\xi) = n(\xi) + r(\xi) \text{ with } n(\xi) = \left\lfloor x(\xi) + \frac{1}{2} \right\rfloor \in \mathbb{N} \text{ and } r(\xi) \in [-1/2, 1/2). \quad (9.20)$$

Lemma 9.15 *The function F defined in* (9.12) *satisfies the following asymptotic behavior*

$$F(e^{-Cr(\xi)}\xi) \sim e^{-An(\xi)^2 - B\xi e^{-Cx(\xi)}} = e^{-An(\xi)^2 - \frac{2A}{C}x(\xi)} \text{ as } \xi \to \infty.$$

Proof We define the function $G : \mathbb{R}^2 \to \mathbb{R}$ by

$$G(x, \xi) = Ax^2 + B\xi e^{-Cx}, \ \forall (x, \xi) \in \mathbb{R}^2. \quad (9.21)$$

Now we first claim that we have:

$$\sum_{n=n(\xi)+1}^{\infty} e^{-G(n, e^{-Cr(\xi)}\xi)} = o\left(e^{-G(n(\xi), e^{-Cr(\xi)}\xi)}\right) \text{ as } \xi \to \infty. \quad (9.22)$$

To prove this property, observe that for $p \geq 1$ we have

$$\begin{aligned}
&G(n(\xi)+p, e^{-Cr(\xi)}\xi) - G(n(\xi), e^{-Cr(\xi)}\xi) \\
&\quad = Be^{-C(n(\xi)+p)}e^{-Cr(\xi)}\xi + A(n(\xi)+p)^2 - Be^{-Cn(\xi)}e^{-Cr(\xi)}\xi - An(\xi)^2 \\
&\quad = B\left(e^{-Cp} - 1\right)e^{-Cx(\xi)}\xi + A\left(2n(\xi)p + p^2\right) \\
&\quad = B\left(e^{-Cp} - 1\right)\frac{2A}{BC}x(\xi) + A\left(2n(\xi)p + p^2\right) \\
&\quad = x(\xi)\left[-\left(1 - e^{-Cp}\right)\frac{2A}{C} + 2Ap\right] - 2A\big(n(\xi) + r(\xi)\big)p + A\left(2n(\xi)p + p^2\right) \\
&\quad = x(\xi)\left[-\left(1 - e^{-Cp}\right)\frac{2A}{C} + 2Ap\right] + A\left(p - r(\xi)\right)^2 - Ar(\xi)^2 \\
&\quad \geq \frac{2A}{C}x(\xi)\left[-\left(1 - e^{-Cp}\right) + Cp\right] + A\left(p - 1/2\right)^2 - A/4.
\end{aligned}$$

From the above estimate we obtain

$$\begin{aligned}
&\sum_{n=n(\xi)+1}^{\infty} e^{-G(n, e^{-Cr(\xi)}\xi) + G(n(\xi), e^{-Cr(\xi)}\xi)} \\
&\qquad \leq e^{-\frac{2A}{C}x(\xi)[C - (1 - e^{-C})]} \sum_{p=1}^{\infty} e^{-A(p-1/2)^2 + A/4} \to 0 \text{ as } \xi \to \infty, \quad (9.23)
\end{aligned}$$

which proves (9.22).

Next we claim that we have

$$\sum_{n=0}^{n(\xi)-1} e^{-G(n, e^{-Cr(\xi)}\xi)} = o\left(e^{-G(n(\xi), e^{-Cr(\xi)}\xi)}\right) \text{ as } \xi \to \infty. \quad (9.24)$$

To see this, note that for $1 \le p \le n(\xi)$ we have

$$
\begin{aligned}
&G(n(\xi)-p, e^{-Cr(\xi)}\xi) - G(n(\xi), e^{-Cr(\xi)}\xi)\\
&\quad = Be^{-C(n(\xi)-p)}e^{-Cr(\xi)}\xi + A(n(\xi)-p)^2 - Be^{-Cn(\xi)}e^{-Cr(\xi)}\xi - An(\xi)^2\\
&\quad = B(e^{Cp}-1)e^{-Cx(\xi)}\xi + A[(n(\xi)-p)^2 - n(\xi)^2]\\
&\quad = B(e^{Cp}-1)e^{-Cx(\xi)}\xi - A[2px(\xi) - p^2 - 2pr(\xi)]\\
&\quad = \frac{2A}{C}x(\xi)[e^{Cp} - 1 - Cp] + A(p^2 - 2pr(\xi))\\
&\quad \ge \frac{2A}{C}x(\xi)[e^{C} - 1 - C] + A(p^2 - p).
\end{aligned}
$$

Therefore we get

$$
\sum_{n=0}^{n(\xi)-1} e^{-G(n,e^{-Cr(\xi)}\xi)+G(n(\xi),e^{-Cr(\xi)}\xi)} \le e^{-\frac{2A}{C}x(\xi)[e^C-1-C]} \sum_{p=0}^{n(\xi)} e^{-A(p^2-p)} \to 0
$$

as $\xi \to \infty$. Coupling (9.22) and (9.24) yields

$$
F\left(e^{-Cr(\xi)}\xi\right) = e^{-G(n(\xi),e^{-Cr(\xi)}\xi)}(1+o(1)) \text{ as } \xi \to \infty.
$$

That completes the proof. □

When C is sufficiently small we obtain a more refined estimate that is useful to understand the large time behavior of the sequence $(P_n(t))_{n\ge0}$.

Lemma 9.16 *Fix $C > 0$ small enough such that*

$$
e^{-C/2}(e^{2C}-1) - 2C > 0 \text{ and } C - e^{C/2}\left(1 - e^{-2C}\right) > 0. \tag{9.25}
$$

Then the function F defined in (9.12) *satisfies*

$$
F(\xi) = \sum_{p=-1,0,1} e^{-A(n(\xi)+p)^2 - B\xi e^{-C(n(\xi)+p)}} + o\left(e^{-An(\xi)^2 - B\xi e^{-Cn(\xi)}}\right) \text{ as } \xi \to \infty.
$$

Proof Recall the definition of the function G in (9.21), let us show that we have:

$$
\sum_{p=n(\xi)+2}^{\infty} e^{-G(p,\xi)} = o\left(e^{-G(n(\xi),\xi)}\right) \text{ as } \xi \to \infty. \tag{9.26}
$$

To prove this property, observe that for $p \ge 2$ we have

$$
\begin{aligned}
G(n(\xi)+p,\xi) &- G(n(\xi),\xi) \\
&= Be^{-C(n(\xi)+p)}\xi + A(n(\xi)+p)^2 - Be^{-Cn(\xi)}\xi - An(\xi)^2 \\
&= B\left(e^{-Cp}-1\right)e^{-Cx(\xi)}e^{Cr(\xi)}\xi + A\left(2n(\xi)p+p^2\right) \\
&= e^{Cr(\xi)}B\left(e^{-Cp}-1\right)\frac{2A}{BC}x(\xi) + A\left(2n(\xi)p+p^2\right) \\
&= x(\xi)\left[-e^{Cr(\xi)}\left(1-e^{-Cp}\right)\frac{2A}{C}+2Ap\right] + A\left(p-r(\xi)\right)^2 - Ar(\xi)^2 \\
&\geq \frac{2A}{C}x(\xi)\left[-e^{C/2}\left(1-e^{-Cp}\right)+Cp\right] + A\left(p-1/2\right)^2 - A/4
\end{aligned}
$$

Now the map $p \mapsto -e^{C/2}\left(1-e^{-Cp}\right)+Cp$ is increasing for $p \geq 1$ so that

$$
\sum_{n=n(\xi)+2}^{\infty} e^{-G(n,\xi)+G(n(\xi),\xi)} \leq e^{-\frac{2A}{C}x(\xi)\left[2C-e^{C/2}\left(1-e^{-2C}\right)\right]} \sum_{p=1}^{\infty} e^{-A(p-1/2)^2+A/4}. \tag{9.27}
$$

Finally thanks to (9.25) we have

$$
2C - e^{C/2}\left(1-e^{-2C}\right) > 0 \text{ for } C > 0 \text{ small enough}
$$

so that

$$
\sum_{p=n(\xi)+2}^{\infty} e^{-G(p,\xi)+G(n(\xi),\xi)} \leq e^{-\frac{2A}{C}x(\xi)\left[2C-e^{C/2}\left(1-e^{-2C}\right)\right]} \sum_{p=1}^{\infty} e^{-A(p-1/2)^2+A/4} \to 0
$$

as $\xi \to \infty$, that is

$$
\sum_{p=n(\xi)+2}^{\infty} e^{-G(p,\xi)} = o\left(e^{-G(n(\xi),\xi)}\right) \text{ as } \xi \to \infty. \tag{9.28}
$$

Next we claim that we have

$$
\sum_{p=0}^{n(\xi)-2} e^{-G(p,\xi)} = o\left(e^{-G(n(\xi),\xi)}\right) \text{ as } \xi \to \infty. \tag{9.29}
$$

To see this, note that for $2 \leq p \leq n(\xi)$ we have

$$
\begin{aligned}
G(n(\xi)-p,\xi)-G(n(\xi),\xi)&\\
&= Be^{-C(n(\xi)-p)}\xi + A(n(\xi)-p)^2 - Be^{-Cn(\xi)}\xi - An(\xi)^2\\
&= B(e^{Cp}-1)e^{-Cx(\xi)}e^{Cr(\xi)}\xi + A[(n(\xi)-p)^2 - n(\xi)^2]\\
&= B(e^{Cp}-1)e^{-Cx(\xi)}e^{Cr(\xi)}\xi - A[2px(\xi) - p^2 - 2pr(\xi)]\\
&= \frac{2A}{C}x(\xi)[e^{Cr(\xi)}e^{Cp} - 1 - Cp] + A(p^2 - 2pr(\xi))\\
&= \frac{2A}{C}x(\xi)[e^{-C/2}e^{Cp} - 1 - Cp] + A(p^2 - 2pr(\xi))\\
&\geq 2Ax(\xi)[e^{-C/2}(e^{C}-1) - C] + A(p^2 - p).
\end{aligned}
$$

Now the map $p \mapsto e^{-C/2}e^{Cp} - 1 - Cp$ is increasing for $p \geq 1$ so that

$$
G(n(\xi)-p,\xi)-G(n(\xi),\xi) \geq 2Ax(\xi)[e^{-C/2}(e^{2C}-1) - 2C] + A(p^2 - p).
$$

Thanks to (9.25) we have

$$
e^{-C/2}(e^{2C}-1) - 2C > 0,
$$

so that we obtain

$$
\sum_{p=0}^{n(\xi)-2} e^{-G(p,\xi)+G(n(\xi),\xi)} \to 0 \text{ as } \xi \to \infty,
$$

and the result follows. □

Now let us prove that when C is small enough, when the time becomes large, at most three variants can survive at the same time. Our precise lemma reads as follows.

Corollary 9.17 *Fix C satisfying (9.25) and let Assumption 9.18 be satisfied. Define for $t \gg 1$, $I(t)$ and $R(t)$ the closest integer part and the fractional part of $Y(t)$ the solution of*

$$
Y(t) = \frac{1}{C}W_0\left(\frac{BC^2}{2A}t\overline{N}(t)\right).
$$

Then the following holds true:

$$
\sum_{p=0}^{I(t)-2} P_p(t) + \sum_{p=I(t)+2}^{\infty} P_p(t) = o\left(P_{I(t)}(t)\right) \text{ as } t \to \infty.
$$

Note that using the notations introduced above (see (9.20)) we have

$$
I(t) = n\left(t\overline{N}(t)\right) \text{ and } R(t) = r\left(t\overline{N}(t)\right).
$$

Remark 9.18 The above corollary means that, at least for C small enough, at most three variants can simultaneously survive in the large time, the variants $I(t)-1$, $I(t)$ and $I(t)+1$.

Proof The proof is a direct consequence of on estimates (9.28) and (9.29). Indeed we have for all $n \geq 0$ and $t > 0$

$$\frac{P_n(t)}{P_{I(t)}(t)} = e^{-A(n^2-I(t)^2)-B(e^{-Cn}-e^{-CI(t)}))t\overline{N}(t)} = e^{-G(n,t\overline{N}(t))+G(I(t),t\overline{N}(t))},$$

and the result follows from these two estimates. □

Proof (Proof of Claim 9.3.1.3.) Now let us show that most of the time, only one variants can survive. To see this, using the same notations as above, note that we have

$$\begin{aligned}\frac{P_{I-1}(t)}{P_{I(t)+1}(t)} &= e^{-A((I(t)-1)^2-(I(t)+1)^2)-B(e^{-C(I(t)-1)}-e^{-C(I(t)+1)}))t\overline{N}(t)}\\ &= e^{-A(-4I(t)-Be^{-CI(t)-CR(t)}e^{CR(t)}(e^{C}-e^{-C}))t\overline{N}(t)}\\ &= e^{4AI(t)-2A/CY(t)e^{CR(t)}(e^{C}-e^{-C}))}\\ &= e^{2AI(t)(2-1/Ce^{CR(t)}(e^{C}-e^{-C}))+O(1)},\end{aligned}$$

while

$$\begin{aligned}\frac{P_{I-1}(t)}{P_{I(t)}(t)} &= e^{-A((I(t)-1)^2-I(t)^2)-B(e^{-C(I(t)-1)}-e^{-CI(t)}))t\overline{N}(t)}\\ &= e^{2AI(t)(1-1/Ce^{CR(t)}(e^{C}-1))+O(1)},\end{aligned}$$

and

$$\begin{aligned}\frac{P_{I+1}(t)}{P_{I(t)}(t)} &= e^{-A((I(t)+1)^2-I(t)^2)-B(e^{-C(I(t)+1)}-e^{-CI(t)}))t\overline{N}(t)}\\ &= e^{-2AI(t)(1+1/Ce^{CR(t)}(e^{-C}-1))+O(1)}.\end{aligned}$$

As a consequence, setting

$$X_1(C) = \frac{1}{C}\ln\frac{C}{\sinh(C)},\ X_2(C) = \frac{1}{C}\ln\frac{C}{e^C-1} \text{ and } X_3(C) = -\frac{1}{C}\ln\frac{1-e^{-C}}{C},$$

for all $\delta > 0$, we have

$$\lim_{\substack{t\to\infty\\ X_1(C)+\delta\leq R(t)}} \frac{P_{I-1}(t)}{P_{I(t)+1}(t)} = 0, \tag{9.30}$$

$$\lim_{\substack{t\to\infty\\ R(t)\leq X_1(C)-\delta}} \frac{P_{I+1}(t)}{P_{I(t)-1}(t)} = 0, \tag{9.31}$$

and

$$\lim_{\substack{t\to\infty \\ X_2(C)+\delta\le R(t)}} \frac{P_{I-1}(t)}{P_{I(t)}(t)} = 0 \text{ and } \lim_{\substack{t\to\infty \\ X_3(C)-\delta\ge R(t)}} \frac{P_{I+1}(t)}{P_{I(t)}(t)} = 0. \tag{9.32}$$

To understand the meaning of the above limits, observe that

$$X_1(C) = -C/6 + O(C^2), \quad X_2(C) = -\frac{1}{2} + O(C) \text{ and } X_3(C) = \frac{1}{2} + O(C) \text{ as } C \to 0.$$

As a consequence, using (9.32), if C is sufficiently small then for some $\delta > 0$ small enough we have

$$P_{I(t)-1}(t) = o\left(P_{I(t)}(t)\right) \text{ for } t \gg 1 \text{ and } -\frac{1}{2} + \delta \le R(t) \le \frac{1}{2},$$
$$P_{I(t)+1}(t) = o\left(P_{I(t)}(t)\right) \text{ for } t \gg 1 \text{ and } -\frac{1}{2} < R(t) \le \frac{1}{2} - \delta.$$

This means that most of the time (when $R(t)$ is close to $-1/2$, only the variant $N(t)$ and $N(t) - 1$ can survive. Using (9.31), if C is sufficiently small then then for some $\delta > 0$ small enough we have

$$P_{I(t)+1}(t) = o\left(P_{I(t)-1}(t)\right) \text{ for } t \gg 1 \text{ and } R(t) \in \left[-\frac{1}{2} + \delta, -\delta\right].$$

For $C > 0$ is small enough, there exists $\delta > 0$ small enough such that the following picture for the survival of the variants holds

$$\sum_{n=0}^{\infty} P_n(t) \sim \begin{cases} P_{I(t)-1}(t) + P_{I(t)}(t) \text{ for } t \gg 1 \text{ and } -1/2 \le R(t) \le -1/2 + \delta, \\ P_{I(t)} \text{ for } t \gg 1 \text{ and } -1/2 + \delta \le R(t) \le 1/2 - \delta, \\ P_{I(t)}(t) + P_{I(t)+1}(t) \text{ for } t \gg 1 \text{ and } 1/2 - \delta < R(t) \le 1/2. \end{cases}$$

This proves Claim 9.3.1.3. □

9.3.2 Replacement dynamics 2: bivalent μ_n

We place ourselves in the case when the sequence μ_n takes only two values, an more precisely we assume that there exist two constants $0 < \mu_2 < \mu_1$ such that $\mu_{2n+1} \equiv \mu_1 > 0$ and $\mu_{2n} \equiv \mu_2$ for all $n \in \mathbb{N}$. Our goal is to give examples of possible behaviors when Assumption 9.8 is not satisfied; in particular, we will not assume that the total mass converges.

Recalling that $\alpha^* := \sup_{n\in\mathbb{N}} \frac{m_n}{\mu_n}$, we can write the total mass as

$$
\begin{aligned}
\sum_{n=0}^{+\infty} P_n(t) &= \sum_{n=0}^{+\infty} P_n^0 e^{m_n \overline{N}(s) - \mu_n t} \\
&= \sum_{n=0}^{+\infty} P_n^0 e^{\left(\frac{m_n}{\mu_n} - \alpha^*\right)\mu_n t\overline{N}(t) + \alpha^*(\mu_n - \mu^*)t\overline{N}(t) + \alpha^*\mu^* t\overline{N}(t) - (\mu_n - \mu^*)t - \mu^* t} \\
&= e^{\mu^* t(\alpha^*\overline{N}(t)-1)} \sum_{n=0}^{+\infty} P_n^0 e^{\mu_n\left(\frac{m_n}{\mu_n} - \alpha^*\right)t\overline{N}(t) + (\mu_n - \mu^*)(\alpha^*\overline{N}(t)-1)t} \\
&= e^{\mu^* t(\alpha^*\overline{N}(t)-1)} \sum_{n=0}^{+\infty} P_n^0 e^{-\mu_n\left(\alpha^* - \frac{m_n}{\mu_n}\right)t\overline{N}(t) - \alpha^*(\mu^* - \mu_n)\left(\overline{N}(t) - \frac{1}{\alpha^*}\right)t} \\
&= e^{\mu_1 t(\alpha^*\overline{N}(t)-1)} \mathcal{F}\big(t, \overline{N}(t)\big), \qquad (9.33)
\end{aligned}
$$

where

$$
\mathcal{F}\big(t, \overline{N}(t)\big) := \sum_{n=0}^{+\infty} P_n^0 e^{-\mu_n\left(\alpha^* - \frac{m_n}{\mu_n}\right)t\overline{N}(t) - \alpha^*(\mu^* - \mu_n)\left(\overline{N}(t) - \frac{1}{\alpha^*}\right)t}. \qquad (9.34)
$$

Since $\mu_{2n+1} = \mu_1$ and $\mu_{2n} = \mu_2$, we set

$$
m_n^1 := m_{2n+1}, \quad P_n^{0,1} := P_{2n+1}^0, \quad m_n^2 := m_{2n}, \quad \text{and } P_{2n}^{0,2} := P_{2n}^0,
$$

so we can rewrite (9.34) as

$$
\begin{aligned}
\mathcal{F}\big(t, \overline{N}(t)\big) &= \sum_{n=0}^{+\infty} P_n^{0,1} e^{-\mu_1\left(\alpha^* - \frac{m_n^1}{\mu_1}\right)t\overline{N}(t)} + \sum_{n=0}^{+\infty} P_n^{0,2} e^{-\mu_2\left(\alpha^* - \frac{m_n^2}{\mu_2}\right)t\overline{N}(t) - (\mu_1 - \mu_2)\left(\alpha^*\overline{N}(t)-1\right)t} \\
&= F_1\big(t\overline{N}(t)\big) + e^{-(\mu_1 - \mu_2)(\alpha^*\overline{N}(t)-1)t} F_2\big(t\overline{N}(t)\big), \qquad (9.35)
\end{aligned}
$$

where

$$
F_1(\xi) := \sum_{n=0}^{+\infty} P_n^{0,1} e^{-\mu_1\left(\alpha^* - \frac{m_n^1}{\mu_1}\right)\xi} \text{ and } F_2(\xi) := \sum_{n=0}^{+\infty} P_n^{0,2} e^{-\mu_2\left(\alpha^* - \frac{m_n^2}{\mu_2}\right)\xi}. \qquad (9.36)
$$

To go a bit further, we notice that taking the logarithm of (9.33) leads to

$$
\mu_1\big(\alpha^*\overline{N}(t) - 1\big)t = -\ln\mathcal{F}\big(t, \overline{N}(t)\big) + \ln\mathcal{P}(t),
$$

so (9.35) becomes

$$
\mathcal{F}\big(t, \overline{N}(t)\big) = F_1\big(t\overline{N}(t)\big) + \left(\frac{\mathcal{F}\big(t, \overline{N}(t)\big)}{\mathcal{P}(t)}\right)^{\frac{\mu_1 - \mu_2}{\mu_1}} F_2\big(t\overline{N}(t)\big). \qquad (9.37)
$$

In the sequel we will assume $\mu_1 = 2\mu_2$. In that case, (9.37) is a second-order polynomial equation in $\sqrt{\mathcal{F}\big(t, \overline{N}(t)\big)}$ which can be inverted to give the following expression of $\mathcal{F}$:

$$\mathcal{F}\left(t,\overline{N}(t)\right)=\frac{1}{4\mathcal{P}(t)}\left(F_2\left(t\overline{N}(t)\right)+\sqrt{F_2\left(t\overline{N}(t)\right)^2+4F_1\left(t\overline{N}(t)\right)\mathcal{P}(t)}\right)^2. \quad (9.38)$$

9.3.2.1 Bivalent example 1: unexpected selection

First let us explain the title of the subsection. Given Proposition 9.9, we can prove that the phenotype that is eventually selected by competition (after the selection for fitness) is the one that maximizes μ_n (in our case, μ_1). In this section we will prove that, for some carefully chosen initial data, selection can make all pathogen expressing the μ_1 phenotype disappear. This strikes us as an unexpected result. We work under the following assumption.

Assumption 9.19 We assume that $\mu_1 = 2\mu_2$, $m_n^1 = \mu_1\alpha^* - B_1e^{-C_1n}$, $P_n^{0,1} = e^{-A_1n^2}$, $m_n^2 = \mu_2\alpha^* - B_2e^{-C_2n}$ and $P_n^{0,2} = e^{-A_2n}$.

Our claim is as follows.

Claim Under Assumption 9.19, we have

$$\lim_{t\to+\infty}\sum_{n=0}^{+\infty}P_n^1(t)=0 \text{ and } \lim_{t\to+\infty}\sum_{n=0}^{+\infty}P_n^2(t)=\frac{\mu^N}{\mu_2\alpha^*}(\mathcal{R}_0-1)>0,$$

wherein $P_n^1(t) = P_{2n+1}(t)$ and $P_n^2(t) = P_{2n}(t)$.

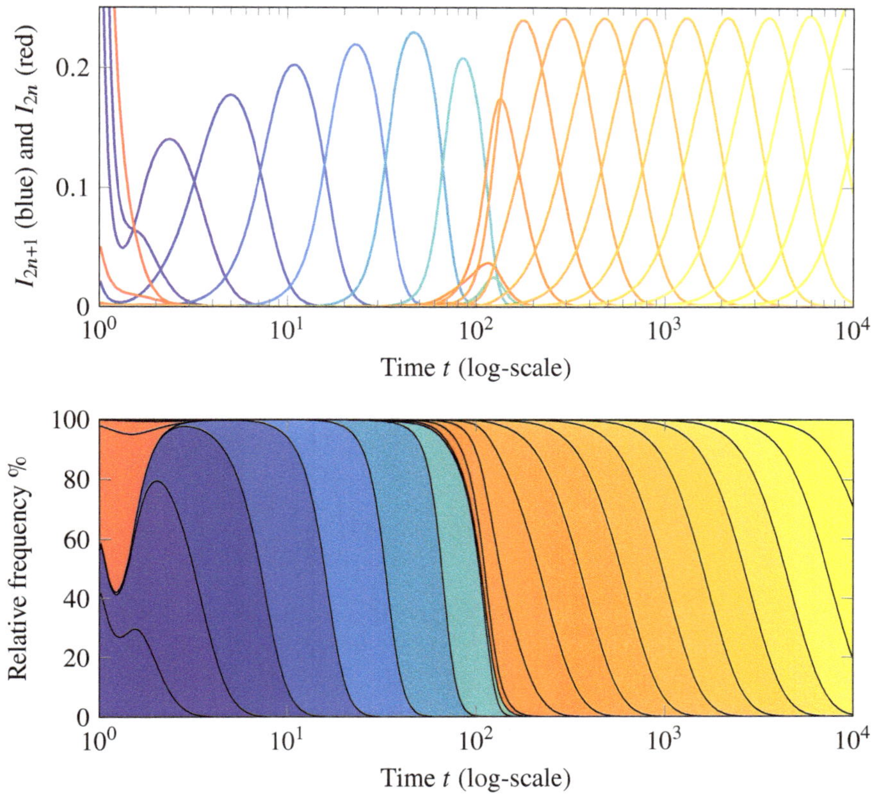

Fig. 9.4: Plots of the solution $P_n(t)$ in the Bivalent case 1. The family $P_n(t)$ is divided in two groups:$P_{2n+1}(t) = P_n^1(t)$ associated with μ_1 which are plotted in a family of cold colors (blue to green), and $P_{2n}(t) = P_n^2(t)$ associated with μ_2 which are plotted in a family of warm colors (red to yellow). Both figures (top and bottom) suggest that the μ_1 family becomes prevalent at first, but the μ_2 family eventually wins the competition (after $t \approx 10^2$). **Top figure:** Value of the density of variants $P_n(t)$. **Bottom figure:** Relative frequencies of the variants as a function of time. **Parameters:** $\beta = 10$, $\mu^N = 10$, $N_0 = 1$, $\alpha^* = 2$, $\mu_1 = 10$, $\mu_2 = 10$, $A_1 = 1$, $A_2 = 3$, $B_1 = B_2 = 10$, $C_1 = C_2 = \frac{1}{2}$. We used a total of 10 variants for the family 1 and 20 variants for the family 2. For interpretation of the colors in the figure(s), the reader is referred to the online version of this article.

Proof (Proof of Claim 9.3.2.1) We rewrite (9.38) as:

$$\mathcal{F}\left(t, \overline{N}(t)\right) = \frac{F_2\left(t\overline{N}(t)\right)^2}{4\mathcal{P}(t)} \left(1 + \sqrt{1 + 4\mathcal{P}(t) \frac{F_1\left(t\overline{N}(t)\right)}{F_2\left(t\overline{N}(t)\right)^2}}\right)^2.$$

We know that $\mathcal{P}(t) \asymp 1$, that $F_1\left(t\overline{N}(t)\right) = O\left(e^{-\frac{A_1}{2C_1}\ln(t\overline{N}(t))^2}\right)$ by Lemma 9.15 and that $F_2\left(t\overline{N}(t)\right) \asymp \left(t\overline{N}(t)\right)^{-\frac{A_2}{C_2}}$ by Claim 9.13. In particular, we obtain

$$4\mathcal{P}(t)\frac{F_1\left(t\overline{N}(t)\right)}{F_2\left(t\overline{N}(t)\right)^2} = O\left(e^{-\frac{A_1}{2C_1}\ln(t\overline{N}(t))^2+\frac{A_2}{C_2}\ln(t\overline{N}(t))}\right) \xrightarrow[t\to+\infty]{} 0,$$

so that

$$\mathcal{F}\left(t,\overline{N}(t)\right) = \frac{F_2\left(t\overline{N}(t)\right)^2}{4\mathcal{P}(t)}(1+o(1))$$

and by (9.33)

$$\mu_1(\alpha^*\overline{N}(t)-1)t = -2\ln\left(F_2(t\overline{N}(t))\right) + O(1).$$

Finally we obtain

$$\sum_{n=0}^{+\infty} P_n^1(t) = e^{\mu_1(\alpha^*\overline{N}(t)-1)t}F_1\left(t\overline{N}(t)\right) = O\left(\frac{F_1\left(t\overline{N}(t)\right)}{F_2\left(t\overline{N}(t)\right)^2}\right) \xrightarrow[t\to+\infty]{} 0.$$

Next we show the convergence of $\sum P_n^2(t)$ to a positive constant. We know that $\sum P_n^1(t) \to 0$ and, by Theorem 9.6, $N(t) \to \frac{1}{\alpha^*}$ and $N'(t) \to 0$. Using the first equation in (9.1a) we have:

$$\begin{aligned}
-\frac{N'(t)-\beta+\mu^N N(t)}{N(t)} &= \sum_{n=0}^{+\infty} m_n P_n(t) = \sum_{n=0}^{+\infty} m_n^1 P_n^1(t) + \sum_{n=0}^{+\infty} m_n^2 P_n^2(t)\\
&= \sum_{n=0}^{+\infty} m_n^2 P_n^2(t) + O\left(\sum_{n=0}^{+\infty} P_n^1(t)\right)\\
&= \sum_{n=0}^{+\infty} \mu_2\left(\frac{m_n^2}{\mu_2}-\alpha^*\right)P_n^2(t) + \sum_{n=0}^{+\infty}\mu_2\alpha^* P_n^2(t) + o(1)\\
&= \mu_2\alpha^*\sum_{n=0}^{+\infty} P_n^2(t) + o(1),
\end{aligned}$$

where we used the fact that $\sum\left(\frac{m_n}{\mu_2}-\alpha^*\right)P_n^2(t) \to 0$, which will justify below. Admitting this fact temporarily, let us finish the argument. We have now

$$\begin{aligned}
\sum_{n=0}^{+\infty} P_n^2(t) = -\frac{1}{\mu_2\alpha^*}\left(\frac{N'(t)-\beta+\mu^N N(t)}{N(t)}\right) + o(1) &= \frac{\mu^N}{\mu_2\alpha^*}\left(\frac{\beta}{\mu^N}\alpha^*-1\right)+o(1)\\
&= \frac{\mu^N}{\mu_2\alpha^*}(\mathcal{R}_0-1)+o(1),
\end{aligned}$$

which is exactly the second part of Claim 9.3.2.1.

There remains to show that $\sum\left(\frac{m_n}{\mu_2}-\alpha^*\right)P_n^2(t)\longrightarrow 0$ as $t\longrightarrow+\infty$. Fix $\varepsilon>0$ arbitrarily. Let $P^\infty:=\frac{2\beta}{\min(\mu^N,\mu_0)}$ so that by Lemma 9.23 we have $\sum P_n^2(t)\le P^\infty$ for t sufficiently large. Let $\mathcal{N}_\varepsilon$ be the set of indices defined by

$$\mathcal{N}_\varepsilon := \left\{n\in\mathbb{N}\,:\,\left|\frac{m_n^2}{\mu_2}-\alpha^*\right|\le\frac{\varepsilon}{2P^\infty}\right\}.$$

Then

$$\sum_{n\in\mathcal{N}_\varepsilon}\left(\frac{m_n}{\mu_2}-\alpha^*\right)P_n^2(t)\le\frac{\varepsilon}{2P^\infty}\sum_{n=0}^{+\infty}P_n^2(t)\le\frac{\varepsilon}{2},$$

for t sufficiently large. On the other hand, for $n\in\mathcal{N}_\varepsilon^c:=\mathbb{N}\backslash\mathcal{N}_\varepsilon$, we have $\alpha^*-\frac{m_n}{\mu_2}\ge\frac{\varepsilon}{2P^\infty}$ so that

$$\begin{aligned}P_n^2(t)=P_n^{0,2}e^{\mu_2\left(\frac{m_n}{\mu_2}-\alpha^*\right)\overline{N}(t)t+\mu_2\left(\alpha^*\overline{N}(t)-1\right)t}&\le P_n^{0,2}e^{-\frac{\mu_2\varepsilon}{2P^\infty}\overline{N}(t)t+\mu_2(\alpha^*\overline{N}(t)-1)t}\\&\le P_n^{0,2}\exp\left(-\frac{\mu_2\varepsilon}{8P^\infty\alpha^*}t\right),\end{aligned}$$

whenever $\overline{N}(t)\ge\frac{1}{2\alpha^*}$ and $|\alpha^*\overline{N}(t)-1|\le\frac{\varepsilon}{8P^\infty\alpha^*}$, which is true for t sufficiently large. Thus

$$\sum_{n\in\mathcal{N}_\varepsilon^c}P_n^2(t)\le e^{-\frac{\mu_2\varepsilon}{8P^\infty\alpha^*}t}\sum_{n\in\mathcal{N}_\varepsilon^c}P_n^{0,2}\xrightarrow[t\to+\infty]{}0,$$

and finally

$$\sum_{n=0}^{+\infty}P_n^2(t)=\sum_{n\in\mathcal{N}_\varepsilon}P_n^2(t)+\sum_{n\in\mathcal{N}_\varepsilon^c}P_n^2(t)\le\varepsilon,$$

for t sufficiently large. This finishes the proof of Claim 9.3.2.1. □

9.3.2.2 Bivalent example 2: alternating persistence

In this subsection we provide an example which shows a very particular asymptotic behavior: the types μ_1 and μ_2 are *both* asymptotically persistent as $t\longrightarrow+\infty$, and become alternatively prevalent in the population. This causes the total mass of infected $\sum P_n(t)$ to fluctuate between two distinct values. This shows, in particular, that an additional assumption (like, for instance, Assumption 9.8) is really necessary to obtain the asymptotic behavior of the mass and that the conclusions of Theorem 9.6 are, in some sense, sharp.

We work under the following assumption.

Assumption 9.20 We assume that $\mu_1=2\mu_2$, $m_n^1=\mu_1\alpha^*-Be^{-Cn}$, $P_n^{0,1}=P_0^1e^{-An^2}$, $m_n^2=\mu_2\alpha^*-\frac{B}{2}e^{-\frac{C}{3}n}$ and $P_n^{0,2}=P_0^2e^{-\frac{A}{18}n^2}$ for some positive constants $A>0$, $B>0$ and $C>0$.

Our claim is as follows.

Claim Under Assumption 9.20, there exist two sequences $t_k^1 \to +\infty$ and $t_k^2 \to +\infty$ as $k \to +\infty$, such that

$$\lim_{k\to+\infty} \sum_{n=0}^{+\infty} P_n^1(t_k^1) = \frac{\mu^N}{\mu_1 \alpha^*}(\mathcal{R}_0 - 1) > 0 \text{ and } \lim_{t\to+\infty} \sum_{n=0}^{+\infty} P_n^2(t_k^1) = 0,$$

and

$$\lim_{k\to+\infty} \sum_{n=0}^{+\infty} P_n^1(t_k^2) = 0 \text{ and } \lim_{k\to+\infty} \sum_{n=0}^{+\infty} P_n^2(t_k^2) = \frac{\mu^N}{\mu_2 \alpha^*}(\mathcal{R}_0 - 1) > 0,$$

wherein $P_n^1(t) = P_{2n+1}(t)$ and $P_n^2(t) = P_{2n}(t)$.

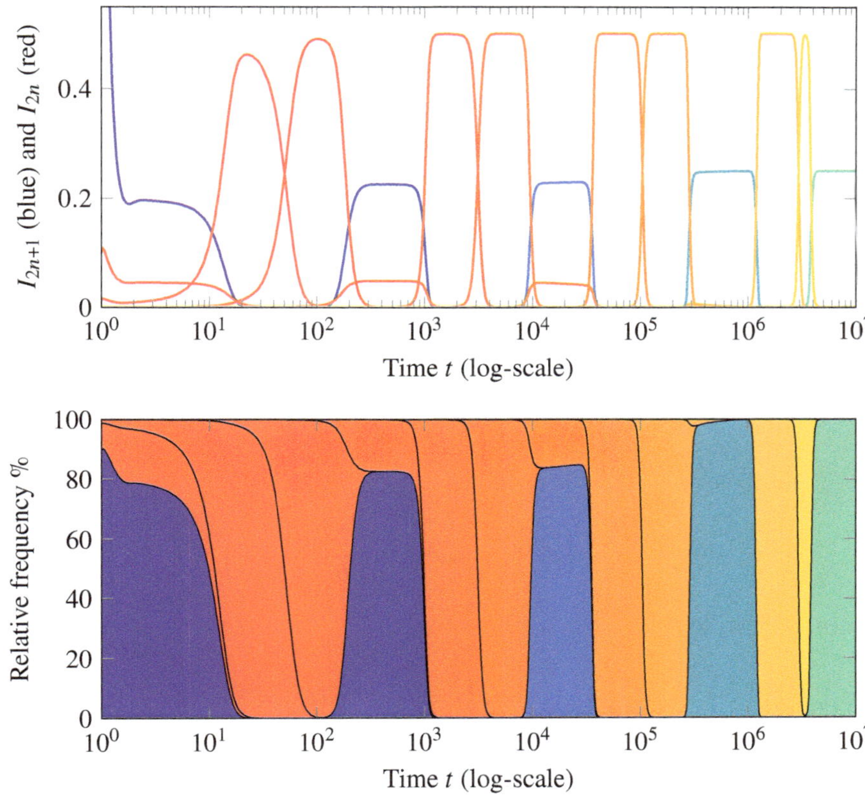

Fig. 9.5: Plots of the solution $P_n(t)$ under Assumption 9.20. The family $P_n(t)$ is divided in two groups:$P_{2n+1}(t) = P_n^1(t)$ associated with μ_1 which are plotted in a family of cold colors (blue to green), and $P_{2n}(t) = P_n^2(t)$ associated with μ_2 which are plotted in a family of warm colors (red to yellow). The top figure suggests that the total pathogen population fluctuates in time. The bottom figure suggests that the μ_1 and μ_2 families become alternatively prevalent. **Top figure:** Value of the density of variants $P_n(t)$. **Bottom figure:** Relative frequencies of the variants as a function of time. The family associated with μ_1 (blue to green) and the family associated with μ_2 (red to yellow) become alternatively dominant. **Parameters:** $\beta = 5$, $\mu^N = 5$, $N_0 = 1$, $\alpha^* = 2$, $\mu_1 = 10$, $\mu_2 = 5$, $P_0^1 = 1$, $P_0^2 = 0.1$, $A = 36$, $B = 2$, $C = 3$. We used a total of 6 variants for the family 1 and 15 variants for the family 2. For interpretation of the colors in the figure(s), the reader is referred to the online version of this article.

Proof (Proof of Claim 9.3.2.2) Recalling (9.36) and Lemma 9.15, we have in this case:

$$F_1(e^{-Cr_1(\xi)}\xi) \sim e^{-An_1(\xi)^2 - \frac{2A}{C}x_1(\xi)},$$
$$F_2(e^{-\frac{C}{3}r_2(\xi)}\xi) \sim e^{-\frac{A}{18}n_2(\xi)^2 - \frac{A}{3C}x_2(\xi)},$$

where

$$x_1(\xi) := n_1(\xi)+r_1(\xi) = \frac{1}{C}W_0\left(\frac{BC^2}{2A}\xi\right) \text{ and } x_2(\xi) := n_2(\xi)+r_2(\xi) = \frac{3}{C}W_0\left(\frac{BC^2}{2A}\xi\right),$$

$n_1(\xi), n_2(\xi) \in \mathbb{N}$ and $r_1(\xi), r_2(\xi) \in \left[-\frac{1}{2}, \frac{1}{2}\right)$. In other words, with $x(\xi) := \frac{1}{C}W_0\left(\frac{BC^2}{2A}\xi\right)$, we have

$$n_1(\xi) = \left\lfloor x(\xi) + \frac{1}{2}\right\rfloor, \qquad n_2(\xi) = \left\lfloor 3x(\xi) + \frac{1}{2}\right\rfloor,$$
$$r_1(\xi) = x(\xi) - n_1(\xi) \in \left(-\frac{1}{2}, \frac{1}{2}\right), \qquad r_2(\xi) = 3x(\xi) - n_2(\xi) \in \left(-\frac{1}{2}, \frac{1}{2}\right).$$

Let ξ_n^1 and ξ_n^2 be defined by the relation

$$\xi_n^1 := \frac{2A}{BC}\left(n - \frac{1}{3}\right)e^{C(n-\frac{1}{3})} \text{ and } \xi_n^2 := \frac{2A}{BC}\left(n + \frac{1}{3}\right)e^{C(n+\frac{1}{3})},$$

so that we have, by definition,

$$x(\xi_n^1) = n - \frac{1}{3}, \quad n_1(\xi_n^1) = n, \quad r_1(\xi_n^1) = -\frac{1}{3}, \quad n_2(\xi_n^1) = 3n - 1, \quad r_2(\xi_n^1) = 0,$$
$$x(\xi_n^2) = n + \frac{1}{3}, \quad n_1(\xi_n^2) = n, \quad r_1(\xi_n^2) = +\frac{1}{3}, \quad n_2(\xi_n^2) = 3n + 1, \quad r_2(\xi_n^2) = 0.$$

Since $W_0(X) = \ln(X) - \ln\ln X + o(1)$ (see [64, (4.19) p.349]), we have for n sufficiently large:

$$\begin{aligned}
x\left(e^{Cr_1(\xi_n^1)}\xi_n^1\right) &= \frac{1}{C}\ln\left(\frac{BC^2}{2A}e^{Cr_1(\xi_n^1)}\xi_n^1\right) - \frac{1}{C}\ln\ln\left(\frac{BC^2}{2A}e^{Cr_1(\xi_n^1)}\xi_n^1\right) + o(1) \\
&= r_1(\xi_n^1) + \frac{1}{C}\ln\left(\frac{BC^2}{2A}\xi\right) - \frac{1}{C}\ln\ln\left(\frac{BC^2}{2A}\xi\right) - \frac{1}{C}\ln\left(1 + \frac{Cr_1(\xi_n^1)}{\ln\left(\frac{BC^2}{2A}\xi\right)}\right) + o(1) \\
&= x(\xi_n^1) + r_1(\xi_n^1) + o(1) = n - \frac{1}{3} - \frac{1}{3} + o(1) = n - \frac{2}{3} + o(1),
\end{aligned}$$

thus for n sufficiently large,

$$n_1\left(e^{Cr_1(\xi_n^1)}\xi_n^1\right) = \left\lfloor x\left(\xi e^{Cr_1(\xi_n^1)}\right) + \frac{1}{2}\right\rfloor = n - 1.$$

Similarly, for n sufficiently large, we have

$$x\left(\xi_n^2 e^{Cr_1(\xi_n^2)}\right) = n + \frac{2}{3} + o(1) \text{ and } n_1\left(e^{Cr_1(\xi_n^2)}\xi_n^2\right) = \left\lfloor x\left(\xi e^{Cr_1(\xi_n^2)}\right) + \frac{1}{2}\right\rfloor = n + 1.$$

Thus, for n sufficiently large, we have

$$
\begin{aligned}
\frac{F_1(\xi_n^1)}{F_2(\xi_n^1)^2} &\sim \exp\Bigg[-An_1\left(e^{Cr_1(\xi_n^1)}\xi_n^1\right)^2 + A\left(\frac{n_2(\xi_n^1)}{3}\right)^2 \\
&\quad -\frac{2A}{C}x_1\left(\xi_n^1 e^{Cr_1(\xi_n^1)}\right) + \frac{2A}{3C}x_2(\xi_n^1)\Bigg] \\
&= \exp\Bigg[A\left(\frac{n_2(\xi_n^1)}{3} - n_1\left(e^{Cr_1(\xi_n^1)}\xi_n^1\right)\right)\left(\frac{n_2(\xi_n^1)}{3} + n_1\left(e^{Cr_1(\xi_n^1)}\xi_n^1\right)\right) \\
&\quad +\frac{2A}{C}\left(\frac{x_2(\xi_n^1)}{3} - x_1\left(e^{Cr_1(\xi_n^1)}\xi_n^1\right)\right)\Bigg] \\
&= \exp\Bigg[A\left(n - \frac{1}{3} - (n-1)\right)\left(n - \frac{1}{3} + n - 1\right) \\
&\quad +\frac{2A}{C}\left(n - \frac{1}{3} + \frac{r_2(\xi_n^1)}{3} - \left(n - 1 + r_1\left(\xi_n^1 e^{Cr_1(\xi)}\right)\right)\right)\Bigg] \\
&= \exp\left[\frac{4A}{3}\left(n - \frac{2}{3}\right) + O(1)\right] \xrightarrow[n\to+\infty]{} +\infty,
\end{aligned}
$$

and similarly,

$$
\begin{aligned}
\frac{F_1(\xi_n^2)}{F_2(\xi_n^2)^2} &\sim \exp\Bigg[-An_1\left(e^{Cr_1(\xi_n^2)}\xi_n^2\right)^2 + A\left(\frac{n_2(\xi_n^2)}{3}\right)^2 \\
&\quad -\frac{2A}{C}x_1\left(\xi_n^2 e^{Cr_1(\xi_n^2)}\right) + \frac{2A}{3C}x_2(\xi_n^2)\Bigg] \\
&= \exp\Bigg[A\left(\frac{n_2(\xi_n^2)}{3} - n_1\left(e^{Cr_1(\xi_n^2)}\xi_n^2\right)\right)\left(\frac{n_2(\xi_n^2)}{3} + n_1\left(e^{Cr_1(\xi_n^2)}\xi_n^2\right)\right) \\
&\quad +\frac{2A}{C}\left(\frac{x_2(\xi_n^2)}{3} - x_1\left(e^{Cr_1(\xi_n^2)}\xi_n^2\right)\right)\Bigg] \\
&= \exp\Bigg[A\left(n + \frac{1}{3} - (n+1)\right)\left(n + \frac{1}{3} + n + 1\right) \\
&\quad +\frac{2A}{C}\left(n + \frac{1}{3} + \frac{r_2(\xi_n^2)}{3} - \left(n + 1 + r_1\left(\xi_n^2 e^{Cr_1(\xi)}\right)\right)\right)\Bigg] \\
&= \exp\left[-\frac{4A}{3}\left(n + \frac{2}{3}\right) + O(1)\right] \xrightarrow[n\to+\infty]{} 0.
\end{aligned}
$$

Now we conclude the proof. Let t_n^1 be a sequence of times such that $t_n^1\overline{N}(t_n^1) = \xi_n^1$. Clearly $t_n^1 \to +\infty$ as $n \to +\infty$. Then, recalling (9.38), we have:

$$\mathcal{F}(t_n^1,\overline{N}(t_n^1)) = \frac{F_1(t_n^1\overline{N}(t_n^1))}{\mathcal{P}(t_n^1)}\left(\frac{F_2(t_n^1\overline{N}(t_n^1))}{\sqrt{F_1(t_n^1\overline{N}(t_n^1))}} + \sqrt{\frac{\mathcal{P}(t_n^1)F_2(t_n^1\overline{N}(t_n^1))^2}{4F_1(t_n^1\overline{N}(t_n^1))}+1}\right)^2$$
$$= \frac{F_1(t_n^1\overline{N}(t_n^1))}{\mathcal{P}(t_n^1)}(1+o(1)),$$

and by (9.33),

$$\mu_1 t_n^1\left(\alpha^*\overline{N}(t_n^1)-1\right) = -\ln F_1((t_n^1\overline{N}(t_n^1)) + O(1).$$

Finally,

$$\sum_{n=0}^{+\infty} P_n^2(t_n^1) = P_0^2 e^{\frac{\mu_2}{\mu_1}\mu_1(\alpha^*\overline{N}(t_n^1)-1)} F_2(t_n^1\overline{N}(t_n^1))$$
$$= P_0^2 e^{\frac{1}{2}\left(-\ln F_1(t_n^1\overline{N}(t_n^1))+O(1)\right)} F_2(t_n^1\overline{N}(t_n^1))$$
$$= O\left(\frac{F_2(\xi_n^1)}{\sqrt{F_1(\xi_n^1)}}\right) \xrightarrow[n\to+\infty]{} 0,$$

and by using (9.1a) together with Theorem 9.6 we obtain

$$\lim_{n\to+\infty}\sum_{n=0}^{+\infty} P_n^1(t_n^1) = \frac{\mu^N}{\mu_1\alpha^*}(\mathcal{R}_0-1),$$

as we did in the proof of Claim 9.3.2.1. The first part of the Claim 9.3.2.2 is proved. Proceeding similarly, let t_n^2 be a sequence of times such that $t_n^2\overline{N}(t_n^2) = \xi_n^1$. Clearly $t_n^2 \to +\infty$ as $n \to +\infty$. We have

$$\mathcal{F}(t_n^2,\overline{N}(t_n^2)) = \frac{F_2(t_n^2\overline{N}(t_n^2))}{4}\left(1+\sqrt{1+\frac{4F_1(t_n^2\overline{N}(t_n^2))}{\mathcal{P}(t_n^2)F_2(t_n^2\overline{N}(t_n^2))^2}}\right)^2$$
$$= \frac{F_2(t_n^2\overline{N}(t_n^2))^2}{4}(1+o(1)),$$

so

$$\mu_1 t_n^2\left(\alpha^*\overline{N}(t_n^2)-1\right) = -2\ln F_2(t_n^2\overline{N}(t_n^2)) + O(1),$$

and

$$\sum_{n=0}^{+\infty} P_n^1(t_n^2) = P_0^1 e^{\mu_1(\alpha^*\overline{N}(t_n^2)-1)} F_1(t_n^2\overline{N}(t_n^2))$$
$$= P_0^1 e^{-2\ln F_1(t_n^2\overline{N}(t_n^2))+O(1)} F_1(t_n^2\overline{N}(t_n^2)) = O\left(\frac{F_1(\xi_n^2)}{F_2(\xi_n^2)^2}\right) \xrightarrow[n\to+\infty]{} 0,$$

so finally

$$\lim_{n\to+\infty}\sum_{n=0}^{+\infty} P_n^2(t_n^2) = \frac{\mu^N}{\mu_2\alpha^*}(\mathcal{R}_0 - 1)\,.$$

This is the second part of Claim 9.3.2.2, and the proof is finished. □

9.4 Discussion

It is a classical result in evolutionary epidemiology [5, 102] known as the 'trade-off hypothesis' that evolution favors variants that maximize the basic reproductive number $\mathcal{R}_0$; here we show once again the robustness of this prediction by considering an infinite number of variants competing for the hosts. Yet, we also go beyond the standard prediction and show that a complexity persists in the asymptotic behavior of the epidemic even in our simplistic model, with many possible outcomes ranging from the simple convergence to a global equilibrium (the case of finite system as in [166], or case i) of Theorem 9.6) to an eternal transient state (the example given in Section 9.3.2.2).

Our examples in section 9.3 shed light on the variety of dynamics that can be observed for the diversity of pathogen variants in the host population. It is likely to be dependent not only on the distribution of the parameters μ_n and $\alpha_n = \frac{m_n}{\mu_n}$, but also on the initial number of infected corresponding to these parameters; depending on the choices we make, we can observe an enhancement, preservation or erosion of the diversity of variants in the population. Example 1 shows the case of an algebraically converging fitness function with an algebraically decreasing initial data. Figure 9.1 (bottom) shows the numerical computation of the relative frequencies of the variants; the number of variants with non-negligible proportion seems to be increasing with time, hence in this case the diversity seems to be increasing with time (enhancement of diversity). In Figure 9.2 (bottom) we show the numerical computations for the case of a fitness function converging exponentially fast with an exponentially decreasing initial data; here the diversity seems to be approximately constant in time, with a constant number of variants that dominate the others (preservation of diversity). Figure 9.3 (bottom) shows a similar fitness function but with Gaussian initial data; here the transition between variants becomes steeper with time, and our analysis in Claim 9.3.1.3 suggests that only one variant dominates all the other for increasingly large periods asymptotically (although this may happen in a larger time frame than the one presented in Figure 9.3). Thus in this case we observe an erosion of the diversity of variants. Those three different types behaviors can be proved analytically, see Claim 9.3.1.1, Claim 9.3.1.2 and Claim 9.3.1.3.

While Figure 9.1, Figure 9.2 and Figure 9.3 focus on the case when μ_n is a constant, in Figure 9.4 and Figure 9.5 we investigate the case when μ_n oscillates between two values and we show that surprising behaviors may occur then. Figure 9.4 illustrates that, when Assumption 9.8 does not hold, we cannot hope to generalize the results of Proposition 9.9: indeed the family of variants with the highest value of μ_n (in blue to green) gets extinct and the family of variants with the lowest value of μ_n (in red to yellow) dominates in this case. This is proved in Claim 9.3.2.1. Figure 9.5 illustrates that in some particular cases, neither family uniformly dominates the population asymptotically, but they both dominate the population alternatively. Claim 9.3.2.2 shows moreover that, in this situation, the total population of infected does not converge to a limit but oscillates between two distinct values.

The assumption that an infinite number of variants exist at the same time corresponds to a creative simplification of reality that allows us to describe possible behaviors occurring in nature. The classical theory for finite systems gives us access to a single type of dynamics: convergence to a unique equilibrium. Yet practical observations in real-life epidemics such as the COVID-19 epidemic suggest that it is not the case [38]. In the end, a description of the succession of infinitely many variants might give a more realistic description of the observed phenomenon than a finite model.

We predict that, when there is equality between different types, the variants that are associated with a fast dynamics are favored in the long run. This may seem counterintuitive since the parameter that is maximized, μ, incidentally corresponds to the inverse of the infection period; however, we conjecture that the behavior that it selected is the one associated with a fast transmission rather than low infection period. This could be checked by splitting the different coefficients associated with predator mortality, prey mortality, and consumption rate in (9.1).

9.5 Proof of the mathematical results

9.5.1 Proof of Proposition 9.4

We prove that problem (9.1) is well posed. Thanks to Assumption 9.1 we observe that the map $F : \mathbb{R}\times\ell^1 \to \mathbb{R}\times\ell^1$ defined by the right hand side of the system of equations (9.1a) is locally Lipschitz on the Banach space $\mathbb{R}\times\ell^1$. Therefore problem (9.1) admits a unique maximal solution $\big(N(t), (P_n(t))_{n\in\mathbb{N}}\big) \in C^1\big([0, T_{\max}), \mathbb{R}\big)\times C^1\big([0, T_{\max}), \ell^1\big)$ for some $T_{\max} > 0$ possibly infinite. Moreover, formula (9.4) holds, i.e.

$$P_n(t) = e^{t\left(m_n\overline{N}(t)-\mu_n\right)}P_n^0,$$

in particular P_n is a non-negative function for all $n \in \mathbb{N}$ and $\big(P_n^0\big)_{n\in\mathbb{N}} \in \ell^1_+ \longmapsto \big(P_n(t)\big)_{n\in\mathbb{N}} \in \ell^1_+$ is continuous for all $t > 0$.

Next, the following a priori estimate proves that the solution $\big(N(t), (P_n(t))_{n\in\mathbb{N}}\big)$ is uniformly bounded in $\mathbb{R}\times\ell^1_+$ so that $T_{\max} = \infty$:

Lemma 9.21 *Suppose that Assumption 9.1 holds true. For all $t \in [0, T_{\max})$,*

$$0 \le N(t) \le N_0 e^{-\mu^N t} + \frac{\beta}{\mu^N}\left(1 - e^{-\mu^N t}\right) \le \max\left(N_0, \frac{\beta}{\mu^N}\right), \tag{9.39}$$

$$N(t) + \sum_{n\in\mathbb{N}} P_n(t) \le \frac{\beta}{\min(\mu^N, \mu_0)} + \left(N_0 + \sum_{n\in\mathbb{N}} P_n^0 - \frac{\beta}{\min(\mu^N, \mu_0)}\right)e^{-\min(\mu^N,\mu_0)t}. \tag{9.40}$$

Proof Using the first equation of (9.1a), we have $N(t) \geq 0$ and due to the positivity of $P_n(t)$ we readily prove the Lemma inequalities for $N(t)$.

Next, adding equations of (9.1a) we remark that

$$\frac{\mathrm{d}}{\mathrm{d}t}\left(N(t) + \sum_{n\in\mathbb{N}} P_n(t)\right) \leq \beta - \mu^N N(t) - \mu_0 \sum_{n\in\mathbb{N}} P_n(t),$$

which yields the inequality for $N(t) + \sum_{n\in\mathbb{N}} P_n(t)$. The Lemma is proved. □

9.5.2 Proof of Proposition 9.5

We first assume that $\mathcal{R}_0 < 1$. Let $\varepsilon > 0$ such that

$$m = 1 - \mathcal{R}_0\left(1 + \varepsilon\frac{\mu^N}{\beta}\right) > 0,$$

there exists $t_0(\varepsilon) > 0$ such that

$$N(t) \leq \frac{\beta}{\mu^N} + \varepsilon.$$

Recall that the equations of (9.1a) for P_n write for all $n \in \mathbb{N}$

$$\frac{\mathrm{d}}{\mathrm{d}t}P_n(t) = m_n P_n N(t) - \mu_n P_n(t),$$

or, by definition of $\mathcal{R}_0$, as

$$\frac{\mathrm{d}}{\mathrm{d}t}P_n(t) \leq \mu_n\left(\mathcal{R}_0\frac{\mu^N}{\beta}N(t) - 1\right)P_n(t).$$

For all $n \in \mathbb{N}$ and $t \geq t_0(\varepsilon)$

$$\frac{\mathrm{d}}{\mathrm{d}t}P_n(t) \leq -m\mu_n P_n(t),$$

therefore

$$0 \leq \sum_{n\in\mathbb{N}} P_n(t) \leq e^{-m\mu_0(t - t_0(\varepsilon))} P_n(t_0(\varepsilon)),$$

which proves that $\lim_{t\to+\infty} \sum_{n=0}^{+\infty} P_n(t) = 0$.

Next, as $N(t)$ is bounded, we consider $\underline{N} := \liminf_{t\to+\infty} N(t)$. Let $(t_k)_{n\geq 0}$ be a sequence that tends to ∞ as $k \to \infty$ and such that $\lim_{k\to\infty} N'(t_k) = 0$ and $\lim_{k\to+\infty} N(t_k) = \underline{N}$. Since (m_n) is bounded, from the following equality for all k

$$\frac{\mathrm{d}}{\mathrm{d}t}N(t_k) = \beta - \mu^N N(t_k) - \sum_{n=0}^{+\infty} m_n N(t_k) P_n(t_k)$$

passing to the limit, we obtain $\liminf_{t\to\infty} N(t) = \frac{\beta}{\mu^N}$. Similarly, $\limsup_{t\to\infty} N(t) = \frac{\beta}{\mu^N}$, therefore

$$\lim_{t\to+\infty} N(t) = \frac{\beta}{\mu^N},$$

which concludes the first part of the proof of the proposition. The second part of Proposition 9.5 will be proved in section 9.5.5, because we need more tools to prove it.

9.5.3 Proof of Theorem 9.6

Before we start the proof of Theorem 9.6, we introduce a few notions that will be useful along this Section. Indeed, in order to get a compactness of the orbits, we need to include our dynamical system in a larger space. We define the distance on $\mathbb{N}$:

$$d(n,m) := \left|\frac{1}{1+n} - \frac{1}{1+m}\right| + |\alpha_n - \alpha_m| + |\mu_n - \mu_m|, \text{ for all } n, m \in \mathbb{N}.$$

We let $\overline{\mathbb{N}}$ be the topological completion of $\mathbb{N}$ for the distance d. It is essentially the smallest closed set for the distance d containing $\mathbb{N}$. Because of Assumption 9.3 the sets $\omega(\alpha)$ and $\omega(\mu)$ are finite and we have, up to a topological isomorphism which we will omit in the rest of the proof,

$$\left(\overline{\mathbb{N}}, d\right) = \left([-K, +\infty) \cap \mathbb{Z}, d\right),$$

wherein we have set $K := \#\omega(\alpha) \times \#\omega(\mu)$,

$$\omega(\alpha) \times \omega(\mu) =: \{(\alpha_{-i}, \mu_{-i}), i = 1, \ldots, K\},$$

and

$$d(n,m) := \left|\frac{\mathbb{1}_{n\geq 0}}{1+n} - \frac{\mathbb{1}_{m\geq 0}}{1+m}\right| + |\alpha_n - \alpha_m| + |\mu_n - \mu_m|, \text{ for all } n, m \in \mathbb{Z} \cap [-K, +\infty).$$

In particular $\overline{\mathbb{N}}$ is Hausdorff and countable, which implies that the Borel σ-algebra is the set of all parts of $\overline{\mathbb{N}}$, and therefore any Borel measure $\nu \in \mathcal{M}(\overline{\mathbb{N}})$ can be represented by a summable sequence:

$$\nu = \sum_{n=-K}^{+\infty} \nu_n \delta_n,$$

where δ_n is the Dirac mass concentrated on $n \in \overline{\mathbb{N}}$ and $(\nu_n)_{n\geq -K}$ is a summable sequence of real numbers. Finally, $\overline{\mathbb{N}}$ is compact for the topology generated by d.

In what follows we will obtain the compactness of the orbit by using the weak-$*$ topology on the space of measures $\mathcal{M}(\overline{\mathbb{N}})$. To mark the difference, we will write

$\mathcal{M}^*(\overline{\mathbb{N}})$, $\mathcal{M}_+^*(\overline{\mathbb{N}})$ instead of $\mathcal{M}(\overline{\mathbb{N}})$, $\mathcal{M}_+(\overline{\mathbb{N}})$ when the space is equipped with the weak-$*$ topology. Recall the topology on $\mathcal{M}(\overline{\mathbb{N}})$ is generated by the norm

$$\|(\nu)_{n\in\overline{\mathbb{N}}}\|_{\mathcal{M}(\overline{\mathbb{N}})} := \sum_{n\in\overline{\mathbb{N}}} |\nu_n|,$$

and that the topology on $\mathcal{M}^*(\overline{\mathbb{N}})$ is that of the weak-$*$ convergence: convergence of a sequence $\nu^n \rightharpoonup \nu$ for this topology holds if, and only if,

$$\sum_{k=-K}^{+\infty} \varphi_k \nu_k^n \xrightarrow[n\to+\infty]{} \sum_{k=-K}^{+\infty} \varphi_k \nu_k,$$

for all $(\varphi_n) \in C(\overline{\mathbb{N}})$ the space of continuous sequences over $\overline{\mathbb{N}}$, which is characterized by

$$\left[(\varphi_k) \in C(\overline{\mathbb{N}})\right] \Longleftrightarrow$$
$$\left[\lim_{j\to+\infty} \varphi_{k_j} = \varphi_{-i} \text{ whenever } \alpha_{k_j} \to \alpha_{-i} \text{ and } \nu_{k_j} \to \nu_{-i} \text{ with } k_j \to +\infty \text{ and } i > 0\right].$$

In what follows we will consider equation (9.1) with an initial data $(P_n)_{n\in\overline{\mathbb{N}}} \in \mathcal{M}_+(\overline{\mathbb{N}})$. Because of our construction of $\overline{\mathbb{N}}$, Assumptions 9.1 and 9.3 need not be adapted to the new framework. Assumption 9.2, however, does. Let us the replace Assumption 9.2 with the following:

Assumption 9.22 We let $N_0 > 0$, $(P_n^0)_{n\in\overline{\mathbb{N}}} \in \mathcal{M}_+(\overline{\mathbb{N}})$ be given and assume that there exists a sequence of indices $n_k \in \overline{\mathbb{N}}$ with

$$P_{n_k}^0 > 0 \text{ and } \lim_{k\to+\infty} \frac{m_{n_k}}{\mu_{n_k}} = \alpha^*.$$

As before the sequence of indices (n_k) need not be strictly monotone and can be eventually stationary.

The following lemma holds true.

Lemma 9.23 *Suppose that Assumption 9.1 holds true. Then we have*

$$0 < \frac{\min(\mu^N, \mu_0)}{\mu^N \beta \min(\mu^N, \mu_0) + m^\infty} \le \liminf_{t\to+\infty} N(t) \le \limsup_{t\to+\infty} N(t) \le \frac{\beta}{\mu^N} < +\infty,$$
$$\limsup_{t\to+\infty} \sum_{n\in\overline{\mathbb{N}}} P_n(t) \le \frac{\beta}{\min(\mu^N, \mu_0)} < +\infty.$$

Proof The upper bounds are proved as in Lemma 9.21. Next we return to the S-component of equation (9.1) and let $\varepsilon > 0$ be given. We have, for t_0 sufficiently large and $t \ge t_0$,

$$N'(t) = \beta - \left(\mu^N + \sum_{n\in\overline{\mathbb{N}}} m_n P_n(t)\right) N(t) \geq \beta - \left(\mu^N + m^\infty \frac{\beta}{\min(\mu^N, \mu_0)} + \varepsilon\right) N(t),$$

therefore

$$\begin{aligned} N(t) &\geq e^{-\left(\mu^N + \frac{\beta m^\infty}{\min(\mu^N,\mu_0)} + \varepsilon\right)(t-t_0)} N(t_0) \\ &\quad + \frac{\beta \min(\mu^N, \mu_0)}{(\mu^N + \varepsilon)\min(\mu^N, \mu_0) + \beta m^\infty}\left(1 - e^{-\left(\mu^N + \frac{\beta m^\infty}{\min(\mu^N,\mu_0)} + \varepsilon\right)(t-t_0)}\right), \end{aligned}$$

so that finally by letting $t \longrightarrow +\infty$ we get

$$\liminf_{t\to+\infty} N(t) \geq \frac{\min(\mu^N, \mu_0)\beta}{(\mu^N + \varepsilon)\min(\mu^N, \mu_0) + \beta m^\infty}.$$

Since $\varepsilon > 0$ is arbitrary we have shown

$$\liminf_{t\to+\infty} N(t) \geq \frac{\min(\mu^N, \mu_0)\beta}{\mu^N \min(\mu^N, \mu_0) + \beta m^\infty}.$$

The Lemma is proved. □

Lemma 9.24 *Suppose that the Assumptions 9.1 and 9.22 hold true. Let* $(N(t), P_i(t))$ *be the corresponding solution of* (9.1). *Then*

$$\limsup_{T\to+\infty} \frac{1}{T}\int_0^T N(t)\mathrm{d}t \leq \frac{1}{\alpha^*}.$$

Proof Let us remark that the second component of (9.1) for any $n \in \mathbb{N}$ can be written as

$$\begin{aligned} P_n(t) &= P_0^n e^{m_n \int_0^t N(s)\mathrm{d}s - \mu_n t}, \\ &= P_0^n \exp\left(\mu_n \int_0^t N(s)\mathrm{d}s \left[\frac{m_n}{\mu_n} - \frac{t}{\int_0^t N(s)\mathrm{d}s}\right]\right). \end{aligned} \tag{9.41}$$

Assume by contradiction that the conclusion of the Lemma does not hold, *i.e.* there exists $\varepsilon > 0$ and a sequence $T_n \longrightarrow +\infty$ such that

$$\frac{1}{T_n}\int_0^{T_n} N(t)\mathrm{d}t \geq \frac{1}{\alpha^*} + \varepsilon \text{ for all } n \in \mathbb{N}.$$

Then

$$\frac{T_n}{\int_0^{T_n} N(t)\mathrm{d}t} \leq \frac{1}{\frac{1}{\alpha^*} + \varepsilon} \leq \alpha^* - \varepsilon',$$

for some $\varepsilon' > 0$. Since $\alpha^* = \sup_n \frac{m_n}{\mu_n}$ there exists some $k \in \mathbb{N}$ with $P_k^0 > 0$ such that $\frac{m_k}{\mu_k} - \alpha^* + \varepsilon' > 0$, and

$$\sum_{i\in\mathbb{N}} P_i(T_n) \geq P_0^k \exp\left(\mu_k \int_0^{T_n} N(s)\mathrm{d}s \left[\frac{m_k}{\mu_k} - \frac{T_n}{\int_0^{T_n} N(s)\mathrm{d}s}\right]\right)$$
$$\geq P_0^k \exp\left(\mu_k \left(\alpha_k - \alpha^* + \varepsilon'\right) \int_0^{T_n} N(s)\mathrm{d}s\right).$$

Since $\int_0^{T_n} N(t)\mathrm{d}t \to +\infty$ when $n \to +\infty$, we have therefore

$$\limsup_{t\to+\infty} \sum_{i\in\overline{\mathbb{N}}} P_i(t) \geq \limsup_{n\to+\infty} \sum_{i\in\overline{\mathbb{N}}} P_i(T_n) = +\infty,$$

which is a contradiction since $P_n(t)$ is bounded in $\mathcal{M}_+(\overline{\mathbb{N}})$ by Lemma 9.23. This completes the proof of the Lemma. □

Remark that, in the Lemma above, Assumption 9.22 is essential. Indeed, were this assumption not true, we could not guarantee that the index k defined in the proof corresponds to a strictly positive P_k^0, hence the contradiction would not be guaranteed either.

The following weak persistence property holds.

Lemma 9.25 *Suppose that Assumptions 9.1 and 9.22 hold true. Let $\left(N(t), P_n(t)\right)$ be the corresponding solution of* (9.1). *Then*

$$\limsup_{t\to+\infty} \sum_{n\in\overline{\mathbb{N}}} P_n(t) \geq \frac{\mu^N}{m^\infty}\left(\mathcal{R}_0 - 1\right) > 0. \tag{9.42}$$

Proof Let us recall that we made the hypothesis that

$$\mathcal{R}_0 = \frac{\beta}{\mu^N}\alpha^* > 1.$$

Assume by contradiction that for t_0 sufficiently large we have

$$\sum_{i\in\overline{\mathbb{N}}} P_i(t) \leq \eta < \frac{\mu^N}{m^\infty}\left(\mathcal{R}_0 - 1\right) \text{ for all } t \geq t_0,$$

with $\eta > 0$.

As a consequence of Lemma 9.24 we have

$$\liminf_{t\to+\infty} N(t) \leq \limsup_{T\to+\infty} \frac{1}{T}\int_0^T N(t)\mathrm{d}t \leq \frac{1}{\alpha^*}. \tag{9.43}$$

Let $\underline{N} := \liminf_{t\to+\infty} N(t)$. Let $(t_n)_{n\geq 0}$ be a sequence that tends to ∞ as $n \to \infty$ and such that $\lim_{n\to+\infty} N'(t_n) = 0$ and $\lim_{n\to+\infty} N(t_n) = \underline{N}$. As $\sum_{i\in\overline{\mathbb{N}}} P_i(t_n) \leq \eta$ for n large enough we deduce from the equality

$$N'(t_n) = \beta - \mu^N N(t_n) - N(t_n) \sum_{i\in\overline{\mathbb{N}}} m_i P_i(t_n),$$

that

$$0 \geq \beta - \mu^N \underline{N} - \underline{N} m^\infty \eta$$

so that

$$\underline{N} \geq \frac{\beta}{\mu^N + m^\infty \eta} > \frac{\beta}{\mu^N \mathcal{R}_0} = \frac{1}{\alpha^*},$$

which contradicts (9.43). □

Proposition 9.26 (Compactness of the orbit and concentration) *Suppose that Assumptions 9.1, 9.22 and 9.3 hold true. Then, the map* $t \mapsto \left(N(t), (P_n(t))_{n\in\overline{\mathbb{N}}}\right) \in \mathbb{R} \times \mathcal{M}_+^*(\overline{\mathbb{N}})$ *is continuous and the corresponding orbit,*

$$O\left(N_0, (P_n^0)_{n\in\overline{\mathbb{N}}}\right) := \left\{\left(N(t), (P_n(t))_{n\in\overline{\mathbb{N}}}\right) \,:\, t \geq 0\right\} \subset \mathbb{R} \times \mathcal{M}_+^*(\overline{\mathbb{N}}),$$

is precompact for the weak-$*$ topology. Moreover if $\mathcal{R}_0 > 1$ and if $t_k \to +\infty$ is an arbitrary sequence along which

$$\liminf_{k\to+\infty} \sum_{n=-K}^{+\infty} P_n(t_k) > 0,$$

then one can extract from (t_k) a subsequence (t_{k_i}) such that the shifted orbits

$$t \mapsto \left(N(t+t_{k_i}), (P_n(t+t_{k_i}))_{n\in\overline{\mathbb{N}}}\right)$$

converge pointwise in $\mathbb{R} \times \mathcal{M}_+^*(\overline{\mathbb{N}})$ to a complete orbit $\left(N^\infty(t), (P_n^\infty(t))_{n\in\overline{\mathbb{N}}}\right)$ that satisfies the following properties:

$$\sum_{n=-K}^{+\infty} P_n^\infty(t) > 0 \text{ and } N^\infty(t) > 0 \text{ for all } t \in \mathbb{R}, \tag{9.44}$$

and

$$\sum_{\{n:\,\alpha_n<\alpha^*\}} P_n^\infty(t) = 0 \text{ for all } t \in \mathbb{R}. \tag{9.45}$$

Finally, the convergence $N(t_{k_i}+t) \to N^\infty(t)$ is locally uniform in $C^1(\mathbb{R})$.

Proof First of all let us remark that

$$P_n(t) = e^{m_n \int_0^t N(s)\mathrm{d}s - \mu_n t} P_n^0 = e^{\mu_n\left(\alpha_n \int_0^t N(s)\mathrm{d}s - t\right)} P_n^0.$$

Since α_n and μ_n are in $C(\overline{\mathbb{N}})$, the map $t \mapsto e^{\mu_n\left(\alpha_n \int_0^t N(s)\mathrm{d}s - t\right)}$ is continuous from $\mathbb{R}^+$ to $C(\overline{\mathbb{N}})$, and therefore $t \mapsto (P_n(t))_{n\in\overline{\mathbb{N}}}$ is continuous from $\mathbb{R}$ to $\mathcal{M}_+^*(\overline{\mathbb{N}})$. Thus $t \to \left(N(t), (P_n(t))_{n\in\overline{\mathbb{N}}}\right)$ is indeed continuous from $\mathbb{R}^+$ to $\mathbb{R} \times \mathcal{M}_+^*(\overline{\mathbb{N}})$. Since $N(t)$ is uniformly bounded and $(P_n(t))_{n\in\overline{\mathbb{N}}}$ is uniformly bounded in total variation norm by Lemma 9.23, it follows from the Prohorov Theorem [32, Vol. II Theorem 8.6.2 p. 202] that the orbit is precompact in $\mathbb{R} \times \mathcal{M}_+^*(\overline{\mathbb{N}})$.

Let t_k be a sequence such that $t_k \to +\infty$. Because $\sum_{n\in\overline{\mathbb{N}}} P_n(t_k)$ is bounded, we can extract from (t_k) a subsequence, still denotes t_k, along which

$$N(t_k) \to N_0^\infty \text{ and } (P_n(t_k))_{n\in\overline{\mathbb{N}}} \rightharpoonup (P_n^{\infty,0})_{n\in\overline{\mathbb{N}}}.$$

We remark that $N'(t+t_k)$ is bounded thanks to Lemma 9.23 and

$$N''(t+t_k) = -\mu^N N'(t+t_k) - N'(t+t_k)\sum_{n\in\overline{\mathbb{N}}} m_n P_n(t+t_k)$$
$$- N(t+t_k)\sum_{n\in\overline{\mathbb{N}}} m_n \left(m_n N(t+t_k) - \mu_n\right) P_n(t+t_k)$$

is also bounded, locally uniformly in t. Thus up to a further extraction and diagonal extraction process, the shifted orbit $N(t+t_k)$ converges locally uniformly in $C^1(\mathbb{R})$. Because

$$P_n(t_k+t) = e^{\mu_n\left(\alpha_n \int_{t_k}^{t+t_k} N(t_k+\sigma)\mathrm{d}\sigma - t\right)} P_n(t_k), \qquad n \in \overline{\mathbb{N}},$$

we can pass to the weak-$*$ limit in the above formula and we get

$$P_n^\infty(t) := e^{\mu_n\left(\alpha_n \int_0^t N^\infty(\sigma)\mathrm{d}\sigma - t\right)} P_n^{\infty,0}, \qquad n\in\overline{\mathbb{N}}, t\in\mathbb{R}, \tag{9.46}$$

so that $(N^\infty(t), (P_n^\infty(t))_{n\in\overline{\mathbb{N}}})$ is a solution of (9.1) with starting from the initial data $(N_0^\infty, (P_n^{\infty,0}))$ and with $t\in\mathbb{R}$: a complete orbit. That $N^\infty(t) > 0$ is a consequence of Lemma 9.23. Since the constant function $n \mapsto 1 \in C(\overline{\mathbb{N}})$ we have

$$\sum_{n\in\overline{\mathbb{N}}} P_n^{\infty,0} = \sum_{n\in\overline{\mathbb{N}}} 1\cdot P_n^{\infty,0} = \lim_{k\to+\infty}\sum_{n\in\overline{\mathbb{N}}} 1\cdot P_n(t_k)$$

so if

$$\liminf_{k\to+\infty}\sum_{n\in\overline{\mathbb{N}}} P_n(t_k) > 0,$$

then we have

$$\sum_{n\in\overline{\mathbb{N}}} P_n^{\infty,0} > 0,$$

and thanks to (9.46), (9.44) is proved.

Next we show the concentration property (9.45). Recalling that

$$\liminf_{t\to+\infty}\frac{1}{t}\int_0^t N(s)\mathrm{d}s \geq \underline{N}$$

for a positive constant $\underline{N} > 0$ and $\limsup_{t\to+\infty}\frac{1}{t}\int_0^t N(s)\mathrm{d}s \leq \frac{1}{\alpha^*}$, we have for t_k sufficiently large $\frac{1}{t_k}\int_0^{t_k} N(s)\mathrm{d}s \leq \frac{1}{\alpha^*} + \frac{\varepsilon \underline{N}}{2\alpha^*}$ and thus

$$\begin{aligned}\sum_{\{n:\alpha_n<\alpha^*-\varepsilon\}} P_n(t_k) &= \sum_{\{n:\alpha_n<\alpha^*-\varepsilon\}} P_n^0 \exp\left[\mu_n t_k\left(\alpha^*\frac{1}{t_k}\int_0^{t_k} N(\sigma)\mathrm{d}\sigma - 1\right.\right.\\ &\qquad \left.\left.-(\alpha^*-\alpha_n)\frac{1}{t_k}\int_0^{t_k} N(\sigma)\mathrm{d}\sigma\right)\right]\\ &\le \sum_{\{n:\alpha_n<\alpha^*-\varepsilon\}} P_n^0 \exp\left[\mu_n t_k\left(\alpha^*\left(\frac{1}{\alpha^*}+\frac{\varepsilon\underline{N}}{2\alpha^*}\right)-1-\varepsilon\underline{N}\right)\right]\\ &= e^{-\mu_0\frac{\varepsilon}{2}\underline{N}t_k}\sum_{n\in\overline{\mathbb{N}}} P_n^0 \xrightarrow[k\to+\infty]{} 0,\end{aligned}$$

Since $\varepsilon > 0$ is arbitrary, (9.45) is proved.

This finishes the proof of Proposition 9.26. □

We are now ready to state our uniform persistence result.

Proposition 9.27 (Uniform persistence) *Suppose that Assumptions 9.1, 9.22 and 9.3 hold true, and that $\mathcal{R}_0 > 1$. Then,*

$$\liminf_{t\to+\infty} \sum_{n\in\overline{\mathbb{N}}} P_n(t) > 0. \tag{9.47}$$

Proof Let us show the uniform persistence property. We adapt the argument of [244, Proposition 3.2] in our non-metric context. Assume by contradiction that (9.47) does not hold. Then, there exists a sequence $t_k \to +\infty$ such that

$$\lim_{k\to+\infty} \sum_{n=-K}^{+\infty} P_n(t_k) = 0. \tag{9.48}$$

By Lemma 9.23 we have $\inf_k N(t_k) > 0$. Because of (9.42), for each k sufficiently large and up to replacing t_k by a subsequence, there exists s_k with $t_{k-1} < s_k < t_k$ such that

$$\sum_{n=-K}^{+\infty} P_n(s_k) = \frac{1}{2}\cdot\frac{\mu^N}{m^\infty}(\mathcal{R}_0-1) \text{ and } \sup_{\sigma\in[s_k,t_k]} \sum_{n=-K}^{+\infty} P_n(\sigma) \le \frac{1}{2}\cdot\frac{\mu^N}{m^\infty}(\mathcal{R}_0-1). \tag{9.49}$$

By Proposition 9.26, there exists a subsequence of s_k, still denoted s_k, such that $N(s_k+t) \to N_0^\infty(t)$ and $(P_n(s_k+t)) \rightharpoonup (P_n^\infty(s_k+t))$, and moreover (9.44) and (9.45) hold for the limit orbit.

Next we show that $(N^\infty(t), (P_n^\infty(t)))$ satisfy the assumptions 9.1, 9.2 and 9.3. The assumptions 9.1 and 9.3 are readily checked since the values of the coefficients α_n and μ_n have not changed (neither have $m_n := \mu_n\alpha_n$). We deduce from (9.45) that

$$\begin{aligned}\sum_{\{n:\alpha_n=\alpha^*\}} P_n^{\infty,0} &= \sum_{n=-K}^{+\infty} P_n^{\infty,0} - \sum_{\{n:\alpha_n<\alpha^*\}} P_n(s_k) = \sum_{n=-K}^{+\infty} P_n^{\infty,0}\\ &= \frac{1}{2}\cdot\frac{\mu^N}{m^\infty}(\mathcal{R}_0-1) > 0,\end{aligned}$$

thus in particular the set $\{n \in \overline{\mathbb{N}} : P_n^{\infty,0} > 0 \text{ and } \alpha_n = \alpha^*\}$ is nonempty. Hence, up to removing the terms for which $P_n^{\infty,0} = 0$, we have

$$\sup_{\{n : P_n^{\infty,0} > 0\}} \frac{m_n}{\mu_n} = \sup_{\{n : P_n^{\infty,0} > 0\}} \alpha_n = \alpha^*,$$

Assumptions 9.2, 9.1 and 9.3 are still satisfied along the subsequence $P_n^{\infty,0} > 0$. In particular we can apply Lemma 9.25 and thus

$$\limsup_{t \to +\infty} \sum_{n=-K}^{+\infty} P_n^\infty(t) \geq \frac{\mu^N}{m^\infty} (\mathcal{R}_0 - 1). \tag{9.50}$$

Now we conclude the argument. Let $T_k := t_k - s_k$. There are two possibilities.

• T_k is bounded. In that case, we further extract a subsequence so that $T_k \to T$. Then we have

$$\sum_{n=-K}^{+\infty} P_n^\infty(T) = \lim_{k \to +\infty} \sum_{n=-K}^{+\infty} P_n(t_k) = 0,$$

and by the uniqueness of the solution to (9.1), we have $P_n^\infty(t) \equiv 0$ for all $t \geq T$. This contradicts (9.50).

• T_k is unbounded. In that case, we further extract a subsequence so that $T_k \to +\infty$. But since

$$\sup_{\sigma \in [s_k, t_k]} \sum_{n=-K}^{+\infty} P_n(\sigma) \leq \frac{1}{2} \cdot \frac{\mu^N}{m^\infty} (\mathcal{R}_0 - 1),$$

we have

$$\sup_{\sigma \in [0, +\infty)} \sum_{n=-K}^{+\infty} P_n^\infty(\sigma) \leq \frac{1}{2} \cdot \frac{\mu^N}{m^\infty} (\mathcal{R}_0 - 1)$$

and this, again, contradicts (9.50).

These contradictions prove that (9.48) cannot hold, hence (9.47) holds. The proof of Proposition 9.27 is finished. □

Lemma 9.28 *Suppose that Assumptions 9.1, 9.22 and 9.3 hold true and let $\big(N(t), (P_n(t))\big)$ be the corresponding solution of* (9.1). *Assume that* $\mathcal{R}_0 > 1$. *Then*

$$\liminf_{T \to +\infty} \frac{1}{T} \int_0^T N(t) \mathrm{d}t \geq \frac{1}{\alpha^*},$$

with α^* given in (9.3).

Proof Assume by contradiction that the conclusion of the Lemma does not hold, *i.e.* there exist $\varepsilon > 0$ and a sequence $t_k \to +\infty$ such that

$$\frac{1}{t_k} \int_0^{t_k} N(t) \mathrm{d}t \leq \frac{1}{\alpha^*} - \varepsilon.$$

Then

$$\sum_{n\in\overline{\mathbb{N}}} P_n(t_k) = \sum_{n\in\overline{\mathbb{N}}} P_n^0 e^{m_n \int_0^{t_k} N(s)\mathrm{d}s - \mu_n t_k} = \sum_{n\in\overline{\mathbb{N}}} P_n^0 e^{\mu_n t_k\left(\frac{m_n}{\mu_n}\frac{1}{t_k}\int_0^{t_k} N(s)\mathrm{d}s - 1\right)}$$
$$\leq \sum_{n\in\overline{\mathbb{N}}} P_n^0 e^{\mu_n t_k\left(\frac{m_n}{\mu_n}\left(\frac{1}{\alpha^*}-\varepsilon\right)-1\right)} \leq \sum_{n\in\overline{\mathbb{N}}} P_n^0 e^{-\varepsilon m_n t_k} \leq e^{-\varepsilon m_0 t_k}\sum_{n\in\overline{\mathbb{N}}} P_n^0$$
$$\xrightarrow[k\to+\infty]{} 0.$$

Therefore

$$\liminf_{t\to+\infty} \sum_{n\in\overline{\mathbb{N}}} P_n(t) \leq \lim_{k\to+\infty} \sum_{n\in\overline{\mathbb{N}}} P_n(t_k) = 0,$$

which is in contradiction with Lemma 9.27. This proves the Lemma. □

Next we give a Lyapunov functional that works for a special case in our model. It is close to the Lyapunov functional given for the Lotka-Volterra case of [166], although in our case it is not possible to factorize $N(t)$ in the first equation of (9.1a), which makes the computations intractable when $\frac{m_n}{\mu_n}$ is not a constant.

Proposition 9.29 (Lyapunov functional) *Let Assumption 9.1 hold true and assume that $\mathcal{R}_0 > 1$. Let $(N^*, P_n^*)_{n\in\overline{\mathbb{N}}}$ be a stationary solution of (9.1a), i.e. $N^* := \frac{1}{\alpha^*}$ and $(P_n^*)_{n\in\overline{\mathbb{N}}}$ satisfies*

$$\sum_{n\in\overline{\mathbb{N}}} \mu_n P_n^* = \frac{\mu^N}{\alpha^*}\left(\mathcal{R}_0 - 1\right).$$

Assume furthermore that $\frac{m_n}{\mu_n} \equiv \alpha^*$ is constant whenever $P_n^* > 0$ and let $\mathcal{N}_+$ be the set of indices for which $P_0^n > 0$. Define $g(x) = x - \ln(x)$, and let

$$D(V) := \left\{(N, (P_n)_n) \,:\, S > 0, P_n = 0 \text{ whenever } P_n^* = 0, \text{ and } \inf_{n\in\mathcal{N}_+} \frac{P_n}{P_n^*} > 0\right\}$$
$$\subset \mathbb{R}\times\ell^\infty\left((P_n^*)^{-1}\right),$$

where $\ell^\infty\left((P_n^*)^{-1}\right)$ is equipped with the norm $\|(\varphi_n)_n\|_{\ell^\infty\left((P_n^*)^{-1}\right)} := \sup_{n\in\overline{\mathbb{N}}}\left|\frac{\varphi_n}{P_n^*}\right|$. Then the functional

$$V(N, (P_n)) := N^* g\left(\frac{N}{N^*}\right) - N^* + \sum_{n\in\mathcal{N}_+}\left[P_n^* g\left(\frac{P_n}{P_n^*}\right) - P_n^*\right]. \tag{9.51}$$

is well-defined and continuous on the open set $D(V)$ for the topology induced by $\mathbb{R}\times\ell^\infty\left((P_n^*)^{-1}\right)$.

Moreover, if $\left(N(t), (P_n(t))_{n\in\mathbb{N}}\right)$ is a solution of (9.1) such that $\left(N(0), (P_n(0))\right) \in D(V)$, then $\left(N(t), (P_n(t))_n\right) \in D(V)$ for all $t \geq 0$, $t \mapsto V(N(t), (P_n(t))_n) \in C^1$ and we have

$$\frac{\mathrm{d}}{\mathrm{d}t} V\left(N(t), (P_n(t))_{n\in\overline{\mathbb{N}}}\right) = -\frac{(N(t)-N^*)^2}{N(t)}\left(-\mu^N - \sum_{n\in\mathcal{N}_+}\alpha\mu_n P_n^*\right). \tag{9.52}$$

Proof First we check that $(N^*, (P_n^*)_{n\in\mathbb{N}})$ is a stationary solution of (9.1). Indeed,

$$\frac{\mathrm{d}}{\mathrm{d}t}(P_n^*) = 0 = \mu_n(\alpha^* N^* - 1)P_n^*,$$

and

$$\frac{\mathrm{d}}{\mathrm{d}t}(N^*) = 0 = \beta - \mu^N N^* - N^* \sum_{n\in\overline{\mathbb{N}}} \alpha^* \mu_n e^{\tau\mu_n} P_n^0.$$

Let $\big(N(t), (P_n(t))_{n\in\mathbb{N}}\big)$ be such that $\big(N(0), (P_n(0))\big) \in D(V)$, we check that $\big(N(t), (P_n(t))\big) \in D(V)$ for all $t \geq 0$. That $N(t) > 0$ is a consequence of Lemma 9.23. Then, we remark that

$$P_n(t) = P_n(0)e^{m_n \int_0^t N(s)\mathrm{d}s - \mu_n t} \geq P_n(0)e^{m_0 \int_0^t N(s)\mathrm{d}s - \mu_\infty t},$$

therefore

$$\inf_{n\in\mathcal{N}_+} \frac{P_n(t)}{P_n^*} \geq e^{m_0 \int_0^t N(s)\mathrm{d}s - \mu_\infty t} \inf_{n\in\mathcal{N}_+} \frac{P_n(0)}{P_n^*} > 0,$$

and we have proved that $\big(N(t), (P_n(t))\big) \in D(V)$ for all $t \geq 0$. The continuity and continuous differentiability of $t \mapsto V\big(N(t), (P_n(t))\big)$ follow from classical arguments.

Next, writing $V_1(t) = N^* g\left(\frac{N(t)}{N^*}\right)$ and $V_2(t) = \sum_{n\in\overline{\mathbb{N}}} P_n^* g\left(\frac{P_n(t)}{P_n^*}\right)$, we have

$$\begin{aligned}
V_1'(t) &= N^* \frac{N'(t)}{N^*} g'\left(\frac{N(t)}{N^*}\right) = \left(\beta - \mu^N N(t) - N(t) \sum_{n\in\overline{\mathbb{N}}} m_n P_n(t)\right)\left(1 - \frac{N^*}{N(t)}\right) \\
&= \left(\beta - \mu^N N(t) - N(t) \sum_{n\in\overline{\mathbb{N}}} m_n P_n(t) - \beta + \mu^N N^* + N^* \sum_{n\in\overline{\mathbb{N}}} m_n P_n^*\right)\left(1 - \frac{N^*}{N(t)}\right) \\
&= -\mu^N \frac{(N(t) - N^*)^2}{N(t)} + \left(N^* \sum_{n\in\overline{\mathbb{N}}} m_n P_n^* - N(t) \sum_{n\in\overline{\mathbb{N}}} m_n P_n(t)\right)\left(1 - \frac{N^*}{N(t)}\right), \\
&= -\mu^N \frac{(N(t) - N^*)^2}{N(t)} + \frac{N(t) - N^*}{N(t)} \sum_{n\in\overline{\mathbb{N}}} \alpha^* \mu_n \big(P_n^* N^* - P_n(t)N(t)\big),
\end{aligned}$$

and

$$\begin{aligned}
V_2'(t) &= \sum_{n\in\overline{\mathbb{N}}} P_n^* \frac{P_n'(t)}{P_n^*} g'\left(\frac{P_n(t)}{P_n^*}\right) = \sum_{n\in\overline{\mathbb{N}}} \mu_n \left(\alpha^* N(t) - 1\right) P_n(t) \left(1 - \frac{P_n^*}{P_n(t)}\right) \\
&= \sum_{n\in\overline{\mathbb{N}}} \mu_n \left(\alpha^* N(t) - 1\right) \left(P_n(t) - P_n^*\right) \\
&= \sum_{n\in\overline{\mathbb{N}}} \mu_n \alpha^* \left(N(t) - N^*\right) \left(P_n(t) - P_n^*\right).
\end{aligned}$$

Recalling $N^* = \frac{1}{\alpha^*}$, we have therefore

$$\begin{aligned}\frac{\mathrm{d}}{\mathrm{d}t}V(N(t),(P_n(t))_n) &= \frac{\mathrm{d}}{\mathrm{d}t}V_1(t) + \frac{\mathrm{d}}{\mathrm{d}t}V_2(t)\\ &= -\mu^N \frac{(N(t)-N^*)^2}{N(t)}\\ &\quad + \frac{N(t)-N^*}{N(t)} \sum_{n\in\overline{\mathbb{N}}} \alpha^*\mu_n \left(P_n^*N^* - P_n(t)N(t) + P_n(t)N(t) - P_n^*N(t)\right)\\ &= -\mu^N \frac{(N(t)-N^*)^2}{N(t)} - \frac{\left(N(t)-N^*\right)^2}{N(t)} \sum_{n\in\overline{\mathbb{N}}} \alpha^*\mu_n P_n^*.\end{aligned}$$

Proposition 9.26 is proved. □

Lemma 9.30 *Suppose that Assumptions 9.1, 9.22 and 9.3 holds true and assume that there exists an index $i \in \overline{\mathbb{N}}$ such that $P_i^0 > 0$ and*

$$\alpha^* = \sup_{n\in\mathbb{N}} \frac{m_n}{\mu_n} = \frac{m_i}{\mu_i}.$$

For $s < t$, let

$$\eta(t;s) := (t-s)\big(\alpha^*\overline{N}(t;s) - 1\big), \text{ where } \overline{N}(t;s) := \frac{1}{t-s}\int_s^t N(\sigma)\mathrm{d}\sigma. \tag{9.53}$$

Then there exists a constant $\overline{\eta} > 0$ such that for any $0 \le s < t$ one has

$$-\overline{\eta} \le \eta(t;s) \le \overline{\eta} < +\infty. \tag{9.54}$$

If moreover $\big(N(t), (P_n(t))_n\big) \in \mathbb{R} \times \mathcal{M}_+(\overline{\mathbb{N}})$ is a uniformly bounded complete orbit such that $\liminf_{t\to-\infty} \sum_{\{k\,:\,\alpha_k=\alpha^*\}} P_n(t) > 0$, then for any $s < t$ one has

$$-\overline{\eta} \le \eta(t;s) \le \overline{\eta} < +\infty. \tag{9.55}$$

Proof Let us write $P_n(t)$ as

$$\begin{aligned}P_n(t) = P_n(s)\exp\left[\mu_n(t-s)\left(\left(\alpha^*\overline{N}(t;s) - 1\right) - \left(\alpha^* - \frac{m_n}{\mu_n}\right)\overline{N}(t;s)\right)\right]\\ = P_n(s)e^{\mu_n\eta(t;s) - (\alpha^*-\alpha_n)\overline{N}(t;s)}.\end{aligned}$$

We claim that $\eta(t;s)$ is uniformly bounded in t, s. Indeed, recalling $\alpha_n = \frac{m_n}{\mu_n}$, we have by Jensen's inequality

$$\exp\left[\sum_{\{n\,:\,\alpha_n=\alpha^*\}} \mu_n\eta(t;s)\frac{P_n(s)}{\sum_{\{k\,:\,\alpha_k=\alpha^*\}} P_k(s)}\right] \le \sum_{\{n\,:\,\alpha_n=\alpha^*\}} e^{\mu_n\eta(t;s)}\frac{P_n(s)}{\sum_{\{k\,:\,\alpha_k=\alpha^*\}} P_k(s)},$$

hence

$$\eta(t;s) \leq \frac{\sum\limits_{\{k:\alpha_k=\alpha^*\}} P_k(s)}{\sum\limits_{\{k:\alpha_k=\alpha^*\}} \mu_k P_k(s)} \ln \left[\sum_{\{n:\alpha_n=\alpha^*\}} e^{\mu_n \eta(t;s)} \frac{P_n(s)}{\sum\limits_{\{k:\alpha_k=\alpha^*\}} P_k(s)} \right]$$

$$= \frac{\sum\limits_{\{k:\alpha_k=\alpha^*\}} P_k(s)}{\sum\limits_{\{k:\alpha_k=\alpha^*\}} \mu_k P_k(s)} \ln \left[\frac{\sum\limits_{\{n:\alpha_n=\alpha^*\}} P_n(t)}{\sum\limits_{\{k:\alpha_k=\alpha^*\}} P_k(s)} \right].$$

Applying Lemma 9.23, the total mass $\sum_{n\in\overline{\mathbb{N}}} P_n(t)$ is bounded above, and by Proposition 9.27 the sum $\sum\limits_{\{k:\alpha_k=\alpha^*\}} P_k(s)$ is bounded below when $s \to +\infty$; therefore there exists $\overline{\eta} < +\infty$, independent of s, such that

$$\eta(t;s) \leq \overline{\eta}.$$

If $\big(N(t), (P_n(t))_n\big) \in \mathbb{R}\times\mathcal{M}_+(\overline{\mathbb{N}})$ is a complete orbit and $\liminf\limits_{t\to-\infty} \sum\limits_{\{k:\alpha_k=\alpha^*\}} P_n(t) > 0$, then there exists an upper bound valid for all $t, s \in \mathbb{R}$.

On the other hand, we claim that $\liminf\limits_{t\to+\infty} \eta(t;s) \geq \underline{\eta}$ for a constant $\underline{\eta}$. Indeed, assume by contradiction that there exists a sequence $t_k \to +\infty$ and $s_k \geq 0$ such that $\eta(t_k;s_k) \to -\infty$. Then

$$\sum_{n\in\overline{\mathbb{N}}} P_n(t) = \sum_{n\in\overline{\mathbb{N}}} P_n(s_k) e^{\mu_n \eta(t_k;s_k) - (\alpha^* - \alpha_n)\overline{N}(t_k;s_k)}$$
$$\leq e^{\mu_0 \eta(t_k;s_k)} \sum_{n\in\overline{\mathbb{N}}} P_n(s_k) \xrightarrow[t\to+\infty]{} 0,$$

which contradicts Proposition 9.27. We have proved (9.54). (9.55) is proved by identical arguments. □

Next we derive a kind of LaSalle principle [206] that shows that complete orbits concentrated on the set $\frac{m_n}{\mu_n} = \alpha^*$ are "almost" stationary.

Lemma 9.31 *Let Assumption 9.1 and 9.3 hold true and assume that* $(P_n^0)_{n\in\overline{\mathbb{N}}}$ *is non-trivial. Assume furthermore that* $\frac{m_n}{\mu_n} \equiv \alpha^*$ *is constant whenever* $P_n^0 > 0$ *and that* $\mathcal{R}_0 > 1$. *Let* $\big(N(t), (P_n(t))_n\big)$ *be a complete orbit of* (9.1) *passing through* (N_0, P_n^0) *at* $t = 0$ *and suppose that*

$$\liminf_{t\to-\infty} \sum_{n\in\overline{\mathbb{N}}} P_n(t) > 0. \tag{9.56}$$

Then we have

$$N(t) \equiv \frac{1}{\alpha^*} \text{ for all } t \in \mathbb{R}.$$

Proof Because of our assumption we have

$$\liminf_{t\to-\infty}\sum_{n\in\overline{\mathbb{N}}} P_n(t) = \liminf_{t\to-\infty}\sum_{\{n:\alpha_n=\alpha^*\}} P_n(t) > 0. \tag{9.57}$$

therefore Lemma 9.30 implies that $\eta(t;s)$, as defined by (9.53), is uniformly bounded. Let $\eta(t) := \eta(t;0)$. We define the distribution:

$$P_n^* := \begin{cases} 0 & \text{if } \frac{m_n}{\mu_n} < \alpha^*, \\ e^{\tau\mu_n}P_n^0, & \text{if } \frac{m_n}{\mu_n} = \alpha^*, \end{cases}$$

where $\tau \geq 0$ is the unique solution of the equation

$$\sum_{n\in\overline{\mathbb{N}}} \mu_n e^{\tau\mu_n}\mathbb{1}_{\alpha_n=\alpha^*}P_n^0 = \frac{\mu^N}{\alpha^*}\left(\mathcal{R}_0 - 1\right).$$

Then (P_n^*) is a stationary distribution with $\inf_{n\in\overline{\mathbb{N}}}\frac{P_n^0}{P_n^*} \geq e^{-|\tau|\mu_0}$, therefore Proposition 9.29 implies that $V\big(N(t),(P_n(t))_n\big)$ defined by (9.51) is well-defined along the orbit and decreasing.

We claim that $V\big(N(t),(P_n(t))_n\big)$ is constant along the orbit. Indeed, let $t_k \to -\infty$ be an aribtrary sequence; since $N(t)$ and $\eta(t)$ are uniformly bounded, we extract from t_k a subsequence, still denoted t_k, such that $N(t_k) \to N^{-\infty}$ and $\eta(t_k) \to \eta^{-\infty}$. Then, we have

$$P_n(t_k) \xrightarrow[k\to+\infty]{} P_n^{-\infty} := P_n^0 e^{\mu_n\eta^{-\infty}}, \text{ in } \mathcal{M}_+(\overline{\mathbb{N}}) \text{ and } \ell^\infty((P_n^*)^{-1}).$$

Moreover $t \mapsto V(N(t),(P_n(t)))$ is decreasing and bounded so there exists V^∞ such that

$$\lim_{t\to-\infty} V(N(t),(P_n(t))) = V^\infty.$$

The shifted orbits $(N(t+t_k),(P_n(t+t_k))_n)$ converge, up to a further extraction, to a complete orbit $(N^{-\infty}(t),(P_n^{-\infty}(t))_n)$; we have $V(N^{-\infty}(t),(P_n^{-\infty}(t))_n) \equiv V^{-\infty}$ for all $t\in\mathbb{R}$, therefore $V'(N^{-\infty}(t),(P_n^{-\infty}(t))_n) = 0$ and, by (9.52),

$$N^{-\infty}(t) \equiv \frac{1}{\alpha^*}, \qquad (N^{-\infty})'(t) \equiv 0, \qquad \text{for all } t\in\mathbb{R},$$

and by the first line in (9.1a),

$$\sum_{n\in\overline{\mathbb{N}}} P^{-\infty}(t) = \beta - \frac{\mu^N}{\alpha^*} \quad\iff\quad \sum_{n\in\overline{\mathbb{N}}} \mu_n e^{\mu_n\eta^{-\infty}}P_n^0 = \frac{\mu^N}{\alpha^*}(\mathcal{R}_0 - 1). \tag{9.58}$$

Since $\eta \mapsto \sum_{n\in\overline{\mathbb{N}}}\mu_n e^{\mu_n\eta}P_n^0$ is strictly increasing, the equation (9.58) has a unique solution which is $\eta = \tau$; therefore $\eta^{-\infty} = \tau$, $P_n^{-\infty} \equiv P_n^*$ and

$$V(N^{-\infty},(P_n^{-\infty})_{n\in\overline{\mathbb{N}}}) = \min_{(N,(P_n))\in D(V)} V(N,(P_n)) = 0.$$

Back to the original complete orbit, by the continuity of V in $\mathbb{R} \times \ell^\infty((P_n^*)^{-1})$ we have that

$$\lim_{k\to+\infty} V(N(t_k), (P_n(t_k))_n) = V(N^{-\infty}, (P_n^{-\infty})_{n\in\overline{\mathbb{N}}}) = \min_{(N,(P_n))\in D(V)} V(N, (P_n)),$$

and since $V(N(t), (P_n(t))_n)$ is decreasing this means that

$$N(t) \equiv \frac{1}{\alpha^*}.$$

Lemma 9.31 is proved. □

We are now in the position to prove Theorem 9.6.

Proof (Proof of Theorem 9.6) Recall that $(P_n)_{n\in\mathbb{N}} \in \ell^1_+$ is given as a sequence over $\mathbb{N}$; without loss of generality, we set

$$\omega(\alpha) \times \omega(\mu) =: \{(\alpha_{-i}, \mu_{-i}) \,:\, i = 1, \ldots, K\},$$

as described as the beginning of Section 9.5.3, and $P^0_{-i} = 0$, so that $(P_n)_{n\in\overline{\mathbb{N}}}$ is well-defined as a member of $\mathcal{M}_+(\overline{\mathbb{N}})$ and the new system is strictly equivalent to the original system for all $t \geq 0$.

We start by dealing with case i) and assume that there is $n_0 \in \mathbb{N}$ such that $\frac{m_{n_0}}{\mu_{n_0}} = \alpha^*$. Thanks to Lemma 9.30 we know that

$$P_n(t) = P_n^0 e^{\mu_n \eta(t;0) - (\alpha_n - \alpha^*)\overline{N}(t;0)}$$

and $\eta(t;0)$ is uniformly bounded for $t \in [0, +\infty)$. Let t_k be a sequence such that $t_k \to +\infty$ and $\eta(t_k;0) \to \eta^\infty \in [0, \overline{\eta}]$. Then

$$P_n(t_k) = P_n^0 e^{\mu_n t_k(\alpha^*\overline{N}(t_k)-1) - \mu_n(\alpha^*-\alpha_n)t_k\overline{N}(t_k)}$$
$$\Longrightarrow P_n^\infty(0) = \begin{cases} 0 & \text{if } \alpha_n < \alpha^*, \\ P_n^0 e^{\mu_n \eta^\infty} & \text{if } \alpha_n = \alpha^*. \end{cases}$$

By Proposition 9.26, we can extract from (t_k) a subsequence, still denoted (t_k), such that $(N(t+t_k), I(t+t_k))$ converges in $\mathbb{R} \times \mathcal{M}_+^*(\overline{\mathbb{N}})$ to a complete orbit $(N^\infty(t), (P_n^\infty(t)))$ with $N^\infty > 0$, and by Proposition 9.27 the limit P_n^∞ is non-trivial. By Proposition 9.26, we have that $P_n^\infty(t) \equiv 0$ whenever $\alpha_n < \alpha^*$; thus $m_n = \alpha^*\mu_n$ whenever $P_n^\infty > 0$. Hence we can apply Lemma 9.31 to show that

$$N^\infty(t) \equiv \frac{1}{\alpha^*} \text{ and } (N^\infty)'(t) \equiv 0.$$

Thus

$$0 = \beta - \mu^N N^\infty - N^\infty \sum_{n\in\overline{\mathbb{N}}} m_n P^\infty(0) \Longrightarrow \sum_{n\in\overline{\mathbb{N}}} m_n P^\infty(0) = \mu^N \left(\frac{\beta}{\mu^N}\alpha^* - 1\right)$$

and finally

$$\sum_{n\in\overline{\mathbb{N}}} \mu_n \mathbb{1}_{\alpha_n=\alpha^*} e^{\mu_n \eta^\infty} P_n^0 = \frac{\mu^N}{\alpha^*}(\mathcal{R}_0 - 1).$$

This equation has a unique solution $\tau \geq 0$, since the left-hand side is a strictly increasing function of η^∞. Thus we have proved that $\eta(t;0)$ converges to this value $\tau \geq 0$ and, finally,

$$\lim_{t\to+\infty} P_n(t) = \lim_{t\to+\infty} P_n^0 e^{\mu_n t(\alpha^*\overline{N}(t)-1)-\mu_n(\alpha^*-\alpha_n)t\overline{N}(t)} = \begin{cases} 0 & \text{if } \alpha_n < \alpha^*, \\ P_n^0 e^{\mu_n \tau} & \text{if } \alpha_n = \alpha^*. \end{cases}$$

This finishes the proof of Theorem 9.6 case i).

Next we deal with case ii) and assume that for all $n \in \mathbb{N}$ we have $\frac{m_n}{\mu_n} < \alpha^*$. Let $t_k \to +\infty$ be arbitrary. By Proposition 9.26 we can extract from t_k a subsequence, still denoted t_k, such that the shifted orbits $\left(N(t+t_k), (P_n(t+t_k))_{n\in\overline{\mathbb{N}}}\right)$ converge pointwise in $\mathbb{R} \times \mathcal{M}_+^*(\overline{\mathbb{N}})$ to a complete orbit $\left(N^\infty(t), (P_n^\infty(t))_{n\in\overline{\mathbb{N}}}\right)$ and the limit $(P_n^\infty)_{n\in\overline{\mathbb{N}}}$ is non-trivial by Proposition 9.27. Because of our assumption that $\frac{m_n}{\mu_n} < \alpha^*$ we have $P_n^\infty = 0$ for all $n \geq 0$ therefore $\frac{m_n}{\mu_n} = \alpha^*$ whenever $P_n^\infty(0) > 0$. Hence we can apply Lemma 9.31 and get

$$N^\infty(t) \equiv \frac{1}{\alpha^*}, \qquad (N^\infty)'(t) \equiv 0, \qquad \text{for all } t \in \mathbb{R}.$$

Because the original sequence is arbitrary, we have proved that

$$N(t) \xrightarrow[t\to+\infty]{} \frac{1}{\alpha^*}, \qquad N'(t) \xrightarrow[t\to+\infty]{} 0, \qquad P_n(t) \xrightarrow[t\to+\infty]{} 0 \text{ for all } n \in \mathbb{N},$$

the fact that the mass does not vanish is a consequence of Proposition 9.27 and the concentration property is a consequence of (9.45) in Proposition 9.26. This finishes the proof of item ii) and ends the proof of Theorem 9.6. □

9.5.4 Proof of Proposition 9.9

Now under the additional assumption 9.8, we prove Proposition 9.9.

Proof (Proof of Proposition 9.9) We decompose the proof in several steps. We let $\overline{N}(t) := \frac{1}{t}\int_0^t N(s)\mathrm{d}s$. Our method is the following: we fix $\varepsilon > 0$ and show that the set of indices $C_\varepsilon := \{n \,:\, \alpha_n \geq \alpha^* - \varepsilon \text{ and } \mu_n \geq \mu^* - \varepsilon\}$ concentrates asymptotically all the mass.

Step 1: We show that

$$\sum_{n=0}^{+\infty} \mathbb{1}_{\alpha_n N(t)<1} P_n(t) \xrightarrow[t\to+\infty]{} 0. \tag{9.59}$$

Indeed since $\overline{N}(t) \to \frac{1}{\alpha^*}$, the function $\mathbb{1}_{\alpha_n N(t)<1} P_n(t)$ converges pointwise to 0 as $t \to +\infty$ because of our assumption that $\frac{m_n}{\mu_n} = \alpha_n < \alpha^*$ for all n (in fact, this function is asymptotically stationary equal to 0 for all fixed $n \in \mathbb{N}$). Moreover we have $\mathbb{1}_{\alpha_n N(t)<1} P_n(t) = \mathbb{1}_{\alpha_n N(t)<1} e^{t\mu_n(\alpha^* \overline{N}(t)-1)} P_n^0 \le P_n^0$, so the sequence is uniformly dominated by $(P_n^0)_{n\in\mathbb{N}} \in \ell_+^1$. By the Lebesgue dominated convergence Theorem, we have therefore

$$\lim_{n\to+\infty} \sum_{n=0}^{+\infty} \mathbb{1}_{\alpha_n N(t)<1} P_n(t) = 0,$$

which proves (9.59).

Step 2: We show that for any $\varepsilon > 0$,

$$\limsup_{t\to+\infty} \sum_{n=0}^{+\infty} \mathbb{1}_{\alpha_n \overline{N}(t)\ge 1} e^{t(\mu^*-\varepsilon)(\alpha_n \overline{N}(t)-1)} P_n^0 < +\infty. \tag{9.60}$$

Indeed let $(A_n)_{n\in\mathbb{N}}$ be a strictly increasing enumeration of the set $\{\alpha_n \,:\, n \in \mathbb{N}\}$. We remark that, by changing the order of summation, we have

$$\sum_{n=0}^{+\infty} \mathbb{1}_{\alpha_n \overline{N}(t)\ge 1} P_n(t) = \sum_{n=0}^{+\infty} \sum_{\{k\,:\,\alpha_k=A_n\}} e^{\mu_n t(A_n \overline{N}(t)-1)} P_k^0,$$

so that, according to Step 1, we have for t sufficiently large

$$\begin{aligned}
\sum_{\{n\,:\,\mu_n\ge\mu^*-\varepsilon\}} P_n(t) &= \sum_{\{n\,:\,\mu_n\ge\mu^*-\varepsilon\}} \left(\mathbb{1}_{\alpha_n\overline{N}(t)<1} + \mathbb{1}_{\alpha_n\overline{N}(t)\ge 1}\right) P_n(t)\\
&= o(1) + \sum_{\{n\,:\,A_n\ge\frac{1}{\overline{N}(t)}\}} \sum_{\{k\,:\,\alpha_k=A_n \text{ and } \mu_k\ge\mu^*-\varepsilon\}} e^{\mu_k t(A_n\overline{N}(t)-1)} P_k^0\\
&\ge o(1) + \sum_{\{n\,:\,A_n\ge\frac{1}{\overline{N}(t)}\}} \sum_{\{k\,:\,\alpha_k=A_n \text{ and } \mu_k\ge\mu^*-\varepsilon\}} e^{(\mu^*-\varepsilon)t(A_n\overline{N}(t)-1)} P_k^0\\
&= o(1) + \sum_{\{n\,:\,A_n\ge\frac{1}{\overline{N}(t)}\}} e^{(\mu^*-\varepsilon)t(A_n\overline{N}(t)-1)} \sum_{\{k\,:\,\alpha_k=A_n \text{ and } \mu_k\ge\mu^*-\varepsilon\}} P_k^0\\
&= o(1) + \sum_{\{n\,:\,A_n\ge\frac{1}{\overline{N}(t)}\}} e^{(\mu^*-\varepsilon)t(A_n\overline{N}(t)-1)} \frac{\sum_{\{k\,:\,\alpha_k=A_n \text{ and } \mu_k\ge\mu^*-\varepsilon\}} P_k^0}{\sum_{\{k\,:\,\alpha_k=A_n\}} P_k^0}\\
&\qquad \times \sum_{\{k\,:\,\alpha_k=A_n\}} P_k^0\\
&\ge o(1) + m \sum_{\{n\,:\,A_n\ge\frac{1}{\overline{N}(t)}\}} \sum_{\{\alpha_k=A_n\}} e^{(\mu^*-\varepsilon)t(A_n\overline{N}(t)-1)} P_k^0,
\end{aligned}$$

wherein $m > 0$ is the constant provided by Assumption 9.8 and the error term $o(1)$ collects terms going to 0 as $t \to +\infty$. Thus

$$\begin{aligned}
&\sum_{\{n:\alpha_n \geq \alpha^* - \varepsilon\}} \mathbb{1}_{\alpha_n \overline{N}(t) \geq 1} e^{t(\mu^* - \varepsilon)(\alpha_n \overline{N}(t) - 1)} P_n^0 \\
&\qquad = \sum_{\{n: A_n \geq \frac{1}{\overline{N}(t)}\}} \sum_{\{\alpha_k = A_n\}} e^{(\mu^* - \varepsilon)t(A_n \overline{N}(t) - 1)} P_k^0 \\
&\qquad \leq o(1) + \frac{1}{m} \sum_{\{n:\mu_n \geq \mu^* - \varepsilon\}} P_n(t) = O(1).
\end{aligned}$$

This proves (9.60).

Step 3: We show that, for all $\varepsilon > 0$,

$$\sum_{\{n:\alpha_n \geq \alpha^* - \varepsilon \text{ and } \mu_n \leq \mu^* - \varepsilon\}} P_n(t) \xrightarrow[t \to +\infty]{} 0. \tag{9.61}$$

Indeed, we have for t sufficiently large

$$\begin{aligned}
&\sum_{\{n:\alpha_n \geq \alpha^* - \varepsilon \text{ and } \mu_n \leq \mu^* - \varepsilon\}} \mathbb{1}_{\alpha_n \overline{N}(t) \geq 1} P_n(t) \\
&= \sum_{n=0}^{+\infty} \mathbb{1}_{A_n \overline{N}(t) \geq 1} \sum_{\{\alpha_k = A_n\}} e^{\mu_n t(A_n \overline{N}(t) - 1)} \mathbb{1}_{\mu_k \leq \mu^* - \varepsilon} P_k^0 \\
&\leq \sum_{n=0}^{+\infty} \mathbb{1}_{A_n \overline{N}(t) \geq 1} \sum_{\{\alpha_k = A_n\}} e^{t(\mu^* - \varepsilon)(A_n \overline{N}(t) - 1)} \mathbb{1}_{\mu_k \leq \mu^* - \varepsilon} P_k^0 \\
&\leq \sum_{n=0}^{+\infty} \mathbb{1}_{A_n \overline{N}(t) \geq 1} e^{-t\frac{\varepsilon}{2}(A_n \overline{N}(t) - 1)} \left(\sum_{\{\alpha_k = A_n\}} P_k^0 \right) e^{t(\mu^* - \frac{\varepsilon}{2})(A_n \overline{N}(t) - 1)} \\
&= \sum_{n=0}^{+\infty} e^{-t\frac{\varepsilon}{2}(\alpha_n \overline{N}(t) - 1)} \mathbb{1}_{\alpha_n \overline{N}(t) \geq 1} e^{t(\mu^* - \frac{\varepsilon}{2})(\alpha_n \overline{N}(t) - 1)} \mathbb{1}_{\mu_n \leq \mu^* - \varepsilon} P_n^0.
\end{aligned}$$

Reducing ε if necessary we may assume that $\frac{2\mu^* - \varepsilon}{\varepsilon} > 1$. Hence by Hölder's inequality we have:

$$\sum_{\{n:\,\alpha_n\geq\alpha^*-\varepsilon \text{ and } \mu_n\leq\mu^*-\varepsilon\}} \mathbb{1}_{\alpha_n\overline{N}(t)\geq 1} P_n(t)$$

$$\leq \left(\sum_{n=0}^{+\infty}\left(e^{-t\frac{\varepsilon}{2}(\alpha_n\overline{N}(t)-1)}\right)^{\frac{\mu^*-\frac{\varepsilon}{2}}{\frac{\varepsilon}{2}}} \mathbb{1}_{\alpha_n\overline{N}(t)\geq 1} e^{t\left(\mu^*-\frac{\varepsilon}{2}\right)\left(\alpha_n\overline{N}(t)-1\right)} P_n^0\right)^{\frac{\frac{\varepsilon}{2}}{\mu^*-\frac{\varepsilon}{2}}}$$

$$\times\left(\sum_{n=0}^{+\infty}\mathbb{1}_{\alpha_n\overline{N}(t)\geq 1} e^{t\left(\mu^*-\frac{\varepsilon}{2}\right)\left(\alpha_n\overline{N}(t)-1\right)} P_n^0\right)^{\frac{\mu^*-\varepsilon}{\mu^*-\frac{\varepsilon}{2}}}$$

$$\leq \left(\sum_{n=0}^{+\infty}\mathbb{1}_{\alpha_n\overline{N}(t)\geq 1} P_n^0\right)^{\frac{\varepsilon}{2\mu^*-\varepsilon}} \left(\sum_{n=0}^{+\infty}\mathbb{1}_{\alpha_n\overline{N}(t)\geq 1} e^{t\left(\mu^*-\frac{\varepsilon}{2}\right)\left(\alpha_n\overline{N}(t)-1\right)} P_n^0\right)^{\frac{2(\mu^*-\varepsilon)}{2\mu^*-\varepsilon}}.$$

Since $\mathbb{1}_{\alpha_n\overline{N}(t)\geq 1} P_n^0$ is uniformly bounded by a summable sequence and converges to 0 for each $n\in\mathbb{N}$, we have by Lebesgue's dominated convergence Theorem:

$$\sum_{n=0}^{+\infty}\mathbb{1}_{\alpha_n\overline{N}(t)\geq 1} P_n^0 \xrightarrow[t\to+\infty]{} 0,$$

and in Step 2 (with ε replaced by $\frac{\varepsilon}{2}$) we proved that

$$\limsup_{t\to+\infty}\sum_{n=0}^{+\infty}\mathbb{1}_{\alpha_n\overline{N}(t)\geq 1} e^{t\left(\mu^*-\frac{\varepsilon}{2}\right)\left(\alpha_n\overline{N}(t)-1\right)} P_n^0 < +\infty,$$

hence

$$\sum_{\{n:\,\alpha_n\geq\alpha^*-\varepsilon \text{ and } \mu_n\leq\mu^*-\varepsilon\}} \mathbb{1}_{\alpha_n\overline{N}(t)\geq 1} P_n(t) \xrightarrow[t\to+\infty]{} 0.$$

We have shown (9.61).

Step 4: We show the convergence of the mass. To do so, we consider the equivalent system set on $\overline{\mathbb{N}}$ by setting $P_{-i}^0 = 0$ for $i\in\{-K,\dots,-1\}$, as in the beginning of Section 9.5.3. Let (t_k) be any sequence such that $t_k\to+\infty$. Thanks to Proposition 9.26 we can extract a subsequence, still denoted t_k, such that $(N(t_k+t), (P_n(t_k+t))_{n\in\overline{\mathbb{N}}})$ converges to a complete orbit $(N^\infty(t), (P_n^\infty(t))_{n\in\overline{\mathbb{N}}})$. Thanks to Theorem 9.6 we know that $P_n^\infty(t)\equiv 0$ whenever $\alpha_n<\alpha^*$ and thanks to Step 3 we know that $P_n^\infty(t)\equiv 0$ whenever $\mu_n<\mu^*$. By Lemma 9.31 we have then $N^\infty(t)\equiv\frac{1}{\alpha^*}$ hence

$$\frac{\mathrm{d}}{\mathrm{d}t}N^\infty(t) = 0 = \beta - \mu^N N^\infty(t) - N^\infty(t)\sum_{n=0}^{+\infty}\mu_n\alpha_n P_n^\infty(t)$$

$$\Longleftrightarrow \frac{1}{\alpha^*}\sum_{n=-K}^{+\infty}\mu^*\alpha^* P_n^\infty(t) = \beta - \frac{\mu^N}{\alpha^*},$$

from which we deduce

$$\sum_{n=-K}^{+\infty} P_n^\infty(t) = \frac{\mu^N}{\alpha^*\mu^*}(\mathcal{R}_0 - 1).$$

Moreover for sequence of indices (n_j) satisfying (9.8), the omega-limit set of the sequence $\left((\alpha_{n_j}, \mu_{n_j})\right)_{j\in\mathbb{N}}$ is the set $\{(\alpha_{-i_1}, \mu_{-i_1}), \ldots, (\alpha_{-i_N}, \mu_{-i_N})\}$ for some $N < K$ and with $\mu_{i_j} < \mu^*, i = 1, \ldots, N$. Thanks to Step 3, we have then

$$\sum_{j=0}^{+\infty} P_{n_j}(t_k + t) \xrightarrow[k\to+\infty]{} \sum_{j=1}^{N} P_{-i_j}^\infty(t) + \sum_{j=0}^{+\infty} P_{n_j}^\infty(t) \equiv 0 \text{ for all } t \in \mathbb{R}.$$

Since the sequence (t_k) is arbitrary, we have indeed proved that $\sum_{n=0}^{+\infty} P_n(t)$ converges and that the limit is given by (9.7), and that for any sequence of indices (n_k) satisfying (9.8), (9.9) holds. Proposition 9.9 is proved. □

9.5.5 Extinction in the case $\mathcal{R}_0 = 1$

We continue the proof of Proposition 9.5 in the limit case $\mathcal{R}_0 = 1$ and we shall show that the infection dies out in this situation.

Proposition 9.32 *Let Assumption 9.1 and 9.3 be satisfied. Then for all initial data $(N_0, (P_n^0)_{n\in\mathbb{N}})$, the solution $(N(t), (P_n(t))_{n\in\mathbb{N}})$ satisfies*

$$\lim_{t\to\infty} N(t) = \frac{\beta}{\mu^N} \text{ and } \lim_{t\to\infty} \sum_{n=0}^{\infty} P_n(t) = 0.$$

Proof To prove this proposition, as in Subsection 9.5.3, we extend the system with $n \in \overline{\mathbb{N}} = \{-K, -K+1, -K+2, \cdots\}$. Fix an arbitrary sequence (t_k) with $t_k \to \infty$ as $k \to \infty$. Then, up to extraction, the sequence of functions $\left(N(t+t_k), (P_n(t+t_k))_{n\in\overline{\mathbb{N}}}\right)$ converges to a complete orbit $\left(N^\infty(t), (P_n^\infty(t))_{n\in\overline{\mathbb{N}}}\right)$ locally uniformly in t with values in $\mathbb{R} \times \mathcal{M}(\overline{\mathbb{N}})$. Now recall that due to Lemma 9.23 one has

$$N^\infty(t) \le \frac{\beta}{\mu^N}, \ \forall t \in \mathbb{R},$$

and $t \mapsto \left(N^\infty(t), (P_n^\infty(t))_{n\in\overline{\mathbb{N}}}\right)$ is solution of

$$\begin{cases} \dfrac{\mathrm{d}}{\mathrm{d}t} N^\infty(t) = \beta - \mu^N N^\infty(t) - N^\infty(t) \displaystyle\sum_{i=-K}^{\infty} m_i P_i^\infty(t) \\ \dfrac{\mathrm{d}}{\mathrm{d}t} P_n^\infty(t) = \mu_n \left(\alpha_n N^\infty(t) - 1\right) P_n^\infty(t), \ \forall n \ge -K \end{cases}$$

Now to complete the proof of the proposition, let us show that

$$\sum_{n\ge -K} P_n^\infty(t) = 0, \ \forall t \in \mathbb{R}.$$

To that aim consider the quantity

$$\ell = \sup_{t\in\mathbb{R}} \sum_{n\geq -K} P_n^\infty(t)$$

and assume by contradiction that $\ell > 0$. Next consider a sequence $(s_p) \subset \mathbb{R}$ such that

$$\ell = \lim_{p\to\infty} \sum_{n\geq -K} P_n^\infty(s_p).$$

As above consider the sequence of functions $(N^\infty(t+s_p), (P_n^\infty(t+s_p))_{n\in\overline{\mathbb{N}}})$ and assume that, possibly along a sub-sequence, it converges to a complete orbit $(\hat{N}(t), (\hat{P}_n(t))_{n\in\overline{\mathbb{N}}})$ locally uniformly in $t \in \mathbb{R}$ with values in $\mathbb{R} \times \mathcal{M}(\overline{\mathbb{N}})$. Hence it becomes a solution for $t \in \mathbb{R}$ of

$$\begin{cases} \dfrac{\mathrm{d}}{\mathrm{d}t}\hat{N}(t) = \beta - \mu^N \hat{N}(t) - \hat{N}(t) \displaystyle\sum_{i=-K}^{\infty} m_i \hat{P}_i(t) \\ \dfrac{\mathrm{d}}{\mathrm{d}t}\hat{P}_n(t) = \mu_n \left(\alpha_n \hat{N}(t) - 1\right) \hat{P}_n(t), \ \forall n \geq -K \end{cases}$$

together with

$$\ell = \sum_{n\geq -K} \hat{P}_n(0) = \sup_{t\in\mathbb{R}} \sum_{n\geq -K} \hat{P}_n(t).$$

Now observe that for all $n \geq -K$ one has

$$\frac{\mathrm{d}}{\mathrm{d}t}\hat{P}_n(t) \leq \mu_n \left(\alpha_n \frac{\beta}{\mu^N} - 1\right) \hat{P}_n(t), \ \forall t \in \mathbb{R}.$$

Hence, since $\mathcal{R}_0 = 1$ we have $\frac{\beta}{\mu^N} = \frac{1}{\alpha^*}$ and $\hat{P}_n(t) \equiv 0$ for all $n \geq -K$ such that $\alpha_n < \alpha^*$. Now set $J = \{n \geq -K : \ \alpha_n = \alpha^*\}$ and the above system of equations reduces to

$$\begin{cases} \dfrac{\mathrm{d}}{\mathrm{d}t}\hat{N}(t) = \beta - \mu^N \hat{N}(t) - \hat{N}(t) \displaystyle\sum_{j\in J} \mu_j \alpha^* \hat{P}_j(t) \\ \dfrac{\mathrm{d}}{\mathrm{d}t}\hat{P}_n(t) = \mu_n \left(\alpha^* \hat{N}(t) - 1\right) \hat{P}_n(t), \ \forall n \in J. \end{cases}$$

Now summing-up the $\hat{P}_n$–components, we get

$$\sum_{n\in J} \frac{\mathrm{d}}{\mathrm{d}t}\hat{P}_n(0) = 0 = (\alpha^* \hat{N}(0) - 1) \sum_{n\in J} \mu_n \hat{P}_n(0),$$

so that, since $\hat{N}(t) \leq \frac{\beta}{\mu^N}$ for all t, we obtain:

$$\hat{N}(0) = \frac{1}{\alpha^*} = \frac{\beta}{\mu^N} \text{ and } \frac{\mathrm{d}}{\mathrm{d}t}\hat{N}(0) = 0.$$

Substituting $t = 0$ into the $\hat{N}$–equation yields

$$0 = \frac{\mathrm{d}}{\mathrm{d}t}\hat{N}(0) = \beta - \mu^N \frac{\beta}{\mu^N} - \frac{\beta}{\mu^N} \sum_{n \in J} \mu_n \alpha^* P_n^\infty(0).$$

This ensures that

$$\alpha^* \left(\inf_{n \geq 0} \mu_n \right) \ell \leq \sum_{n \in J} \mu_n \alpha^* P_n^\infty(0) = 0,$$

a contradiction that proves that $\ell = 0$ and completes the proof of the proposition. □

References

[1] Aguiar, M., Ballesteros, S., Kooi, B.W., Stollenwerk, N.: The role of seasonality and import in a minimalistic multi-strain dengue model capturing differences between primary and secondary infections: complex dynamics and its implications for data analysis. J. Theoret. Biol. **289**, 181–196 (2011). DOI 10.1016/j.jtbi.2011.08.043

[2] Aguiar, M., Kooi, B., Stollenwerk, N.: Epidemiology of dengue fever: a model with temporary cross-immunity and possible secondary infection shows bifurcations and chaotic behaviour in wide parameter regions. Math. Model. Nat. Phenom. **3**(4), 48–70 (2008). DOI 10.1051/mmnp:2008070

[3] Aguiar, M., Kooi, B.W., Rocha, F., Ghaffari, P., Stollenwerk, N.: How much complexity is needed to describe the fluctuations observed in dengue hemorrhagic fever incidence data? Ecological Complexity **16**, 31–40 (2013). DOI 10.1016/j.ecocom.2012.09.001

[4] Ai, S., Du, Y., Peng, R.: Traveling waves for a generalized Holling-Tanner predator-prey model. J. Differential Equations **263**(11), 7782–7814 (2017). DOI 10.1016/j.jde.2017.08.021

[5] Alizon, S., Hurford, T.A., Mideo, N., Van Baalen, M.: Virulence evolution and the trade-off hypothesis: history, current state of affairs and the future. J. Evol. Biol. **22**(2), 245–259 (2009). DOI 10.1111/j.1420-9101.2008.01658.x

[6] Arditi, R., Saiah, H.: Empirical evidence of the role of heterogeneity in ratio-dependent consumption. Ecology **73**(5), 1544–1551 (1992). DOI 10.2307/1940007

[7] Ardito, A., Ricciardi, P.: Lyapunov functions for a generalized Gause-type model. J. Math. Biol. **33**(8), 816–828 (1995). DOI 10.1007/BF00187283

[8] Armbruster, D., Guckenheimer, J., Holmes, P.: Kuramoto-Sivashinsky dynamics on the center-unstable manifold. SIAM J. Appl. Math. **49**(3), 676–691 (1989). DOI 10.1137/0149039. URL `https://doi.org/10.1137/0149039`

[9] Armstrong, R.A., McGehee, R.: Competitive exclusion. Am. Nat. **115**(2), 151–170 (1980). DOI 10.1086/283553. URL `https://doi-org.docelec.u-bordeaux.fr/10.1086/283553`

A. Ducrot et al., *Differential Equations and Population Dynamics II*, Lecture Notes on Mathematical Modelling in the Life Sciences, https://doi.org/10.1007/978-3-031-91409-6

[10] Arnold, L., Jones, C., Mischaikow, K., Raugel, G.: Dynamical systems, *Lecture Notes in Mathematics*, vol. 1609. Springer-Verlag, Berlin (1995). DOI 10.1007/BFb0095237. URL https://doi.org/10.1007/BFb0095237. Lectures given at the Second C.I.M.E. Session held in Montecatini Terme, June 13–22, 1994

[11] Arnold, V.I.: Geometrical methods in the theory of ordinary differential equations, *Grundlehren der mathematischen Wissenschaften [Fundamental Principles of Mathematical Sciences]*, vol. 250, second edn. Springer-Verlag, New York (1988). DOI 10.1007/978-1-4612-1037-5. URL https://doi.org/10.1007/978-1-4612-1037-5. Translated from the Russian by Joseph Szücs [József M. Szűcs]

[12] Aronson, D.G., Weinberger, H.F.: Multidimensional nonlinear diffusion arising in population genetics. Adv. in Math. **30**(1), 33–76 (1978). DOI 10.1016/0001-8708(78)90130-5. URL https://doi.org/10.1016/0001-8708(78)90130-5

[13] Ashwin, P., Mei, Z.: Normal form for Hopf bifurcation of partial differential equations on the square. Nonlinearity **8**(5), 715–734 (1995). URL http://stacks.iop.org/0951-7715/8/715

[14] Avez, A.: Calcul différentiel. Collection Maîtrise de Mathématiques Pures. [Collection of Pure Mathematics for the Master's Degree]. Masson, Paris (1983)

[15] Babin, A.V., Pilyugin, S.Y.: Continuous dependence of attractors on the shape of domain. Zap. Nauchn. Sem. S.-Peterburg. Otdel. Mat. Inst. Steklov. (POMI) **221**, 58–66, 254 (1995). DOI 10.1007/BF02355582. URL https://doi.org/10.1007/BF02355582

[16] Babin, A.V., Vishik, M.I.: Unstable invariant sets of semigroups of nonlinear operators and their perturbations. Uspekhi Mat. Nauk **41**(4(250)), 3–34, 239 (1986)

[17] Ball, J.: Saddle point analysis for an ordinary differential equation in a banach space, and an application to dynamic buckling of a beam. In: Nonlinear elasticity, pp. 93–160. Elsevier (1973)

[18] Barles, G., Mirrahimi, S., Perthame, B.: Concentration in Lotka-Volterra parabolic or integral equations: a general convergence result. Methods Appl. Anal. **16**(3), 321–340 (2009). DOI 10.4310/MAA.2009.v16.n3.a4. URL https://doi-org.docelec.u-bordeaux.fr/10.4310/MAA.2009.v16.n3.a4

[19] Barles, G., Perthame, B.: Concentrations and constrained Hamilton-Jacobi equations arising in adaptive dynamics. In: Recent developments in nonlinear partial differential equations, *Contemp. Math.*, vol. 439, pp. 57–68. Amer. Math. Soc., Providence, RI (2007). DOI 10.1090/conm/439/08463. URL https://doi-org.docelec.u-bordeaux.fr/10.1090/conm/439/08463

[20] Barnsley, M.F.: Fractals everywhere, second edn. Academic Press Professional, Boston, MA (1993). Revised with the assistance of and with a foreword by Hawley Rising, III

[21] Bates, P.W., Jones, C.K.R.T.: Invariant manifolds for semilinear partial differential equations. In: Dynamics reported, Vol. 2, *Dynam. Report. Ser. Dynam. Systems Appl.*, vol. 2, pp. 1–38. Wiley, Chichester (1989)

[22] Bates, P.W., Lu, K., Zeng, C.: Existence and persistence of invariant manifolds for semiflows in Banach space, vol. 645. American Mathematical Soc. (1998)

[23] Bates, P.W., Lu, K., Zeng, C.: Approximately invariant manifolds and global dynamics of spike states. Invent. Math. **174**(2), 355–433 (2008). DOI 10.1007/s00222-008-0141-y. URL `https://doi.org/10.1007/s00222-008-0141-y`

[24] Baurmann, M., Gross, T., Feudel, U.: Instabilities in spatially extended predator-prey systems: spatio-temporal patterns in the neighborhood of Turing-Hopf bifurcations. J. Theoret. Biol. **245**(2), 220–229 (2007). DOI 10.1016/j.jtbi.2006.09.036. URL `https://doi.org/10.1016/j.jtbi.2006.09.036`

[25] Benaïm, M., Hirsch, M.W.: Asymptotic pseudotrajectories and chain recurrent flows, with applications. J. Dynam. Differential Equations **8**(1), 141–176 (1996). DOI 10.1007/BF02218617. URL `https://doi.org/10.1007/BF02218617`

[26] Berkson, J.: Application of the logistic function to bio-assay. Journal of the American statistical association **39**(227), 357–365 (1944)

[27] Berryman, A.A.: The orgins and evolution of predator-prey theory. Ecology **73**(5), 1530–1535 (1992). DOI 10.2307/1940005

[28] Bertoni, S.: Periodic solutions for non-linear equations of structured populations. Journal of Mathematical Analysis and Applications **220**(1), 250–267 (1998). DOI 10.1006/jmaa.1997.5878. URL `https://www.sciencedirect.com/science/article/pii/S0022247X97958789`

[29] Beverton, R.J., Holt, S.J.: On the dynamics of exploited fish populations, vol. 11. Springer Science & Business Media (2012). DOI 10.1007/978-94-011-2106-4

[30] Bibikov, Y.N.: Local theory of nonlinear analytic ordinary differential equations, *Lecture Notes in Mathematics*, vol. 702. Springer-Verlag, Berlin-New York (1979)

[31] Billotti, J.E., LaSalle, J.P.: Dissipative periodic processes. Bull. Amer. Math. Soc. **77**, 1082–1088 (1971). DOI 10.1090/S0002-9904-1971-12879-3. URL `https://doi.org/10.1090/S0002-9904-1971-12879-3`

[32] Bogachev, V.I.: Measure theory. Vol. I, II. Springer-Verlag, Berlin (2007). DOI 10.1007/978-3-540-34514-5. URL `http://dx.doi.org/10.1007/978-3-540-34514-5`

[33] Bouin, E., Calvez, V., Meunier, N., Mirrahimi, S., Perthame, B., Raoul, G., Voituriez, R.: Invasion fronts with variable motility: phenotype selection, spatial sorting and wave acceleration. C. R. Math. Acad. Sci. Paris **350**(15-16), 761–766 (2012). DOI 10.1016/j.crma.2012.09.010. URL `http://dx.doi.org/10.1016/j.crma.2012.09.010`

[34] Bowen, R.: ω-limit sets for axiom A diffeomorphisms. J. Differential Equations **18**(2), 333–339 (1975). DOI 10.1016/0022-0396(75)90065-0

[35] Bramson, M.: Convergence of solutions of the Kolmogorov equation to travelling waves. Mem. Amer. Math. Soc. **44**(285), iv+190 (1983). DOI 10.1090/memo/0285. URL https://doi.org/10.1090/memo/0285
[36] Brauer, F.: Backward bifurcations in simple vaccination models. J. Math. Anal. Appl. **298**(2), 418–431 (2004). DOI 10.1016/j.jmaa.2004.05.045. URL https://doi.org/10.1016/j.jmaa.2004.05.045
[37] Brezis, H.: Functional analysis, Sobolev spaces and partial differential equations. Universitext. Springer, New York (2011)
[38] Brüssow, H.: Covid-19: Omicron – the latest, the least virulent, but probably not the last variant of concern of sars-cov-2. Microb. Biotechnol. **15**(7), 1927–1939 (2022). DOI 10.1111/1751-7915.14064
[39] Burie, J.B., Djidjou-Demasse, R., Ducrot, A.: Asymptotic and transient behaviour for a nonlocal problem arising in population genetics. European J. Appl. Math. **31**(1), 84–110 (2020). DOI 10.1017/s0956792518000487. URL https://doi.org/10.1017/s0956792518000487
[40] Butler, G.J., Wolkowicz, G.S.K.: A mathematical model of the chemostat with a general class of functions describing nutrient uptake. SIAM J. Appl. Math. **45**(1), 138–151 (1985). DOI 10.1137/0145006. URL https://doi-org.docelec.u-bordeaux.fr/10.1137/0145006
[41] Cantrell, R.S., Cosner, C.: Practical persistence in ecological models via comparison methods. Proc. Roy. Soc. Edinburgh Sect. A **126**(2), 247–272 (1996). DOI 10.1017/S0308210500022721. URL https://doi.org/10.1017/S0308210500022721
[42] Cantrell, R.S., Cosner, C.: Spatial ecology via reaction-diffusion equations. Wiley Series in Mathematical and Computational Biology. John Wiley & Sons, Ltd., Chichester (2003). DOI 10.1002/0470871296. URL https://doi.org/10.1002/0470871296
[43] Cantrell, R.S., Cosner, C., Hutson, V.: Permanence in ecological systems with spatial heterogeneity. Proc. Roy. Soc. Edinburgh Sect. A **123**(3), 533–559 (1993). DOI 10.1017/S0308210500025877. URL https://doi.org/10.1017/S0308210500025877
[44] Carr, J.: Applications of centre manifold theory, *Applied Mathematical Sciences*, vol. 35. Springer-Verlag, New York-Berlin (1981)
[45] Carvalho, A.N., Langa, J.A., Robinson, J.C.: Attractors for infinite-dimensional non-autonomous dynamical systems, *Applied Mathematical Sciences*, vol. 182. Springer, New York (2013). DOI 10.1007/978-1-4614-4581-4. URL https://doi.org/10.1007/978-1-4614-4581-4
[46] Cheban, D.N.: Nonautonomous dynamics—nonlinear oscillations and global attractors. Springer Monographs in Mathematics. Springer, Cham ([2020] ©2020). DOI 10.1007/978-3-030-34292-0. URL https://doi.org/10.1007/978-3-030-34292-0
[47] Cheng, K.S.: Uniqueness of a limit cycle for a predator-prey system. SIAM J. Math. Anal. **12**(4), 541–548 (1981). DOI 10.1137/0512047. URL https://doi.org/10.1137/0512047

[48] Cheng, K.S., Hsu, S.B., Lin, S.S.: Some results on global stability of a predator-prey system. Journal of Mathematical Biology **12**(1), 115–126 (1982). DOI 10.1007/BF00275207

[49] Chicone, C., Latushkin, Y.: Center manifolds for infinite-dimensional nonautonomous differential equations. J. Differential Equations **141**(2), 356–399 (1997). DOI 10.1006/jdeq.1997.3343. URL `https://doi.org/10.1006/jdeq.1997.3343`

[50] Chicone, C., Latushkin, Y.: Evolution semigroups in dynamical systems and differential equations, *Mathematical Surveys and Monographs*, vol. 70. American Mathematical Society, Providence, RI (1999). DOI 10.1090/surv/070. URL `https://doi.org/10.1090/surv/070`

[51] Chitty, H.: The snowshoe rabbit enquiry, 1946-48. The Journal of Animal Ecology pp. 15–20 (1950). DOI 10.2307/1567

[52] Cholewa, J.W., Dlotko, T.: Global attractors in abstract parabolic problems, *London Mathematical Society Lecture Note Series*, vol. 278. Cambridge University Press, Cambridge (2000). DOI 10.1017/CBO9780511526404. URL `https://doi.org/10.1017/CBO9780511526404`

[53] Chow, S.N., Hale, J.K.: Methods of bifurcation theory, *Grundlehren der Mathematischen Wissenschaften*, vol. 251. Springer-Verlag, New York-Berlin (1982)

[54] Chow, S.N., Li, C.Z., Wang, D.: Normal forms and bifurcation of planar vector fields. Cambridge University Press, Cambridge (1994). DOI 10.1017/CBO9780511665639. URL `https://doi.org/10.1017/CBO9780511665639`

[55] Chow, S.N., Lin, X.B., Lu, K.: Smooth invariant foliations in infinite-dimensional spaces. J. Differential Equations **94**(2), 266–291 (1991). DOI 10.1016/0022-0396(91)90093-O. URL `https://doi.org/10.1016/0022-0396(91)90093-O`

[56] Chow, S.N., Liu, W., Yi, Y.: Center manifolds for invariant sets. J. Differential Equations **168**(2), 355–385 (2000). DOI 10.1006/jdeq.2000.3890. URL `https://doi.org/10.1006/jdeq.2000.3890`. Special issue in celebration of Jack K. Hale's 70th birthday, Part 2 (Atlanta, GA/Lisbon, 1998)

[57] Chow, S.N., Lu, K.: C^k centre unstable manifolds. Proc. Roy. Soc. Edinburgh Sect. A **108**(3-4), 303–320 (1988). DOI 10.1017/S0308210500014682. URL `https://doi.org/10.1017/S0308210500014682`

[58] Chow, S.N., Lu, K.: Invariant manifolds for flows in Banach spaces. J. Differential Equations **74**(2), 285–317 (1988). DOI 10.1016/0022-0396(88)90007-1. URL `https://doi.org/10.1016/0022-0396(88)90007-1`

[59] Chow, S.N., Lu, K., Shen, Y.Q.: Normal forms for quasiperiodic evolutionary equations. Discrete Contin. Dynam. Systems **2**(1), 65–94 (1996). DOI 10.3934/dcds.1996.2.65. URL `https://doi.org/10.3934/dcds.1996.2.65`

[60] Chu, J., Liu, Z., Magal, P., Ruan, S.: Normal forms for an age structured model. J. Dynam. Differential Equations **28**(3-4), 733–761 (2016). DOI 10.1007/s10884-015-9500-8. URL `https://doi.org/10.1007/s10884-015-9500-8`

[61] Chueshov, I.D.: Introduction to the theory of infinite-dimensional dissipative systems. University Lectures in Contemporary Mathematics. AKTA, Kharkiv (1999)

[62] Cinquin, O., Demongeot, J.: Positive and negative feedback: striking a balance between necessary antagonists. J. Theoret. Biol. **216**(2), 229–241 (2002). DOI 10.1006/jtbi.2002.2544. URL https://doi.org/10.1006/jtbi.2002.2544

[63] Conley, C.: Isolated invariant sets and the Morse index, *CBMS Regional Conference Series in Mathematics*, vol. 38. American Mathematical Society, Providence, RI (1978)

[64] Corless, R.M., Gonnet, G.H., Hare, D.E.G., Jeffrey, D.J., Knuth, D.E.: On the Lambert W function. Adv. Comput. Math. **5**(4), 329–359 (1996). DOI 10.1007/BF02124750. URL https://doi.org/10.1007/BF02124750

[65] Cosner, C.: Persistence (permanence), compressivity, and practical persistence in some reaction-diffusion models from ecology. In: Comparison methods and stability theory (Waterloo, ON, 1993), *Lecture Notes in Pure and Appl. Math.*, vol. 162, pp. 101–115. Dekker, New York (1994)

[66] Cosner, C., DeAngelis, D.L., Ault, J.S., Olson, D.B.: Effects of spatial grouping on the functional response of predators. Theoretical population biology **56**(1), 65–75 (1999). DOI 10.1006/tpbi.1999.1414

[67] Crandall, M.G., Rabinowitz, P.H.: Bifurcation from simple eigenvalues. J. Functional Analysis **8**, 321–340 (1971). DOI 10.1016/0022-1236(71)90015-2. URL https://doi.org/10.1016/0022-1236(71)90015-2

[68] Crandall, M.G., Rabinowitz, P.H.: Bifurcation, perturbation of simple eigenvalues and linearized stability. Arch. Rational Mech. Anal. **52**, 161–180 (1973). DOI 10.1007/BF00282325. URL https://doi.org/10.1007/BF00282325

[69] Crandall, M.G., Rabinowitz, P.H.: The Hopf bifurcation theorem in infinite dimensions. Arch. Rational Mech. Anal. **67**(1), 53–72 (1977). DOI 10.1007/BF00280827. URL https://doi.org/10.1007/BF00280827

[70] Cushing, J.M.: Two species competition in a periodic environment. J. Math. Biol. **10**(4), 385–400 (1980). DOI 10.1007/BF00276097

[71] Cushing, J.M.: Bifurcation of time periodic solutions of the McKendrick equations with applications to population dynamics. Comput. Math. Appl. **9**(3), 459–478 (1983). DOI 10.1016/0898-1221(83)90060-3. URL https://doi.org/10.1016/0898-1221(83)90060-3. Hyperbolic partial differential equations

[72] Da Prato, G., Lunardi, A.: Stability, instability and center manifold theorem for fully nonlinear autonomous parabolic equations in Banach space. Arch. Rational Mech. Anal. **101**(2), 115–141 (1988). DOI 10.1007/BF00251457. URL https://doi.org/10.1007/BF00251457

[73] Day, T., Gandon, S.: Applying population-genetic models in theoretical evolutionary epidemiology. Ecol. Lett. **10**(10), 876–888 (2007). DOI 10.1111/j.1461-0248.2007.01091.x. URL pub/DayGandon2007.pdf

[74] De Wit, A., Lima, D., Dewel, G., Borckmans, P.: Spatiotemporal dynamics near a codimension-two point. Physical Review E **54**(1), 261 (1996). DOI 10.1103/PhysRevE.54.261

[75] Deimling, K.: Nonlinear functional analysis. Springer-Verlag, Berlin (1985). DOI 10.1007/978-3-662-00547-7. URL https://doi.org/10.1007/978-3-662-00547-7

[76] Dell'Antonio, G.F., D'Onofrio, B.: Construction of a center-unstable manifold for C^1-flows and an application to the Navier-Stokes equation. Arch. Rational Mech. Anal. **93**(2), 185–201 (1986). DOI 10.1007/BF00279959. URL https://doi.org/10.1007/BF00279959

[77] Demazure, M.: Bifurcations and catastrophes. Universitext. Springer-Verlag, Berlin (2000). DOI 10.1007/978-3-642-57134-3. URL https://doi.org/10.1007/978-3-642-57134-3. Geometry of solutions to nonlinear problems, Translated from the 1989 French original by David Chillingworth

[78] Demongeot, J., Glade, N., Forest, L.: Liénard systems and potential-Hamiltonian decomposition. I. Methodology. C. R. Math. Acad. Sci. Paris **344**(2), 121–126 (2007). DOI 10.1016/j.crma.2006.10.016. URL https://doi.org/10.1016/j.crma.2006.10.016

[79] Desvillettes, L., Jabin, P.E., Mischler, S., Raoul, G.: On selection dynamics for continuous structured populations. Commun. Math. Sci. **6**(3), 729–747 (2008)

[80] Deuflhard, P., Röblitz, S.: A guide to numerical modelling in systems biology, *Texts in Computational Science and Engineering*, vol. 12. Springer, Cham (2015). DOI 10.1007/978-3-319-20059-0. URL https://doi.org/10.1007/978-3-319-20059-0

[81] Diekmann, O., van Gils, S.A.: Invariant manifolds for Volterra integral equations of convolution type. J. Differential Equations **54**(2), 139–180 (1984). DOI 10.1016/0022-0396(84)90156-6. URL https://doi.org/10.1016/0022-0396(84)90156-6

[82] Diekmann, O., van Gils, S.A.: The center manifold for delay equations in the light of suns and stars. In: Singularity theory and its applications, Part II (Coventry, 1988/1989), *Lecture Notes in Math.*, vol. 1463, pp. 122–141. Springer, Berlin (1991). DOI 10.1007/BFb0085429. URL https://doi.org/10.1007/BFb0085429

[83] Diekmann, O., van Gils, S.A., Verduyn Lunel, S.M., Walther, H.O.: Delay equations, *Applied Mathematical Sciences*, vol. 110. Springer-Verlag, New York (1995). DOI 10.1007/978-1-4612-4206-2. URL https://doi.org/10.1007/978-1-4612-4206-2. Functional, complex, and nonlinear analysis

[84] Diekmann, O., Heesterbeek, J.A.P.: Mathematical epidemiology of infectious diseases. Wiley Series in Mathematical and Computational Biology. John Wiley & Sons, Ltd., Chichester (2000). Model building, analysis and interpretation

[85] Ding, W., Huang, W.: Traveling wave solutions for some classes of diffusive predator-prey models. J. Dynam. Differential Equations **28**(3-4), 1293–1308

(2016). DOI 10.1007/s10884-015-9472-8. URL https://doi.org/10.1007/s10884-015-9472-8

[86] van den Driessche, P., Watmough, J.: A simple SIS epidemic model with a backward bifurcation. J. Math. Biol. **40**(6), 525–540 (2000). DOI 10.1007/s002850000032. URL https://doi.org/10.1007/s002850000032

[87] Ducrot, A., Fu, X., Magal, P.: Turing and Turing-Hopf bifurcations for a reaction diffusion equation with nonlocal advection. J. Nonlinear Sci. **28**(5), 1959–1997 (2018). DOI 10.1007/s00332-018-9472-z. URL https://doi.org/10.1007/s00332-018-9472-z

[88] Ducrot, A., Giletti, T., Matano, H.: Spreading speeds for multidimensional reaction-diffusion systems of the prey-predator type. Calc. Var. Partial Differential Equations **58**(4), Paper No. 137, 34 (2019). DOI 10.1007/s00526-019-1576-2. URL https://doi.org/10.1007/s00526-019-1576-2

[89] Ducrot, A., Griette, Q., Liu, Z., Magal, P.: Differential equations and population dynamics I. Introductory approaches. Lecture Notes on Mathematical Modelling in the Life Sciences. Springer, Cham ([2022] ©2022). DOI 10.1007/978-3-030-98136-5. URL https://doi.org/10.1007/978-3-030-98136-5. With forewords by Jacques Demongeot and Glenn Webb

[90] Ducrot, A., Kang, H., Magal, P.: Hopf bifurcation theorem for second-order semi-linear Gurtin-MacCamy equation. J. Evol. Equ. **22**(3), Paper No. 72, 40 (2022). DOI 10.1007/s00028-022-00833-3. URL https://doi.org/10.1007/s00028-022-00833-3

[91] Ducrot, A., Kang, H., Magal, P.: A short proof for Hopf bifurcation in Gurtin-MacCamy's population dynamics model. Proc. Amer. Math. Soc. **151**(8), 3561–3575 (2023). DOI 10.1090/proc/15892. URL https://doi.org/10.1090/proc/15892

[92] Ducrot, A., Langlais, M., Magal, P.: Multiple travelling waves for an *SI*-epidemic model. Netw. Heterog. Media **8**(1), 171–190 (2013). DOI 10.3934/nhm.2013.8.171. URL https://doi.org/10.3934/nhm.2013.8.171

[93] Ducrot, A., Madec, S.: Singularly perturbed elliptic system modeling the competitive interactions for a single resource. Math. Models Methods Appl. Sci. **23**(11), 1939–1977 (2013). DOI 10.1142/S021820251350022X. URL https://doi.org/10.1142/S021820251350022X

[94] Ducrot, A., Magal, P.: A center manifold for second order semilinear differential equations on the real line and applications to the existence of wave trains for the Gurtin-McCamy equation. Trans. Amer. Math. Soc. **372**(5), 3487–3537 (2019). DOI 10.1090/tran/7780. URL https://doi.org/10.1090/tran/7780

[95] Dunbar, S.R.: Travelling wave solutions of diffusive Lotka-Volterra equations. J. Math. Biol. **17**(1), 11–32 (1983). DOI 10.1007/BF00276112. URL https://doi.org/10.1007/BF00276112

[96] Dunbar, S.R.: Traveling wave solutions of diffusive Lotka-Volterra equations: a heteroclinic connection in $\mathbf{R}^4$. Trans. Amer. Math. Soc. **286**(2), 557–594 (1984). DOI 10.2307/1999810. URL https://doi.org/10.2307/1999810

[97] Dunbar, S.R.: Traveling waves in diffusive predator-prey equations: periodic orbits and point-to-periodic heteroclinic orbits. SIAM J. Appl. Math. **46**(6), 1057–1078 (1986). DOI 10.1137/0146063. URL https://doi.org/10.1137/0146063

[98] Dushoff, J., Huang, W., Castillo-Chavez, C.: Backwards bifurcations and catastrophe in simple models of fatal diseases. J. Math. Biol. **36**(3), 227–248 (1998). DOI 10.1007/s002850050099. URL https://doi.org/10.1007/s002850050099

[99] Eckmann, J.P., Epstein, H., Wayne, C.E.: Normal forms for parabolic partial differential equations. Ann. Inst. H. Poincaré Phys. Théor. **58**(3), 287–308 (1993). URL http://www.numdam.org/item?id=AIHPA_1993__58_3_287_0

[100] Eckmann, J.P., Ruelle, D.: Ergodic theory of chaos and strange attractors. Rev. Modern Phys. **57**(3), 617–656 (1985). DOI 10.1103/RevModPhys.57.617. URL https://doi.org/10.1103/RevModPhys.57.617

[101] Elton, C., Nicholson, M.: The ten-year cycle in numbers of the lynx in canada. The Journal of Animal Ecology pp. 215–244 (1942). DOI 10.2307/1358

[102] Ewald, P.W.: Host-parasite relations, vectors, and the evolution of disease severity. Annual Review of Ecology and Systematics **14**(1), 465–485 (1983). DOI 10.1146/annurev.es.14.110183.002341

[103] Faou, E., Grébert, B., Paturel, E.: Birkhoff normal form for splitting methods applied to semilinear Hamiltonian PDEs. I. Finite-dimensional discretization. Numer. Math. **114**(3), 429–458 (2010). DOI 10.1007/s00211-009-0258-y. URL https://doi.org/10.1007/s00211-009-0258-y

[104] Faou, E., Grébert, B., Paturel, E.: Birkhoff normal form for splitting methods applied to semilinear Hamiltonian PDEs. II. Abstract splitting. Numer. Math. **114**(3), 459–490 (2010). DOI 10.1007/s00211-009-0257-z. URL https://doi.org/10.1007/s00211-009-0257-z

[105] Faria, T.: Normal forms and Hopf bifurcation for partial differential equations with delays. Trans. Amer. Math. Soc. **352**(5), 2217–2238 (2000). DOI 10.1090/S0002-9947-00-02280-7. URL https://doi.org/10.1090/S0002-9947-00-02280-7

[106] Faria, T.: Normal forms for semilinear functional differential equations in Banach spaces and applications. II. Discrete Contin. Dynam. Systems **7**(1), 155–176 (2001). DOI 10.3934/dcds.2001.7.155. URL https://doi.org/10.3934/dcds.2001.7.155

[107] Faria, T., Huang, W., Wu, J.: Smoothness of center manifolds for maps and formal adjoints for semilinear FDEs in general Banach spaces. SIAM J. Math. Anal. **34**(1), 173–203 (2002). DOI 10.1137/S0036141001384971. URL https://doi.org/10.1137/S0036141001384971

[108] Faria, T., Magalhães, L.T.: Normal forms for retarded functional-differential equations and applications to Bogdanov-Takens singularity. J. Differential Equations **122**(2), 201–224 (1995). DOI 10.1006/jdeq.1995.1145. URL https://doi.org/10.1006/jdeq.1995.1145

[109] Faria, T., Magalhães, L.T.: Normal forms for retarded functional-differential equations with parameters and applications to Hopf bifurcation. J. Differential

Equations **122**(2), 181–200 (1995). DOI 10.1006/jdeq.1995.1144. URL https://doi.org/10.1006/jdeq.1995.1144

[110] Feng, X., Ruan, S., Teng, Z., Wang, K.: Stability and backward bifurcation in a malaria transmission model with applications to the control of malaria in China. Math. Biosci. **266**, 52–64 (2015). DOI 10.1016/j.mbs.2015.05.005. URL https://doi.org/10.1016/j.mbs.2015.05.005

[111] Fenichel, N.: Geometric singular perturbation theory for ordinary differential equations. J. Differential Equations **31**(1), 53–98 (1979). DOI 10.1016/0022-0396(79)90152-9. URL https://doi.org/10.1016/0022-0396(79)90152-9

[112] Fisher, M.E., Goh, B.S.: Stability results for delayed-recruitment models in population dynamics. J. Math. Biol. **19**(1), 147–156 (1984). DOI 10.1007/BF00275937. URL https://doi.org/10.1007/BF00275937

[113] Fisher, R.A.: The wave of advance of advantageous genes. Annals of Eugenics **7**(4), 355–369 (1937). DOI 10.1111/j.1469-1809.1937.tb02153.x. URL http://dx.doi.org/10.1111/j.1469-1809.1937.tb02153.x

[114] Foias, C., Hoang, L., Olson, E., Ziane, M.: On the solutions to the normal form of the Navier-Stokes equations. Indiana Univ. Math. J. **55**(2), 631–686 (2006). DOI 10.1512/iumj.2006.55.2830. URL https://doi.org/10.1512/iumj.2006.55.2830

[115] Freedman, H.I., So, J.W.H.: Persistence in discrete semidynamical systems. SIAM J. Math. Anal. **20**(4), 930–938 (1989). DOI 10.1137/0520062. URL https://doi.org/10.1137/0520062

[116] Gallay, T.: A center-stable manifold theorem for differential equations in Banach spaces. Comm. Math. Phys. **152**(2), 249–268 (1993). URL http://projecteuclid.org/euclid.cmp/1104252409

[117] Gardner, R., Jones, C.K.R.T.: Stability of travelling wave solutions of diffusive predator-prey systems. Trans. Amer. Math. Soc. **327**(2), 465–524 (1991). DOI 10.2307/2001812. URL https://doi.org/10.2307/2001812

[118] Gardner, R.A.: Existence of travelling wave solutions of predator-prey systems via the connection index. SIAM J. Appl. Math. **44**(1), 56–79 (1984). DOI 10.1137/0144006. URL https://doi.org/10.1137/0144006

[119] Gause, G.: The struggle for existence. Williams and Wilkins, Baltimore (1934)

[120] Gilpin, M.E.: Do hares eat lynx? The American Naturalist **107**(957), 727–730 (1973). DOI 10.1086/282870

[121] Ginzburg, L.R., Akcakaya, H.R.: Consequences of ratio-dependent predation for steady-state properties of ecosystems. Ecology **73**(5), 1536–1543 (1992). DOI 10.2307/1940006

[122] Gobbino, M., Sardella, M.: On the connectedness of attractors for dynamical systems. J. Differential Equations **133**(1), 1–14 (1997). DOI 10.1006/jdeq.1996.3166. URL https://doi.org/10.1006/jdeq.1996.3166

[123] Goh, B.S.: Global stability in two species interactions. J. Math. Biol. **3**(3-4), 313–318 (1976). DOI 10.1007/BF00275063. URL https://doi.org/10.1007/BF00275063

[124] Goh, B.S.: Global stability in many-species systems. The American Naturalist **111**(977), 135–143 (1977). DOI 10.1086/283144

[125] Goh, B.S.: Global stability in a class of prey-predator models. Bull. Math. Biol. **40**(4), 525–533 (1978). DOI 10.1016/S0092-8240(78)80053-6. URL https://doi.org/10.1016/S0092-8240(78)80053-6

[126] Golubitsky, M., Stewart, I., Schaeffer, D.G.: Singularities and groups in bifurcation theory. Vol. II, *Applied Mathematical Sciences*, vol. 69. Springer-Verlag, New York (1988). DOI 10.1007/978-1-4612-4574-2. URL https://doi.org/10.1007/978-1-4612-4574-2

[127] Griette, Q.: Singular measure traveling waves in an epidemiological model with continuous phenotypes. Trans. Amer. Math. Soc. **371**(6), 4411–4458 (2019). DOI 10.1090/tran/7700

[128] Guckenheimer, J., Holmes, P.: Nonlinear oscillations, dynamical systems, and bifurcations of vector fields, *Applied Mathematical Sciences*, vol. 42. Springer-Verlag, New York (1983). DOI 10.1007/978-1-4612-1140-2. URL https://doi.org/10.1007/978-1-4612-1140-2

[129] Guckenheimer, J., Oster, G., Ipaktchi, A.: The dynamics of density dependent population models. J. Math. Biol. **4**(2), 101–147 (1977). DOI 10.1007/BF00275980. URL https://doi.org/10.1007/BF00275980

[130] Guo, S., Wu, J.: Bifurcation theory of functional differential equations, *Applied Mathematical Sciences*, vol. 184. Springer, New York (2013). DOI 10.1007/978-1-4614-6992-6. URL https://doi.org/10.1007/978-1-4614-6992-6

[131] Gutierrez, A.: Physiological basis of ratio-dependent predator-prey theory: the metabolic pool model as a paradigm. Ecology **73**(5), 1552–1563 (1992). DOI 10.2307/1940008

[132] Hadamard, J.: Sur l'itération et les solutions asymptotiques des équations différentielles. Bull. Soc. Math. France **29**, 224–228 (1901)

[133] Hadeler, K.P., van den Driessche, P.: Backward bifurcation in epidemic control. Math. Biosci. **146**(1), 15–35 (1997). DOI 10.1016/S0025-5564(97)00027-8. URL https://doi.org/10.1016/S0025-5564(97)00027-8

[134] Hale, J.K.: Integral manifolds of perturbed differential systems. Ann. of Math. (2) **73**, 496–531 (1961). DOI 10.2307/1970314. URL https://doi.org/10.2307/1970314

[135] Hale, J.K.: Ordinary differential equations, *Pure and Applied Mathematics*, vol. Vol. XXI. Wiley-Interscience [John Wiley & Sons], New York-London-Sydney (1969)

[136] Hale, J.K.: Ordinary differential equations, second edn. Robert E. Krieger Publishing Co., Inc., Huntington, NY (1980)

[137] Hale, J.K.: Flows on centre manifolds for scalar functional-differential equations. Proc. Roy. Soc. Edinburgh Sect. A **101**(3-4), 193–201 (1985). DOI 10.1017/S030821050002076X. URL https://doi.org/10.1017/S030821050002076X

[138] Hale, J.K.: Asymptotic behavior of dissipative systems, *Mathematical Surveys and Monographs*, vol. 25. American Mathematical Society, Providence, RI

(1988). DOI 10.1090/surv/025. URL https://doi.org/10.1090/surv/025

[139] Hale, J.K.: Dissipation and attractors. In: International Conference on Differential Equations, Vol. 1, 2 (Berlin, 1999), pp. 622–637. World Sci. Publ., River Edge, NJ (2000)

[140] Hale, J.K.: Functional differential equations. In: Analytic Theory of Differential Equations: The Proceedings of the Conference at Western Michigan University, Kalamazoo, from 30 April to 2 May 1970, pp. 9–22. Springer (2006)

[141] Hale, J.K., Koçak, H.: Dynamics and bifurcations, *Texts in Applied Mathematics*, vol. 3. Springer-Verlag, New York (1991). DOI 10.1007/978-1-4612-4426-4. URL https://doi.org/10.1007/978-1-4612-4426-4

[142] Hale, J.K., Lopes, O.: Fixed point theorems and dissipative processes. J. Differential Equations **13**, 391–402 (1973). DOI 10.1016/0022-0396(73)90025-9. URL https://doi.org/10.1016/0022-0396(73)90025-9

[143] Hale, J.K., Raugel, G.: Lower semicontinuity of the attractor for a singularly perturbed hyperbolic equation. J. Dynam. Differential Equations **2**(1), 19–67 (1990). DOI 10.1007/BF01047769. URL https://doi.org/10.1007/BF01047769

[144] Hale, J.K., Verduyn Lunel, S.M.: Introduction to functional-differential equations, *Applied Mathematical Sciences*, vol. 99. Springer-Verlag, New York (1993). DOI 10.1007/978-1-4612-4342-7. URL https://doi.org/10.1007/978-1-4612-4342-7

[145] Hale, J.K., Waltman, P.: Persistence in infinite-dimensional systems. SIAM J. Math. Anal. **20**(2), 388–395 (1989). DOI 10.1137/0520025. URL https://doi.org/10.1137/0520025

[146] Hamel, F., Roques, L.: Fast propagation for KPP equations with slowly decaying initial conditions. J. Differential Equations **249**(7), 1726–1745 (2010). DOI 10.1016/j.jde.2010.06.025. URL https://doi.org/10.1016/j.jde.2010.06.025

[147] Han, M., Yu, P.: Normal forms, Melnikov functions and bifurcations of limit cycles, *Applied Mathematical Sciences*, vol. 181. Springer, London (2012). DOI 10.1007/978-1-4471-2918-9. URL https://doi.org/10.1007/978-1-4471-2918-9

[148] Haragus, M., Iooss, G.: Local bifurcations, center manifolds, and normal forms in infinite-dimensional dynamical systems. Universitext. Springer-Verlag London, Ltd., London; EDP Sciences, Les Ulis (2011). DOI 10.1007/978-0-85729-112-7. URL https://doi.org/10.1007/978-0-85729-112-7

[149] Hardin, G.: The competitive exclusion principle: An idea that took a century to be born has implications in ecology, economics, and genetics. science **131**(3409), 1292–1297 (1960). DOI 10.1126/science.131.3409.1292

[150] Harrison, G.W.: Global stability of predator-prey interactions. J. Math. Biol. **8**(2), 159–171 (1979). DOI 10.1007/BF00279719. URL https://doi.org/10.1007/BF00279719

[151] Hassard, B.D., Kazarinoff, N.D., Wan, Y.H.: Theory and applications of Hopf bifurcation, *London Mathematical Society Lecture Note Series*, vol. 41. Cambridge University Press, Cambridge-New York (1981)
[152] Hastings, A.: Population biology: concepts and models. Springer Science & Business Media (1997). DOI 10.1007/978-1-4757-2731-9
[153] Henry, D.: Geometric theory of semilinear parabolic equations, *Lecture Notes in Mathematics*, vol. 840. Springer-Verlag, Berlin-New York (1981)
[154] Hill, A.V.: The possible effects of the aggregation of the molecules of haemoglobin on its dissociation curves. J. Physiol. **40**, 4–7 (1910)
[155] Hirsch, M.W., Pugh, C.C., Shub, M.: Invariant manifolds. Bull. Amer. Math. Soc. **76**, 1015–1019 (1970). DOI 10.1090/S0002-9904-1970-12537-X. URL https://doi.org/10.1090/S0002-9904-1970-12537-X
[156] Hirsch, M.W., Smith, H.L., Zhao, X.Q.: Chain transitivity, attractivity, and strong repellors for semidynamical systems. J. Dynam. Differential Equations **13**(1), 107–131 (2001). DOI 10.1023/A:1009044515567. URL https://doi.org/10.1023/A:1009044515567
[157] Hofbauer, J.: An index theorem for dissipative semiflows. Rocky Mountain J. Math. **20**(4), 1017–1031 (1990). DOI 10.1216/rmjm/1181073059. URL https://doi.org/10.1216/rmjm/1181073059. Geoffrey J. Butler Memorial Conference in Differential Equations and Mathematical Biology (Edmonton, AB, 1988)
[158] Hofbauer, J., Sigmund, K.: Evolutionary games and population dynamics. Cambridge University Press, Cambridge (1998). DOI 10.1017/CBO9781139173179. URL https://doi.org/10.1017/CBO9781139173179
[159] Hofbauer, J., So, J.W.H.: Uniform persistence and repellors for maps. Proc. Amer. Math. Soc. **107**(4), 1137–1142 (1989). DOI 10.2307/2047679. URL https://doi.org/10.2307/2047679
[160] Holling, C.S.: The components of predation as revealed by a study of small-mammal predation of the european pine sawfly. The canadian entomologist **91**(5), 293–320 (1959). DOI 10.4039/Ent91293-5
[161] Holling, C.S.: Some characteristics of simple types of predation and parasitism. The canadian entomologist **91**(7), 385–398 (1959). DOI 10.4039/Ent91385-7
[162] Hopf, E.: Abzweigung einer periodischen Lösung von einer stationären eines Differential systems. Ber. Verh. Sächs. Akad. Wiss. Leipzig Math.-Nat. Kl. **95**(1), 3–22 (1943)
[163] Hopf, E.: Selected Works of Eberhard Hopf with Commentaries: With Commentaries, vol. 17. American Mathematical Soc. (2002)
[164] Hosono, Y.: Traveling waves for the Lotka-Volterra predator-prey system without diffusion of the predator. Discrete Contin. Dyn. Syst. Ser. B **20**(1), 161–171 (2015). DOI 10.3934/dcdsb.2015.20.161. URL https://doi.org/10.3934/dcdsb.2015.20.161
[165] Hsu, C.H., Yang, C.R., Yang, T.H., Yang, T.S.: Existence of traveling wave solutions for diffusive predator-prey type systems. J. Differential Equations

252(4), 3040–3075 (2012). DOI 10.1016/j.jde.2011.11.008. URL https://doi.org/10.1016/j.jde.2011.11.008
[166] Hsu, S.B.: Limiting behavior for competing species. SIAM J. Appl. Math. **34**(4), 760–763 (1978). DOI 10.1137/0134064
[167] Hsu, S.B.: On global stability of a predator-prey system. Math. Biosci. **39**(1-2), 1–10 (1978). DOI 10.1016/0025-5564(78)90025-1. URL https://doi.org/10.1016/0025-5564(78)90025-1
[168] Hsu, S.B., Hubbell, S.P., Waltman, P.: Competing predators. SIAM J. Appl. Math. **35**(4), 617–625 (1978). DOI 10.1137/0135051. URL https://doi.org/10.1137/0135051
[169] Hsu, S.B., Hubbell, S.P., Waltman, P.: Competing predators. SIAM J. Appl. Math. **35**(4), 617–625 (1978). DOI 10.1137/0135051. URL https://doi.org/10.1137/0135051
[170] Hsu, S.B., Hubbell, S.P., Waltman, P.: A contribution to the theory of competing predators. Ecological Monographs **48**(3), 337–349 (1978). DOI 10.2307/2937235. URL https://esajournals.onlinelibrary.wiley.com/doi/abs/10.2307/2937235
[171] Hsu, S.B., Liu, Z., Magal, P.: A Holling predator-prey model with handling and searching predators. SIAM J. Appl. Math. **80**(4), 1778–1795 (2020). DOI 10.1137/20M1316007. URL https://doi.org/10.1137/20M1316007
[172] Hsu, S.B., Smith, H.L., Waltman, P.: Competitive exclusion and coexistence for competitive systems on ordered Banach spaces. Trans. Amer. Math. Soc. **348**(10), 4083–4094 (1996). DOI 10.1090/S0002-9947-96-01724-2. URL https://doi.org/10.1090/S0002-9947-96-01724-2
[173] Huang, J., Lu, G., Ruan, S.: Existence of traveling wave solutions in a diffusive predator-prey model. J. Math. Biol. **46**(2), 132–152 (2003). DOI 10.1007/s00285-002-0171-9. URL https://doi.org/10.1007/s00285-002-0171-9
[174] Huang, W.: Traveling wave solutions for a class of predator-prey systems. J. Dynam. Differential Equations **24**(3), 633–644 (2012). DOI 10.1007/s10884-012-9255-4. URL https://doi.org/10.1007/s10884-012-9255-4
[175] Huang, W.: A geometric approach in the study of traveling waves for some classes of non-monotone reaction-diffusion systems. J. Differential Equations **260**(3), 2190–2224 (2016). DOI 10.1016/j.jde.2015.09.060. URL https://doi.org/10.1016/j.jde.2015.09.060
[176] Hupkes, H.J., Verduyn Lunel, S.M.: Center manifold theory for functional differential equations of mixed type. J. Dynam. Differential Equations **19**(2), 497–560 (2007). DOI 10.1007/s10884-006-9055-9. URL https://doi.org/10.1007/s10884-006-9055-9
[177] Hutson, V.: The stability under perturbations of repulsive sets. J. Differential Equations **76**(1), 77–90 (1988). DOI 10.1016/0022-0396(88)90064-2. URL https://doi.org/10.1016/0022-0396(88)90064-2
[178] Hutson, V.: The existence of an equilibrium for permanent systems. Rocky Mountain J. Math. **20**(4), 1033–1040 (1990). DOI 10.1216/rmjm/1181073060. URL https://doi.org/10.1216/rmjm/1181073060. Ge-

offrey J. Butler Memorial Conference in Differential Equations and Mathematical Biology (Edmonton, AB, 1988)

[179] Hutson, V., Mischaikow, K.: An approach to practical persistence. J. Math. Biol. **37**(5), 447–466 (1998). DOI 10.1007/s002850050137. URL `https://doi.org/10.1007/s002850050137`

[180] Hutson, V., Moran, W.: Persistence of species obeying difference equations. J. Math. Biol. **15**(2), 203–213 (1982). DOI 10.1007/BF00275073. URL `https://doi.org/10.1007/BF00275073`

[181] Hutson, V., Schmitt, K.: Permanence and the dynamics of biological systems. Mathematical biosciences **111**(1), 1–71 (1992). DOI 10.1016/0025-5564(92)90078-B

[182] Iannelli, M.: Mathematical theory of age-structured population dynamics. Giardini editori e stampatori in Pisa (1995)

[183] Iooss, G.: Existence et stabilité de la solution périodique secondaire intervenant dans les problèmes d'evolution du type Navier-Stokes. Arch. Rational Mech. Anal. **47**, 301–329 (1972). DOI 10.1007/BF00281637. URL `https://doi.org/10.1007/BF00281637`

[184] Iooss, G.: Bifurcation of maps and applications, *North-Holland Mathematics Studies*, vol. 36. North-Holland Publishing Co., Amsterdam-New York (1979)

[185] Iooss, G., Joseph, D.D.: Elementary stability and bifurcation theory, second edn. Undergraduate Texts in Mathematics. Springer-Verlag, New York (1990). DOI 10.1007/978-1-4612-0997-3. URL `https://doi.org/10.1007/978-1-4612-0997-3`

[186] Iudovich, V.I.: The onset of auto-oscillations in a fluid. J. Appl. Math. Mech. **35**, 587–603 (1971). DOI 10.1016/0021-8928(71)90053-0. URL `https://doi.org/10.1016/0021-8928(71)90053-0`

[187] Jabin, P.E., Raoul, G.: On selection dynamics for competitive interactions. J. Math. Biol. **63**(3), 493–517 (2011). DOI 10.1007/s00285-010-0370-8

[188] Jackson, P., Jackson, A.F., Dallet, R., Devitre, D., De Crem, J.: Les félins: [toutes les espèces du monde]. Delachaux et Niestlé (1996)

[189] Joseph, D.D., Sattinger, D.H.: Bifurcating time periodic solutions and their stability. Arch. Rational Mech. Anal. **45**, 79–109 (1972). DOI 10.1007/BF00253039. URL `https://doi.org/10.1007/BF00253039`

[190] Kelley, A.: The stable, center-stable, center, center-unstable, unstable manifolds. J. Differential Equations **3**, 546–570 (1967). DOI 10.1016/0022-0396(67)90016-2. URL `https://doi.org/10.1016/0022-0396(67)90016-2`

[191] Kielhöfer, H.: Bifurcation theory, *Applied Mathematical Sciences*, vol. 156. Springer-Verlag, New York (2004). DOI 10.1007/b97365. URL `https://doi.org/10.1007/b97365`. An introduction with applications to PDEs

[192] Kokubu, H.: Normal forms for parametrized vector fields and its application to bifurcations of some reaction diffusion equations. Japan J. Appl. Math. **1**(2), 273–297 (1984). DOI 10.1007/BF03167061. URL `https://doi.org/10.1007/BF03167061`

[193] Kolmogorov, A.: Théorie générale des systèmes dynamiques de la mécanique classique. Séminaire Janet. Mécanique analytique et mécanique céleste **1**, 1–20 (1957)
[194] Kolmogorov, A.N., Petrovski, I.G., Piskunov, N.S.: Étude de l'équation de la diffusion avec croissance de la quantité de matière et son application à un problème biologique. Bull. Univ. Moskow, Ser. Internat., Sec. A **1**, 1–25 (1937)
[195] Kostova, T., Li, J.: Oscillations and stability due to juvenile competitive effects on adult fertility. Computers & Mathematics with Applications **32**(11), 57–70 (1996). DOI 10.1016/S0898-1221(96)00197-6
[196] Kribs-Zaleta, C.M., Velasco-Hernández, J.X.: A simple vaccination model with multiple endemic states. Mathematical biosciences **164**(2), 183–201 (2000). DOI 10.1016/S0025-5564(00)00003-1
[197] Krisztin, T.: An invariance principle of Lyapunov-Razumikhin type and compartmental systems. In: World Congress of Nonlinear Analysts '92, Vol. I–IV (Tampa, FL, 1992), pp. 1371–1379. de Gruyter, Berlin (1996)
[198] Krylov, N., Bogoliubov, N.: The application of methods of nonlinear mechanics to the theory of stationary oscillations. Publication 8 of the Ukrainian Academy of Sciences, Kiev (1934)
[199] Kuang, Y., Freedman, H.I.: Uniqueness of limit cycles in Gause-type models of predator-prey systems. Math. Biosci. **88**(1), 67–84 (1988). DOI 10.1016/0025-5564(88)90049-1. URL `https://doi.org/10.1016/0025-5564(88)90049-1`
[200] Kuznetsov, Y.A.: Elements of applied bifurcation theory, *Applied Mathematical Sciences*, vol. 112, second edn. Springer-Verlag, New York (1998)
[201] Kuznetsov, Y.A.: Elements of applied bifurcation theory, *Applied Mathematical Sciences*, vol. 112, third edn. Springer-Verlag, New York (2004). DOI 10.1007/978-1-4757-3978-7. URL `https://doi.org/10.1007/978-1-4757-3978-7`
[202] Lang, S.: Real analysis, second edn. Addison-Wesley Publishing Company, Advanced Book Program, Reading, MA (1983)
[203] Lang, S.: Real and functional analysis, *Graduate Texts in Mathematics*, vol. 142, third edn. Springer-Verlag, New York (1993). DOI 10.1007/978-1-4612-0897-6. URL `https://doi.org/10.1007/978-1-4612-0897-6`
[204] Larson, D.A.: Transient bounds and time-asymptotic behavior of solutions to nonlinear equations of Fisher type. SIAM J. Appl. Math. **34**(1), 93–103 (1978). DOI 10.1137/0134008. URL `https://doi.org/10.1137/0134008`
[205] LaSalle, J.: The extent of asymptotic stability. Proc. Nat. Acad. Sci. U.S.A. **46**, 363–365 (1960). DOI 10.1073/pnas.46.3.363. URL `https://doi.org/10.1073/pnas.46.3.363`
[206] LaSalle, J.: Some extensions of Liapunov's second method. IRE Trans. **CT-7**, 520–527 (1960)
[207] LaSalle, J.P.: The stability of dynamical systems. Regional Conference Series in Applied Mathematics. Society for Industrial and Applied Mathematics,

Philadelphia, PA (1976). With an appendix: "Limiting equations and stability of nonautonomous ordinary differential equations" by Z. Artstein

[208] Lau, K.S.: On the nonlinear diffusion equation of Kolmogorov, Petrovsky, and Piscounov. J. Differential Equations **59**(1), 44–70 (1985). DOI 10.1016/0022-0396(85)90137-8. URL https://doi.org/10.1016/0022-0396(85)90137-8

[209] Leonov, G.A.: Attractors, limit cycles and homoclinic orbits of low dimensional quadratic systems. Analytical methods. Can. Appl. Math. Q. **17**(1), 121–159 (2009)

[210] Leonov, G.A.: General existence conditions of homoclinic trajectories in dissipative systems. Lorenz, Shimizu-Morioka, Lu and Chen systems. Phys. Lett. A **376**(45), 3045–3050 (2012). DOI 10.1016/j.physleta.2012.07.003. URL https://doi.org/10.1016/j.physleta.2012.07.003

[211] Leonov, G.A., Kuznetsov, N.V.: On differences and similarities in the analysis of Lorenz, Chen, and Lu systems. Appl. Math. Comput. **256**, 334–343 (2015). DOI 10.1016/j.amc.2014.12.132. URL https://doi.org/10.1016/j.amc.2014.12.132

[212] Levin, S.A., Segel, L.A.: Pattern generation in space and aspect. SIAM Rev. **27**(1), 45–67 (1985). DOI 10.1137/1027002. URL https://doi.org/10.1137/1027002

[213] Li, B.: Global asymptotic behavior of the chemostat: general response functions and different removal rates. SIAM J. Appl. Math. **59**(2), 411–422 (1999). DOI 10.1137/S003613999631100X. URL https://doi-org.docelec.u-bordeaux.fr/10.1137/S003613999631100X

[214] Li, H., Xiao, H.: Traveling wave solutions for diffusive predator-prey type systems with nonlinear density dependence. Comput. Math. Appl. **74**(10), 2221–2230 (2017). DOI 10.1016/j.camwa.2017.06.048. URL https://doi.org/10.1016/j.camwa.2017.06.048

[215] Li, T.Y., Yorke, J.A.: Period three implies chaos. Amer. Math. Monthly **82**(10), 985–992 (1975). DOI 10.2307/2318254. URL https://doi.org/10.2307/2318254

[216] Lichtenberg, A.J., Lieberman, M.A.: Regular and stochastic motion, *Applied Mathematical Sciences*, vol. 38. Springer-Verlag, New York (1983). DOI 10.1007/978-1-4757-4257-2. URL https://doi.org/10.1007/978-1-4757-4257-2

[217] Lin, X.D., So, J.W.H., Wu, J.H.: Centre manifolds for partial differential equations with delays. Proc. Roy. Soc. Edinburgh Sect. A **122**(3-4), 237–254 (1992). DOI 10.1017/S0308210500021090. URL https://doi.org/10.1017/S0308210500021090

[218] Liu, L.F., Cohen, J.E.: Equilibrium and local stability in a logistic matrix model for age-structured populations. J. Math. Biol. **25**(1), 73–88 (1987). DOI 10.1007/BF00275889. URL https://doi.org/10.1007/BF00275889

[219] Liu, Z., Li, N.: Stability and bifurcation in a predator-prey model with age structure and delays. J. Nonlinear Sci. **25**(4), 937–957 (2015). DOI 10.1007/s00332-015-9245-x. URL https://doi.org/10.1007/s00332-015-9245-x

[220] Liu, Z., Magal, P.: Bogdanov-Takens bifurcation in a predator-prey model with age structure. Z. Angew. Math. Phys. **72**(1), Paper No. 4, 24 (2021). DOI 10.1007/s00033-020-01434-1. URL `https://doi.org/10.1007/s00033-020-01434-1`

[221] Liu, Z., Magal, P., Ruan, S.: Hopf bifurcation for non-densely defined Cauchy problems. Z. Angew. Math. Phys. **62**(2), 191–222 (2011). DOI 10.1007/s00033-010-0088-x. URL `https://doi.org/10.1007/s00033-010-0088-x`

[222] Liu, Z., Magal, P., Ruan, S.: Center-unstable manifolds for nondensely defined Cauchy problems and applications to stability of Hopf bifurcation. Can. Appl. Math. Q. **20**(2), 135–178 (2012)

[223] Liu, Z., Magal, P., Ruan, S.: Normal forms for semilinear equations with non-dense domain with applications to age structured models. J. Differential Equations **257**(4), 921–1011 (2014). DOI 10.1016/j.jde.2014.04.018. URL `https://doi.org/10.1016/j.jde.2014.04.018`

[224] Liu, Z., Magal, P., Ruan, S.: Oscillations in age-structured models of consumer-resource mutualisms. Discrete Contin. Dyn. Syst. Ser. B **21**(2), 537–555 (2016). DOI 10.3934/dcdsb.2016.21.537. URL `https://doi.org/10.3934/dcdsb.2016.21.537`

[225] Liu, Z., Tang, H., Magal, P.: Hopf bifurcation for a spatially and age structured population dynamics model. Discrete Contin. Dyn. Syst. Ser. B **20**(6), 1735–1757 (2015). DOI 10.3934/dcdsb.2015.20.1735. URL `https://doi.org/10.3934/dcdsb.2015.20.1735`

[226] Lobry, J., Flandrois, J., Carret, G., Pave, A.: Monod's bacterial growth model revisited. Bulletin of Mathematical Biology **54**, 117–122 (1992). DOI 10.1007/BF02458623

[227] Lorenz, E.N.: Deterministic nonperiodic flow. J. Atmospheric Sci. **20**(2), 130–141 (1963). DOI 10.1175/1520-0469(1963)020<0130:DNF>2.0.CO;2. URL `https://doi.org/10.1175/1520-0469(1963)020<0130:DNF>2.0.CO;2`

[228] Lorz, A., Perthame, B.: Long-term behaviour of phenotypically structured models. Proc. R. Soc. Lond. A **470**(2167), 20140089, 10 (2014). DOI 10.1098/rspa.2014.0089

[229] Lotka, A.J.: Contribution to the theory of periodic reactions. The Journal of Physical Chemistry **14**(3), 271–274 (1910). DOI 10.1021/j150111a004

[230] Lotka, A.J.: The growth of mixed populations: Two species competing for a common food supply. J. Wash. Acad. Sci **22**(16/17), 461–469 (1932). URL `http://www.jstor.org/stable/24530449`

[231] Lu, K., Wang, Q., Young, L.S.: Strange attractors for periodically forced parabolic equations. Mem. Amer. Math. Soc. **224**(1054), vi+85 (2013). DOI 10.1090/S0065-9266-2012-00669-1. URL `https://doi.org/10.1090/S0065-9266-2012-00669-1`

[232] Luo, A.C.J.: Bifurcation and stability in nonlinear dynamical systems, *Nonlinear Systems and Complexity*, vol. 28. Springer, Cham ([2019] ©2019). DOI 10.1007/978-3-030-22910-8. URL `https://doi.org/10.1007/978-3-030-22910-8`

[233] Lyapunov, A.: Problème général de la stabilité du mouvement. Ann. Fac. Sci. Toulouse Sci. Math. Sci. Phys. (2) **9**, 203–474 (1907). URL http://www.numdam.org/item?id=AFST_1907_2_9__203_0

[234] Ma, Z., Magal, P.: Global asymptotic stability for Gurtin-MacCamy's population dynamics model. Proc. Amer. Math. Soc. **152**(2), 765–780 (2024). DOI 10.1090/proc/15629. URL https://doi.org/10.1090/proc/15629

[235] Magal, P.: Mutation and recombination in a model of phenotype evolution. J. Evol. Equ. **2**(1), 21–39 (2002). DOI 10.1007/s00028-002-8078-x

[236] Magal, P.: Perturbation of a globally stable steady state and uniform persistence. J. Dynam. Differential Equations **21**(1), 1–20 (2009). DOI 10.1007/s10884-008-9127-0. URL https://doi.org/10.1007/s10884-008-9127-0

[237] Magal, P., McCluskey, C.: Two-group infection age model including an application to nosocomial infection. SIAM J. Appl. Math. **73**(2), 1058–1095 (2013). DOI 10.1137/120882056

[238] Magal, P., McCluskey, C.C., Webb, G.F.: Lyapunov functional and global asymptotic stability for an infection-age model. Appl. Anal. **89**(7), 1109–1140 (2010). DOI 10.1080/00036810903208122

[239] Magal, P., Ruan, S.: Center manifolds for semilinear equations with non-dense domain and applications to Hopf bifurcation in age structured models. Mem. Amer. Math. Soc. **202**(951), vi+71 (2009). DOI 10.1090/S0065-9266-09-00568-7. URL https://doi.org/10.1090/S0065-9266-09-00568-7

[240] Magal, P., Ruan, S.: Theory and applications of abstract semilinear Cauchy problems, *Applied Mathematical Sciences*, vol. 201. Springer, Cham (2018). DOI 10.1007/978-3-030-01506-0. URL https://doi.org/10.1007/978-3-030-01506-0. With a foreword by Glenn Webb

[241] Magal, P., Seydi, O.: Persistence of a normally hyperbolic manifold for a system of non densely defined Cauchy problems. J. Differential Equations **267**(5), 2950–3008 (2019). DOI 10.1016/j.jde.2019.03.035. URL https://doi.org/10.1016/j.jde.2019.03.035

[242] Magal, P., Thieme, H.R.: Eventual compactness for semiflows generated by nonlinear age-structured models. Commun. Pure Appl. Anal. **3**(4), 695–727 (2004). DOI 10.3934/cpaa.2004.3.695. URL https://doi.org/10.3934/cpaa.2004.3.695

[243] Magal, P., Webb, G.F.: Mutation, selection, and recombination in a model of phenotype evolution. Discrete Contin. Dynam. Systems **6**(1), 221–236 (2000). DOI 10.3934/dcds.2000.6.221

[244] Magal, P., Zhao, X.Q.: Global attractors and steady states for uniformly persistent dynamical systems. SIAM J. Math. Anal. **37**(1), 251–275 (2005). DOI 10.1137/S0036141003439173. URL https://doi.org/10.1137/S0036141003439173

[245] Malchow, H., Petrovskii, S.V., Venturino, E.: Spatiotemporal patterns in ecology and epidemiology. Chapman & Hall/CRC Mathematical and Computational Biology Series. Chapman & Hall/CRC, Boca Raton, FL (2008). Theory, models, and simulation, With 1 CD-ROM (Windows, Macintosh and UNIX)

[246] Marsden, J.: The Hopf bifurcation for nonlinear semigroups. Bull. Amer. Math. Soc. **79**, 537–541 (1973). DOI 10.1090/S0002-9904-1973-13191-X. URL `https://doi.org/10.1090/S0002-9904-1973-13191-X`

[247] Marsden, J.E., McCracken, M.: The Hopf bifurcation and its applications, *Applied Mathematical Sciences*, vol. Vol. 19. Springer-Verlag, New York (1976). With contributions by P. Chernoff, G. Childs, S. Chow, J. R. Dorroh, J. Guckenheimer, L. Howard, N. Kopell, O. Lanford, J. Mallet-Paret, G. Oster, O. Ruiz, S. Schecter, D. Schmidt and S. Smale

[248] Martin Jr., R.H.: Nonlinear operators and differential equations in Banach spaces. Pure and Applied Mathematics. Wiley-Interscience [John Wiley & Sons], New York-London-Sydney (1976)

[249] McGehee, R., Armstrong, R.A.: Some mathematical problems concerning the ecological principle of competitive exclusion. J. Differential Equations **23**(1), 30–52 (1977). DOI 10.1016/0022-0396(77)90135-8. URL `https://doi.org/10.1016/0022-0396(77)90135-8`

[250] McKean, H.P., Shatah, J.: The nonlinear Schrödinger equation and the nonlinear heat equation reduction to linear form. Comm. Pure Appl. Math. **44**(8-9), 1067–1080 (1991). DOI 10.1002/cpa.3160440817. URL `https://doi.org/10.1002/cpa.3160440817`

[251] Meixner, M., De Wit, A., Bose, S., Schöll, E.: Generic spatiotemporal dynamics near codimension-two Turing-Hopf bifurcations. Phys. Rev. E (3) **55**(6), 6690–6697 (1997). DOI 10.1103/PhysRevE.55.6690. URL `https://doi.org/10.1103/PhysRevE.55.6690`

[252] Metz, J.A., Diekmann, O.: The dynamics of physiologically structured populations, vol. 68. Springer (2014). DOI 10.1007/978-3-662-13159-6

[253] Meyer, K.R., Hall, G.R.: Introduction to Hamiltonian dynamical systems and the N-body problem, *Applied Mathematical Sciences*, vol. 90. Springer-Verlag, New York (1992). DOI 10.1007/978-1-4757-4073-8. URL `https://doi.org/10.1007/978-1-4757-4073-8`

[254] Meyer-Nieberg, P.: Banach lattices. Universitext. Springer-Verlag, Berlin (1991). DOI 10.1007/978-3-642-76724-1. URL `https://doi-org.docelec.u-bordeaux.fr/10.1007/978-3-642-76724-1`

[255] Michaelis, L., Menten, M.L.: Die kinetik der invertinwirkung. Biochemische Zeitschrift **49**, 333–369 (1913)

[256] Mielke, A.: A reduction principle for nonautonomous systems in infinite-dimensional spaces. J. Differential Equations **65**(1), 68–88 (1986). DOI 10.1016/0022-0396(86)90042-2. URL `https://doi.org/10.1016/0022-0396(86)90042-2`

[257] Mielke, A.: Normal hyperbolicity of center manifolds and Saint-Venant's principle. Arch. Rational Mech. Anal. **110**(4), 353–372 (1990). DOI 10.1007/BF00393272. URL `https://doi.org/10.1007/BF00393272`

[258] Mischaikow, K., Smith, H., Thieme, H.R.: Asymptotically autonomous semiflows: chain recurrence and Lyapunov functions. Trans. Amer. Math. Soc. **347**(5), 1669–1685 (1995). DOI 10.2307/2154964. URL `https://doi.org/10.2307/2154964`

[259] Mischaikow, K., Smith, H., Thieme, H.R.: Asymptotically autonomous semiflows: chain recurrence and Lyapunov functions. Trans. Amer. Math. Soc. **347**(5), 1669–1685 (1995). DOI 10.2307/2154964. URL https://doi.org/10.2307/2154964
[260] Monod, J.: Recherches sur la croissance des cultures bactériennes. Hermann Paris (1942). Thèse de doctorat, Université de Paris
[261] Monod, J.: La technique de culture continue: théorie et applications. Ann. Inst. Pasteur **79**, 390–410 (1950). URL https://gallica.bnf.fr/ark:/12148/bpt6k6211277h/f34.item
[262] Monod, J., Wyman, J., Changeux, J.P.: On the nature of allosteric transitions: a plausible model. Journal of Molecular Biology **12**(1), 88–118 (1965)
[263] Murray, J.D.: Mathematical biology, *Biomathematics*, vol. 19. Springer-Verlag, Berlin (1989). DOI 10.1007/978-3-662-08539-4. URL https://doi.org/10.1007/978-3-662-08539-4
[264] Murray, J.D.: Mathematical biology. I An introduction, *Interdisciplinary Applied Mathematics*, vol. 17, third edn. Springer-Verlag, New York (2002)
[265] Murray, J.D.: Mathematical biology. II Spatial models and biomedical applications, *Interdisciplinary Applied Mathematics*, vol. 18, third edn. Springer-Verlag, New York (2003)
[266] Nakanishi, K., Schlag, W.: Invariant manifolds around soliton manifolds for the nonlinear Klein-Gordon equation. SIAM J. Math. Anal. **44**(2), 1175–1210 (2012). DOI 10.1137/11082720X. URL https://doi.org/10.1137/11082720X
[267] Nikolenko, N.V.: The method of Poincaré normal forms in problems of integrability of equations of evolution type. Uspekhi Mat. Nauk **41**(5(251)), 109–152, 263 (1986)
[268] Novick, A., Szilard, L.: Description of the chemostat. Science **112**(2920), 715–716 (1950). DOI 10.1126/science.112.2920.715
[269] Nussbaum, R.D.: Some asymptotic fixed point theorems. Trans. Amer. Math. Soc. **171**, 349–375 (1972). DOI 10.2307/1996387. URL https://doi.org/10.2307/1996387
[270] Perko, L.: Differential equations and dynamical systems, *Texts in Applied Mathematics*, vol. 7, third edn. Springer-Verlag, New York (2001). DOI 10.1007/978-1-4613-0003-8. URL https://doi.org/10.1007/978-1-4613-0003-8
[271] Perron, O.: über Stabilität und asymptotisches Verhalten der Integrale von Differentialgleichungssystemen. Math. Z. **29**(1), 129–160 (1929). DOI 10.1007/BF01180524. URL https://doi.org/10.1007/BF01180524
[272] Perthame, B.: Transport equations in biology. Frontiers in Mathematics. Birkhäuser Verlag, Basel (2007)
[273] Pliss, V.A.: A reduction principle in the theory of stability of motion. Izv. Akad. Nauk SSSR Ser. Mat. **28**, 1297–1324 (1964)
[274] Poincaré, H.: Les méthodes nouvelles de la mécanique céleste: Méthodes de MM. Newcomb, Glydén, Lindstedt et Bohlin. 1893, vol. 2. Gauthier-Villars it fils (1893)

[275] Prüss, J.: On the qualitative behaviour of populations with age-specific interactions. Comput. Math. Appl. **9**(3), 327–339 (1983). DOI 10.1016/0898-1221(83)90020-2. URL https://doi.org/10.1016/0898-1221(83)90020-2. Hyperbolic partial differential equations

[276] Qin, Y., Su, K.: Attractors for nonlinear autonomous dynamical systems. Current Natural Sciences. EDP Sciences, Les Ulis; Science Press, Beijing ([2022] ©2022). DOI 10.1051/978-2-7598-2703-9. URL https://doi.org/10.1051/978-2-7598-2703-9

[277] Raoul, G.: Long time evolution of populations under selection and vanishing mutations. Acta Appl. Math. **114**(1-2), 1–14 (2011). DOI 10.1007/s10440-011-9603-0

[278] Rasmussen, M.: Attractivity and bifurcation for nonautonomous dynamical systems, *Lecture Notes in Mathematics*, vol. 1907. Springer, Berlin (2007)

[279] Raugel, G.: Global attractors in partial differential equations. In: Handbook of dynamical systems, Vol. 2, pp. 885–982. North-Holland, Amsterdam (2002). DOI 10.1016/S1874-575X(02)80038-8. URL https://doi.org/10.1016/S1874-575X(02)80038-8

[280] Ricker, W.E.: Stock and recruitment. Journal of the Fisheries Board of Canada **11**(5), 559–623 (1954). DOI 10.1139/f54-039

[281] Ricker, W.E.: Computation and interpretation of biological statistics of fish populations. Bull. Fish. Res. Bd. Can. **191**, 1–382 (1975)

[282] Robinson, C.: Stability theorems and hyperbolicity in dynamical systems. Rocky Mountain J. Math. **7**(3), 425–437 (1977). DOI 10.1216/RMJ-1977-7-3-425. URL https://doi.org/10.1216/RMJ-1977-7-3-425

[283] Robinson, J.C.: Infinite-dimensional dynamical systems. Cambridge Texts in Applied Mathematics. Cambridge University Press, Cambridge (2001). DOI 10.1007/978-94-010-0732-0. URL https://doi.org/10.1007/978-94-010-0732-0. An introduction to dissipative parabolic PDEs and the theory of global attractors

[284] Rosenzweig, M.L.: Paradox of enrichment: destabilization of exploitation ecosystems in ecological time. Science **171**(3969), 385–387 (1971). DOI 10.1126/science.171.3969.385

[285] Rosenzweig, M.L., MacArthur, R.H.: Graphical representation and stability conditions of predator-prey interactions. The American Naturalist **97**(895), 209–223 (1963). DOI 10.1086/282272

[286] Rothe, F.: Convergence to travelling fronts in semilinear parabolic equations. Proc. Roy. Soc. Edinburgh Sect. A **80**(3-4), 213–234 (1978). DOI 10.1017/S0308210500010258. URL https://doi.org/10.1017/S0308210500010258

[287] Roussarie, R.: Bifurcations of planar vector fields and Hilbert's sixteenth problem. Modern Birkhäuser Classics. Birkhäuser/Springer, Basel (1998). DOI 10.1007/978-3-0348-8798-4. URL https://doi.org/10.1007/978-3-0348-8798-4. [2013] reprint of the 1998 edition [MR1628014]

[288] Rovinsky, A., Menzinger, M.: Interaction of Turing and Hopf bifurcations in chemical systems. Phys. Rev. A (3) **46**(10), 6315–6322 (1992). DOI 10.

1103/PhysRevA.46.6315. URL https://doi.org/10.1103/PhysRevA.46.6315

[289] Ruan, S.: Turing instability and travelling waves in diffusive plankton models with delayed nutrient recycling. IMA J. Appl. Math. **61**(1), 15–32 (1998). DOI 10.1093/imamat/61.1.15. URL https://doi.org/10.1093/imamat/61.1.15

[290] Ruan, S., Wang, W.: Dynamical behavior of an epidemic model with a nonlinear incidence rate. J. Differential Equations **188**(1), 135–163 (2003). DOI 10.1016/S0022-0396(02)00089-X. URL https://doi.org/10.1016/S0022-0396(02)00089-X

[291] Ruan, S., Zhao, X.Q.: Persistence and extinction in two species reaction-diffusion systems with delays. J. Differential Equations **156**(1), 71–92 (1999). DOI 10.1006/jdeq.1998.3599. URL https://doi.org/10.1006/jdeq.1998.3599

[292] Saltzman, B.: Finite amplitude free convection as an initial value problem—i. Journal of atmospheric sciences **19**(4), 329–341 (1962). DOI 10.1175/1520-0469(1962)019<0329:FAFCAA>2.0.CO;2

[293] Sattinger, D.H.: Bifurcation of periodic solutions of the Navier-Stokes equations. Arch. Rational Mech. Anal. **41**, 66–80 (1971). DOI 10.1007/BF00250178. URL https://doi.org/10.1007/BF00250178

[294] Scarpellini, B.: Center manifolds of infinite dimensions. I. Main results and applications. Z. Angew. Math. Phys. **42**(1), 1–32 (1991). DOI 10.1007/BF00962056. URL https://doi.org/10.1007/BF00962056

[295] Scheel, A.: Radially symmetric patterns of reaction-diffusion systems. Mem. Amer. Math. Soc. **165**(786), viii+86 (2003). DOI 10.1090/memo/0786. URL https://doi.org/10.1090/memo/0786

[296] Schreiber, S.J.: Criteria for C^r robust permanence. J. Differential Equations **162**(2), 400–426 (2000). DOI 10.1006/jdeq.1999.3719. URL https://doi.org/10.1006/jdeq.1999.3719

[297] Sell, G.R.: The Floquet problem for almost periodic linear differential equations. In: Ordinary and partial differential equations (Proc. Conf., Univ. Dundee, Dundee, 1974), *Lecture Notes in Math.*, vol. Vol. 415, pp. 239–251. Springer, Berlin-New York (1974)

[298] Sell, G.R., You, Y.: Dynamics of evolutionary equations, *Applied Mathematical Sciences*, vol. 143. Springer-Verlag, New York (2002). DOI 10.1007/978-1-4757-5037-9. URL https://doi.org/10.1007/978-1-4757-5037-9

[299] Seydel, R.: Practical bifurcation and stability analysis, *Interdisciplinary Applied Mathematics*, vol. 5, third edn. Springer, New York (2010). DOI 10.1007/978-1-4419-1740-9. URL https://doi.org/10.1007/978-1-4419-1740-9

[300] Sharkovsky, O.M.: Co-existence of cycles of a continuous mapping of the line into itself. Ukrain. Mat. Ž. **16**, 61–71 (1964)

[301] Sharkovsky, O.M., Kolyada, S.F., Sivak, A.G., Fedorenko, V.V.: Dynamics of one-dimensional maps, *Mathematics and its Applications*, vol. 407. Kluwer Academic Publishers Group, Dordrecht (1997). DOI 10.1007/978-94-015-8897-3. URL https://doi.org/10.1007/978-94-015-8897-3. Trans-

lated from the 1989 Russian original by Sivak, P. Malyshev and D. Malyshev and revised by the authors

[302] Shatah, J.: Normal forms and quadratic nonlinear Klein-Gordon equations. Comm. Pure Appl. Math. **38**(5), 685–696 (1985). DOI 10.1002/cpa.3160380516. URL `https://doi.org/10.1002/cpa.3160380516`

[303] Sherratt, J.A., Smith, M.J., Rademacher, J.D.: Locating the transition from periodic oscillations to spatiotemporal chaos in the wake of invasion. Proceedings of the National Academy of Sciences **106**(27), 10890–10895 (2009). DOI 10.1073/pnas.0900161106

[304] Shi, H.B., Ruan, S.: Spatial, temporal and spatiotemporal patterns of diffusive predator-prey models with mutual interference. IMA J. Appl. Math. **80**(5), 1534–1568 (2015). DOI 10.1093/imamat/hxv006. URL `https://doi.org/10.1093/imamat/hxv006`

[305] Shil'nikov, A.L.: On bifurcations of the Lorenz attractor in the Shimizu-Morioka model. Phys. D **62**(1-4), 338–346 (1993). DOI 10.1016/0167-2789(93)90292-9. URL `https://doi.org/10.1016/0167-2789(93)90292-9`. Homoclinic chaos (Brussels, 1991)

[306] Shil'nikov, L.P., Shilnikov, A.L., Turaev, D.V., Chua, L.O.: Methods of qualitative theory in nonlinear dynamics. Part I, *World Scientific Series on Nonlinear Science. Series A: Monographs and Treatises*, vol. 4. World Scientific Publishing Co., Inc., River Edge, NJ (1998). DOI 10.1142/9789812798596. URL `https://doi.org/10.1142/9789812798596`. With the collaboration of Sergey Gonchenko (Sections 3.7 and 3.8), Oleg Sten'kin (Section 3.9 and Appendix A) and Mikhail Shashkov (Sections 6.1 and 6.2)

[307] Shimizu, T., Morioka, N.: On the bifurcation of a symmetric limit cycle to an asymmetric one in a simple model. Phys. Lett. A **76**(3-4), 201–204 (1980). DOI 10.1016/0375-9601(80)90466-1. URL `https://doi.org/10.1016/0375-9601(80)90466-1`

[308] Siegel, C.L.: Uber die normalform analytischer differentialgleichungen in der nahe einer gleichgewichtslosung. Matematika **5**(2), 119–128 (1961)

[309] Siegel, C.L., Moser, J.K.: Lectures on celestial mechanics. Classics in Mathematics. Springer-Verlag, Berlin (1995). Translated from the German by C. I. Kalme, Reprint of the 1971 translation

[310] Sijbrand, J.: Properties of center manifolds. Trans. Amer. Math. Soc. **289**(2), 431–469 (1985). DOI 10.2307/2000247. URL `https://doi.org/10.2307/2000247`

[311] Smith, H., Zhao, X.Q.: Microbial growth in a plug flow reactor with wall adherence and cell motility. J. Math. Anal. Appl. **241**(1), 134–155 (2000). DOI 10.1006/jmaa.1999.6642. URL `https://doi.org/10.1006/jmaa.1999.6642`

[312] Smith, H.L.: Competitive coexistence in an oscillating chemostat. SIAM J. Appl. Math. **40**(3), 498–522 (1981). DOI 10.1137/0140042. URL `https://doi.org/10.1137/0140042`

[313] Smith, H.L.: Monotone dynamical systems, *Mathematical Surveys and Monographs*, vol. 41. American Mathematical Society, Providence, RI (1995). An introduction to the theory of competitive and cooperative systems

[314] Smith, H.L., Thieme, H.R.: Dynamical systems and population persistence, *Graduate Studies in Mathematics*, vol. 118. American Mathematical Society, Providence, RI (2011). DOI 10.1090/gsm/118. URL https://doi.org/10.1090/gsm/118
[315] Smith, H.L., Waltman, P.: Perturbation of a globally stable steady state. Proc. Amer. Math. Soc. **127**(2), 447–453 (1999). DOI 10.1090/S0002-9939-99-04768-1. URL https://doi.org/10.1090/S0002-9939-99-04768-1
[316] Smith, H.L., Zhao, X.Q.: Dynamics of a periodically pulsed bio-reactor model. J. Differential Equations **155**(2), 368–404 (1999). DOI 10.1006/jdeq.1998.3587. URL https://doi.org/10.1006/jdeq.1998.3587
[317] Smith, H.L., Zhao, X.Q.: Robust persistence for semidynamical systems. In: Proceedings of the Third World Congress of Nonlinear Analysts, Part 9 (Catania, 2000), vol. 47, pp. 6169–6179 (2001). DOI 10.1016/S0362-546X(01)00678-2. URL https://doi.org/10.1016/S0362-546X(01)00678-2
[318] Strogatz, S.H.: Nonlinear dynamics and chaos, second edn. Westview Press, Boulder, CO (2015). With applications to physics, biology, chemistry, and engineering
[319] Stumpf, E.: On a differential equation with state-dependent delay: a center-unstable manifold connecting an equilibrium and a periodic orbit. J. Dynam. Differential Equations **24**(2), 197–248 (2012). DOI 10.1007/s10884-012-9245-6. URL https://doi.org/10.1007/s10884-012-9245-6
[320] Swart, J.H.: Hopf bifurcation and the stability of nonlinear age-dependent population models. Comput. Math. Appl. **15**(6-8), 555–564 (1988). DOI 10.1016/0898-1221(88)90280-5. URL https://doi.org/10.1016/0898-1221(88)90280-5. Hyperbolic partial differential equations. V
[321] Tang, H., Liu, Z.: Hopf bifurcation for a predator-prey model with age structure. Appl. Math. Model. **40**(2), 726–737 (2016). DOI 10.1016/j.apm.2015.09.015. URL https://doi.org/10.1016/j.apm.2015.09.015
[322] Tang, Y., Huang, D., Ruan, S., Zhang, W.: Coexistence of limit cycles and homoclinic loops in a SIRS model with a nonlinear incidence rate. SIAM J. Appl. Math. **69**(2), 621–639 (2008). DOI 10.1137/070700966. URL https://doi.org/10.1137/070700966
[323] Taylor, P.A., Williams, P.J.L.: Theoretical studies on the coexistence of competing species under continuous-flow conditions. Canadian Journal of Microbiology **21**(1), 90–98 (1975). DOI 10.1139/m75-013. PMID: 1116041
[324] Temam, R.: Infinite-dimensional dynamical systems in mechanics and physics, *Applied Mathematical Sciences*, vol. 68, second edn. Springer-Verlag, New York (1997). DOI 10.1007/978-1-4612-0645-3. URL https://doi.org/10.1007/978-1-4612-0645-3
[325] Thieme, H.R.: Convergence results and a Poincaré-Bendixson trichotom for asymptotically autonomous differential equations. J. Math. Biol. **30**(755–763 (1992). DOI 10.1007/BF00173267. URL https://doi.org/1 1007/BF00173267

[326] Thieme, H.R.: Persistence under relaxed point-dissipativity (with application to an endemic model). SIAM J. Math. Anal. **24**(2), 407–435 (1993). DOI 10.1137/0524026. URL https://doi.org/10.1137/0524026

[327] Thieme, H.R.: Asymptotically autonomous differential equations in the plane. Rocky Mountain J. Math. **24**(1), 351–380 (1994). DOI 10.1216/rmjm/1181072470. URL https://doi.org/10.1216/rmjm/1181072470. 20th Midwest ODE Meeting (Iowa City, IA, 1991)

[328] Thieme, H.R.: Asymptotically autonomous differential equations in the plane. II. Stricter Poincaré/Bendixson type results. Differential Integral Equations **7**(5-6), 1625–1640 (1994)

[329] Thieme, H.R.: Uniform weak implies uniform strong persistence for non-autonomous semiflows. Proc. Amer. Math. Soc. **127**(8), 2395–2403 (1999). DOI 10.1090/S0002-9939-99-05034-0. URL https://doi.org/10.1090/S0002-9939-99-05034-0

[330] Thieme, H.R.: Uniform persistence and permanence for non-autonomous semiflows in population biology. Math. Biosci. **166**(2), 173–201 (2000). DOI 10.1016/S0025-5564(00)00018-3. URL https://doi.org/10.1016/S0025-5564(00)00018-3

[331] Thieme, H.R.: Mathematics in population biology. Princeton Series in Theoretical and Computational Biology. Princeton University Press, Princeton, NJ (2003)

[332] Thieme, H.R.: Global stability of the endemic equilibrium in infinite dimension: Lyapunov functions and positive operators. J. Differential Equations **250**(9), 3772–3801 (2011). DOI 10.1016/j.jde.2011.01.007

[333] Turchin, P.: Complex population dynamics: a theoretical/empirical synthesis, *Monographs in Population Biology*, vol. 35. Princeton University Press, Princeton, NJ (2003)

[334] Turyn, L.: A center-unstable manifold theorem for parametrically excited surface waves. SIAM J. Math. Anal. **27**(1), 241–257 (1996). DOI 10.1137/0527013. URL https://doi.org/10.1137/0527013

[335] Uchiyama, K.: The behavior of solutions of some nonlinear diffusion equations for large time. J. Math. Kyoto Univ. **18**(3), 453–508 (1978). DOI 10.1215/kjm/1250522506. URL https://doi.org/10.1215/kjm/1250522506

Van Minh, N., Wu, J.: Invariant manifolds of partial functional differential equations. J. Differential Equations **198**(2), 381–421 (2004). DOI 10.1016/j. .2003.10.006. URL https://doi.org/10.1016/j.jde.2003.10.006

lerbauwhede, A.: Invariant manifolds in infinite dimensions. In: Dynam- infinite-dimensional systems (Lisbon, 1986), *NATO Adv. Sci. Inst. Ser. ut. Systems Sci.*, vol. 37, pp. 409–420. Springer, Berlin (1987)

uwhede, A.: Centre manifolds, normal forms and elementary bifur- Dynamics reported, Vol. 2, *Dynam. Report. Ser. Dynam. Systems* pp. 89–169. Wiley, Chichester (1989)

de, A., van Gils, S.A.: Center manifolds and contractions Banach spaces. J. Funct. Anal. **72**(2), 209–224 (1987). -1236(87)90086-3. URL https://doi.org/10.1016/ 086-3

[340] Vanderbauwhede, A., Iooss, G.: Center manifold theory in infinite dimensions. In: Dynamics reported: expositions in dynamical systems, *Dynam. Report. Expositions Dynam. Systems (N.S.)*, vol. 1, pp. 125–163. Springer, Berlin (1992)

[341] Verhulst, P.F.: Notice sur la loi que la population poursuit dans son accroissement. Correspondance Mathématique et Physique **vol.X**, 113–121 (1838)

[342] Volpert, A.I., Volpert, V.A., Volpert, V.A.: Traveling wave solutions of parabolic systems, *Translations of Mathematical Monographs*, vol. 140. American Mathematical Society, Providence, RI (1994). DOI 10.1090/mmono/140. URL `https://doi.org/10.1090/mmono/140`. Translated from the Russian manuscript by James F. Heyda

[343] Volterra, V.: Fluctuations in the abundance of a species considered mathematically 1. Nature (1926). DOI 10.1038/118558a0

[344] Volterra, V.: Variations and fluctuations of the number of individuals in animal species living together. ICES Journal of Marine Science **3**(1), 3–51 (1928)

[345] Volterra, V.: Leçons sur la théorie mathématique de la lutte pour la vie. Gauthier-Villars et cie. (1931)

[346] Waltman, P.: A brief survey of persistence in dynamical systems. In: Delay differential equations and dynamical systems (Claremont, CA, 1990), *Lecture Notes in Math.*, vol. 1475, pp. 31–40. Springer, Berlin (1991). DOI 10.1007/BFb0083477. URL `https://doi.org/10.1007/BFb0083477`

[347] Wang, D.: An introduction to the normal form theory of ordinary differential equations. Adv. in Math. (China) **19**(1), 38–71 (1990)

[348] Webb, G.F.: Compactness of bounded trajectories of dynamical systems in infinite-dimensional spaces. Proc. Roy. Soc. Edinburgh Sect. A **84**(1-2), 19–33 (1979). DOI 10.1017/S0308210500016930. URL `https://doi.org/10.1017/S0308210500016930`

[349] Webb, G.F.: Theory of nonlinear age-dependent population dynamics, *Monographs and Textbooks in Pure and Applied Mathematics*, vol. 89. Marcel Dekker, Inc., New York (1985)

[350] Wiggins, S.: Global bifurcations and chaos, *Applied Mathematical Sciences*, vol. 73. Springer-Verlag, New York (1988). DOI 10.1007/978-1-4612-1042-9. URL `https://doi.org/10.1007/978-1-4612-1042-9`. Analytical methods

[351] Wiggins, S.: Normally hyperbolic invariant manifolds in dynamical systems, *Applied Mathematical Sciences*, vol. 105. Springer-Verlag, New York (1994). DOI 10.1007/978-1-4612-4312-0. URL `https://doi.org/10.1007/978-1-4612-4312-0`. With the assistance of György Haller and Igor Mezić

[352] Wiggins, S.: Introduction to applied nonlinear dynamical systems and chaos, *Texts in Applied Mathematics*, vol. 2, second edn. Springer-Verlag, New York (2003)

[353] Winfree, A.T.: The geometry of biological time, *Biomathematics*, vol. 8. Springer-Verlag, Berlin-New York (1980)

[354] Wolkowicz, G.S.K., Lu, Z.Q.: Global dynamics of a mathematical model of competition in the chemostat: general response functions and differential death rates. SIAM J. Appl. Math. **52**(1), 222–233 (1992). DOI 10.1137/0152012

[355] Wolkowicz, G.S.K., Xia, H.: Global asymptotic behavior of a chemostat model with discrete delays. SIAM J. Appl. Math. **57**(4), 1019–1043 (1997). DOI 10.1137/S0036139995287314

[356] Wu, J.: Theory and applications of partial functional-differential equations, *Applied Mathematical Sciences*, vol. 119. Springer-Verlag, New York (1996). DOI 10.1007/978-1-4612-4050-1. URL `https://doi.org/10.1007/978-1-4612-4050-1`

[357] Wu, J., Zou, X.: Traveling wave fronts of reaction-diffusion systems with delay. J. Dynam. Differential Equations **13**(3), 651–687 (2001). DOI 10.1023/A:1016690424892. URL `https://doi.org/10.1023/A:1016690424892`

[358] Xiao, D., Ruan, S.: Global analysis of an epidemic model with nonmonotone incidence rate. Math. Biosci. **208**(2), 419–429 (2007). DOI 10.1016/j.mbs.2006.09.025. URL `https://doi.org/10.1016/j.mbs.2006.09.025`

[359] Yang, F., Freedman, H.I.: Competing predators for a prey in a chemostat model with periodic nutrient input. J. Math. Biol. **29**(8), 715–732 (1991). DOI 10.1007/BF00160188. URL `https://doi.org/10.1007/BF00160188`

[360] Yi, Y.: A generalized integral manifold theorem. J. Differential Equations **102**(1), 153–187 (1993). DOI 10.1006/jdeq.1993.1026. URL `https://doi.org/10.1006/jdeq.1993.1026`

[361] Yukon: Data for Yukon Snowshoe-Hare (accessed December 1, 2020). URL `https://yukon.ca/en/snowshoe-hare`

[362] Zanolin, F.: Permanence and positive periodic solutions for Kolmogorov competing species systems. Results Math. **21**(1-2), 224–250 (1992). DOI 10.1007/BF03323081. URL `https://doi.org/10.1007/BF03323081`

[363] Zehnder, E.: A simple proof of a generalization of a theorem by C. L. Siegel. In: Geometry and topology (Proc. III Latin Amer. School of Math., Inst. Mat. Pura Aplicada CNPq, Rio de Janeiro, 1976), *Lecture Notes in Math.*, vol. Vol. 597, pp. 855–866. Springer, Berlin-New York (1977)

[364] Zehnder, E.: C. L. Siegel's linearization theorem in infinite dimensions. Manuscripta Math. **23**(4), 363–371 (1977/78). DOI 10.1007/BF01167695. URL `https://doi.org/10.1007/BF01167695`

[365] Zhang, Z.F.: Proof of the uniqueness theorem of limit cycles of generalized Liénard equations. Appl. Anal. **23**(1-2), 63–76 (1986). DOI 10.1080/00036818608839631. URL `https://doi.org/10.1080/00036818608839631`

[366] Zhao, X.Q.: Uniform persistence and periodic coexistence states in infinite-dimensional periodic semiflows with applications. Canad. Appl. Math. Quart. **3**(4), 473–495 (1995)

[367] Zhao, X.Q.: Asymptotic behavior for asymptotically periodic semiflows with applications. Comm. Appl. Nonlinear Anal. **3**(4), 43–66 (1996)

[368] Zhao, X.Q.: Convergence in asymptotically periodic tridiagonal competitive-cooperative systems. Canad. Appl. Math. Quart. **6**(3), 287–301 (1998). Ge-

offrey J. Butler Memorial Conference in Differential Equations and Mathematical Biology (Edmonton, AB, 1996)

[369] Zhao, X.Q.: Uniform persistence in processes with application to nonautonomous competitive models. J. Math. Anal. Appl. **258**(1), 87–101 (2001). DOI 10.1006/jmaa.2000.7361. URL https://doi.org/10.1006/jmaa.2000.7361

[370] Zhao, X.Q.: Dynamical systems in population biology, *CMS Books in Mathematics/Ouvrages de Mathématiques de la SMC*, vol. 16. Springer-Verlag, New York (2003). DOI 10.1007/978-0-387-21761-1. URL https://doi.org/10.1007/978-0-387-21761-1

[371] Zhao, X.Q.: Persistence in almost periodic predator-prey reaction-diffusion systems. In: Dynamical systems and their applications in biology (Cape Breton Island, NS, 2001), *Fields Inst. Commun.*, vol. 36, pp. 259–268. Amer. Math. Soc., Providence, RI (2003)

[372] Zhao, X.Q.: Dynamical systems in population biology, second edn. CMS Books in Mathematics/Ouvrages de Mathématiques de la SMC. Springer, Cham (2017). DOI 10.1007/978-3-319-56433-3. URL https://doi.org/10.1007/978-3-319-56433-3

[373] Zhao, X.Q., Hutson, V.: Permanence in Kolmogorov periodic predator-prey models with diffusion. Nonlinear Anal. **23**(5), 651–668 (1994). DOI 10.1016/0362-546X(94)90244-5. URL https://doi.org/10.1016/0362-546X(94)90244-5

GPSR Compliance
The European Union's (EU) General Product Safety Regulation (GPSR) is a set of rules that requires consumer products to be safe and our obligations to ensure this.

If you have any concerns about our products, you can contact us on

ProductSafety@springernature.com

In case Publisher is established outside the EU, the EU authorized representative is:

Springer Nature Customer Service Center GmbH
Europaplatz 3
69115 Heidelberg, Germany

www.ingramcontent.com/pod-product-compliance
Ingram Content Group UK Ltd.
Pitfield, Milton Keynes, MK11 3LW, UK
UKHW021639180926
475650UK00001B/36

9783031914119